AF332951

Direct Methods of Solving Crystal Structures

NATO ASI Series

Advanced Science Institutes Series

A series presenting the results of activities sponsored by the NATO Science Committee, which aims at the dissemination of advanced scientific and technological knowledge, with a view to strengthening links between scientific communities.

The series is published by an international board of publishers in conjunction with the NATO Scientific Affairs Division

A	**Life Sciences**	Plenum Publishing Corporation
B	**Physics**	New York and London
C	**Mathematical and Physical Sciences**	Kluwer Academic Publishers
D	**Behavioral and Social Sciences**	Dordrecht, Boston, and London
E	**Applied Sciences**	
F	**Computer and Systems Sciences**	Springer-Verlag
G	**Ecological Sciences**	Berlin, Heidelberg, New York, London,
H	**Cell Biology**	Paris, Tokyo, Hong Kong, and Barcelona
I	**Global Environmental Change**	

Recent Volumes in this Series

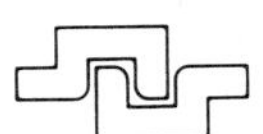

Series B: Physics

Direct Methods of Solving Crystal Structures

Edited by

Henk Schenk

University of Amsterdam
Amsterdam, The Netherlands

Plenum Press
New York and London
Published in cooperation with NATO Scientific Affairs Division

Proceedings of a NATO Advanced Study Institute/
18th Course of the International School of Crystallography
on Direct Methods of Solving Crystal Structures,
held April 18–29, 1990, in Erice, Sicily, Italy

Library of Congress Cataloging-in-Publication Data

Direct methods of solving crystal structures / edited by Henk Schenk.
      p.   cm. -- (NATO ASI series. Series B, Physics ; v. 274)
    Sponsored by NATO Scientific Affairs Division.
    "Proceedings of a NATO Advanced Study Institute/18th Course of the
International School of Crystallography on Direct Methods of Solving
Crystal Structures, held April 18-29, 1990, in Erice, Sicily,
Italy"--Copyright p.
    Includes bibliographical references and index.
    ISBN 0-306-44040-7
    1. Crystallography--Congresses.   I. Schenk, H. (Henk) II. North
Atlantic Treaty Organization. Scientific Affairs Division.
III. NATO Advanced Study Institute/18th Course of the International
School of Crystallography on Direct Methods of Solving Crystal
Structures (1990 : Erice, Italy) IV. Series.
QD921.D527 1991
  548--dc20   91-29312
                                                          CIP

ISBN 0-306-44040-7

PREFACE

This book of the proceedings of the 1990 NATO Advanced Study Institute on
"Direct Methods for Solving Crystal Structures" held in April in the Ettore Majorana
Centre in Erice, Sicily, provides a complete account of the activities of the Institute. The
ASI was intended to guide the participants to the state of the art of Direct Methods by
bringing together all leading experts in the field and selected participants. Introductory
lectures and advanced tutorials for small groups have been mixed in the right proportions
to achieve the optimal transfer of expert knowledge. In the end an anonymous inquiry
proved undoubtedly the success of the School and the value of its content. Therefore I
am confident that this book, reflecting the content of the ASI and speaking for itself,
will prove a valuable addition to the literature on Direct Methods.

The 1990 NATO ASI formed part of two series of advanced schools: it was the
18th course of the International School of Crystallography of the Ettore Majorana Centre
in Erice, Sicily, and the tenth ASI on Direct Methods. Previous ASIs on Direct Methods
were held in:

1970	Parma, Italy
1971	York, England
1974	Erice, Italy
1975	York, England
1976	Sao Carlos, Brazil
1978	Erice, Italy
1980	York, England
1984	Erice, Italy
1985	Madras, India.

York and Erice are prominent places in this list and it is therefore appropriate to
mention in particular Professors Michael Woolfson and Lodovico Riva di Sanseverino for
their continuous efforts to ensure the success of the ASIs. I would also like to
acknowledge the fact that many of the schools were NATO Advanced Study Institutes.

As Director of the School I wish to express my sincere thanks to Professors Paola
Spadon and Lodovico Riva di Sanseverino who were responsible for the local organisation
of the ASI, assisted by many other scientists and non-scientists and in cooperation with
the management and personnel of the very well equiped Ettore Majorana Centre in the
acient town of Erice. Good organisation is the most important condition for the fruitful
exchange of knowledge and their organisation was excellent. In preparing the program of
the ASI I had the valuable assistance of many of the lecturers, in particular of Professors
Herbert Hauptman and Michael Woolfson. Next I would I like to thank all contributors
to this proceedings, a few of whom had to do a hard job they were replacing lecturers
unable to attend the School at very late notice. For completeness one extra paper was
added to the volum about probability theory and I thank René Peschar for writing it.
Last but not least I would like to thank Dinie Hartgers, Ewald Holzhaus, Ineke Baas,
Chris Kyriakidis and Yuan–Fang Wang and in particular Huib Schenk who assisted me in
editing this proceedings and compiling the index.

On behalf of the Organizing Committee of the 18th Course of the International School of Crystallography at the Erice Ettore Majorana Centre I gratefully acknowledge a number of institutions for support. In the first place of course I wish to mention the NATO Scientific Affairs Division which provided the most significant single financial contribution to the ASI and cooperated very efficiently in the organisation. In addition in alphabetical order the following organisations made contributions: Azienda Autonoma di Turismo di Palermo e Monreale, Banco di Sicilia at Palermo, Biopolymer Research Centre CNR Padova, the British Council at Roma and London, Cassa di Risparmio della Provincia Siciliane at Palermo, the Commission of the European Communities (Bridge office) at Bruxelles, the Committee On Science and Technology in Developing countries (COSTED), the Ettore Majorana Centre for Scientific Culture, the International Union of Crystallography, the International Council of Scientific Unions, the Italian African Institute in Rome, the Italian National Research Council and the National Science Foundation in Washington.

Since most of the leading experts in Direct Methods lectured at the ASI and agreed to share their exclusive knowledge with small groups of participants, in the worksessions the Advanced Study Institute was a great success and was at the most advanced level imaginable. For myself as director it was a pleasure simply to be involved.

Henk Schenk

CONTENTS

AN ELEMENTARY INTRODUCTION TO DIRECT METHODS

H. Schenk

Laboratory for Crystallography, University of Amsterdam
Nieuwe Achtergracht 166, 1018 WV Amsterdam
The Netherlands

Introduction

In a diffraction experiment intensities I_{hkl} are measured whereas $F_{hkl} = |F_{hkl}|\exp(i\varphi_{hkl})$ are necessary to image the electron density. Now $|F_{hkl}|$ can be calculated straightforward from I_{hkl} but the relative phases φ_{hkl} are lost in the experiment and cause the so called phase problem. Direct methods try to evaluate phases φ_{hkl} "directly" from the measured diffraction intensities I_{hkl} by using relationships among the phases, relationships, whose values are based on the intensities only. Roughly it can be stated that, since the crystal structure can be described by a limited number of parameters (the positions of the atoms) and since many more intensities can be measured, relationships among the structure factors F_{hkl}, and thus among the phases φ_{hkl}, must exist. Direct Methods identify and use these phase relationships to solve the phase problem.

In a nutshell a Direct Method proceeds as follows: In the first step as many phase relationships as possible are collected, the origin is fixed by specifying the phases of a few suitable reflections numerically and then, using the phase relationships, new phases are calculated. In general, however, it will not be possible to phase all strong reflections and hence a few more starting reflections are selected, which act as unknowns (symbols, ambiguities) and from which new phases can be calculated using the phase relationships (the so-called phase extension). This process generally develops like a snow ball, provided a good choice of origin-defining reflections and unknowns has been made. Finally, when most of the strong reflections have got a phase, the numerical values of the unknowns are evaluated and using a Fourier summation (expression 1) an image of the structure is produced.

In this intoductory chapter this process will be explored in more detail and also some practical exercises will be introduced. Therefore, paragraphs in this chapter will deal with successively the origin of the phase problem, weak and strong structure factors, normalized structure factors, the physical meaning of the triplet relation, the origin definition in the centrosymmetric triclinic space group P1bar (for typographical reasons to be referred to in this chapter as P$\underline{1}$) and the process of phase propagation. Moreover, in the Advanced Study Institute the snow ball-like phase propagation will be carried out by hand for a structure in the same spacegroup.

The Phase Problem

The electron density $\rho(x,y,z)$, where x, y and z define a position in the unit cell in fractions of the cell edges, is given by the Fourier summation

$$\rho(x,y,z) = \sum_h \sum_k \sum_l |F_{hkl}| \cos[2\pi(hx+ky+lz)+\varphi_{hkl}] \qquad (1)$$

This expression shows that the electron density ρ is a superposition of planar density waves of the cosine form with amplitudes $|F_{hkl}|$, with a spatial direction and a wavelength defined by hkl, and with the position of the maxima of the planar waves with respect to the origin determined by the so-called phases φ_{hkl}. In the two upper rows of Fig. 1 a number of these planar density waves is depicted with their phases φ_{hkl} all equal to 0, and their $|F_{hkl}|$'s all equal. In order to image the electron density ρ of a particular structure one needs to know the $|F_{hkl}|$-values and the corresponding phases φ_{hkl} and is then able to calculate the Fourier summation 1. This process is depicted in the final row of Fig. 1, where the electron density of naphtalene is given and a few of the most important terms of expression 1 for that structure.

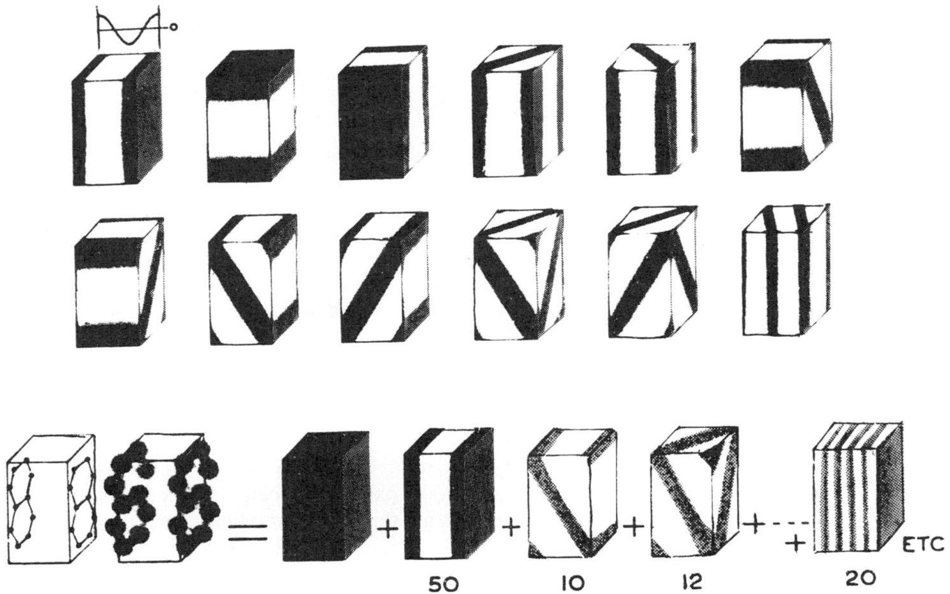

Fig. 1. Fourier components and the Fourier summation of naphtalene.

The central problem in a crystal structure determination follows from the fact that only intensities $I_{hkl} \approx |F_{hkl}|^2$ can be measured by means of X-ray diffraction, whereas for each term of expression 1 not only $|F_{hkl}|$ but also the phase φ_{hkl} must be known in order to obtain the electron density $\rho(x,y,z)$. Unless in very exceptional cases, these phases cannot be measured and therefore they have to be deducted in one way or another. This is generally referred to as the phase problem and solving the phase problem is the most important and sometimes most troublesome step of a crystal structure determination. To illustrate the phase problem it is recalled that in the final row of Fig. 1 the structure factor amplitudes were used in a Fourier summation 1 with their correct phases and that as a result the structure arises. However, if these phases were not known and taken arbitrarily to be equal to 0, then the Fourier summation 1 would show only one large maximum at the origin and indicate nothing of the structure at all. Another possibility would be to take phases at random, and then a completely grey image would result from expression 1, again not showing anything of the structure.

The phase problem thus is a serious problem; without solving it no image of the structure will be obtained. Even when the structurefactors are all set to 1 or taken from another crystal the desired structure will show up in a Fourier summation, provided the correct phases have been used. However, wrong phases and correct amplitudes reveal no structure; the phase problem thus is a very serious problem indeed. Therefore methods to evaluate phases play a central role in crystal structure determinations. The so–called Direct Methods constitute one approach to solve the phase problem, Patterson Methods another. In protein crystallography two important methods to determine phases are based on isomorphous replacement and anomalous dispersion respectively. In this Advanced Study Institute the emphasis lies on Direct Methods whilst all other methods will be dealt with to the extent necessary to understand their interaction with Direct Methods. Patterson methods will have their place in the advanced section of the Institute.

Strong and weak structure factors

If in a crystal structure the atoms lie in the neighbourhood of the set of planes H (H is a compact notation for the indices hkl), as indicated in Fig. 2a, the reflection has a large intensity I_{hkl}. This intensity is small when the atoms are randomly distributed with respect to the same planes (Fig. 2b). This follows also directly from the structure factor expression

$$F_H = |F_H|\exp(i\varphi_H) = \sum_{j=1}^{N} f_j \exp[2\pi i(hx_j + ky_j + lz_j)] \qquad (2)$$

because a large structure factor F_H will only be found if the atoms lie near positions for which $hx_j + ky_j + lz_j \simeq$ constant modulo 1 for all j.

Conversely, a strong intensity of H implies that the electron density will peak in planar regions which lie d_H apart and a weak intensity that the electron density is scattered with respect of the planes H. The choice of the origin with respect to these planar regions of electron density defines the phase φ_H in expression 2. Large and small structure factor amplitudes thus may be used to predict where in the unit cell electron density can approximately be expected.

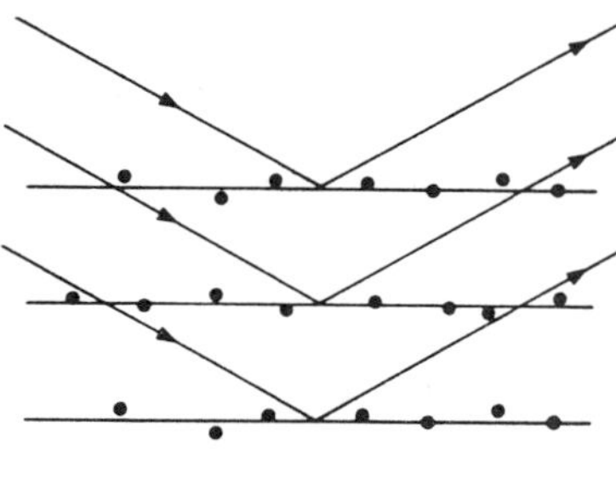

Fig. 2a

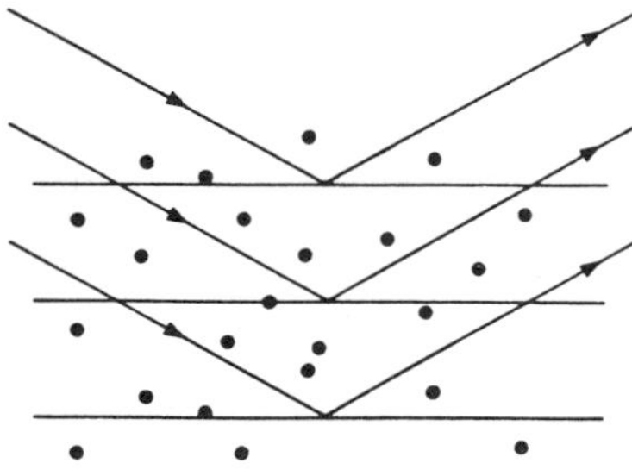

Fig. 2b

Normalized structure factors

The amplitudes of different structure factors F_H cannot be compared directly since the form factor decreases with increasing reflection angle θ. Therefore the so called normalized structure factors are used in direct methods instead. They are defined as:

$$|E_H|^2 = |F_H|^2 / (\sum_{j=1}^{N} f_j^2) \qquad (3)$$

and have the property that $< |E_H|^2 > = 1$ for all values of θ, because the expectation value of $|F_H|$ at given θ is expressed by:

$$<|F_H|^2>_\theta = \sum_{j=1}^{N} f_j^2(\theta) \tag{4}$$

The structure factor expression in terms of the normalized structure factor is then:

$$E_H = \frac{1}{(\sum_{j=1}^{N} f_j^2)^{1/2}} \sum_{j=1}^{N} f_j \exp 2\pi i (hx_j + ky_j + lz_j) \tag{5}$$

If the form factor f_j has the same shape for all atoms ($f_j = Z_j f$), expression 1 can be written as

$$E_H = \frac{1}{(\sum_{j=1}^{N} Z_j^2)^{1/2}} \sum_{j=1}^{N} Z_j \exp 2\pi i (hx_j + ky_j + lz_j) \tag{6}$$

This is clearly the structure factor formula of a point atom structure, because no θ-dependent factors are present any more. It follows easily from expression 6 that for a structure of N equal atoms the maximum possible value of E_H is equal to $N^{1/2}$.

In the early beginning of Direct Methods also unitary structure factors U_H have been used extensively. For an equal atom structure the relation between U and E is given by $U_H = N^{-1/2} E_H$. Hence the maximum value of U_H is equal to unity.

<u>The triplet relation</u>

If two reflections H and K are both strong then the electron density is likely to be found in the neighbourhood of two sets of equidistant planes H and K at the same time, that is to say the electron density will be found near the lines of intersection of the planes H and K as indicated in projection in Fig. 3a. Geometrically the reflections H=K and −H−K fit on the grid formed by H and K. Now furthermore, a large $|E|$ for the reflection −H−K implies that the electron density will also peak in planes lying d_{-H-K} apart. It is therefore most likely that these planes run through the lines of intersection of the planes H and K, in other words that the three sets of planes have their lines of intersection in common (Fig. 3b).

The origin O has not been fixed yed and by choosing it at an arbitrary point in

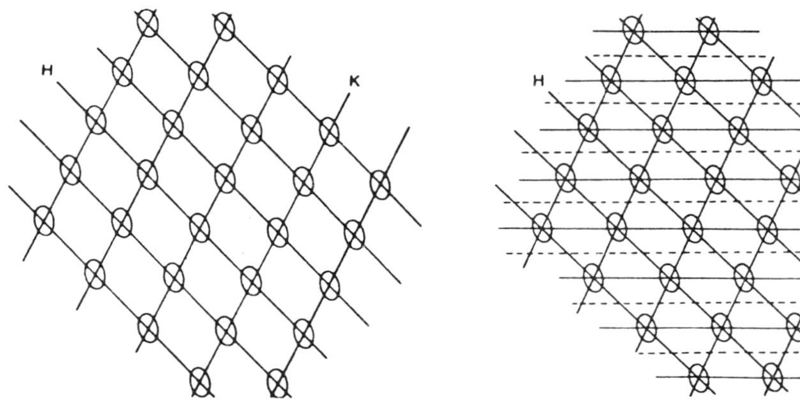

Fig. 3a   Fig. 3b

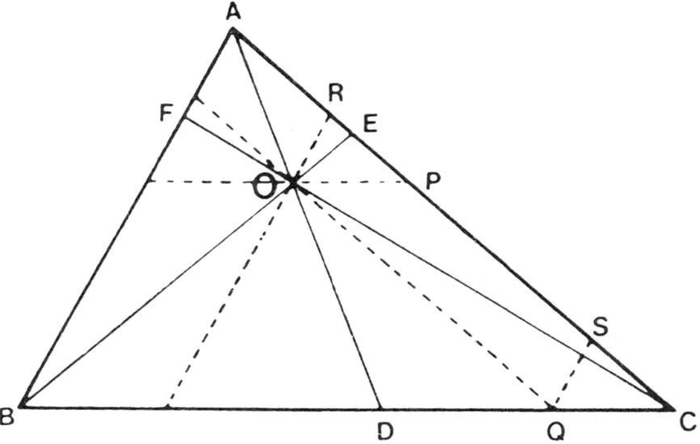

Fig. 4.   In an arbitrary triangle ABC an origin O is chosen arbitrarily. Theorem:
AO/AD + BO/BE + CO/CF = 2.
Proof: AO/AD = AP/AC; CO/CF = CR/AC; BO/BE = BQ/BC = AS/AC; because
RP = SC,   AP + CR + AS = 2AC

fig. 3b the triplet phase relationship

$$\varphi_H + \varphi_K + \varphi_{-H-K} = 0 \quad (\text{modulo } 2\pi) \tag{7}$$

follows easily from a planimetric theorem (Schenk, 1979), as illustrated in Fig. 4. Since
the origin has arbitrary been chosen, expression 7 is completely general; relations of this
type are usually called structure invariants, although a better name would be origin
invariants. Most probably the name structure invariant has been given by the early
researchers in the field because these invariants carry structure information as is shown
clearly in this paragraph.

In Fig. 3b the ideal situation is sketched; of course a small shift of the planes of
largest density of −H−K does not affect the reasoning given above. However, the most
unlikely position for these planes is the one indicated by the broken lines in Fig. 3b
because then the planes −H−K of largest electron density just clear the lines of
intersection of H and K. Therefore the triplet relation has a probability character and
this is emphasized by writing it as

$$\varphi_H + \varphi_K + \varphi_{-H-K} \simeq 0 \tag{8}$$

in which $\simeq$ means that the most probable value of 7 is 0. This relation holds more
likely the larger the structure factors of H, K and −H−K or the larger the value of the
product of the normalized structure factors $E_3 = N^{-1/2} |E_H E_K E_{-H-K}|$. The triplet relation
and its probabilistic basis will be dealt with extensively in next chapters.

Origin definition in P$\bar{1}$

For a phase determination in a particular space group always the restrictive rules
are employed as resulting from the conventional choice of the unit cell. In space group
P$\underline{1}$ the only symmetry element is a centre of symmetry, and then the convention as
given in the International Tables for X−ray Crystallography states that the unit cell has
to be chosen with its origin at one of the available centres (see fig. 5). As a result all
phases will have one out of two possible values only, 0 or π. In the unit cell eight
different centres of symmetry are present; they all are differently surrounded by the
electron density and lie half of the unit cell apart from each other. If one centre has
been chosen as origin, the other centres then have fractional coordinates $(1/2,0,0)$,
$(0,1/2,0)$, $(0,0,1/2)$, $(1/2,1/2,0)$, $(1/2,0,1/2)$, $(0,1/2,1/2)$ and $(1/2,1/2,1/2)$. Of course for
the description of the structure it does not matter which one of the eight centres has
been fixed as origin and thus one is free to choose any of them. In Direct Methods

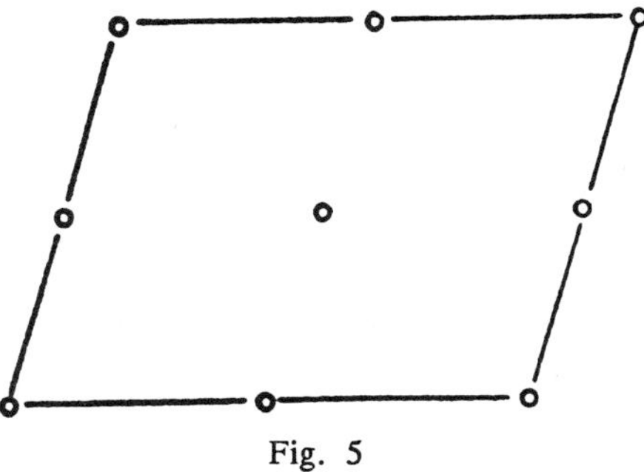

Fig. 5

generally such equivalent origin possibilities are referred to as permissable origins.

The problem now arises how to select one of the centres as the origin and firstly this process will be demonstrated in two dimensions. In fact all to be done is to ensure that the different centres of symmetry will be surrounded by different electron densities and that can be achieved by assigning numerical phases to a few terms in expression 1. In Fig. 6 the planes of reflection 210 are drawn. In case the phase of 210 is chosen to be 0 the maxima of the electron–density wave run through the centres 1 and 2 and the minima through 3 and 4. In case the sign of 210 is chosen to be π, the centres 3 and 4 are found at maxima and 1 and 2 at minima of the density wave. Thus the choice of either $\varphi_{210}=0$ or $\varphi_{210}=\pi$ breaks the group of four identically surrounded centres up into two groups: 1 and 2 have a different electron density than 3 and 4. By making a next choice, in which 1 and 3 form the one group and 2 and 4 the other, we can ensure that the electron density at all 4 centres in the projection is different. For instance this can be done with reflection (120) as is shown in Fig. 7. Thus by choosing $\varphi_{210}=0$ and $\varphi_{120}=0$ the four centres of symmetry get a different electron density and hence the origin has been fixed in the two– dimensional unit cell.

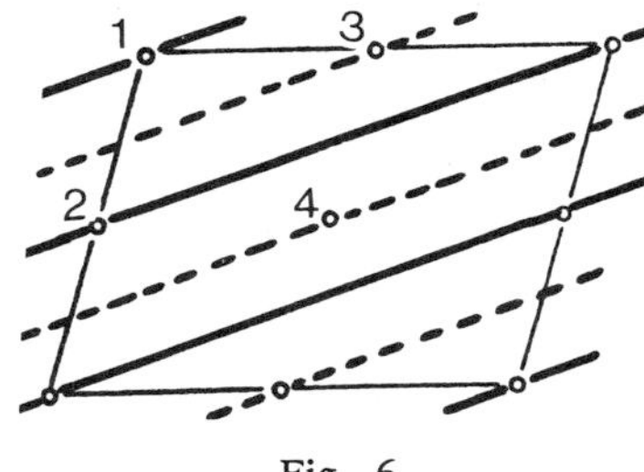

Fig. 6

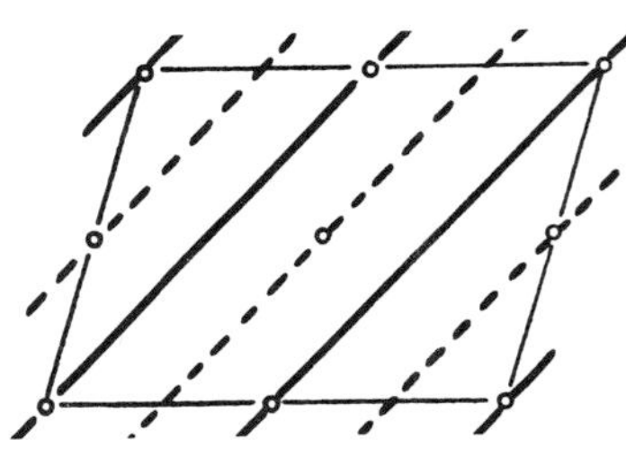

Fig. 7

It can also be seen easily that reflection 220 cannot be used for the origin definition at all. For this reflection all centres of symmetry lie either on the maximum of the density wave or in the minimum (see Fig. 8). The phases of these reflections are the same whatever origin is chosen and therefore unsuitable for origin selection.

Reflection 110 is, on the other hand, an alternative origin fixing reflexion, because it splits the centres again into two groups, now 1 and 4 in one and 2 and 3 in the other group (see Fig. 9). Thus the choice $\varphi_{110}=0$ and $\varphi_{210}=0$ equally well fixes the origin as $\varphi_{120}=0$ and $\varphi_{110}=0$ or as the above mentioned couple $\varphi_{120}=0$ and $\varphi_{210}=0$. Of course, in two dimensions not more than two reflections can be chosen to fix the origin; as soon as that has been done, all other reflections must be phased relative to the origin–fixing phases.

The above reasoning can be generalized, because all reflections with h even and k odd contribute similar to the electron density as reflection 210, all reflections with h odd and k even similar as 120, and all reflections with h odd and k odd similar as 110. This implies that any reflection in the above three groups can be used to fix the origin and not just the example reflections.

Real crystals are three dimensional and therefore three reflections can be chosen following a similar reasoning as for two dimensions. There are now eight centres of symmetry with different electron density surroundings. Therefore the process of fixing the origin consists of choosing the phases of three reflections, whose contributions to the Fourier summation 1 have to be such that the eight centres will be surrounded by a different electron density. Without going through the details again, the final procedure is given here, using the eight parity groups eee, ooe, eoe, oee, eeo, oeo, eoo, and ooo where the e's and o's stand for even and odd respectively, and the position of the letter reflects the index h, k and l respectively.

step 1:   No origin fixing with eee reflections.
step 2:   Choose origin defining reflection 1 from any of the 7 remaining parity groups. This parity group is used now and cannot be used again.
step 3:   Choose reflection 2 from any of the 6 remaining parity groups. This parity group cannot be used again and also the parity group of the sum of the indices of reflection 1 and 2 cannot be chosen any longer. (e.g. reflection 1 is 324, reflection 2 is 410, then used are oee, eoe, but also ooe (334+410=734).
step 4:   Choose reflection 3 from one of the remaining 4 parity groups.
step 5:   Assign phases (e.g. $\varphi=0$) to the three reflections.

When the three reflections are chosen such that they have a large $|E|$-value and many phase relationships with large reliabilities, then the origin will be optimally fixed and the successive phase propagation will follow an optimal path.

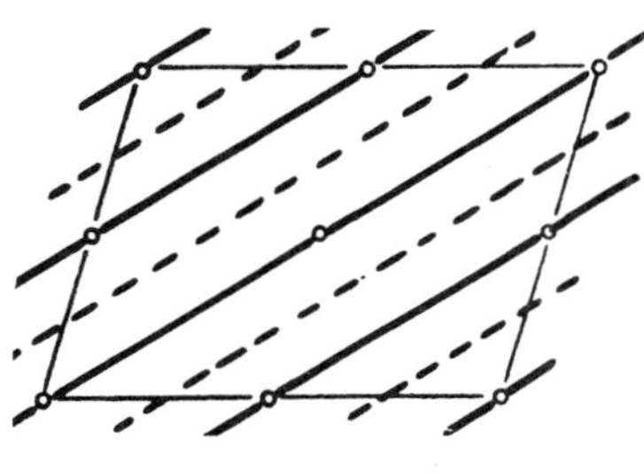

Fig. 8

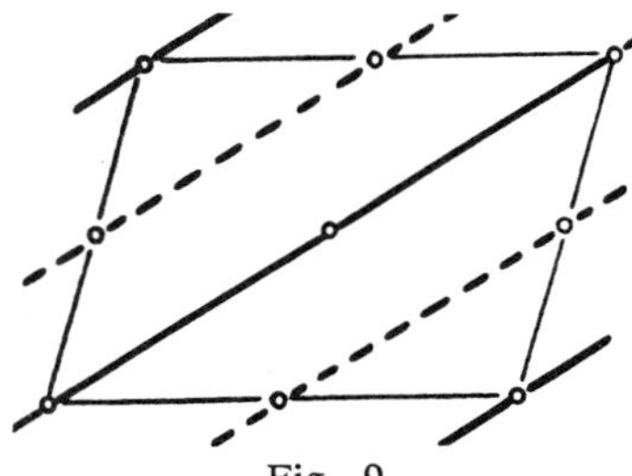

Fig. 9

Phase propagation

As stated before, in general it will not be possible to phase all reflections just from the origin defining phases. In space group P1 it is even impossible, because then the final phases will all be equal to 0, and the expectation is that approximately half of them will be π and the other half 0. So at least a fourth reflection, whose phase has to be π, must be added to the three origin defining ones in order to solve the structure.

However, in most cases four phases in the starting set will again not be enough to phase all reflections and more starting reflections will be needed. How this can be done will be explained in later chapters. For now it suffices to know that it is possible to construct starting sets consisting from more reflections than the origin defining reflections alone which are capable to access the other phases through the use of phase relationships such as the triplet relation 8 by means of a process called phase propagation or phase extension. Of course, the phases of the not–origin defining reflections are unknown and they are referred to as unknown symbols or ambiguities.

The phase propagation is carried out either symbolically (symbolic addition) or numerically (multi solution). In this introduction the latter path will be followed, later also symbolic addition will be dealt with. Therefore, the starting set gets first numerical values. This implies of course that ambiguities will get a number of different trial values, i.e. in centrosymmetric structures each ambiguity gets two different trial values, 0 or π. Thus, given a starting set with 3 origin defining reflections and 4 unknown extra reflections, there are $2^4 = 16$ starting points for the phase extension, and hence, in the multisolution approach, the phase extension has to be carried out 16 times. The actual phase extension uses in general just triplet relationships 8. When two of the reflections in 8 are known, the third one can be calculated and successively used in other triplets for calculating new phases again, this process being the actual phase propagation.

In order to carry out such a phase propagation for a structure by hand as will be done in the ASI, triplet relations for the strongest reflections and a pre–determined starting set are necessary. Different phase propagations will start from different starting points (different numerical values for the same starting phases), so simulating a multisolution approach. In a centrosymmetric structure it is easier to work with signs in stead of phases. A phase of 0 is then represented by a sign of +1 and a phase of π by a sign of −1 and the triplet relation 8 is at the same time replaced by

$$S_H S_K S_{-H-K} \simeq +1 \tag{9}$$

or $S_H = S_K S_{-H-K}$ because $S_H = S_{-H}$. In the propagation process new signs are simply calculated by using expression 9.

<u>References</u>

Schenk, H. (1979). J. Chem. Ed. 56, 383.

STRUCTURE FACTOR ALGEBRA: STRUCTURE INVARIANTS AND SEMINVARIANTS

George M. Sheldrick

Institut für Anorganische Chemie der Universität
Tammannstraße 4, D-3400 Göttingen, Federal Republic of Germany

1. INTRODUCTION

Structure factor algebra (Bertaut, 1956) provides a general method of incorporating space group symmetry into crystallographic programs. We shall consider some elementary applications, along the lines described by S.R. Hall at the 1970 Advanced Study Institute on Direct and Patterson Methods at Parma.

A convenient method for defining space group symmetry, employed in many program systems, is to specify the general equivalent positions in terms of fractional coordinates x, y, z, as listed in Volume A of *International Tables*. This is equivalent to a 3x3 rotation matrix R and a translation vector t for each equivalent position. There are simpler methods of indicating the lattice type and whether a centre of symmetry is present, so we shall assume throughout a set of R and t which do not include these operations. If the structure is centrosymmetric, we shall also assume that the origin is at a centre of symmetry. For the mth equivalent position:

$$\mathbf{x}_m = R_m \cdot \mathbf{x} + \mathbf{t}_m$$

which is a shorthand for:

Direct Methods of Solving Crystal Stuctures
Edited by H. Schenk, Plenum Press, New York, 1991

$$x_m = R_{11}\, x + R_{12}\, y + R_{13}\, z + t_1$$
$$y_m = R_{21}\, x + R_{22}\, y + R_{23}\, z + t_3$$
$$z_m = R_{31}\, x + R_{32}\, y + R_{33}\, z + t_3$$

By convention, the first equivalent position is always x, y, z.

Examples:

$P\bar{1}$: No need to specify centre of symmetry, so $m = 1$ only, with $R_{11} = R_{22} = R_{33} = 1$, all other elements of R and all elements of t are zero.

$C2$: Only the equivalent positions x, y, z and $-x, y, -z$ are specified, since the other two are generated by the C lattice. $m = 1$: $R_{11} = 1$, $R_{22} = 1$, $R_{33} = 1$; $m = 2$: $R_{11}=-1$, $R_{22}=1$, $R_{33}=-1$; rest of R and t zero.

$P4_1$: Equivalent positions and the corresponding R and t are:

$$m = 1 \qquad x,\ y,\ z \qquad R = \begin{pmatrix} 1 & 0 & 0 \\ 0 & 1 & 0 \\ 0 & 0 & 1 \end{pmatrix} \qquad t = \begin{pmatrix} 0 \\ 0 \\ 0 \end{pmatrix}$$

$$m = 2 \qquad -x,\ -y,\ 1/2+z \qquad R = \begin{pmatrix} -1 & 0 & 0 \\ 0 & -1 & 0 \\ 0 & 0 & 1 \end{pmatrix} \qquad t = \begin{pmatrix} 0 \\ 0 \\ 0.5 \end{pmatrix}$$

$$m = 3 \qquad -y,\ x,\ 1/4+z \qquad R = \begin{pmatrix} 0 & -1 & 0 \\ 1 & 0 & 0 \\ 0 & 0 & 1 \end{pmatrix} \qquad t = \begin{pmatrix} 0 \\ 0 \\ 0.25 \end{pmatrix}$$

$$m = 4 \qquad y,\ -x,\ 3/4+z \qquad R = \begin{pmatrix} 0 & 1 & 0 \\ -1 & 0 & 0 \\ 0 & 0 & 1 \end{pmatrix} \qquad t = \begin{pmatrix} 0 \\ 0 \\ 0.75 \end{pmatrix}$$

2. EQUIVALENT REFLECTIONS AND PHASE SHIFTS

Equivalent reflections determine the Laue symmetry of the diffraction pattern: they have the same intensity as each other, but not necessarily the same phase angles. Their indices may be generated as follows:

$$h_m = R_{11}\ h + R_{21}\ k + R_{31}\ \ell$$

$$k_m = R_{12}\ h + R_{22}\ k + R_{32}\ \ell$$

$$\ell_m = R_{13}\ h + R_{23}\ k + R_{33}\ \ell$$

Note that the order of multiplication is *different* from that in the expression for x_m, y_m and z_m. We shall refer to the reflection $h = (h,k,\ell)$ (which corresponds to $m = 1$) as the *prime* reflection; it may be chosen anywhere in reciprocal space. The *phase shift* of the *equivalent* reflection $h_m = (h_m, k_m, \ell_m)$ is the phase angle in radians which must be added to the phase angle ϕ of the prime reflection to give ϕ_m :

$$\phi_m = \phi + \text{'phase shift'} = \phi - 2\pi h \cdot t_m = \phi - 2\pi(ht_1 + kt_2 + \ell t_3)$$

In the absence of anomalous dispersion, Friedel's law requires that $F_{-m} = F_m$ and $\phi_{-m} = -\phi_m$, where $-m$ refers to the reflection $(-h_m, -k_m, -\ell_m)$, which enables us to generate the remaining equivalent reflections. Thus in $P4_1$ (see above) the reflection $(1,2,3)$ will have the following equivalents, all with the same F and E values:

	h_m	ϕ_m	$-h_m$	ϕ_{-m}
$m = 1$	1 2 3	$+\phi$	-1 -2 -3	$-\phi$
$m = 2$	-1 -2 3	$+\phi-3\pi$	1 2 -3	$-\phi+3\pi$
$m = 3$	2 -1 3	$+\phi-3\pi/2$	-2 1 -3	$-\phi+3\pi/2$
$m = 4$	-2 1 3	$+\phi-9\pi/2$	2 -1 -3	$-\phi+9\pi/2$

[because of the 2π degeneracy, $+9\pi/2$ is equivalent to $+\pi/2$ etc.].

3. REFLECTIONS WITH RESTRICTED PHASE ANGLES

In centrosymmetric structures reflection phase angles are restricted to 0 or π (assuming that the centre of symmetry is at the origin). Similarly in non-centrosymmetric structures, reflections which belong to centrosymmetric projections are restricted to two values which differ by π.

To derive the allowed values of the restricted phases, we need to find an equivalent reflection of which the Friedel opposite coincides with the prime reflection, i.e. $-h_m = h$. The phase angle of $-h_m$, calculated as described in section 2, must be the same as that of h since they are the same reflection:

$$-[\phi-2\pi(ht_1+kt_2+\ell t_3)] = \phi$$

therefore $$2\phi = 2\pi(ht_1+kt_2+\ell t_3) \; [+2n\pi \text{ with } n \text{ integral}]$$

and $$\phi = \pi(ht_1+kt_2+\ell t_3) + n\pi$$

Thus, because of the 2π degeneracy in the value of 2ϕ, there will be two possible values of ϕ (the phase of the *prime* reflection) which differ by π. Note that it is possible for equivalent reflections to have different phase restrictions.

Example: consider the reflection $(0,1,6)$ in $P2_12_12_1$:

equivalent position		h_m	ϕ_m	$-h_m$	ϕ_{-m}
$m = 1$	$x,\ y,\ z$	0 1 6	$+\phi$	0 -1 -6	$-\phi$
$m = 2$	$\tfrac{1}{2}-x,\ -y,\ \tfrac{1}{2}+z$	0 -1 6	$+\phi-6\pi$	0 1 -6	$-\phi+6\pi$
$m = 3$	$\tfrac{1}{2}+x,\ \tfrac{1}{2}-y,\ -z$	0 -1 -6	$+\phi-\pi$	0 1 6	$-\phi+\pi$
$m = 4$	$-x,\ \tfrac{1}{2}+y,\ \tfrac{1}{2}-z$	0 1 -6	$+\phi-7\pi$	0 -1 6	$-\phi+7\pi$

The Friedel opposite generated from $m = 3$ gives the prime reflection, so equating the phase angles:

$$\phi_{-3} = -\phi+\pi = \phi \qquad\qquad \text{thus } \phi = (\pi/2)+n\pi = \pi/2 \text{ or } 3\pi/2.$$

4. SYSTEMATIC ABSENCES

A reflection will be systematically absent when any equivalent reflection has identical indices to those of the prime reflection, and has a phase shift from it which is not a multiple of 2π. If F (and E) were not zero, we would be in the impossible situation of having two different phase angles for the same reflection. By way of example, consider $(0,0,2)$ and $(0,0,4)$ in $P4_1$:

	h_m			ϕ_m		h_m			ϕ_m
$m = 1$	0	0	2	ϕ		0	0	4	ϕ
$m = 2$	0	0	2	$\phi - 2\pi$		0	0	4	$\phi - 4\pi$
$m = 3$	0	0	2	$\phi - \pi$		0	0	4	$\phi - 2\pi$
$m = 4$	0	0	2	$\phi - 3\pi$		0	0	4	$\phi - 6\pi$

(0,0,2) is systematically absent but (0,0,4) is allowed. For many direct methods procedures, it is a sensible precaution to eliminate systematic absences from the data set.

5. STATISTICAL WEIGHTS

As will be discussed in section A10 of this course, the normalised structure factors are derived by:

$$E_h^2 = F_h^2 \; / \; \epsilon \cdot \Sigma f_h^2$$

where the summation is over all atoms in the cell, and f is the product of scattering factor and temperature factor for one atom. ϵ is a factor which is needed to make the mean E^2 equal to unity for any class of reflection; it corrects e.g. for systematic absences and the resulting redistribution in intensity. Thus in the above example of the (0,0,4) reflection in $P4_1$, since 3 out of every 4 reflections of the type $(0,0,\ell)$ are systematically absent, ϵ is 4. There is a simple algorithm for finding ϵ: in the noncentrosymmetric space groups, ϵ is the number of equivalent reflections for which h_m = h; in centrosymmetric space groups, ϵ is the number of equivalent reflections for which h_m = h or -h. The prime reflection is included in this sum, so ϵ is 1 for *general* reflections.

In the case of $P4$, this algorithm gives ϵ = 4 for all $(0,0,\ell)$ reflections; this time none of them are systematically absent. The intensity gained (in a statistical sense) by the $(0,0,\ell)$ reflections has been lost by other reflections close to the c* axis in reciprocal space; the effect arises because there is a minimum distance an atom can lie from a 4-fold axis. A similar effect arises for mirror planes etc. (Wilson, 1964). In principle the reflections bordering on c* should have $\epsilon < 1$, but in practice they are rarely corrected for this.

Giacovazzo (1974a,b) has argued convincingly that additional statistical weights should be applied to phase relations. For triplet phase relations in $P2_12_12_1$, the weights are 0, 1, $2^{\frac{1}{2}}$ or 2 depending on the types of reflections involved. The resulting modified tangent formula appears to give more accurate phase estimates than the normal formula. These weights are derived by structure factor algebra, but the calculations are too complicated to discuss here.

6. THE EFFECT OF ORIGIN AND ENANTIOMORPH CHOICE ON PHASES

In $P4_1$, the whole structure may be pushed parallel to the 4_1 axis without violating the space group: we have simply chosen to define the origin of the unit cell at a different point relative to the structure. Adding $\Delta\mathbf{z}$ to the fractional coordinates of all atoms will multiply all terms in the structure factor sum by:

$$exp(2\pi i h \cdot \mathbf{x}) = exp(2\pi i \ell.\Delta z)$$

which will add $2\pi\ell.\Delta z$ onto all phase angles. Thus the phases of reflections with $\ell = 0$ are unchanged: they only determine the projected structure. Possible phase shifts of this type must be taken into account when comparing two phase sets to see if they are 'similar', i.e. correspond to the same structure.

As a general rule, reversing the signs of all phase angles converts a non-centrosymmetric structure into its enantiomorph; for certain pairs of space groups (e.g. $P4_1$ and $P4_3$) the space group is converted into its enantiomorph in the process. The rule implies that inversion of the whole structure through a point at the origin produces the enantiomeric structure, which is true for the origin chosen in *International Tables* for all space groups except *Fdd2*, $I4_1$, $I4_122$, $I4_1md$, $I4_1cd$, $I\bar{4}2d$ and $F4_132$ (Parthé & Gelato, 1984). For example in *Fdd2*, inversion in the origin gives the same structure, whereas inversion in a point at $(1/8,1/8,0)$ gives the opposite enantiomer. The latter changes the phases of the reflections $(h,k,0)$ with $h+k = 4n+2$ (which have restricted values of 0 or π) from ϕ to $\pi-\phi$.

7. STRUCTURE SEMINVARIANTS

In $P4_1$ there are 4_1 axes at $(0,0,z)$ and $(\frac{1}{2},\frac{1}{2},z)$ and diad axes at $(0,\frac{1}{2},z)$ and $(\frac{1}{2},0,z)$. The only move of the structure in the **ab** plane which leaves the expressions for the equivalent positions unchanged is $(\frac{1}{2},\frac{1}{2},0)$. Since we move every atom, all terms in the structure factor sum are multiplied by $exp(2\pi ih.\mathbf{x}) = exp(\pi i(h+k))$. The reflections with $h+k$ odd will then change phase by π, and those with $h+k$ even will have unchanged phases. Since the phases of the reflections $(h,k,0)$ were unchanged by moving the structure along the **c** axis, the phases of $(h,k,0)$ with $(h+k)$ *even* are unchanged for *any* translation of the structure relative to the origin of the unit-cell which *leaves the symmetry elements unchanged*. These phases are examples of *structure seminvariants*.

8. STRUCTURE INVARIANTS

Adding a vector $\Delta\mathbf{x}$ to the fractional coordinates of all atoms will change the sum of the phase angles $\phi_\mathbf{h} + \phi_\mathbf{k} + \phi_\mathbf{l}$ of a triplet phase relation by:

$$(2\pi\mathbf{h} \cdot \Delta\mathbf{x}) + (2\pi\mathbf{k} \cdot \Delta\mathbf{x}) + (2\pi\mathbf{l} \cdot \Delta\mathbf{x}) = 2\pi(\mathbf{h}+\mathbf{k}+\mathbf{l}) \cdot \Delta\mathbf{x}$$

which is identically zero since $\mathbf{h} + \mathbf{k} + \mathbf{l}$ is zero. The triplet phase sum is thus *invariant* with respect to the position of the structure relative to the origin of the unit-cell (whether or not the positions of the symmetry elements change). The triplet phase sum is an example of a *structure invariant*. By an analogous argument the quartet phase sum $\phi_\mathbf{h} + \phi_\mathbf{k} + \phi_\mathbf{l} + \phi_\mathbf{m}$ with $\mathbf{h} + \mathbf{k} + \mathbf{l} + \mathbf{m} = 0$ is also a *structure invariant*.

The concepts of *structure invariants* and *seminvariants* were established in a series of epic papers by Karle and Hauptman (e.g. 1953, 1956). Tables of seminvariants for all space groups may be found in *International Tables*, Vol. IV (for a correction see Lessinger & Wondratschek, 1975), and improved tables for centred lattices are given by Hall (1982). The identification of seminvariants by means of structure factor algebra has been discussed by Gramaccioli & Zechmeister (1972).

9. PHASE RELATIONS AND SYMMETRY

The phase problem is essentially solved if we can obtain approximately correct phases for a list of independent prime reflections with E greater than some limit, say 1.2. We shall refer to this list as the *unique* list; although the reflections could be chosen anywhere in reciprocal space, it is convenient to choose the octant with h, k and $\ell \geqslant 0$ in the orthorhombic case etc. The remaining reflections can be generated from this list as described in section 2. Phase relations are found by searching the whole of reciprocal space, and then converting to relations between reflections in the unique list by applying the appropriate phase shifts and sign changes. The most widely used phase relations are *triplet phase relations* for which:

$$h + k + 1 = 0 \quad \text{and} \quad E_h, \ E_k \text{ and } E_1 \text{ are all large.}$$

We can find all such relations if h is in the unique list and k and 1 are allowed to be equivalent reflections of those in the unique list. We should need to store the following information for such a phase relation:

$$n_h, \ \pm n_k, \ \pm n_1, \ \phi_o \quad \text{and} \quad weight$$

n are code numbers of reflections in the unique list, the $\pm$ signs are applied to the appropriate phases, and ϕ_o is the combined phase shift.

10. SIGMA-1 RELATIONS AS SPECIAL TRIPLE PHASE RELATIONS

In the course of searching for the triplet phase relations, a number of relations are found of the type:

$$h = k' + k'' \quad (E_h, \ E_{k'} = E_{k''} \text{ all large})$$

where k' and k" are equivalent reflections generated from the same prime reflection, with $\phi_{k'} = \phi + s'$ and $\phi_{k''} = -\phi + s''$. Since ϕ cancels in this special case,

$$\phi_h \approx \phi_{k'} + \phi_{k''} = s'+s''$$

and we have an estimate for the phase of ϕ_h, *which must be a seminvariant*. To combine all the indications for a particular ϕ_h, we sum the real and imaginary contributions:

$$A = \sum W.cos(s'+s'') \qquad\qquad B = \sum W.sin(s'+s'')$$

where the weight $[W = \sigma_3\sigma_2^{-3/2}E_h(E_k^{\,2}-1)]$ is less than the weight of a normal triplet phase relation because **k'** and **k"** are not independent. Finally we project the result along γ (the first of the two allowed phase angles for h), and apply the centrosymmetric probability formula:

$$P = \tfrac{1}{2} + \tfrac{1}{2}.tanh[(A.cos\gamma +B.sin\gamma)/2]$$

where P is the probability that $\phi_h = \gamma$. The probability that $\phi_h = \gamma+\pi$ is then $1-P$. This (highly oversimplified) approach may be used for all space groups, but does not find every so-called $\sum_1$ indication in some high symmetry space groups. It is also possible to have $\phi_{k'} = \phi+s'$ and $\phi_{k''} = \phi+s''$ in the above example. If the phase of ϕ_k is restricted to two values which differ by π (centrosymmetric structure or projection), then $2\phi_k$ is known, and we can add the indication to the summations for ϕ_h.

Note that the (E^2-1) term enables the $\sum_1$ formula to give (weak) indications of phase shifts of π in *symmorphic* space groups such as $P\bar{1}$ of *C2*, unlike the triplet phase relation.

EXERCISES TO A2: STRUCTURE FACTOR ALGEBRA

$P2_12_12_1$	x, y, z	$P4_12_12$	x, y, z
	$(\tfrac{1}{2})$+x, $(\tfrac{1}{2})$-y, -z		$(\tfrac{1}{2})$-y, $(\tfrac{1}{2})$+x, $(\tfrac{1}{4})$+z
	$(\tfrac{1}{2})$-x, -y, $(\tfrac{1}{2})$+z		y, x, -z
	-x, $(\tfrac{1}{2})$+y, $(\tfrac{1}{2})$-z		$(\tfrac{1}{2})$-x, $(\tfrac{1}{2})$+y, $(\tfrac{1}{4})$-z
			-x, -y, $(\tfrac{1}{2})$+z
$P3_1$	x, y, z		$(\tfrac{1}{2})$+y, $(\tfrac{1}{2})$-x, (3/4)+z
	-y, x-y, (1/3)+z		-y, -x, $(\tfrac{1}{2})$-z
	y-x, -x, (2/3)+z		$(\tfrac{1}{2})$+x; $(\tfrac{1}{2})$-y, (3/4)-z

1. In the space group $P4_12_12$, the reflection $(1,1,3)$ generates the following set of equivalents (remember to transpose R!): $h_1 = 1,1,3$; $h_2 = 1,-1,3$; $h_3 = 1,1,-3$ $h_4 = -1,1,-3$; $h_5 = -1,-1,3$; $h_6 = -1,1,3$; $h_7 = -1,-1,-3$; $h_8 = 1,-1,-3$. Since no two of these are identical, the ϵ-value is 1 and the reflection cannot be systematically absent. However $h_1 = h_{-7}$, so $\phi = -(\phi-2\pi(3\ \frac{1}{2}))$, thus $\phi = \pi/2$ or $3\pi/2$.

Use this approach for the reflections $(0,5,0)$, $(3,-3,0)$ and $(2,0,3)$ in this space group to establish (a) whether they are systematically absent, and if not (b) the ϵ-value and (c) the phase restrictions, if any.

2. Why is the triplet phase relation:

$$\phi_{(3,2,0)} + \phi_{(-3,0,-4)} + \phi_{(0,-2,4)} \simeq 0$$

in the space group $P2_12_12_1$ *useless* for direct methods? (Hint: are the reflection phases restricted?). Is this three phase sum a structure invariant or seminvariant (or neither)? Supplementary question for advanced students: it is however very useful in the *experimental determination of absolute structure:* why?

3. If $E_{(h,k,0)}$ is very large in the space group $P3_1$, are there preferred values for the corresponding phase? Is this reflection a structure invariant or seminvariant (or neither) when (a) $h-k = 3n$, (b) $h-k \neq 3n$?

REFERENCES

Bertaut, E.F. (1956). Acta Cryst., 9, 769-770

Giacovazzo, C. (1974a,b). Acta Cryst., **A30**, 626-630 and 631-634

Gramaccioli, C.M. & Zechmeister, K. (1972). Acta Cryst., **A28**, 154-158

Hall, S.R. (1982). Acta Cryst., **A38**, 874-875

Hauptman, H. & Karle, J. (1953). *Solution of the Phase Problem I. The Centrosymmetric Crystal*. A.C.A. Monograph No. 3. Pittsburgh: Polycrystal Book Service

Hauptman, H. & Karle, J. (1956). Acta Cryst., 9, 45-55

Lessinger, L. & Wondratschek, H. (1975). Acta Cryst., **A31**, 521

Parthé, E. & Gelato, L.M. (1984). Acta Cryst., **A40**, 169-183

Wilson, A.J.C. (1964). Acta Cryst., 17, 1591-1592

STRUCTURE INVARIANTS, SEMINVARIANTS AND ORIGIN DEFINITION

H. Krabbendam

Laboratorium voor Kristal- en Structuurchemie
Bijvoet Centrum voor Biomoleculair Onderzoek
Rijksuniversiteit Utrecht, The Netherlands

1. INTRODUCTION

In a crystal a system of basic vectors **a**, **b**, **c** is chosen that complies with the crystal symmetry. These basic vectors represent the translational symmetry of the crystal; this means that in the three-dimensional lattice, generated by the basic vectors, the lattice points represent positions that are translational equivalent, i.e. points that have identical surroundings in the same orientation. In order to define a crystal structure completely it is sufficient to specify the atomic positions within one translational unit, the unit cell, of the lattice. However, the lattice, and thus the origin of the unit cell, can be moved freely, in a parallel fashion, within the crystal; in order to specify the atomic coordinates unambiguously, the origin of the unit cell must be defined in a unique way. Though the position of the origin can be chosen anywhere, some positions are preferred, dependent on the symmetry of the crystal. An example is space group $P\bar{1}$, in which one of the eight different positions of the centres of symmetry is usually chosen as the origin position (Fig. 1.1). The choice of origin is not trivial, as the eight positions are not equivalent (do not have the same surroundings).

The crystal structure, i.e. the elctron-density function, is found from the observed structure-factor amplitudes by a Fourier summation, provided the structure-factor phases have been

Direct Methods of Solving Crystal Stuctures
Edited by H. Schenk, Plenum Press, New York, 1991

estimated (the phase problem). However, the structure-factor phases depend on the choice of the origin of the unit cell and also on the choice of the enantiomorph (the phases change sign upon inversion of the atomic coordinates in non-centrosymmetric space groups). Before a Fourier summation can be performed one must be sure that all phases refer to the same origin and the same enantiomorph. One of the concerns in the application of phase-determining methods is to ascertain that this is the case.

The most important phase-determining methods are the "direct methods". In this paper some of the principles that are important to these methods, and that have all to do with origin and enantiomorph definition, will be dealt with; these principles include: phase modulation upon shift of origin, the occurrence of quantities that are independent on the choice of origin whatsoever (structure invariant phase sums), or of quantities that are independent on the choice of origin from a limited number of possible positions, as in $P\bar{1}$ (structure seminvariant phases or phase sums), and origin/ and enantiomorph defintion by the specification of the values of a limited number of well-chosen structure-factor phases (dependent on the space group). The latter subject, origin fixation, will be dealt with on the basis of some examples; a fuller treatment can be found in (1), (2), (3) (see also references therein).

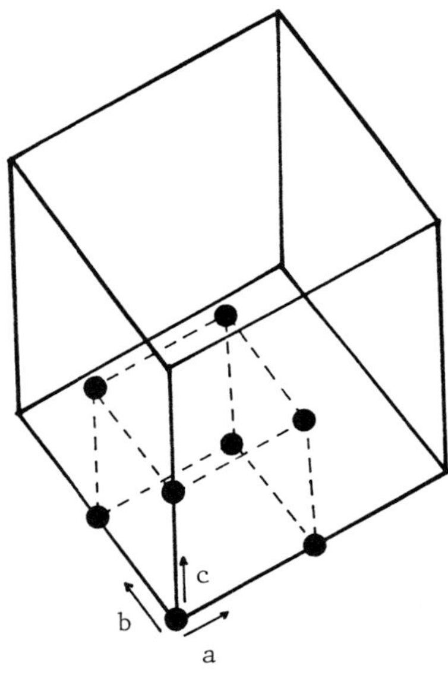

Fig. 1.1. The eight positions of the inversion
centres in space group $P\bar{1}$.

22

2. PHASE MODULATION

If the origin of the unit cell is shifted from a point 0 to a point 0' over a vector $\mathbf{t}$ (Fig. 2.1) then a quantitive expression that gives the change in the structure-factor phase $\phi_\mathbf{h}$ can be found as follows:

If $F_\mathbf{h}$ is defined relative to the original origin 0 and $F_\mathbf{h}'$ relative to the new origin 0', then

$$F_\mathbf{h} = \sum_{i=1}^{N} f_i \, \exp\{2\pi\mathbf{h}.\mathbf{r}_i\} \tag{2.1}$$

and

$$F_\mathbf{h}' = \sum_{i=1}^{N} f_i \, \exp\{2\pi i \, \mathbf{h}.(\mathbf{r}_i-\mathbf{t})\}.$$

$$= (\sum_{i=1}^{N} f_i \, \exp 2\pi i \, \mathbf{h}.\mathbf{r}_i) \, \exp\{-2\pi i \mathbf{h}\cdot\mathbf{t}\}$$

$$= F_\mathbf{h} \, \exp(-2\pi i \mathbf{h}.\mathbf{t}) \tag{2.2}$$

$$= |F_\mathbf{h}| \, \exp\{-2\pi i \, \mathbf{h}.\mathbf{t} + i\phi_\mathbf{h}\}$$

so $\quad |F_\mathbf{h}'| \; = \; |F_\mathbf{h}|$ $\hfill$ (2.3)

and $\quad \phi_\mathbf{h}' = -2\pi\mathbf{h}.\mathbf{t} + \phi_\mathbf{h}$ $\hfill$ (2.4)

So the conclusion is that the structure-factor amplitude is independent of the choice of origin (not surprising, as it is an observed quantity) and the phase is modulated by an amount $-2\pi\mathbf{h}.\mathbf{t}$, if $\mathbf{t}$ the shiftvector is from the old to the new origin.

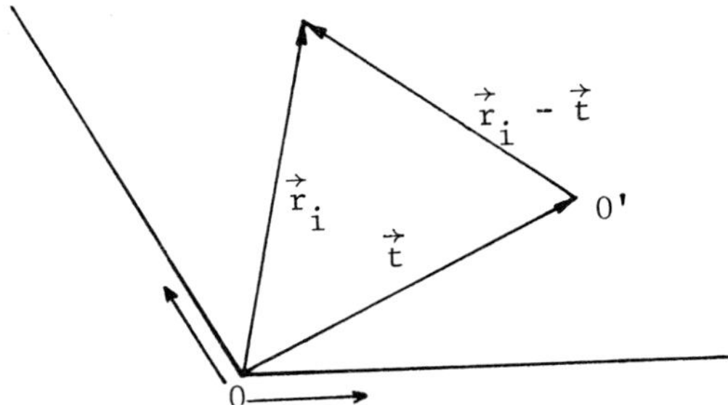

Fig. 2.1. Origin shift from 0 to 0'.

3. STRUCTURE INVARIANTS

Structure invariants are quantities that are independent on the choice of origin, where-ever the origin may happen to be. The products of structure factors

$$F_{h_1}^{a_1} \, F_{h_2}^{a_2} \, F_{h_3}^{a_3} \, \ldots \ldots \tag{3.1}$$

and, consequently, the sum of phases

$$a_1 \phi_{h_1} + a_2 \phi_{h_2} + a_3 \phi_{h_3} + \ldots \ldots \tag{3.2}$$

are structure invariants, provided

$$a_1 \mathbf{h}_1 + a_2 \mathbf{h}_2 + a_3 \mathbf{h}_3 + \ldots \equiv 0 \tag{3.3}$$

Prove

If $F_\mathbf{h}$ and $F'_\mathbf{h}$ are the structure factors that correspond to the origins 0 and 0' respectively, and $\mathbf{t}$ is the vector between 0 and 0' (Fig. 2.1) then, using (2.2):

$$F_{h_1}^{'a_1} F_{h_2}^{'a_2} F_{h_3}^{'a_3} \ldots = F_{h_1}^{a_1} F_{h_2}^{a_2} F_{h_3}^{a_3} \ldots \exp\{-2\pi i (a_1 h_1 + a_2 h_2 + a_3 h_3 + \ldots)\}$$

$$= F_{h_1}^{a_1} F_{h_2}^{a_2} F_{h_3}^{a_3} \ldots$$

if (3.3) is valid.

24

Important structure invariants in direct methods are

(i) the triple product:
   $F_hF_kF_{-h-k} = |F_hF_kF_{-h-k}| \exp\{i(\phi_h+\phi_k+\phi_{-h-k})\}$, where $\phi_h+\phi_k+\phi_{-h-k}$
   is called the triplet invariant or three-phase invariant.
(ii) the quartet $F_hF_kF_lF_{-h-k-l}$, of which the phase $\phi_h+\phi_k+\phi_l+\phi_{-h-k-l}$
   is called the quartet invariant of four-phase invariant

4. ALLOWED ORIGINS

In many spacegroups preferred positions of the origin present themselves quite logically: in centrosymmetric space groups the natural choice of the origin is on one of the centres of symmetry, in P2 one can choose between positions on one of the four two-fold axes, in P4 there are two sets of positions that one can choose from, i.e. either on one of the two fourfold axes or on one of the two twofold axes. To each choice (for a special class of origins) there corresponds a special shape of the expressions for the equivalent positions and, consequently, for the shape of the structure-factor expression. The different classes of allowed origins are called equivalence classes (e.g. in P4 there are two equivalence classes, one is the class of fourfold axes and the other is the class of twofold axes). Shifts of origin from one allowed origin position to another within one class let the shape of the structure-factor expression unchanged.

Example

In $P\bar{1}$ the origin can be chosen anywhere if the coordinates of the equivalent positions are simply chosen as x,y,z; then the structure-factor expression is in the most general form (2.1). If the equivalent positions are defined to be x,y,z and $\bar{x},\bar{y},\bar{z}$, with the corresponding structure-factor expression:

$$F_h = 2\sum_{j=1}^{N/2} F_h \cos2\pi\mathbf{h}.\mathbf{r}_j \tag{4.1}$$

then the allowed positions of the origin are the eight centres of symmetry (see paragraph 1); these eight centres of symmetry form an equivalence class (the only one in this space group).

In order to arrive at a general procedure to derive the allowed origin positions within an equivalence class of some specified space group, we will first give an example and then generalize the result.

In P2 the point group operation that represents the rotation around the two-fold axis is given by the matrix

$$C = \begin{pmatrix} \bar{1} & & \\ & 1 & \\ & & \bar{1} \end{pmatrix}$$

This means that, by the two-fold operation, a point P, with coordinates x,y,z relative to a two-fold axis at O_1 (Fig. 4.1), is mapped to a point P_1 with coordinates $(\bar{x},y,\bar{z})$, as is shown by

$$\begin{pmatrix} -1 & & \\ & 1 & \\ & & -1 \end{pmatrix} \begin{pmatrix} x \\ y \\ z \end{pmatrix} = \begin{pmatrix} -x \\ y \\ -z \end{pmatrix}$$

or, if the points P and P_1 have position vectors $\mathbf{r}$ and $\mathbf{r}_1$ respectively:

$$C\mathbf{r} = \mathbf{r}_1$$

If there is a two-fold axis at another point O_2 as well, and the vector between O_1 and O_2 is $\mathbf{d}$, then the position of P relative to O_2 is $\mathbf{r}-\mathbf{d}$. Rotation of P through the two-fold axis at O_2 results in a point P_2 with a position vector $C(\mathbf{r}-\mathbf{d})$. Now the following observation is essential: P_1 and P_2 have the same surroundings as P, only rotated through 180°, so P_1 and P_2 have the same surroundings in the same orientation; this means that P_1 and P_2 are translationally equivalent, so the vector between P_2 and P_1 must be a lattice vector (indicated as $\mathbf{m}(m_1,m_2,m_3)$). From Fig. 4.1. it then follows that

$$C\mathbf{r} = \mathbf{d} + C(\mathbf{r}-\mathbf{d}) + \mathbf{m} \tag{4.2}$$

or

$$(C - I)\mathbf{d} = \mathbf{m} \tag{4.3}$$

where I is the unit matrix. Written in components:

$$\begin{pmatrix} \bar{2} & 0 & 0 \\ 0 & 0 & 0 \\ 0 & 0 & \bar{2} \end{pmatrix} \begin{pmatrix} d_1 \\ d_2 \\ d_3 \end{pmatrix} = \begin{pmatrix} m_1 \\ m_2 \\ m_3 \end{pmatrix}$$

or

$$-2d_1 = m_1$$
$$0d_2 = m_2$$
$$-2d_3 = m_3$$

For different lattice vectors (m_1, m_2, m_3) the following solutions are obtained:

m_1	m_2	m_3	d_1	d_2	d_3
0	0	0	0	y	0
1	0	0	1/2	y	0
0	0	1	0	y	1/2
1	0	1	1/2	y	1/2

$$(4.4)$$

where y can have any value. The results agree with what we know about the four possible positions of the two-fold axes in P2.

Equations (4.2) and (4.3) are generally valid for any point-group operation in any space group; only, in order to find all allowed origins in an equivalence class the complete independent set of point group operations C_i valid for that class must be applied, leading to a number of equations:

$$(C_i - I)\ \mathbf{d} = \mathbf{m} \tag{4.5}$$

form which $\mathbf{d}$ can be solved.

5. STRUCTURE SEMINVARIANTS

Structure factors or products of structure factors, or structure-factor phases or sums of structure-factor phases, whose values are insensitive to origin shifts $\mathbf{d}$ between allowed origins of an equivalence class, are called structure seminvariants.

If an (allowed) origin shift is applied then a seminvariant
phase $\phi_{\mathbf{h}_S}$ should remain unchanged, i.e. the phase-shift should
be zero (modulus 2π). The phase shift is $-2\pi\,\mathbf{h}_S.\mathbf{d}$ (see (2.4)),
so the requirement for $\phi_{\mathbf{h}_S}$ to be a seminvariant is

$$\mathbf{h}_S.\mathbf{d} = n \quad (n \text{ is an integer}) \tag{5.1}$$

for any of the allowed phase shifts $\mathbf{d}$.

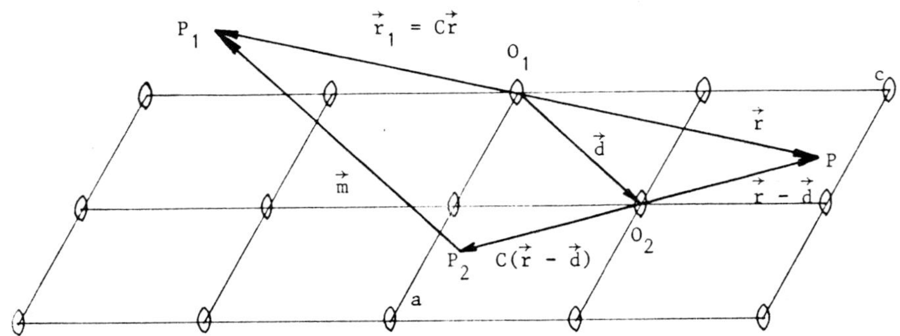

Fig. 4.1. Derivation of the allowed origin positions in P2 (equation 4.3)

The solution to (5.1) will be a vector $\mathbf{h}_S(h_S, k_S, l_S)$ of which
the components h_S, k_S, l_S will be subject to certain restrictions.

Example

In $P2_1$ it is sufficient to regard the allowed origin shifts $(0,y,0)$,
$(1/2,y,0)$, $(0,y,1/2)$; the shift $(1/2,y,1/2)$ does not give anything
new. Application of (5.1) gives:

$$(h_S\ k_S\ l_S) \begin{pmatrix} 0 \\ y \\ 0 \end{pmatrix} = k_S y = n_1 \rightarrow k_S = 0$$

$$(h_S\ k_S\ l_S) \begin{pmatrix} 1/2 \\ y \\ 0 \end{pmatrix} = 1/2\,h_S + k_S y = 1/2\,h_S = n_2$$

$$h_S = 2n_2 \text{ i.e. } h_S = 0 \bmod (2)$$

28

$$(h_s \; k_s \; l_s) \begin{pmatrix} 0 \\ y \\ 1/2 \end{pmatrix} = 1/2 \; l_s + k_s y = 1/2 \; l_s = n_3$$

$$l_s = 2n_3 \text{ i.e. } l_s = 0 \bmod (2)$$

The conclusion is that in P2 the phase ϕ_{gog} (g = "gerade", i.e. even) is a structure seminvariant. In the same way it is easy to prove that $\phi_{uku} + \phi_{u\bar{k}u}$ is a structure seminvariant (a so-called two-phase structure seminvariant); here u means "ungerade", i.e. odd.

Example

In $P3_1$ the structure-factor expression can be chosen such that the origin is on one of the three 3_1 axes; the allowed origin shifts are (0 0 z), (2/3 -1/3 z) or (1/3 -2/3 z), for any z; these allowed shifts can be derived by application of the procedure described in paragraph 4 or by inspection of fig. 5.1. Application of (5.1) gives

$$(h_s \; k_s \; l_s) \begin{pmatrix} 0 \\ 0 \\ z \end{pmatrix} = -l_s z = n_1 \rightarrow l_s = 0 \bmod 0$$

$$(h_s \; k_s \; l) \begin{pmatrix} 2/3 \\ 1/3 \\ z \end{pmatrix} = 2/3 \; h_s + 1/3 \; k_s + l_s z = n_2$$

$$(h_s \; k_s \; l_s) \begin{pmatrix} 1/3 \\ 2/3 \\ z \end{pmatrix} = 1/3 \; h_s + 2/3 \; k_s + l_s z = n_3$$

This system of equations reduces to

$$l_s = 0 \bmod 0$$
$$2h_s + k_s = 0 \bmod 3$$
$$h_s + 2k_s = 0 \bmod 3$$

Addition of the two last equations results in $h_s - k_s = 0 \bmod 3$. An example of a seminvariant structure factor is $F_{1\bar{2}0}$; the wave crests of the corresponding density wave are drawn in Fig. 5.1; it is seen that the phase of this density wave does not change if the origin is shifted from one allowed position to another.

It is seen from these examples that there is a correspondence between the restrictions on the allowed positions of the origin and the restrictions on the Laue indices h_s, k_s, l_s of the seminvariant phase. In P3$_1$ the allowed origins are restricted to the plane **a-b, c** and, correspondingly, the restrictions on h_s, k_s, l_s are two-dimensional by nature, i.e. $h_s - k_s \equiv 0 \bmod 3$ and $1 = 0 \bmod 0$. In P2$_1$ the allowed origins are restricted to positions on one of the four screw axes, but not to one plane or one line; this corresponds with the fact that

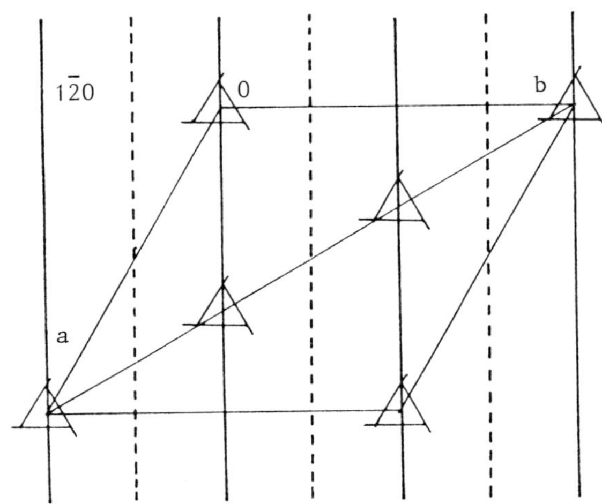

Fig. 5.1. The allowed origins in P3$_1$ are at (0 0 z), (2/3 - 1/3 z), (1/3 - 2/3 z), all lying in the plane **a-b, c**.

the restrictions on h_s, k_s, l_s have a three-dimensional character, i.e. $h_s \equiv 0 \bmod 2$, $k_s \equiv 0 \bmod 2$, $l_s \equiv 0 \bmod 2$. In order to have a concise way to express these restrictions the notions of "seminvariant vector" and "seminvariant modulus" have been introduced: in P3$_1$ the seminvariant vector is (h-k, l) and the seminvariant modulus is (3.0); in P2$_1$ the seminvariant vector is (h,k,l) and the seminvariant modulus is (2,0,2). Tables of seminvariant vectors and moduli for all space groups can be found in (4).

6. ORIGIN DEFINITION IN P1: THE PRIMITIVITY CONDITION

In P1 the origin can be chosen anywhere in the unit cell. In order to fix the position of the origin regard three density waves corresponding to the structure factors $F_{h_1 k_1 l_1}$, $F_{h_2 k_2 l_2}$, $F_{h_3 k_3 l_3}$ respectively; the relative positions of the three waves are uniquely determined by the electron density function $\rho(\mathbf{r})$ (the Fourier expansion of $\rho(\mathbf{r})$ is unique). The three density waves form a fixed frame of reference. By choosing the values of their phases (i.e. the distance of the wave crests to the origin) the position of the origin is fixed (be it on an unknown position as long as $\rho(\mathbf{r})$ is unknown). However, if one of the chosen density waves has several maxima within the unit cell , then its phase can be measured from any of the maxima, and several positions of the origin will be possible. If, for instance, two density waves have only one maximum within the unit cell, and the third wave has two maxima, then the volume of the cell spanned by the three density waves will be one half of the volume of the unit cell, i.e. the volume of the parallelopipedum in reciprocal space spanned by the reciprocal vectors (h_1,k_1,l_1), (h_2,k_2,l_2), (h_3,k_3,l_3) will be twice the volume V^* of the reciprocal cell. In this case two positions for the origin are possible, and the origin is not fixed uniquely (see Fig. 6.1).

Suppose we intend to fix the origin by choosing the values of three phases $\phi_{\mathbf{H}_i}$, with $\mathbf{h}_i = h_i a^* + k_i b^* + l_i C^*$ (i = 1,2,3). Then the volume $(V^*)'$ of the cell in reciprocal space spanned by the three reciprocal vectors $\mathbf{h}_i$ is:

$$(V^*)' = \mathbf{H}_1 . [\mathbf{H}_2 \mathbf{x} \mathbf{H}_3]$$

$$= (h_1 a^* + k_1 b^* + l_1 c^*) . [(h_2 a^* + k_2 b^* + l_2 c^*)$$

$$x\ (h_3 a^* + k_3 b^* + l_3 c^*)] \tag{6.1}$$

After some calculation the result is:

$$(V^*)' = \begin{vmatrix} h_1 & k_1 & l_1 \\ h_2 & k_2 & l_2 \\ h_3 & k_3 & l_3 \end{vmatrix} V^* \tag{6.2}$$

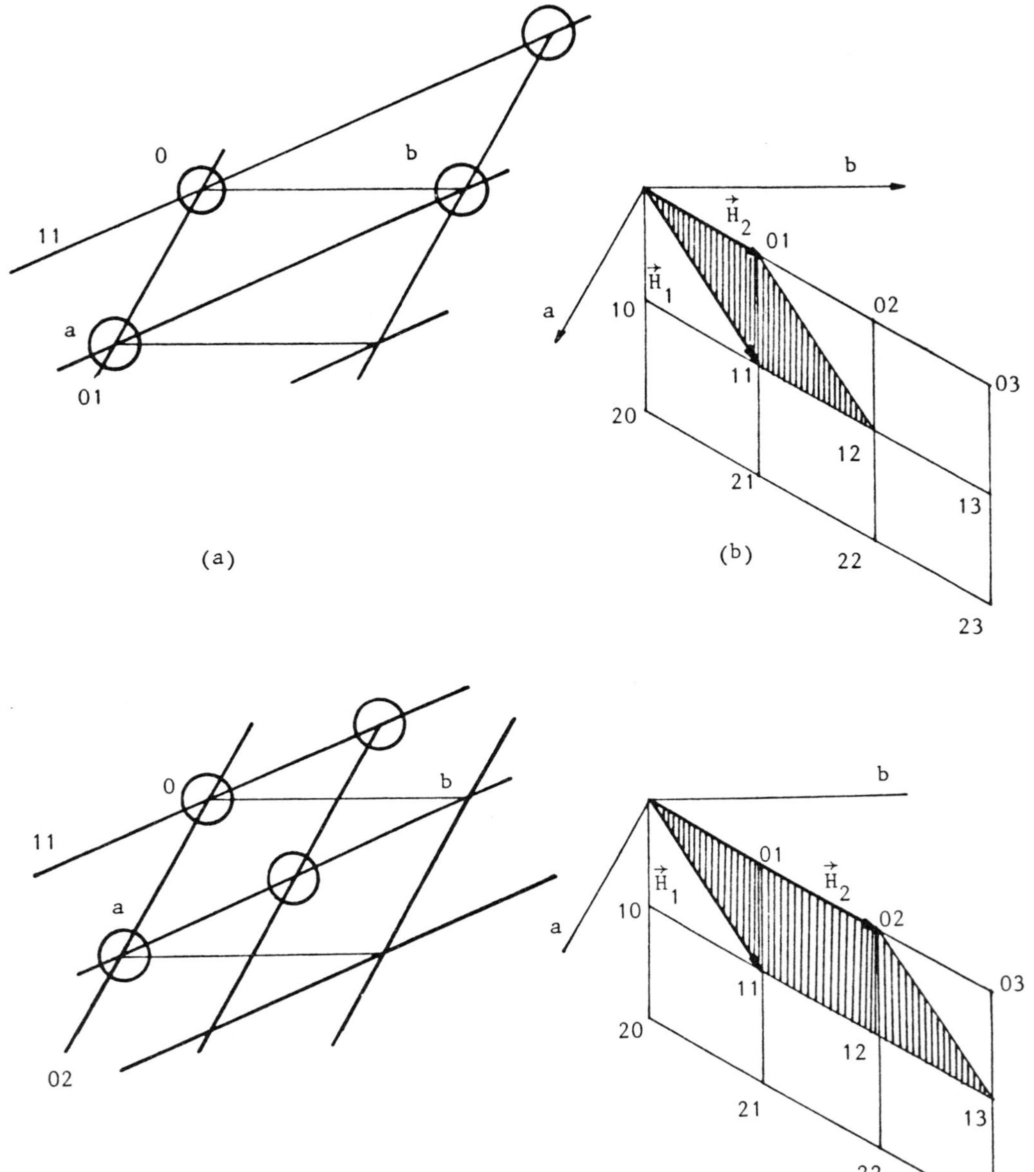

Fig. 6.1. Primitivity condition in plane group p1.
    (a) The density waves (01) and (11) (with ϕ_{01} and ϕ_{11} both chosen as 0) define the origin uniquely.
    (b) $\mathbf{H}_1 = (0,1)$, $\mathbf{H}_2 = (1,1)$ span a volume in reciprocal space equal to V*. Note that the determinant is -1.
    (c) The density waves (02) and (11) allow two possible positions for the origin.
    (d) $\mathbf{H}_1 = (0,2)$, $\mathbf{H}_2 = (1,1)$ span a volume in reciprocal space equal to 2V*. Note that the determinant is -2.

For a unique definition of the origin the requirement is that
$(V^*)' = V^*$. Then (6.2) gives

$$\begin{vmatrix} h_1 & k_1 & l_1 \\ h_2 & k_2 & l_2 \\ h_3 & k_3 & l_3 \end{vmatrix} = \pm 1 \qquad (6.3)$$

Expression (6.3) is known as the primitivity condition.

7. ORIGIN DEFINITION IN P$\bar{1}$

In P$\bar{1}$ the seminvariant vector is (h,k,l) and the
seminvariant modulus is $(2,2,2)$, so the phase ϕ_{ggg} is a structure
seminvariant. As the value of the phases from the ggg class of
phases are insensitive to a shift of origin from one allowed
position to another, phases of this class can not be used to fix
the origin. In order to see what phases can define the origin we
first notice that the eight allowed origins (centres of
symmetry) are in a three-dimensional arrangement (not restricted
to a line or a plane), as is exemplified by the fact that the
seminvariant vector is a three-dimensional vector. So it is
reasonable to assume that three phases can be given values 0 or
π to fix the origin. To begin with we take three phases with
simple indices: ϕ_{100}, ϕ_{010}, ϕ_{111}; the indices fulfill the
primitivity condition:

$$\begin{vmatrix} 1 & 0 & 0 \\ 0 & 1 & 0 \\ 1 & 1 & 1 \end{vmatrix} = +1$$

As is shown in Fig. 7.1, the origin is uniquely fixed by giving
each of the three phases one of the values 0 or π. In Fig. 7.2
the 100 density wave is replaced by the 300 density wave; as it
appears, nothing is changed, the origin is fixed at the same
position as it was before. Further trials show that replacing
the index 1 by any other odd number and 0 by any even number
does not have any influence on the results, as far as origin
fixing is concerned. The conclusion is that in P$\bar{1}$, because of
the restricted number of allowed origin positions at special
positions, the primitivity condition (6.3) may be modified into

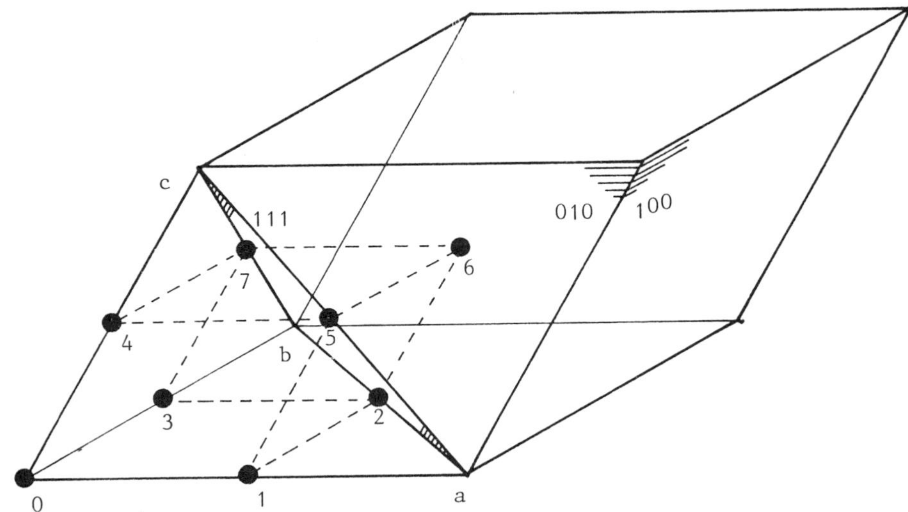

Fig. 7.1. In P$\bar{1}$ the density waves (100), (010), (111) are all given a
phase equal to 0. These waves select origins [0,3,4,7],
[0,1,4,5], [0,2,5,7] respectively; the origin is fixed on 0 (the
only origin common to all groups).

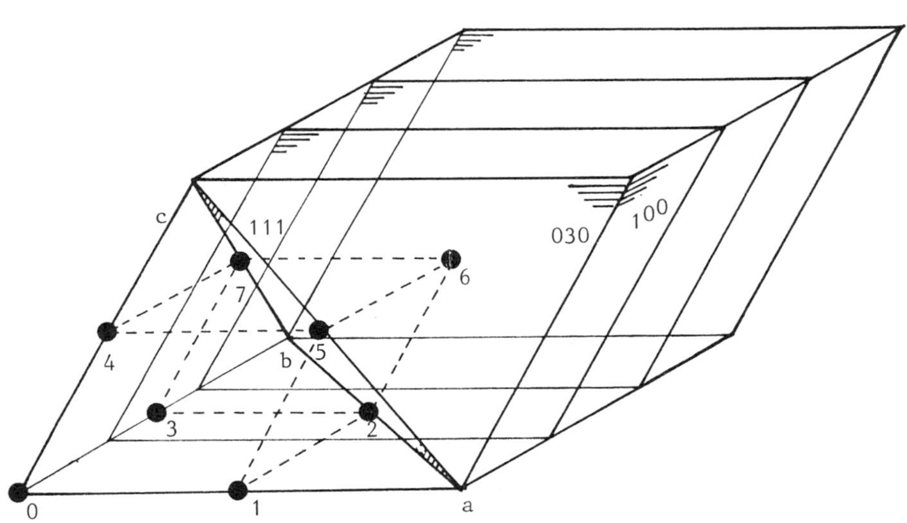

Fig. 7.2. In P$\bar{1}$ the density waves (100), (030), (111) behave as (100),
(010), (111) (Fig. 7.1); the origin is fixed on 0.

$$\begin{vmatrix} h_1 & k_1 & l_1 \\ h_2 & k_2 & l_2 \\ h_3 & k_3 & l_3 \end{vmatrix} \bmod(2,2,2) = \pm 1 \qquad\qquad (7.1)$$

In order to define some rules that ensure that (7.1) is fulfilled, it is practical to introduce the notion of "parity-groups" for the Laue-indices. These parity groups are: ggg, ggu, gug, ugg, guu, ugu, uug, uuu (g=even, u=odd). The rules are: (i) do not choose a phase from parity group ggg (a structure seminvariant) (ii) do not choose two reflections from the same parity group (the determinant will contain two equal rows, giving it the value 0) (iii) do not choose three reflections of which the indices add up to ggg (these reflections are linearly dependent).

Example

The phases ϕ_{321}, ϕ_{435}, ϕ_{227} define the origin uniquely; they belong to the parity groups ugu, guu, ggu respectively, which are all different and do not add up to ggg. The primitivity condition is fulfilled:

$$\begin{vmatrix} 3 & 2 & 1 \\ 4 & 3 & 5 \\ 2 & 2 & 7 \end{vmatrix} \bmod(2,2,2) = \begin{vmatrix} 1 & 0 & 1 \\ 0 & 1 & 1 \\ 0 & 0 & 1 \end{vmatrix} = +1$$

The phases ϕ_{321}, ϕ_{435}, ϕ_{312} do not define the origin uniquely: they belong to the parity groups ugu, guu, uug respectively, which are all different but which happen to add up to ggg. The primitivity condition is not fulfilled:

$$\begin{vmatrix} 3 & 2 & 1 \\ 4 & 3 & 5 \\ 3 & 1 & 2 \end{vmatrix} \bmod(2,2,2) = \begin{vmatrix} 1 & 0 & 1 \\ 0 & 1 & 1 \\ 1 & 1 & 0 \end{vmatrix} = -2$$

and two possible positions for the origin remain.

8. ORIGIN DEFINITION IN SOME NON-CENTRIC SPACE GROUPS

(i) $P2_12_12_1$

The origin is chosen, by convention, midway between each pair of non-intersecting screw-axes. The result is that there are eight possible positions for the origin, the situation is identical to that in $P\bar{1}$: the seminvariant vector is (h,k,l) and the seminvariant modulus is $(2,2,2)$, as in $P\bar{1}$. The special phase-restricted groups, and the values the phases can assume, are as follows:

ogg	$0,\pi$	gog	$0,\pi$	ggo	$0,\pi$
ouu	$\pm\pi/2$	uog	$0,\pi$	ugo	$\pm\pi/2$
oug	$\pm\pi/2$	gou	$\pm\pi/2$	guo	$0,\pi$
ogu	$0,\pi$	uou	$\pm\pi/2$	uuo	$\pm\pi/2$

Example

The phase-values $\phi_{023}=0$, $\phi_{102}=0$, $\phi_{011}=+\frac{\pi}{2}$ define the origin uniquely: the indices are chosen from different parity groups (ogu, uog, ouu) which do not add up to ggg.

(ii) Pm

In Pm, a monoclinic space group, there are two mirror planes, 1/2b apart, perpendicular to **b**. The allowed origins are situated in either of the mirror planes. The choice between the mirror planes can be made by choosing $\phi_{ouo} = 0$ or π. The location of the origin in the chosen mirror plane is free; the situation is a two-dimensional analog of P1; the position of the origin in the chosen mirror plane is determined by fixing the values of two phases, $\phi_{h_2k_2l_2}$ and $\phi_{h_3k_3l_3}$, of which the indices obey the two-dimensional primitivity condition

$$\begin{vmatrix} h_2 & l_2 \\ h_3 & l_3 \end{vmatrix} = \pm\, 1$$

This condition is identical to the requirement

$$\begin{vmatrix} (0 & k_1 & 0) \\ h_2 & k_2 & l_2 \\ h_3 & k_3 & l_3 \end{vmatrix} \bmod(020) = \pm\, 1$$

as can be seen by working out the determinant. The seminvariant modulus in Pm is (020); this modulus is only applied to the indices in parentheses.

(iii) $P3_1$

In $P3_1$ there are three 3_1-axes, all perpendicular to the line $(\mathbf{a}-\mathbf{b})$ (see Fig. 5.1). The allowed origins are restricted to positions on the three screw-axes, and are thus restricted to the plane $\mathbf{a-b},\mathbf{c}$. The problem of origin fixing is now a two-dimensional problem, and only 2 phases need to be specified to fix the origin. This is reflected by the fact that the seminvariant vector $(h-k,l)$ is a two-dimensional vector. The three screw-axes are $1/3|\mathbf{a-b}|$ apart, and this is what makes the seminvariant modulus to be $(3,0)$ (the position along the screw-axis is free). On the line $\mathbf{a-b}$ the origin is fixed by giving the phase $\phi_{h_1k_10}$ with $h_1-k_1 = 1 \bmod 3$ (not a seminvariant) its proper value. The position of the origin along the screw-axis is fixed by giving the phase $\phi_{h_3k_31}$ some value (as far as the c-axis is concerned, this is a one-dimensional analog to P1, so only $l_3=1$ will do). The problem is that the proper value of $\phi_{h_1k_10}$ is not known (it is not a phase with restricted values); the only thing one can do is to give it some value (which will fix the origin in the plane $\mathbf{a-b},\mathbf{c}$ close to one of the three screw axes) and adjust its phase later in a phase refinement process.

9. ENANTIOMORPH DEFINITION

A crystal structure posessing space group symmetry not containing a S_n axis has one of two possible enantiomorphic forms A or B. As the diffraction intensities are insensitive to the choice between A and B (if anomalous diffraction effects are ignored), no information about the actual enantiomorphic form can be obtained by ordinary diffraction methods. Nevertheless, during a direct method's phasing process, going from estimated invariant phases to an estimate of structure-factor phases, a choice between A and B is made, either as an inherent part of the phase-determining process or as a consequence of our intervention beforehand. The reason for us to fix the enantiomorph (which one is irrelevant) beforehand is that we want to try to avoid a result in which one part of the structure-factor phases correspond to A and another part to B, i.e. the situation in which parts of both enantiomorphs will appear in the resulting electron density function.

Generally, except in some odd space groups, the transition from A to B can be made by reversing the signs of the atomic coordinates; this goes together with a change in sign of all structure-factor phases. So the solution of the problem of enantiomorph fixing seems to be easy: our choice of origin fixing phases (of which at least one must be far from 0 and π) will fix the enantiomorph on, say, A; reversion of the signs of the origin fixing phases will change the enantiomorph to B. Unfortunately this is not true; a change of sign of the origin fixing phases changes either the enantiomorph to B, keeping the origin fixed at, say, O_1, or changes the origin from O_1 to O_2 and keeps the enantiomorph fixed at A:

$$A(O_1) \text{ or } B(O_2) \quad \underset{\Longleftarrow \Longrightarrow}{\overset{\phi \rightarrow -\phi}{}} \quad B(O_1) \text{ or } A(O_2)$$

The only way to fix the enaniomorph is by choosing the sign of an invariant (triplet, quartet) phase or of a seminvariant phase (not close to 0 or π). In some space groups, using only phases with restricted values, this amounts to choosing the phase of an extra reflection, apart from the origin fixing phases.

Example

In $P2_12_12_1$ the seminvariant vector is (h,k,l), the seminvariant modulus is $(2,2,2)$, so the seminvariant phase is ϕ_{ggg}. The origin can be fixed by choosing the values of three restricted phases, e.g. ogu $(0,\pi)$, uog $(0,\pi)$, ouu $(\pm \frac{\pi}{2})$ (see paragraph 8). Choose, in this sequence, the phase values $0,0,+\pi/2$, and call the chosen origin $(0,0,0)$. Reversion of the signs of the origin fixing phases to $0,0,-\pi/2$ can be brought about either by changing the enantiomorph and keeping the origin fixed on $(0,0,0)$, or by keeping the enantiomorph fixed and changing the origin to $(0,1/2,0)$. Next choose the sign of a fourth restricted phase, chosen such that its sign does not change upon a sift of origin from $(0,0,0)$ to $(0,1/2,0)$. Any phase with indices from parity group ugo will have this property (choice from $\pm\pi/2$). Specifying this phase value as e.g. $+\pi/2$ will fix the enantiomorph, as this phase is insensitive to the special change of origin that was applied:

	reflection	enantiomorph A		enantiomorph B	
		$0,0,0$	$01/20$	$0,0,0$	$0,1/2,0$
origin	ogu	0	0	0	0
fixing	uog	0	0	0	0
phases	ouu	$+\pi/2$	$-\pi/2$	$-\pi/2$	$+\pi/2$
	ugo	$+\pi/2$	$+\pi/2$	$-\pi/2$	$-\pi/2$

Note that (uog)+(ugo) = (ggg) is a seminvariant; as $\phi_{uog}+\phi_{ugo} = \pm\pi/2$, and ϕ_{uog} is fixed, the choice of ϕ_{ugo} (from $\pm\pi/2$) fixes the phase of the seminvariant and thus defines the enantiomorph.

REFERENCES

1.   C. Giacovazzo, "Direct Methods in Crystallography", Academic Press, London, New York, Toronto, Sydney, San Francisco (1980).

2.   D. Rogers, in "Theory and Practice of Direct Methods in Crystallography", eds. M.F.C. Ladd & R.A. Palmer, Plenum Press, New York and London (1980).

3.   H.A. Hauptman, "Crystal Structure Determination. The Role of the Cosine Seminvariants", Plenum Press, New York and London (1972).

4.   J. Karle, in International Tables for X-ray Crystallography", eds. J.A. Ibers and W.C. Hamilton, The Kynoch Press, Birmingham (1974).

THE SAYRE EQUATION

Michael M. Woolfson

Physics Department
University of York
York YO1 5DD, U.K.

A landmark in direct methods was the simultaneous publication of three papers by Sayre (1952), Cochran (1952) and Zachariasen (1952). They all dealt with the same ultimate topic, the sign relationship

$$s(\mathbf{h})\ s(\mathbf{k})\ s(\mathbf{h+k})\ \approx\ +1 \quad , \qquad (1)$$

from different points of view but in the paper by Sayre there was derived a powerful and elegant equation which gave exact relationships between structure factors. While exact equations, in particular cases, also resulted from the determinantal inequalities of Karle and Hauptman (1950) the Sayre equation had the advantage that it linked the structure factors in a very simple way.

Sayre considered the electron density associated with a structure containing equal resolved atoms. The electron density and the structure factors are related by Fourier transformation thus

$$\rho(\mathbf{r})\ \xleftrightarrow{\ F\ T\ }\ F(\mathbf{h}) \qquad . \qquad (2)$$

Next Sayre considered the squared electron density, the Fourier transformation of this being given by

Direct Methods of Solving Crystal Stuctures
Edited by H. Schenk, Plenum Press, New York, 1991

$$\rho(\mathbf{r})^2 \quad \xleftrightarrow{\text{F T}} \quad G(\mathbf{h}) \qquad . \qquad (3)$$

The usual structure factor equation may be written as

$$F(\mathbf{h}) \;=\; \sum_{i=1}^{N} f \, \exp(2\pi i \mathbf{h}.\mathbf{r}_j) \qquad (4)$$

where f is the scattering factor for the equal atoms. Similarly we may write

$$G(\mathbf{h}) \;=\; \sum_{i=1}^{N} g \, \exp(2\pi i \mathbf{h}.\mathbf{r}_j) \qquad (5)$$

where g is the scattering factor for the squared-density atoms which are again equal and resolved. From equations (4) and (5)

$$F(\mathbf{h}) \qquad \frac{f}{g} G(\mathbf{h}) \qquad . \qquad (6)$$

We now apply the convolution theorem which states that

$$\text{F T } (\rho_1 \times \rho_2) \;=\; \text{F T } (\rho_1) \; * \; \text{F T } (\rho_2) \qquad (7)$$

where * indicates convolution. From this we find

$$G(\mathbf{h}) = \text{F T } (\rho \times \rho) = F(\mathbf{h}) \; * \; F(\mathbf{h}) \qquad (8)$$

or, taking account of the periodic nature of ρ,

$$G(\mathbf{h}) \;=\; \frac{1}{V} \sum_{\mathbf{k}} F(\mathbf{k}) \; F(\mathbf{h}-\mathbf{k}) \qquad . \qquad (9)$$

Combining equations (6) and (9) gives the Sayre equation

$$F(\mathbf{h}) \;=\; \frac{f}{gV} \sum_{\mathbf{k}} F(\mathbf{k}) \; F(\mathbf{h}-\mathbf{k}) \qquad (10)$$

The values of g may be found from

$$g \;=\; f \; * \; f \qquad (11)$$

although in a particular case a simple analytical form for f
and for ρ may be employed to simplify the calculation of g.

In the original paper Sayre argued from equation (10)
that, for a centrosymmetric structure, if $F(\mathbf{h})$ was a large
structure factor then a large term in the summation would
tend to have the same sign - which then leads to
relationship (1). However, it should be noted that
equation (10) is applicable to both centrosymmetric and
non-centrosymmetric structures so that it has implications
for general phases as well.

NON-EQUAL ATOMS

If the atoms of a structure are not equal, but still
resolved, then equation (9) is still valid although the
structure factor of the squared structure on the left-hand
side is a completely unknown quantity. Later we shall see,
by reference to this equation, that some techniques of phase
determination have a tendency, if not counteracted in some
way, to give the phases of the squared rather than those of
the structure itself. This manifests itself through
density maps in which heavier atoms tend to be
overemphasized.

It was shown by Woolfson (1958) that a modified
equation could be produced which would be valid for a
structure containing two types of atom. To derive this
equation one considers the electron density, ρ, its square,
ρ^2, and its cube, ρ^3, with corresponding scattering factors
f_s, g_s and h_s for the two kinds of atom, s=1 and s=2.
Constants $A(\mathbf{h})$ and $B(\mathbf{h})$ can always be found such that

$$f_1(\mathbf{h}) \;=\; A(\mathbf{h})\,g_1(\mathbf{h}) \;+\; B(\mathbf{h})\,h_1(\mathbf{h})$$

and
$$f_2(\mathbf{h}) \;=\; A(\mathbf{h})\,g_2(\mathbf{h}) \;+\; B(\mathbf{h})\,h_2(\mathbf{h}) \quad . \tag{12}$$

From this we find

$$F(\mathbf{h}) \;=\; A(\mathbf{h})\,G(\mathbf{h}) \;+\; B(\mathbf{h})\,H(\mathbf{h}) \tag{13}$$

where H(h) is the Fourier transform of ρ^3.   This leads to
the equation

$$F(h) = \frac{A(h)}{V} \sum_{k} F(k)\ F(h-k) + \frac{B(h)}{V^2} \sum_{k} \sum_{l} F(k)\ F(l)\ F(h-k-l).$$

$$(14)$$

This equation has found very little utility; unless
there is a very heavy atom, which will usually assist
structure solution as it happens, then the simplicity and
convenience of the Sayre equation, compared with equation
(14), outweighs any errors in its application.

APPLICABILITY

It is usually assumed that all the relationships used
in direct methods are dependent on the non-negativity of
electron density.   While this is true for some
relationships, such as the Karle Hauptman determinants, it
is not true for all and in particular it is not a necessary
condition for the Sayre equation.   If normalized structure
factors, E's, are used then as long as the resolution of the
data is high enough the 'atoms' will have diffraction
ripples, corresponding to negative electron density, but
will still be equal and resolved.   For that reason it is
valid to write the Sayre equation in the form

$$E(h) = \frac{f}{gV} \sum_{k} E(k)\ E(h-k) \quad . \qquad (15)$$

Actually it has been found that the Sayre equation holds
reasonably well over a very large range of conditions
(Shiono, 1989).   For example, with the structure of Avian
Pancreatic Polypeptide (Glover et. al., 1983) containing 36
amino acids (+80 H_2O), space group C2 at 2.2Å resolution,
the residual, comparing the magnitudes of the two sides of
equation (15), is 0.136 while the mean difference in phase
between the two sides is 30.2°.   This is a pointer towards
the application of the Sayre equation to the solution of
very large structures - even proteins.

44

REFERENCES

Cochran, W., 1952, A relation between the signs of structure factors. *Acta Cryst.*, **5**, 65-67.

Glover, I., Haneef, I., Pitts, J., Wood, S., Moss, D., Tickle, I. and Blundell, T., 1983, Conformational flexibility in a small globular hormone: X-ray analysis of Avian Pancreatic Polypeptide at 0.98Å resolution. *Biopolymers,* **22**, 293-304.

Karle, J. and Hauptman, H., 1950, The phases and magnitudes of the structure factors. *Acta Cryst.*, **3**, 181-187.

Sayre, D., 1952, The squaring method: a new method for phase determination. *Acta Cryst.*, **5**, 60-65.

Shiono, M., 1989, An investigation of the Sayre equation. D.Phil. Thesis, University of York.

Woolfson, M.M., 1958, An equation between structure factors for structures containing unequal or overlapped atoms. I. The equation and its properties. *Acta Cryst.*, **11**, 277-283.

Zachariasen, W.H., 1952, A new analytical method for solving crystal structures. *Acta Cryst.*, **5**, 68-73.

THE COCHRAN DISTRIBUTION

Michael M. Woolfson

Physics Department
University of York
York YO1 5DD, U.K.

INTRODUCTION

All the earlier published work using direct methods, right up to the end of the 1950's, involved the application of inequality relationships, usually of the Harker-Kasper type (Harker & Kasper, 1948) and/or the triple sign relationship for solving either centrosymmetric structures or centrosymmetric projections of non-centrosymmetric structures. In fact some of the very earliest theoretical work pointed towards the possibility of applying direct methods to non-centrosymmetric structures as well. The determinant inequalities, developed by Karle & Hauptman (1950), which will later be described in greater detail, could restrict general phases in the non-centrosymmetric case and Sayre's equation (Sayre, 1952) could be applied to any equal-atom structure regardless of space group and therefore could put some constraints on the phases of the involved structure factors.

Written in the usual way, Sayre's equation appears as

$$F(\mathbf{h}) \; = \; \frac{f}{gV} \; \sum_{\mathbf{k}} F(\mathbf{k}) \; F(\mathbf{h}-\mathbf{k}) \quad , \tag{1}$$

Direct Methods of Solving Crystal Stuctures
Edited by H. Schenk, Plenum Press, New York, 1991

where f and g are the atomic scattering factors for normal atoms and squared atoms respectively and V the volume of the unit cell. However, we recall that $F(\bar{h}) = F(h)^*$, where $*$ denotes *complex conjugate*, and multiplying both sides of equation (1) by this quantity gives

$$|F(h)|^2 = \frac{f}{gV} \sum_{k} F(\bar{h})\ F(k)\ F(h-k) \quad .\qquad (2)$$

Since the left-hand side is real we may derive two equations from equation (2) by equating the real and imaginary parts of the two sides, viz.

$$|F(h)|^2 = \frac{f}{gV} \sum_{k} |F(h)\ F(k)\ F(h-k)|\ \cos\{\phi(h)-\phi(k)-\phi(h-k)\} \quad (3a)$$

and

$$0 \quad = \frac{f}{gV} \sum_{k} |F(h)\ F(k)\ F(h-k)|\ \sin\{\phi(h)-\phi(k)-\phi(h-k)\} \quad . \quad (3b)$$

From equation (3a) we may deduce that if $|F(h)|^2$ is large and if $|F(h)\ F(k)\ F(h-k)|$ is also large then $\cos\{\phi(h)-\phi(k)-\phi(h-k)\}$ should be as large as possible. The condition for this is that

$$\Phi_3(h,k) \ = \ \phi(h)-\phi(k)-\phi(h-k) \ \approx \ 0 \quad (\text{modulo } 2\pi) \qquad (4)$$

which is interpreted as saying that the quantity $\Phi_3(h,k)$ should be close to some multiple of 2π.

THE COCHRAN DISTRIBUTION

While the three-phase relationship (4) was implicit in various theoretical approaches it was first explicitly stated and theoretically examined by Cochran (1955). If the magnitudes of the three involved structure factors and the atomic contents of the unit cell are known then the probability density of the quantity $\Phi_3(h,k)$ is given by

$$P\{\Phi_3(h,k)\} \ = \ \frac{1}{2\pi\ I_0(\kappa)} \ \exp\ \{\ \kappa\ \cos\ \Phi_3(h,k)\} \qquad (5)$$

where
$$\kappa = 2 \; \frac{\sigma_3}{\sigma_2^{3/2}} \; | \; E(h) \; E(k) \; E(h-k) | \qquad (6)$$

σ^n is defined in terms of the atomic numbers z_i for the N atoms in the unit cell as

$$\sigma^n = \sum_{i=1}^{N} z_i^n \quad , \qquad (7)$$

$I_0(\kappa)$ is a modified Bessel function of order 0 and the E's are normalized structure factors. In the case of an equal atom structure the quantity $\sigma_3/\sigma_2^{3/2}$ equals $N^{-1/2}$.

To see what this means in terms of constraint on the three-phase relationship we consider an equal atom structure with N = 256 with each of the three normalized structure magnitudes equal to 2.0. This gives κ = 1 and the corresponding Cochran distribution is shown in Figure 1.

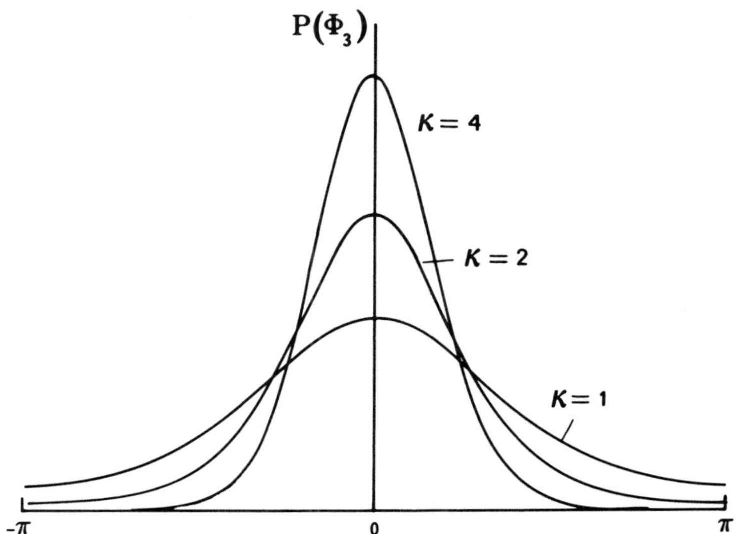

Fig.1   The Cochran distributions for κ = 1, 2 and 4.

The distributions for κ = 2 and 4 are also shown in the figure. It is clear that the higher is the value of κ the more the probability density is concentrated around zero. A measure of the tightness of a distribution is its variance

(square of the standard deviation) and Karle & Karle (1966)
derived a formula giving the variance as a function of κ
which involves an infinite series in terms of modified Bessel
functions of increasing order,

$$V(\kappa) \;=\; \frac{\pi^2}{3} \;+\; 4\sum_{t=1}^{\infty} \frac{(-1)^t}{t^2} \; \frac{I_t(\kappa)}{I_0(\kappa)} \tag{8}$$

$V(\kappa)$ and $\sigma(\kappa)$ as functions of κ are shown in Figure 2.
It will be seen that, for $\kappa = 4$ for example, the standard
deviation of about 30° constrains the triple-phase
relationship quite tightly.

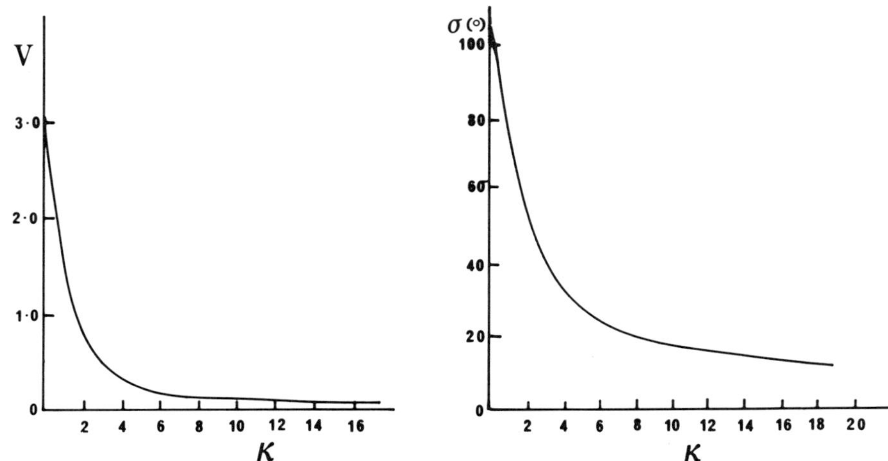

Fig.2   The variance and standard deviation of a triple
          -phase relationship as a function of κ.

USING THE COCHRAN DISTRIBUTION

Relationship (4) can be put into the form
$$\phi(\mathbf{h}) \approx \phi(\mathbf{k}) + \phi(\mathbf{h-k}) \tag{9}$$
which gives an estimate of the value of $\phi(\mathbf{h})$ if the values of
the phases on the right-hand side are known.   Indeed, if the
precise values were known for the phases on the right-hand

side then $V(\kappa)$ would be the variance of the estimate of $\phi(\mathbf{h})$. This indicates that, just as in the centrosymmetric case, a knowledge of a small number of phases might be used in a chain process, using equation (9), to derive phase estimates for a large set of structure factors. However, although in retrospect that seems blindingly obvious it took several years for that process to become a reality. When Karle & Karle (1964) showed how to use equation (9), together with the tangent formula which is described later, to solve the non-centrosymmetric structure of L-arginine dihydrate then the era of modern direct methods was really launched.

REFERENCES

Cochran, W., 1955, Relations between the phases of structure factors. *Acta Cryst.*, **8**, 473-478.

Harker, D. and Kasper, J.S. ,1948, Phases of Fourier coefficients directly from crystal diffraction data. *Acta Cryst.*, **1**, 70-75.

Karle, J. and Hauptman, H., 1950, The phases and magnitudes of the structure factors. *Acta Cryst.*, **3**, 181-187.

Karle, I.L. and Karle, J., 1964, The application of the symbolic addition method to the structure of L-arginine dihydrate. *Acta Cryst.*, **17**, 835-841.

Karle, J. and Karle, I.L.., 1966, The symbolic addition procedure for phase determination for centrosymmetric and noncentrosymmetric crystals. *Acta Cryst.*, **21**, 849-859.

Sayre, D., 1952, The squaring method; a new method of phase determination. *Acta Cryst.*, **5**, 60-65.

DETERMINANTAL INEQUALITIES LEAD TO TRIPLET AND QUARTET RELATIONSHIPS

Herbert A. Hauptman

Medical Foundation of Buffalo, Inc.
73 High Street
Buffalo, N.Y. 14203 U.S.A.

1.   THE DETERMINANTAL INEQUALITIES

A necessary and sufficient condition that the electron density
function in a crystal be non-negative everywhere is that the system of
determinantal inequalities hold:

$$
\begin{vmatrix}
E_0 & E_{\bar{H}_1} & E_{\bar{H}_2} & \cdots\cdots & E_{\bar{H}_n} \\
E_{H_1} & E_0 & E_{H_1 - H_2} & \cdots\cdots & E_{H_1 - H_n} \\
E_{H_2} & E_{H_2 - H_1} & E_0 & \cdots\cdots & E_{H_2 - H_n} \\
\cdot & \cdot & \cdot & & \cdot \\
\cdot & \cdot & \cdot & & \cdot \\
\cdot & \cdot & \cdot & & \cdot \\
E_{H_n} & E_{H_n - H_1} & E_{H_n - H_2} & \cdots\cdots & E_0
\end{vmatrix} \geq 0
\tag{1}
$$

where the (complex) normalized structure factor E_H is defined by

$$
E_H = |E_H| \exp\left(i\phi_H\right) = \frac{1}{\sigma_2^{1/2}} \sum_{j=1}^{N} Z_j \exp\left(-2\pi i H \cdot r_j\right),
\tag{2}
$$

H is an arbitrary reciprocal lattice vector, Z_j is the atomic number and
r_j the position vector of the atom labeled j, N is the number of atoms
in the unit cell, and

$$
\sigma_2 = \sum_{j=1}^{N} Z_j^2 .
\tag{3}
$$

For simplicity and without any loss of generality, only the equal atom
case is considered here, so that

$$
E_H = \frac{1}{N^{1/2}} \sum_{j=1}^{N} \exp\left(-2\pi i H \cdot r_j\right)
\tag{4}
$$

Direct Methods of Solving Crystal Stuctures
Edited by H. Schenk, Plenum Press, New York, 1991

and

$$E_0 = N^{1/2} \, . \tag{5}$$

Since the magnitudes $|E|$ are presumed to be known, the system of inequalities (1) serves, in favorable cases, to restrict the possible values of the phases ϕ.

2. TRIPLETS

Replacing H_1 by H and H_2 by K in (1), the case $n = 2$ reduces to

$$\begin{vmatrix} N^{1/2} & E_{\bar{H}} & E_{\bar{K}} \\ E_H & N^{1/2} & E_{H-K} \\ E_K & E_{K-H} & N^{1/2} \end{vmatrix} \geq 0 \tag{6}$$

or, after some simplification, to

$$2\left|E_H E_K E_{H+K}\right| \cos\left(\phi_H + \phi_K + \phi_{-H-K}\right) \geq$$
$$\sqrt{N}\left(|E_H|^2 + |E_K|^2 + |E_{H+K}|^2\right) - N^{3/2} \tag{7}$$

from which it is clear that, provided that $|E_H|$, $|E_K|$ and $|E_{H+K}|$ are sufficiently large,

$$\cos\left(\phi_H + \phi_K + \phi_{-H-K}\right) > 0. \tag{8}$$

For example, if

$$|E_H| = |E_K| = \sqrt{\frac{N}{3}} \, , \quad |E_{H+K}| = \sqrt{\frac{N}{2}} \, , \tag{9}$$

then (7) reduces to

$$\cos\left(\phi_H + \phi_K + \phi_{-H-K}\right) \geq \frac{1}{4}\sqrt{2} \sim 0.354. \tag{10}$$

If, on the other hand, the three magnitudes $|E_H|$, $|E_K|$ and $|E_{H+K}|$ are small, the inequality (7) yields no useful information.

The results briefly summarized here are clarified and extended by the probabilistic theory of the triplets which leads to the result that

$$\cos\left(\phi_H + \phi_K + \phi_{-H-K}\right) \sim 1 \tag{11}$$

provided that the three magnitudes $|E_H|$, $|E_K|$ and $|E_{H+K}|$ are sufficiently large.

3. THE NEGATIVE QUARTETS

Replacing H_1 by H, H_2 by K, and H_3 by L in (1), the case $n = 3$ becomes

$$\begin{vmatrix} N^{1/2} & E_{\bar{H}} & E_{\bar{K}} & E_{\bar{L}} \\ E_H & N^{1/2} & E_{H-K} & E_{H-L} \\ E_K & E_{-H+K} & N^{1/2} & E_{K-L} \\ E_L & E_{-H+L} & E_{-K+L} & N^{1/2} \end{vmatrix} \geq 0. \tag{12}$$

Under the assumption that

$$|E_{H-K}| = |E_L| = 0 \tag{13}$$

(12) reduces to

$$\begin{vmatrix} N^{1/2} & E_{\bar{H}} & E_{\bar{K}} & 0 \\ E_H & N^{1/2} & 0 & E_{H-L} \\ E_K & 0 & N^{1/2} & E_{K-L} \\ 0 & E_{-H+L} & E_{-K+L} & N^{1/2} \end{vmatrix} \geq 0 \tag{14}$$

or, after simplification, to

$$2\left|E_H E_K E_{H-L} E_{K-L}\right| \cos\left(\phi_H + \phi_{\bar{K}} + \phi_{-H+L} + \phi_{K-L}\right) \leq$$

$$N^2 - N\left(|E_H|^2 + |E_K|^2 + |E_{H-L}|^2 + |E_{K-L}|^2\right)$$

$$+ \left|E_H E_{K-L}\right|^2 + \left|E_K E_{H-L}\right|^2 \tag{15}$$

in which the linear combination of four phases

$$\phi_H + \phi_{\bar{K}} + \phi_{-H+L} + \phi_{K-L} \tag{16}$$

is seen to be a structure invariant (quartet) since

$$H - K + (-H+L) + (K-L) = 0. \tag{17}$$

Clearly, if the four magnitudes $|E_H|$, $|E_K|$ and $|E_{H-L}|$, and $|E_{K-L}|$ are sufficiently large

$$\cos\left(\phi_H + \phi_{\bar{K}} + \phi_{-H+L} + \phi_{K-L}\right) < 0. \tag{18}$$

For example, if

$$|E_H| = |E_K| = |E_{H-L}| = |E_{K-L}| = \sqrt{\frac{N}{3}}, \tag{19}$$

then (15) reduces to

$$\cos\left(\phi_H + \phi_{\overline{K}} + \phi_{-H+L} + \phi_{K-L}\right) \leq -\frac{1}{2} \tag{20}$$

on the assumption, it must be recalled, that Eq. (13) hold. If, on the other hand, the four magnitudes $|E_H|$, $|E_K|$, $|E_{H-L}|$, and $|E_{K-L}|$ are all small, the inequality (15) yields no useful information.

These results are extended and made more precise by the probabilistic theory of the quartets which, briefly summarized, states that, on the assumption that the four "main terms"

$$|E_H|, \quad |E_K|, \quad |E_L|, \quad |E_{H+K+L}| \tag{21}$$

are all large,

$$\cos\left(\phi_H + \phi_K + \phi_L + \phi_{-H-K-L}\right) \approx 1 \tag{22}$$

or

$$\cos\left(\phi_H + \phi_K + \phi_L + \phi_{-H-K-L}\right) \approx -1 \tag{23}$$

according as the three "cross terms"

$$|E_{H+K}|, \quad |E_{K+L}|, \quad |E_{L+H}| \tag{24}$$

are all large or all small, respectively. In fact the probabilistic theory of the quartets goes even further and, in favorable cases, yields estimates of the cosines

$$\cos\left(\phi_H + \phi_K + \phi_L + \phi_{-H-K-L}\right) \tag{25}$$

in terms of all seven magnitudes

$$|E_H|, \quad |E_K|, \quad |E_L|, \quad |E_{H+K+L}|, \quad |E_{H+K}|, \quad |E_{K+L}|, \quad |E_{L+H}|. \tag{26}$$

THE CALCULATION OF NORMALIZED STRUCTURE FACTORS AND OF

TRIPLET INVARIANTS

Davide Viterbo

Dipartimento di Chimica
Universitá della Calabria
I-87030 Arcavacata di Rende (Cs), Italy

Statistical analysis of structure factor amplitudes; unitary and normalized structure factors

The statistical analysis of the observed structure factor moduli gives very useful indications on the presence or not of those symmetry elements which do not give rise to systematic absences. It also allows to obtain an estimate both of the scale factor by which one has to multiply the measured data to scale them to their absolute value, and of the temperature factor. Finally it is a basic step in the calculation of unitary and normalized structure factors, which will be seen to be useful quantities when using direct methods.

The theoretical probability distributions of structure amplitudes was first derived by Wilson (1949) under the hypothesis that the atomic positions are random variables with uniform distribution throughout the unit cell (i.e. all points in the cell have the same probability of hosting an atom).

Their form depends on whether the crystal possesses (centric distribution) or not (acentric distribution) an inversion center.

For the two cases we have

$$P_{\bar{1}}(|F|) = \frac{\sqrt{2}}{\sqrt{\pi\Sigma}} \exp\{-|F|^2/2\Sigma\} \quad \text{(centric)} \qquad (1)$$

$$P_1(|F|) = \frac{2|F|}{\Sigma} \exp\{-|F|^2/\Sigma\} \quad \text{(acentric)} \qquad (2)$$

where

$$\Sigma = \sum_{j=1}^{N} f_j^2 = \langle|F|^2\rangle/\epsilon_{\mathbf{h}} \qquad (3)$$

and $\epsilon_{\mathbf{h}}$ is a factor depending on the specific symmetry of the weighted reciprocal lattice. Σ being a function of the atomic scattering factors will not have a constant value in all the reciprocal space, but it will decrease for increasing values of $\sin\Theta/\lambda$. This is a problem in the practical use of distributions (1) and (2). In order to overcome this problem the <u>unitary structure factors</u>

$$|U_{\mathbf{h}}| = \frac{|F_{\mathbf{h}}|}{\sum\limits_{j=1}^{N} f_j} \tag{4}$$

and the <u>normalized structure factors</u>

$$|E_{\mathbf{h}}| = \frac{|F_{\mathbf{h}}|}{\sqrt{<|F|^2>}} = \frac{|F_{\mathbf{h}}|}{\sqrt{\epsilon_{\mathbf{h}}\Sigma}} \tag{5}$$

are introduced. In the same way by which $<|F|^2>$ was defined in (3), we have

$$<|U_{\mathbf{h}}|^2> = \epsilon_{\mathbf{h}}\sum_{j=1}^{N}[f_j/\Sigma\ f_j]^2 = \epsilon_{\mathbf{h}}\Sigma_U \tag{6}$$

and in the case of all equal atoms $\Sigma_U = 1/N$. From (5) we can immediately derive

$$<|E_{\mathbf{h}}|^2> = 1 \tag{7}$$

Both U and E are independent of the scattering angle Θ and correspond to idealized point atom structures.

The centric and acentric distributions, when expressed in terms of normalized structure factors, become

$$P_{\bar{1}}(|E|) = \frac{\sqrt{2}}{\sqrt{\pi}}\ \exp\{-|E|^2/2\} \tag{8}$$

$$P_1(|E|) = 2|E|\ \exp\{-|E|^2\} \tag{9}$$

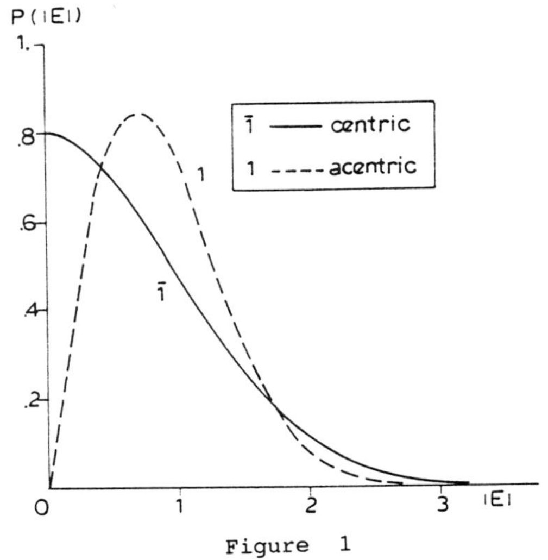

Figure 1

which are completely independent of the structure complexity. They are represented in the two curves of Figure 1, which are quite different; the centric distribution foresees a higher percent of reflexions with extreme intensity (weak and strong reflexions), with respect to the acentric one, showing a maximum corresponding to intermediate values of the intensities. The comparison of the distribution of the observed amplitudes with the theoretical distributions of Figure 1 will allow to establish the presence or not of the inversion centre in the crystal under consideration.

From the distributions it is possible to derive the theoretical mean values of several functions of $|E|$ or the theoretical percent of $|E|>t$. Some of these values are reported in Table 1 and may be compared with the corresponding experimental values to verify whether the structure is centrosymmetric or not.

TABLE 1

function	Theoretical		Experimental for AZOS
	centro	non—centro	
$<\|E\|^2>$	1.000	1.000	1.000
$<\|E\|^2-1>$	0.968	0.736	0.942
$<\|E\|>$	0.798	0.886	0.797
% $\|E\|>1.0$	31.7	36.8	32.7
% $\|E\|>2.0$	4.6	1.8	3.8
% $\|E\|>3.0$	0.3	0.01	0.4

Wilson plot

So far we have implicitly assumed that the observed structure moduli are on an absolute scale, but in general the values of $|F_h|^2_{obs}$, obtained from the intensities, are on a relative scale. Assuming that the thermal motion is isotropic and equal for all the atoms, we may write

$$|F_h|^2_{obs} = K \ |{}^OF_h|^2 \ exp\{-2Bs^2\} \qquad (10)$$

where K is the scale factor , $|{}^OF_h|$ is the structure amplitude in absolute scale for atoms at rest, B is the overall isotropic temperature factor and $s=sin\Theta/\lambda$.

Using the results of the statistical analysis of the intensities, Wilson (1942) proposed a method to derive the values of K and B.

Let us consider a set of observed intensities falling within a restricted range of $\underline{s}$, such that within this range the decrease of the $\underline{f}$'s with $\underline{s}$ may be neglected. The average value of both sides of (10) will be

$$<|F_h|^2_{obs}>_s = K <|{}^oF_h|^2>_s \exp\{-2B<s^2>\}$$

$$= K \, \Sigma_s \exp\{-2B<s^2>\} \qquad (11)$$

from which

$$\ln [<|F_h|^2_{obs}>_s/\Sigma_s] = \ln K - 2B<s^2> \qquad (12)$$

where $<s^2>$ is the mean value of $\sin^2\Theta/\lambda^2$ in the considered interval and $\Sigma_s = \sum_{j=1}^{N} {}^of_j^2$ is computed using the tabulated values of the atomic scattering factors for atoms at rest for $\underline{s} = \sqrt{<s^2>}$. Dividing the reciprocal lattice into several intervals of $\underline{s}$, (12) tells us that a linear relation will exist between $\ln[<|F_{obs}|^2>_s/\Sigma_s]$ and $<s^2>$ and that a plot of these values obtained from the experimental data can be interpolated by the best straight line passing through them. The intercept of the line on the vertical axis will give us $\underline{\ln K}$ and its slope the value of $\underline{2B}$.

Figure 2 is an example of such <u>Wilson plot</u> for a typical small organic structure (<u>p</u>-carboxyphenylazoxycyanide-dimethyl sulphoxide (Calleri et Al., 1976), AZOS hereinafter); the numerical values of the terms appearing in equation (12), obtained from 1908 observed reflexions, are given in Table 2. The main reason for the deviations of the experimental points from the straight line is the breakdown of the condition of equipropbability of all atomic positions, assumed in deriving (1) and (2); in fact, the presence of structural regularities, such as the phenyl hexagon in AZOS, is in contrast with this assumption. The dotted line in Figure 2, obtained by least-square fitting, has a slope of -9.20 and an intercept of 0.72. From these values we can derive that B=4.60 $\mathring{A}^2$ and K=2.05.

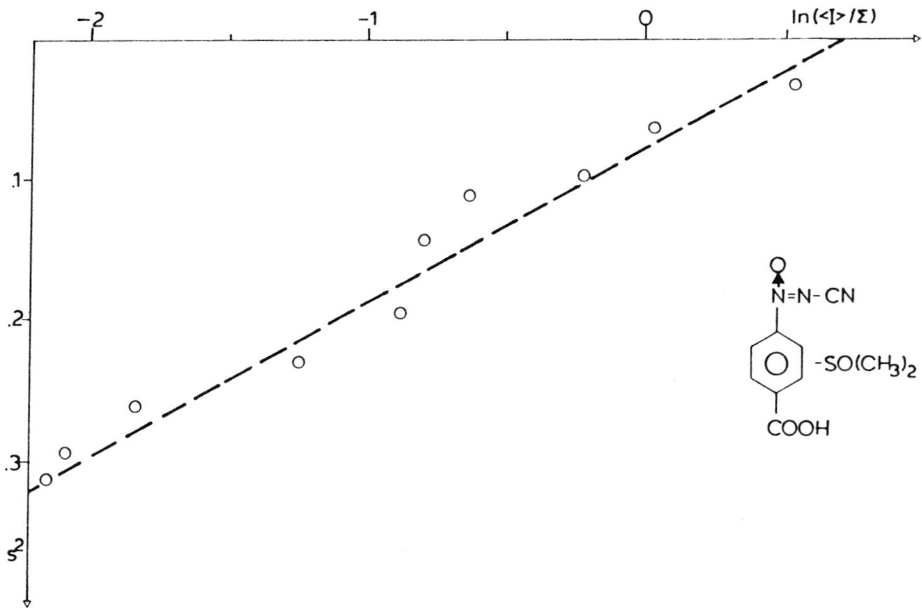

Figure 2

TABLE 2

	s^2 interval	N	$<s^2>$	$<\lvert F\rvert^2>_s$	Σ_s	$<\lvert F\rvert^2>_s/\Sigma_s$	$\ln[<\lvert F\rvert^2>_s/\Sigma_s]$
1	0.000-0.066	82	0.033	3737	2200	1.699	0.530
2	0.033-0.098	124	0.066	1360	1323	1.028	0.027
3	0.066-0.131	156	0.098	821	1021	0.804	-0.218
4	0.098-0.164	182	0.131	420	791	0.531	-0.633
5	0.131-0.196	210	0.164	293	647	0.453	-0.792
6	0.164-0.229	231	0.196	234	563	0.417	-0.876
7	0.196-0.262	248	0.229	146	508	0.288	-1.245
8	0.229-0.295	269	0.262	73	451	0.162	-1.821
9	0.262-0.327	272	0.295	52	410	0.127	-2.066
10	0.295-0.327	134	0.311	46	393	0.118	-2.138

The normalized structure factors can then be calculated by means of the relation

$$\lvert E_{\mathbf{h}}\rvert \;=\; \left[\frac{\lvert F_{\mathbf{h}}\rvert^2_{\text{obs}}}{K\,\exp\{-2Bs^2\}\,\epsilon_{\mathbf{h}}\,\Sigma\,{}^{\text{o}}f_j^{\,2}}\right]^{\frac{1}{2}} \qquad (13)$$

These values are then rescaled so that the normalization condition (7) is satisfied. Some attention should be given to the evaluation of the $\epsilon_{\mathbf{h}}$ factor which according to Stewart & Karle (1976) may be obtained for each reflexion by applying structure-factor algebra. It is in fact given by the number of times the transformation $\mathbf{h}' = \mathbf{h}R_s$ yields an identity $\mathbf{h}' \equiv \mathbf{h}$ as $\mathbf{R}_s$ varies over all rotation matrices of the space group; in the case of non-primitive cells this number should be multiplied by the multiplicity of the cell.

The last column in Table 1 gives the experimental values of several statistical indicators based on the $\lvert E\rvert$ values of AZOS (space group $P2_1/a$), which confirm the presence of an inversion centre.

In order to take into account the deviations from the ideal linear trend due to structural regularities, Karle & Hauptman (1953) proposed the K-curve approach by which the experimental curve is fitted by a monotonically decreasing function. But, in their extensive analysis, Hall & Subramanian (1982 a,b) and Subramanian & Hall (1982) have shown that the best E values are obtained by using the simple linear interpolation.

Special attention should be given to weak reflections. A method for a correct treatment of weak measured reflexions has been proposed by French & Wilson (1978). As suggested by Vickovic & Viterbo (1979), if the measurement of weak reflections is skipped, before carrying out Wilson statistics they should be introduced with a properly estimated value.

Use of *a-priori* information

Main (1976) proposed the following classification of the possible types of *a-priori* structural information:

1) randomly positioned atoms;

2) randomly positioned and randomly oriented fragment;
3) randomly positioned but correctly oriented fragment;
4) correctly positioned and correctly oriented fragment.

By introducing the group scattering factor $g_i(\mathbf{h})$ (3) becomes

$$<|F_{\mathbf{h}}|^2> = \sum_{i=1}^{m} |g_i(\mathbf{h})|^2 \qquad (14)$$

for $\underline{m}$ groups in the unit cell. Main (1976) derived the form of the right-hand side of (14) for the four considered cases:

1) In this case (14) is identical to (3), *i.e.*

$$\sum_{i=1}^{m} |g_i(\mathbf{h})|^2 = \epsilon_{\mathbf{h}} \sum_{j=1}^{N} f_j(\mathbf{h})^2 \quad ; \qquad (15)$$

2) in this case the group scattering factor is expressed using Debye (1915) scattering equation

$$|g_i(\mathbf{h})|^2 = \epsilon_{\mathbf{h}} \sum_{j=1}^{n_i} \sum_{k=1}^{n_i} f_j(\mathbf{h}) f_k(\mathbf{h}) \frac{\sin 4\pi s d_{jk}}{4\pi s d_{jk}} \quad ; \qquad (16)$$

where d_{jk} is the distance between the j-th and k-th atom in the fragment;

3) the group scattering factor becomes

$$|g_i(\mathbf{h})|^2 = \sum_{s=1}^{P} \sum_{j=1}^{n_i} \sum_{k=1}^{n_i} f_j(\mathbf{h}) f_k(\mathbf{h}) \exp\{2\pi i \mathbf{h} \mathbf{R}_s (\mathbf{r}_j - \mathbf{r}_k)\} \qquad (17)$$

where P is the number of point-group symmetry operators and $\mathbf{r}_j - \mathbf{r}_k$ are the interatomic vectors of the oriented fragment; note that the $\epsilon_{\mathbf{h}}$ factor is not applied because symmetry is already accounted for.

4) the group scattering factor is nothing but the contribution of the n_i known atoms of the fragment to the structure factor, $F_M(\mathbf{h})$; its square will be

$$|g_i(\mathbf{h})|^2 = \sum_{j=1}^{M} \sum_{k=1}^{M} f_j(\mathbf{h}) f_k(\mathbf{h}) \exp\{2\pi h i (\mathbf{r}_j - \mathbf{r}_k)\} \qquad (18)$$

where $M = n_i \times$ (space group order) is the number of known atoms in the cell (for simplicity atoms in special positions are not considered).

Pseudotranslational symmetry

An other type of *a-priori* information is the knowledge of the type of pseudotranslational symmetry which may be present in a crystal structure.

We will say that a structure possesses <u>pseudosymmetry</u> when a non-negligible part of the atoms approximately satisfy a higher symmetry than that of the whole structure. It can be noted that, in the special case in which these atoms exactly satisfy a higher symmetry, they will identify a supergroup of the space group to which the whole structure belongs.

An important type of pseudosymmetry is the pseudotranslational symmetry occurring when a non-negligible amount of the electron density, say $\varrho_p(\mathbf{r})$ is repeated by a pseudotranslation (in the sense that it does not necessarily correspond to a crystallographic translation) $\mathbf{u}$, i.e. $\varrho_p(\mathbf{r}) \approx \varrho_p(\mathbf{r+u})$. The vector $\mathbf{u}$ may be expressed as

$$\mathbf{u} = \frac{\nu_1}{\mu_1}\,\mathbf{a} + \frac{\nu_2}{\mu_2}\,\mathbf{b} + \frac{\nu_3}{\mu_3}\,\mathbf{c} \tag{19}$$

where $\mathbf{a}$, $\mathbf{b}$, $\mathbf{c}$ are the unit-cell vectors and ν_i and μ_i are pairs of integers with no common factor, such that $0 \le \nu_i \le \mu_i$ for $i=1,2,3$.

Buerger (1956, 1959) suggested that such structures should be treated as the sum of a substructure, $\varrho_p(\mathbf{r})$ and the complement structure, $\varrho_c(\mathbf{r}) = \varrho(\mathbf{r}) - \varrho_p(\mathbf{r})$.

A two-dimensional example of pseudotranslational symmetry is shown in Figure 3, where the two heavy atoms 1 and 1' are related by a centering translation, $\mathbf{u}=\mathbf{a}/2+\mathbf{b}/2$, while the light atoms occupy general unrelated positions: the heavy atoms form the substructure and the light atoms the complement structure. If the structure only contained the heavy atoms, the unit cell would be centered and the reflections with h+k=2n+1 would be systematically absent. For the structure of Figure 3, only the light atoms will contribute to these reflections (superstructure reflections), which will therefore be systematically weaker than the reflections with h+k=2n (substructure reflections) to which both heavy and light atoms contribute.

Because of the presence of pseudotranslational symmetry, imposing certain relations between subsets of atomic positions, the assumption that the atomic coordinates are independent random variables is violated.

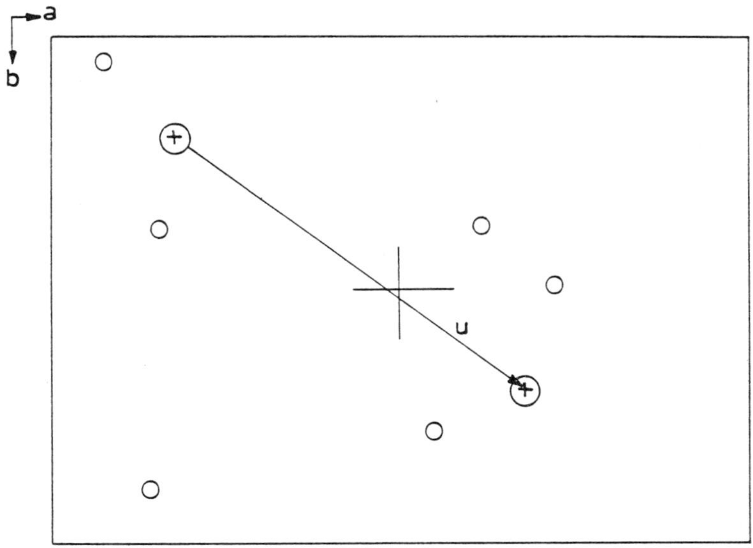

Figure 3

In the reciprocal space this is revealed by a nonuniform value of the average normalized intensity, $<|E_h|^2>$, when the reflections are grouped in different ways. As we have seen, the reflections have to be divided into the set of substructure reflections with high mean intensity and the set of substructure reflections, with low mean intensity.

It has been shown by Cascarano, Giacovazzo and Luic (1985 a,b) that, for a crystal possessing r independent pseudotranslation vectors u_i (i=1 to r), the average intensity is given by

$$<|F_h|^2> = \epsilon_h \ (\alpha_h \ \Sigma_p + \Sigma_q) \qquad (20)$$

with $\quad \Sigma_p = \sum_{j=1}^{p} f_j^2 \quad$ and $\quad \Sigma_q = \sum_{j=1}^{q} f_j^2$

where p is the total number of atoms in the unit-cell related by pseudotranslation and q is the total number of atoms in the cell not affected by pseudotranslation. In (20)

$$\alpha_h = \gamma_h \ (n_1 n_2 \ldots n_r)/m \qquad (21)$$

where m is the general order of the space group, n_i is the index of the i-th pseudotranslation (*i.e.* the smallest integer for which $n_i u_i$ is a lattice vector) and γ_h is the number of times the algebraic congruences

$$hR_s u_i \equiv 0 \ (mod \ 1) \quad for \ i=1 \ to \ r \qquad (22)$$

are simultaneously satisfied when R_s varies over the m rotation matrices of the space group. The maximum value of is m when $\alpha_h = n_1 n_2 \ldots n_r$. If no pseudotranslation occurs ($p=0$) then (20) reduces to (3). Reflexions with $\alpha_h = 0$ are superstructure reflexions for which $<|F_h|^2> = \epsilon_h \Sigma_q$.

Pseudotranslational symmetry may be recognized *via* the unequal distributions of the normalized structure factors E' computed without taking pseudotranslation into account. Considering pseudotranslations with index $n<16$ and with $\mu_i \leq 4$, then 64 different vectors u_i may be generated. From (22) the corresponding sets of substructure reflexions are derived; their indices obey the conditions

$$\frac{\nu_1}{\mu_1} h \ + \ \frac{\nu_2}{\mu_2} k \ + \ \frac{\nu_3}{\mu_3} l \equiv 0 \ (mod \ 1) \ . \qquad (23)$$

For each set the number of reflection and the average normalized intensity are calculated.

If all atoms have similar unitary scattering factors, then

$$<|E'_h|^2> = (\alpha_h \ \Sigma_p + \Sigma_q)/\Sigma \approx (\alpha_h p + q)/N \qquad (24)$$

and

$$q/N \approx (<|E'_h|^2> - \alpha)/(1-\alpha) \qquad (25)$$

where the average is carried out by varying h over an homogeneous set of reflexions with the same $\alpha_h = \alpha$. When the set of superstructure reflexions (with $\alpha=0$) is used, then (25) reduces to

$$q/N \approx <|E'_h|^2>_{sup} \qquad (26)$$

which may be used to estimate $\underline{p}$ and $\underline{q}$. For each of the 64 types of pseudotranslation the ratio

$$< \left| E'_{\mathbf{h}} \right|^2 >_{sub} / < \left| E'_{\mathbf{h}} \right|^2 >_{sup} \qquad\qquad (27)$$

is calculated; the actual pseudotranslation is indicated by the set with the highest value of this ratio.

An example of the application of this method to the structure of freieslebanite (Ito & Novacki, 1974) (PbAgSbS$_3$, space group P2$_1$/a, Z=4) is shown in Table 3, where the high value of the ratio (27) for the set no. 53 indicates the existence of a pseudotranslation of type $\mathbf{u} = \mathbf{a}/2$ + $\mathbf{b}/3$; a value of $\underline{p}$/N=0.55 is also estimated.

Once the pseudotranslational symmetry has been identified, renormalization can be accomplished combining (5) with (20); it is expected that the new E's will loose the hypercentric character and this is clearly shown in Figure 4, where N(z) is a cumulative distribution function; the theoretical trends of N(z) for centro- and non-centrosymmetric structures may be derived from (8) and (9) respectively.

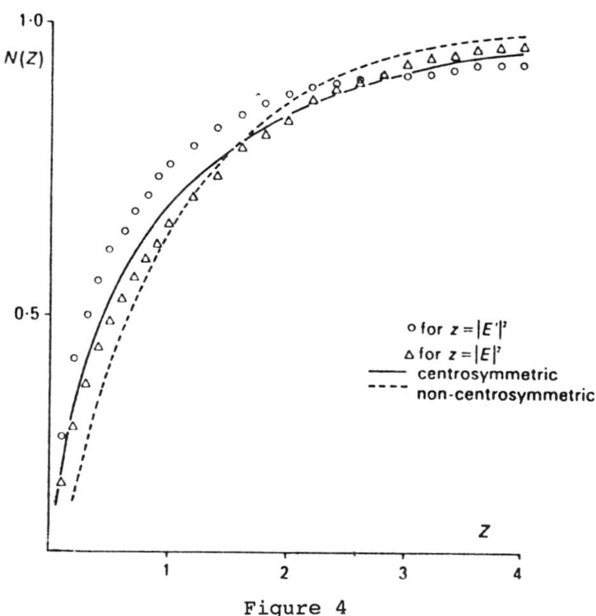

Figure 4

This statistical method for identifying the type of psuedotranslational symmetry and the related sub- and superstructure reflections has been recently improved (Cascarano, Giacovazzo & Luic, 1988) and other similar methods have been proposed (Fan Hai-Fu, Yao Jia-Xing & Qian Jin-Zi, 1988).

TABLE 3

$\langle |E|^2 \rangle$ values for the reflexions devided into different classes.
The asterisk indicates the correct pseudotranslation.

	Class		N	$\langle |E|^2 \rangle$		Class		N	$\langle |E|^2 \rangle$
1)	all		2000	1.000	2)	h	= 2n	1026	1.665
3)	k	= 2n	978	.960	4)	l	= 2n	1006	1.057
5)	h+ k+ l	= 2n	1010	.990	6)	h+ k	= 2n	1024	.917
7)	h+ l	= 2n	996	1.031	8)	k+ l	= 2n	1008	.986
9)	h	= 3n	649	1.055	10)	k	= 3n	631	2.022
11)	l	= 3n	671	.976	12)	h+ k	= 3n	1128	.835
13)	h+ l	= 3n	662	.970	14)	k+ l	= 3n	1126	.810
15)	h+ k+ l	= 3n	1121	.787	16)	h+ k+2l	= 3n	1129	.803
17)	h+2k+ l	= 3n	1121	.787	18)	2h+ k+ l	= 3n	1129	.803
19)	h+2k	= 3n	1128	.835	20)	h+2l	= 3n	654	.964
21)	k+2l	= 3n	1126	.810	22)	l	= 4n	496	1.051
23)	k	= 4n	458	.923	24)	h	= 4n	512	1.685
25)	h+ k	= 4n	768	.707	26)	h+ l	= 4n	488	1.024
27)	k+ l	= 4n	760	.964	28)	h+ k+ l	= 4n	764	.989
29)	2h+2k+ l	= 4n	506	1.070	30)	2h+ k+2l	= 4n	484	.966
31)	h+2k+2l	= 4n	508	2.402	32)	2h+ k+ l	= 4n	758	.965
33)	h+2k+ l	= 4n	500	1.012	34)	h+ k+2l	= 4n	772	.680
35)	h+2k	= 4n	512	1.664	36)	h+2l	= 4n	508	1.618
37)	k+2l	= 4n	492	.958	38)	2h+ k	= 4n	510	.852
39)	2h+ l	= 4n	492	1.051	40)	2k+ l	= 4n	506	1.084
41)	3h+3k+ l	= 4n	758	.979	42)	3h+ k+3l	= 4n	764	.989
43)	h+3k+3l	= 4n	758	.979	44)	h+2k+3l	= 4n	510	1.000
45)	h+3k+2l	= 4n	772	.680	46)	3h+ k+2l	= 4n	772	.680
47)	h+3k	= 4n	768	.707	48)	h+3l	= 4n	490	1.027
49)	k+3l	= 4n	760	.964	50)	3k+2l	= 6n	325	.926
51)	2k+3l	= 6n	313	2.107	52)	2h+3k	= 6n	323	.949
53)	3h+2k	= 6n	339	3.684 *	54)	3h+2l	= 6n	343	1.631
55)	2h+3l	= 6n	323	1.104	56)	2h+2k+3l	= 6n	570	.883
57)	3h+2k+3l	= 6n	313	2.081	58)	3h+3k+2l	= 6n	345	.896
59)	4k+3l	=12n	157	2.057	60)	4h+3l	=12n	159	1.079
61)	4h+3k	=12n	151	.842	62)	3k+4l	=12n	153	.852
63)	3h+4k	=12n	171	3.685	64)	3h+4l	=12n	169	1.593

Setting up triplet invariants

Since the most important relations are those estimated with the highest reliability, we will only set up triplets

$$\phi_h \approx \phi_k + \phi_{h-k} \qquad (28)$$

relating reflections with large $|E|$. The minimum $|E|$ value is chosen in such a way that the number of employed reflections is approximately equal to ten times the number of atoms in the asymmetric unit and usually lies between 1.2 and 1.6.

For each reflection h all pairs of reflections k and $h-k$ with large $|E|$ must be found; the set of all relationships obtained in this way is often called the Σ_2 listing. The triplet search must take into account the reciprocal lattice symmetry and the relations between the phases of equivalent reflections.

If all the indices are reconducted within the same independent portion of the reciprocal lattice, each reflection h will have to be combined with all vectors $\pm kR_i$, obtained by varying R_i over all rotation matrices of the space group not related by an inversion centre; if also the $|E|$ value of reflection $h\pm kR_i$ is large then the triplet is retained. In general this third reflection will not be in the considered independent portion of reciprocal lattice, and it can be expressed as $\pm HR_j$. The phase relationship will then become

$$\phi_h \approx \pm\phi_{kR_i} \pm \phi_{HR_j} \approx \pm\phi_k \pm \phi_H - 2\pi(kT_i - HT_j) \qquad (29)$$

Most computer programs store triplet relations in the form

$$n_h, \ \pm n_k, \ \pm n_H, \ \ \phi, \ G \qquad (30)$$

where n_h, n_k, n_H represent the sequential numbers of the corresponding reflections in the sorted list, with the sign indicated by (29), $\phi = 2\pi(kT_i + HT_j)$ and G is the parameter of Cochran's (1955) probability distribution given by

$$G_{hk} = 2\sigma_3\sigma_2^{-3/2}|E_h E_k E_{h-k}| \ \ \text{with} \ \sigma_n = \sum_{j=1}^{N} Z_j^n \qquad (31)$$

where Z_j is the atomic number of the j-th atom.

For centrosymmetric structures (29) reduces to

$$S(_h) \approx S(_k)S(_H)(-1)^{2(kT_i - HT_j)} \qquad (32)$$

and in (30) the signs are of no use, while G is the argument of the *tanh* sign probability formula derived by Cochran & Woolfson (1955) and is equal to half the value given in (31).

References

- Buerger M.J. (1956), Proc. Natl. Acad. Sci. USA, **42**, 776.

- Buerger M.J. (1959), "Vector Space and its Application in Crystal Structure Investigation", J. Whiley, New York.

- Calleri M., Chiari G., Chiesi Villa A., Gaetani Manfredotti A., Guastini C. & Viterbo D. (1976), Acta Crystallogr., **B32**, 1032.

- Cascarano G., Giacovazzo C. & Luic M. (1985), Acta Crystal-logr., **A41**, 544.

- Cascarano G., Giacovazzo C. & Luic M. (1985), in "Structure & Statistics in Crystallography", Editor A.J.C. Wilson, pp. 67-77, Adenine Press, Guilderland NY.

- Cascarano G., Giacovazzo C. & Luic M. (1988), Acta Crystal-logr., **A44**, 176.

- Chochran W. (1955), Acta Crystallogr., **8**, 473.

- Chochran W. & Woolfson M.M. (1955), Acta Crystallogr., **8**, 1.

- Debye P. (1915), Ann. Phys. Lpz., **46**, 809.

- Fan Hai-Fu, Yao Jia-Xing & Qian Jin-Zi (1988), Acta Crystallogr., **A44**, 688.

- French S. & Wilson K. (1978), Acta Crystallogr., **A34**, 517.

- Hall S.R. & Subramanian V. (1982), Acta Crystallogr., **A38**, 590.

- Hall S.R. & Subramanian V. (1982), Acta Crystallogr., **A38**, 598.

- Ito T. & Novacki W. (1974), Z. Kristallogr., **137**, 399.

- Karle J. & Hauptman H. (1953), Acta Crystallogr., **6**, 473.

- Main P. (1976), in "Crystallographic Computing Techniques", Editor F.R. Ahmed, pp. 97-105, Munksgaard, Copenhagen.

- Stewart J.M. & Karle J. (1976), Acta Crystallogr., **A32**, 1005.

- Subramanian V. & Hall S.R. (1982), Acta Crystallogr., **A38**, 577.

- Vickovic I. & Viterbo D. (1979), Acta Crystallogr., **A35**, 500.

- Wilson A.J.C. (1949), Acta Crystallogr., **2**, 318.

- Wilson A.J.C. (1942), Nature, **150**, 152.

Exercises

1 - Compute the scale and temperature factor from the Wilson plot of Figure 2.

2 - Derive the reflexion multiplicity $\underline{m}$ and the symmetry factor ϵ for reflexions of types

$$hkl, \ 0kl, \ h0l, \ hk0, \ h00, \ 0k0, \ 001$$

assuming that the space group is $P2_1/c$ with general equivalent positions

$$x,y,z; \ \ \bar{x},\bar{y},\bar{z}; \ \ \bar{x},\tfrac{1}{2}+y,\tfrac{1}{2}-z; \ x,\tfrac{1}{2}-y,\tfrac{1}{2}+z.$$

3 - Using Debye' formula (16), derive what is the average bond distance
     contributing to the positive hump at $s^2=0.21\text{Å}^{-2}$ and to the negative
     humps at $s^2=0.11$ and 0.29Å^{-2} in Figure 2. Consider that the function
     $\sin x/x$ has the first maximum at $x=7.73$ and the first and second
     minimum at $x=4.49$ and $x=10.90$

STARTING SET DETERMINATON

Paul T. Beurskens

Crystallography Laboratory, Toernooiveld
6525 ED Nijmegen, The Netherlands

1. Introduction

For most practical applications of direct methods it is convenient to distinguish the following stages:
(a) the initiation: how to get started.
(b) the middle stage: phase generation and phase refinement.
(c) the final selection of the most probable phase set.
In this chapter we discuss some practical aspects of (a) and (b) with emphasis on the determination of the starting set. But how to choose starting reflections depends on what we plan to do with them!

The most important phase relationship

$$\varphi(\mathbf{h}+\mathbf{h}') \approx \varphi(\mathbf{h}) + \varphi(\mathbf{h}') \tag{1}$$

or

$$\varphi(\mathbf{h}) \approx \varphi(\Sigma_{\mathbf{k}} E(\mathbf{h}-\mathbf{k})\, E(\mathbf{k})) \tag{2}$$

has been derived, used, and discussed by many authors, with varying notation and nomenclature. The relationship is 'probably' or 'approximately' valid for strong reflections, i.e. reflections with large |E| values. Expression (1) is the sum-of-angle formula.The most popular form of (2) now is called the tangent formula (Karle and Hauptman, 1956). It is most powerful in the middle stage of any phasing procedure. If many phases are known, application of (1) leads to multiple phase indications for many reflections, which in fact implies that we are applying (2). In the initial stage, however, not many phases are known, and the application of single phase indications (1) may lead to phase errors. Even one single aberrant relation, i.e. (1) with a phase error of π, may cause failure to solve the structure. Thus special care is needed at the initial stage of the phasing procedure.

2. The sigma-formulae

The method proposed by Hauptman and Karle (1953) may serve as an example of a straightforward procedure. A number of formulae, denoted sigma-1, sigma-2, sigma-3, etc. were derived for centrosymmetric structures.
The sigma-2 relation is given by the sum-of-angle formula (1) and its summation (2).
The now well-known sigma-1 relation, for instance for space group P2/m:

$$S(0,2k,0) \approx S(\Sigma_h \Sigma_l (|E(h,k,l)|^2 - 1))$$

where S means 'sign of', may be used to find the sign of a structure seminvariant from experimental data. $|E(0,2k,0)|$ must be large, but notice however that weak reflections (having small $|E(h,k,l)|$ values) are used in the summations.
The sigma-3 relation gives an expression for a two-phase seminvariant, e.g. for space group P2/m:

$$S(h', k', l')\, S(h'', k'', l'') \approx S(\Sigma_k(|E(h,k,l)|^2 - 1))$$

with $h' - h'' = 2h,\ k' - k'' = 0,\ l' - l'' = 2l$.

Alternatively, if $S(h', k', l')$ is known, the sigma-3 relation can be used to find $S(h'', k'', l'')$.

The sign determination procedure consists of (a) the assignment of origin fixing signs (e.g. 3 for space group P2/m), (b) the use of sigma-1 to determine the signs of some seminvariants, (c) the use of sigma-3 (and perhaps other sigma -formulae) together with the results from (a) and (b) to find a few more signs, and finally (d) the use of eq. (2) (i.e. sigma-2) to determine as many signs as possible, for reflections with large $|E|$ values. The question of failure or success is mainly dependent on whether weak sign indications are accepted or not.

These early concepts have evolved considerably. The sigma-1 relation is still used in many computerised procedures. The seminvariants are extensively employed (e.g. Giacovazzo, 1977). The sigma-3 relation has been used as a figure of merit in the final stage (e.g. Beurskens and Van den Hark, 1975). The sigma-5 relation is now known as a quartet (e.g. Schenk, 1974).

3. Symbolic Addition

The very successful symbolic addition method (Karle and Karle, 1963) used letter symbols to represent phases. The phasing procedure begins with (a) the assignment of some origin specifications, (b) if possible the determination of some phases with the sigma-1 formula, and (c) if possible the determination of some more phases with the sum-of-angle formula (1). Typically about 5 - 10 phases are obtained this way.

Then all possible strong triplets (three reflections entering eq. 1) are inspected, and one reflection occurring frequently in combination with the results from (a), (b) and (c) is selected. This reflection is given the symbol A, i.e. the symbol A represents the phase of this reflection. This symbol will be used in (1) to generate more phases. If not all strong reflections can be reached by successive application of (1), it may be necessary to assign the symbol B to another selected reflection. Later a third (C) or even a fourth choice (D) may become necessary. At the end of the phase generation process (almost) all strong reflections have their phases expressed in terms of these symbols.

In the final stage of the symbolic addition procedure the symbolic phases have to be converted into numerical values. In principle one may try out all possible numerical values for each of the letter symbols; for instance by using all permutations of $\pm\frac{1}{4}\pi, \pm\frac{3}{4}\pi$. In practice, one may find relations between the symbols. For instance, if the phase of one reflection is found as A from one triplet (1) and as B from a different triplet (2) than the symbolic relation A = B may be used to reduce the number of possible solutions. In addition, other relations (e.g. the sigma-3) or properties (e.g. not all phases equal to 0) may be used as figures of merit to reduce the number of solutions.

The route in the phase generation process is not fixed. If the Fourier syntheses of one of more solutions do not lead to the elucidation of the structure, one may restart the procedure using another route, for instance use a different set of origin determining phases. The success of the procedure depends heavily on the skill of the crystallographer who is doing the work, and nowadays students can learn much about direct methods by reading the early literature.

4. Symbolic phase correlation

Significant improvement of direct methods techniques was made possible by the use of computers. In order to overcome the step-by-step assignment of symbolic phases, where each step in the procedure may introduce phase errors, we designed a procedure which employed many phases simultaneously (Beurskens, 1963, 1964; Beurskens and Van den Hark, 1980). In the initial steps, the phases of a large number of reflections are expressed in terms of some 10 - 26 symbols. Many correlations (relations between the symbols) are found and may be used to eliminate symbols. Thus while increasing the number of participating triples (1), the number of unknown symbols is gradually reduced. Alternatively, the entire ensemble of symbolic relations may be analysed in a multisolution fashion.

Procedure (Fig. 1) and definitions:
H is the primary set of reflections **h** of which the phase is known, either absolute (e.g. by origin choice, or sigma-1) or expressed in terms of one or more symbols. The selection of the initial symbol assignments is merely based on the $|E|$-value, which should be as large as possible, but there should be no relation (1) among the initial assignments.

H+H' denotes any possible vector sum **h+h'**, using also symmetry related reflections.

K is the secondary set of reflections **k**, selected from H+H', which are to be used as 'temporarily' accepted reflections.

H+K and K+K' are the numerous reflections for which symbolic phases have been obtained from H and K.

For 10-30 reflections in set H, and 20-200 reflections in set K, many thousands of reflections will have their phases expressed by a phase symbol. The number of such phase-indications is a multiple of the number of unique reflections! Therefore inspection of the results will give a large number of symbolic phase relations.

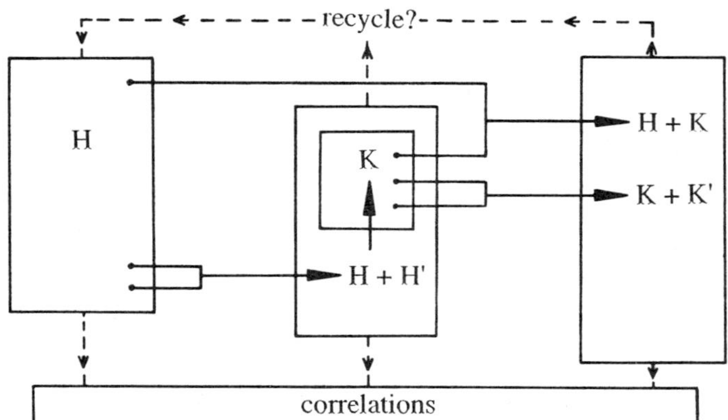

Fig. 1. Diagram showing the cyclic phase correlation procedure.

Example. In the primary set H: reflection h_1, h_2 and h_3 with symbols A, B and C, respectively; in the secondary set K: reflection $k_1 = h_1 + h_2$, symbol A + B; in the resulting set H + K: reflection $k_1 - h_3$, symbol A + B - C. Suppose that $h_1 + h_2 - h_3 = -h_1$, we then have the symbolic relation $A + B - C \approx -A$.

If we allow the symbols to take the values $\pm\frac{1}{4}\pi$, $\pm\frac{3}{4}\pi$ only, then for only 10 symbols there exist over one million permutations. Nevertheless, the analysis of the symbols is straightforward and fast (Beurskens and Prick, 1981). Finally, we end up with an ordered list of possible solutions, each with numerical phase values for the symbols, and a consistency criterion based on eq. (2). How to find the correct solution from the list of possible solutions is not a trivial matter. It requires the use of other information, such as the use of weak reflections, which is outside the scope of this chapter.

Some points are worth mentioning.

- The generation of symbolic phases from sets H and K is directly related to the existence of quadruples (De Vries, 1965; Viterbo and Woolfson, 1973) among the triples that are used and, consequently, strengthened quartets (Schenk, 1974) play an important role in the phase correlation procedure (Van den Hark, 1976).

- If the primary set does not contain any origin and enantiomorph choice, the n-fold origin ambiguity is expressed by an n-fold multiplicity of identical solutions. Any one of the possible solutions then implies the specification of the origin (and enantiomorph).

- The concept of primary and secondary reflections has also been used in various multisolution and random phasing methods (e.g. Declercq et al., 1975).

5. Convergence mapping

Various multisolution techniques use more-or-less arbitrarily specified phases as starting phases for a phase generation (1) and tangent refinement (2) procedure. By repeating the entire refinement procedure for all possible permutations of the originally specified starting phases, one hopes to encounter the correct solution among many false solutions.

One way of optimising this procedure, originally introduced in the MULTAN system, is the very successful convergence procedure (Germain et al., 1970). The aim of the convergence mapping is to find a starting set which contains origin (and enantiomorph) fixing phases, as few as possible permutational phases, and which enables the generation of all other phases by (2), while avoiding 'weak links' (the use of single phase indications (1)).

Table 1. Bottom of convergence map (example)	Table 2. Starting set (used in Table 1)	Table 3. Improved convergence map (refl. 2 added to starting set)
11 = 2 + 7, 12 + 5, 23 + 26	Sigma-1: 8, 26	11 = 2 + 7, 12 + 5, 23 + 26
2 = 10 + 6	Origin: 7, 12, 23	10 = 6 + 12, 3 + 23, 6 + 2
10 = 6 + 12, 3 + 23	Enantiomorph:	6 = 7 + 5, 3 + 12
6 = 7 + 5, 3 + 12	Permutation: 3, 5	

One example is given in Table 1. This table gives the bottom of a convergence map. One reflection is indicated by a sequential number: increasing numbers indicate reflections with smaller $|E|$-values.

During the development of the convergence map the selection of the origin, enantiomorph and permutation reflections takes place: the results for this example are given in Table 2, together with possible other *a priori* known phase information for this example: two phases of seminvariants determined by the sigma-1 relation.

The bottom line of Table 1 gives: the phase of reflection 6 is obtained through two relationships (1), i.e. via reflections 7 and 5, and via reflection 3 and 12. We ignore here any symmetry, so 7 and 5 could imply negative phases or numerical terms (for instance: $\varphi(6) = \varphi(7) - \varphi(5) + \pi$). So the bottom line tells us that reflection 6 is obtained by two strong interactions (1).

The second to last line of Table 1 tells us that reflection 10 is also obtained by two interactions. Then reflection 2 is obtained by just one interaction. Thereafter reflection 11 is obtained by three triples (1). And so on.

In this particular example the line "2 = 10 + 6" is a weak link. A serious phase error is introduced if this single indication happens to be incorrect. It would have been better to define reflection 2 as an additional permutation reflection (at the cost of a severe increase of computer time!). Table 1 would then be changed to Table 3.

The converging map is developed as follows:
The sigma-2 list, which is a complete list of interactions (1) for all participating reflections (i.e. usually $|E|$ larger than a fixed value, for instance 1.20), is used to estimate the 'importance' α of each of the reflections. The reflection with the largest α is the reflection for which the phase can be determined by (2) with the highest reliability (many contributors, with large $|E|$ values).

The reflections with $\alpha = 0$ cannot be obtained by (1). These reflections cannot be phased, and they will be removed definitively from the sigma-2 listing.

Let us consider the reflection with the lowest α ($\alpha > 0$). This reflection can be phased by other reflections, but it is the reflection that is least useful for the phasing of other reflections. It will be removed from the sigma-2 listing, for the purpose of finding the more important reflections. Later, having found a good starting set, the complete sigma-2 listing is restored. The α values for all reflections are now recalculated from the shortened sigma-2 listing.

What about the reflection with the next lowest α? It will be removed from the sigma-2 listing, unless this reflection is essential in the phasing process. This means that we should not remove reflections which cannot be phased by the remaining reflections! For instance, the remaining reflections must contain origin (and enantiomorph) defining reflections. In fact, they are selected by this criterion: any reflection which is essential to origin (and enantiomorph) fixation is not removed from the list.

While the least useful reflections are removed, ensuring that the origin (and enantiomorph) defining reflections are kept, the selection of permutation reflections is considered. That is, when one reflection cannot be phased by any relation (1) this reflection will be used as a permutation reflection. (This reflection will be used with several different numerical phases, or it will be given a symbolic phase.) Moreover, when a reflection cannot be phased by more than one or two relations (1), it is a 'weak link' and this reflection may also be used as a permutation reflection. An example of the end of such a procedure is given in Tables 1 and 2 (or 3).

The main result of the convergence procedure is the selection of the starting reflections. It is customary to use the convergence map from the bottom upwards as the pathway for the phase generation. This, however, is not the best possible pathway. A divergence mapping is used in SIMPLE (Overbeek and Schenk, 1978) which tries to find the best possible pathway from the starting set obtained by the convergence procedure.

6. Conclusions

New formulae and new strategies for phase determination, applied to problems of increasing complexity, continuously require ingenuity in finding and using the starting set reflections. But consider Sheldrick (1980), in discussing the multisolution philosophy: 'If you try to solve a crystal structure many different ways, one of them will probably work. Why it works and others fail may not be obvious.'

This is true for direct methods in general, as well as for practical computer applications.
What matters is : did you solve the phase problem?

References

Beurskens, P.T., 1963, Technical report on *Sign Correlation by the Sayre Equation*. The Crystallography Laboratory, University of Pittsburgh, Pennsylvania.

Beurskens, P.T., 1964, A computer procedure for the systematic application of Sayre's equation for solving the phase problem for the analysis of centrosymmetric structures, Acta Cryst., 17:462.

Beurskens, P.T., 1965, Thesis, University of Utrecht, The Netherlands.

Beurskens, P.T., and Prick, P.A.J., 1981, A fast algorithm for determining phase values for symbols from weighted symbol relations in direct methods, Acta Cryst., A37:180.

Beurskens, P.T., and van den Hark, Th.E.M., 1975, Application of sigma-3 and pair-relationships in polar space groups, Acta Cryst., A31:S14.

Beurskens, P.T., and van den Hark, Th.E.M., 1980, Phase correlation with calculated cosine invariants for routine structure analysis, in *Theory and practice of direct methods in crystallography*, edited by M.F.C Ladd and R.A. Palmer, Plenum Press, New York.

Declercq, J.P., Germain, G., and Woolfson, M.M., 1975, On the application of phase relationships to complex structures VIII. an extension of the magic-integer approach, Acta Cryst., A31:367.

De Vries, A., 1965, The correlation method. A systematic use of the interrelationship between sign relations, Acta Cryst., 18:473.

Germain, G., Main, P., and Woolfson, M.M., 1970, On the application of phase relationships to complex structures II. Getting a good start, Acta Cryst., B26:274.

Giacovazzo, C., 1977, A general approach to phase relationships : the method of representations, Acta Cryst., A33:933.

Hauptman, H., and Karle, J., 1953, Solution of the phase problem, I. The centrosymmetric crystal, A.C.A. Monograph No. 3.

Karle, J., and Hauptman, H., 1956, A theory of phase determination for the four types of non-centrosymmetric space group 1P222 2P22 3P$_1$2 3P$_2$2, Acta Cryst., 9:635.

Karle, I.L., and Karle, J., 1963, An application of a new phase determination procedure to the structure of cyclo(hexaglycyl) hemihydrate, Acta Cryst., 16:969.

Schenk, 1974, On the use of negative quartets, Acta Cryst., A30:477.

Sheldrick, G.M., 1980, Multisolution experiences, in *Direct methods in crystallography*, N.A.T.O. Advanced Study Institute, Summerschool York.

Overbeek, A.R. and Schenk, H., 1978, in *Computing in Crystallography*, edited by H. Schenk, R. Olthof-Hazekamp, H. van Koningsveld and G.C.Bassi; Delft: Delft University Press.

Van den Hark, Th.E.M., 1976, Thesis, University of Nijmegen, The Netherlands.

Viterbo, D., and Woolfson, M.M., 1973, The determination of structure invarients I. Quadrupoles and their uses, Acta Cryst., A29:205.

SYMBOLIC ADDITION

Henk Schenk

Laboratory for Crystallography, University of Amsterdam
Nieuwe Achtergracht 129, 1018 WV Amsterdam
The Netherlands

Introduction

Symbolic Addition was the first Direct Method for routine structure determination (Karle & Karle, 1966). Moreover, when automated Symbolic Addition leads to very fast-running programs; structures are solved in a fraction of the time necessary to carry out a complete multisolution approach (Peschar, 1991; Schenk and Kiers, 1983; Overbeek and Schenk, 1978; Schenk, 1971). For an Advanced Study Instute there is one more advantage: the procedures used in Symbolic Addition enable easy insight in the actual processes going on in Direct Methods and as a result it is of great educational value too.

In Symbolic Addition key roles are played by the triplet relation

$$\varphi_H \approx \varphi_K + \varphi_{H-K} \tag{1}$$

for large values of the quantity

$$E_3 = N^{-1/2} |E_H E_K E_{H-K}| \tag{2}$$

and its centrosymmetric analogue

$$S_H \approx S_K S_{H-K} \tag{3}$$

with S_i is the sign of reflection i. Expression 3 is also more probable true when E_3 has larger values. When more than one relation 1 or 3 are available these two expressions have their analogues in:

$$\varphi_H = \frac{\sum_K E_3 (\varphi_K + \varphi_{H-K})}{\sum_K E_3} \tag{4}$$

and

$$S_H = A \sum_K E_3 S_K S_{H-K} \tag{5}$$

One more important expression is used frequently when in the final stage of a Symbolic Addition numerical phases have to be refined. This is carried out by means of the tangent formula, or rather, because phases must be known on the interval $-\pi \leqslant \varphi < \pi$ and a tangent expression is only uniquely determined in the interval $-\pi/2 \leqslant \varphi < \pi/2$, by its exponential analogue:

$$\exp i\varphi_H = \frac{\displaystyle\sum_K E_3 \exp i(\varphi_H + \varphi_{H-K})}{\left| \displaystyle\sum_K E_3 \exp i(\varphi_H + \varphi_{H-K}) \right|} \tag{6}$$

Whereas expression 6 gives directly the phases itself, the tangent formula, being itself unique on the interval 0 to π only, is therefore always used together with the signs of nominator and denominator in order to produce the phases. The latter procedure is the equivalent of using expression 6 directly (Schenk, 1972). All expressions 1, 3, 5 and 6 are extensively dealt with in other chapters of this book.

In historical perpesctive the use of symbols in the applications of Direct Methods is nearly as old as the methods themselves. In the same year when Harker and Kasper (1948) derived the first inequalities, Gilles (1948) was the first to use symbols as unknowns to determine the phases of a structure by means of inequalities. Also Zachariasen (1952) and Rumanova (1954) used the same approach to solve structures. However, the real break–through of symbols in phase determinations, is due to Karle and Karle (1963 and 1966), when they first solved a noncentrosymmetric structure and later described a standard procedure for the application of the Symbolic Addition.

A Symbolic–Addition method starts with collecting n phase relationships for m unknown phases where $m \ll n$. Then origin–defining phases are selected, and a number of other unknown phases is given a symbolic phase. These starting reflections are chosen such, that in principle all other phases can be reached from this set via the phase relationships. Then a phase propagation is carried out, using throughout symbolic phases. With the extended phase set as basis, the numerical values of the symbols are then determined through Figures of Merit. In most cases an additional phase extension using expression 6 is carried out, followed by a Fourier summation, which will reveal the structure in case the relationships donot carry to much errors and the phase extension has been carried out such that no error propagation has taken place.

Although alternative methods like the Symbolic Addition get more interest these days, the majority of structures are still solved by multisolution methods in which the phase propagation is carried out as many times as there are different starting points for the starting set. This explains why automated programs based on Symbolic Addition are so much faster: they apply the phase extension only once for all solutions at the same time by employing symbols and a second time for the selected numerical values of the unknown symbols.

It is the object of this chapter to introduce the reader in more detail to the Symbolic–addition methods and its potentials.

The Elements of a Symbolic Addition

A complete Symbolic Addition will start from a file with hkl indices and structure factor magnitudes, and ends with a roughly refined structure. The elements of such a procedure are the following:
a. Calculation of the normalized structure factors.
b. Generation of the phase relationships (triplets etc.).
c. Starting set determination.
d. Symbolic phase extension.
e. Determination of the approximate numerical values of the symbols.
f. Numerical phase refinement.
g. Calculation of an E–Fourier synthesis and its interpretation.
h. Fast structure refinement.

A brief description of the different steps is as follows: In step a the experimental structure-factor data are corrected for thermal motion (Wilson plot, Debije curve, K curve, see chapter on Multan), brought to absolute scale, and finally converted into normalized structure factors $|E|$. Then in step b all necessary phase relationships are collected. For example in the program system SIMPEL (Kiers and Schenk, 1983) not only triplet relations are generated, but also positive and negative quartets, Σ_1 relationships, Harker-Kasper relations, and quintets. Step c is carried out either by means of a convergence procedure or during the phase extension (step d) itself. In order to make sure that the final symbolic phases are as correct as possible, the phase extension step d has to be performed very carefully; in particular the first stages of this process must be free from errors because errors propagate usually as easily as correct phases. In the Optimal Symbolic Addition process (see chapter of Peschar, this book) in order to make sure that no errors can be made many different sets of symbolic phases are extended simultaneously. The numerical values of the symbols are determined in step e by means of a variety of figures of merit, some of which can be generally used whilst others are reliable in a limited number of space groups only. Then in most cases the tangent refinement 6 or the triplet refinement 5 is used (step f) in order to get at least 10 phased reflections per atom in the asymmetric unit for the calculation of a well-resolved E-Fourier synthesis. In step g this E-map is calculated and interpreted. The last step involves a quick least-squares refinement. In SIMPEL this is carried out by means of a diagonal least-squares routine with individual isotropic thermal parameters with facilities for automatic error correction (Shi and Schenk, 1989).

In this chapter steps c, d and e will be dealt with in more detail, for steps a, f and g the reader is referred to other chapters in this book. It may be noted that in step f the triplet refinement 4 is preferred for centrosymmetric space groups, because of computer efficiency.

<u>Starting Set Determination</u>

If one wants to start a Symbolic Addition with a complete starting set (origin and symbol-defining reflections), a careful analysis of the system of phase relationships is necessary. In many cases a convergence procedure (Germain, Main and Woolfson, 1970) will be sufficient to determine this starting set. In this procedure starting from the 200 highest $|E|$ values and all associated triplets or, as is done in SIMPEL, from the 30 to 50 strongest reflections and all their triplets and quartets, the phase, which is determined worst by means of the phase relationships, is identified and removed from the set of phases. At the same time the associated phase relationships are removed from the relation list. Then the next reflection which is related worst to the other reflection through the relationsips is identified and removed. This recipe is followed until no reflection is left in the reflection list and no relation in the relation list. All removed reflections, which share at the time of removal no relationsip with the reflections still remaining in the reflection set, belong to the starting set. After finishing the procedure these reflections can be used either for defining the origin or for assigning a symbolic phase.

However, in some structure determinations this starting set does not suffice, because a number of strong reflections may not be phased accurately in the course of the phase extension procedure since they are linked via weak Σ_2 relationships only to the set of phases reliably phased from the starting set reflections. In SIMPEL this is remedied by using a divergence (or accessability) procedure (Overbeek, 1985) This divergence procedure starts from the convergence starting set, adopt a strict acceptance scheme and establishes all phases reliably accessible from the starting set. Now weak links will result in groups of phases which are not accessible, and additional symbolic phases can be assigned in order to improve the starting set.

The determination of the starting set in centrosymmetric and noncentrosymmetric space groups is very similar. The only essential difference is that in the latter phase-restricted reflections are preferable to fix the origin. If the space group has restricted reflections, which do not enter in the stronger Σ_2 relations, it is advisable to delay the origin fixing to a later stage and to assign one or more extra symbols. The phase-extension procedure is then carried out, and finally from the symbolic phases of the restricted reflections a symbol or symbolic sum is deduced which fixes the origin (Overbeek, 1985).

One more difference appears in noncentrosymmetric Symbolic Addition: it is necessary to fix the enantiomorph. Sometimes this can be done by assigning a value to a special projection reflection as in space group $P2_12_12_1$, but in many cases the enantiomorph is most strongly fixed when it is done with a reflection or group of reflections which have no interaction with the origin defining reflections. When an appropriate reflection is not obviously available at this stage of the phase determination, the choice of enantiomorph can also be delayed till after the phase extension step.

The Handling of Symbols in the Phase Propagation

For phase propagation in centrosymmetric structures expressions 3 and 5 are used. Thus if $S_K=a$ and $S_{H-K}=bc$, and $S_H=S_K*S_{H-K}$ forms a reliable triplet, then $S_H=a*bc=abc$. Of course, whatever the sign of a the value of a^2 will be always +1, and thus symbolic signs can be easily reduced, e.g. $a^3bc^2d^4=ab$. In case more than one phase indication are available for a particular reflection they must be calculated all and only in case they all lead to the same symbolic sign the final sign is evident. When one is forced (by educators or circumstances) to apply Symbolic Addition by hand, this receipe will de sufficient.

In noncentrosymmetric space groups expressions 1 and 4 are used. Now if the phases are $\varphi_K=a+2b$ and $\varphi_{H-K}=-b+c+\pi$ and the triplet relation $\varphi_H=\varphi_K+\varphi_{H-K}$ is reliable then the resulting phase is given by $\varphi_H=a+b+c+\pi$. In noncentrosymmetric structures some phases (also symbolic phases) may be restricted and for those symbols or symbol combinations again only one multiple may occur. E.g. if b is either 0 or π, then $2b=0$ and thus $a+2b$ reduces to a. Therefore, when b occurs and is is phase-restricted, then it will take the form +b only because when it occurs in other combinations these can be reduced by adding or subtracting 2b. Another example: in case symbol c is equal to $\pm \pi/2$ then the phases can be reduced by adding or subtracting as many times the quantity $2c+\pi$ untill the coefficient of c equals 1. Also in the noncentrosymmetric case all different phase indications should be identified and calculated.

Symbolic Phase Extension

In the symbolic phase propagation the number of erroneous phases needs to be kept to a minimum, and therefore strict rules are necessary to accept new signs from the symbolic sign indications calculated by means of the phase relations. In centrosymmetric cases several authors suggest the use of the hyperbolic tangent formulae (Cochran and Woolfson, 1955) for the triplets

$$P_+(S_HS_KS_{H-K}) = \frac{1}{2} + \frac{1}{2}\tanh(|E_3|S_HS_KS_{H-K}) \tag{7}$$

and for the sign of a reflection when more than one indication is present from different phase relationships

$$P_+(S_H) = \frac{1}{2} + \frac{1}{2}\tanh(\sum_K |E_3|S_KS_{H-K}) \tag{8}$$

However, expression 7 underestimates the true probabilities in many structures, and overestimates them in others (Schenk, 1973). Moreover, in Symbolic-addition procedures the available terms for the evaluation of expression 8 are never a random selection of all possible terms, and therefore $P_+(S_H)$ calculated with 8 in most cases overestimates the true probability. In view of this, flexible but strict criteria are to be preferred, and e.g. in the program system SIMPEL83 the acceptance criteria are chosen relative to the actual E_3 values (Overbeek, 1985); a new phase based on a single relationship is accepted only if the triple product of the relation belongs to the strongest E_3-values, and for double and multiple phase indications the lower limit for ΣE_3 is increased by some 30% for each extra indication. Furthermore, as soon as different symbolic-phase indications are found for a reflection, this reflection is not accepted as a phase which is allowed to generate new ones. This is because in that case the symbolic-phase relation resulting from the two phase indications will propagate, and in the end a very strong relation is found which may dominate the Σ_2-consistency figure of merit. In case the relation is correct this does not matter, but if not, the correct set of phases may not be found at all because it was lost in the phase-extension process.

In the noncentrosymmetric case the variance of expressions 1 and 4 may be used, however, only in case variances can be calculated which reflect the true uncertainties in the phase relationships and resulting phases rather a similar scheme should be adopted as given for the centrosymmetric case. Its philosophy is very pragmatic indeed and based on two ideas: 1. in case it is possible to solve a structure by Direct Methods at least a few phase relations with the highest E_3-values can be trusted, and 2. to avoid eror propagation the acceptance of new phases has to be very conservative.

In many cases it will be necessary to relax the strict acceptance criteria in the final stages of the extension, in order to find a sufficiently large number of phases. Generally this relaxation does not introduce many erroneous phases, because at that stage there is already a vast number of error-free symbolic ones.

It is also possible to incorporate the starting set determination (step c) in the phase extension procedure. In this case first the origin-defining reflections are chosen, and then they are used in the phase-extension process using the strict acceptance criteria described before. When no new phases are found any more, a symbol is assigned to a reflection which enters in many Σ_2 relations with the already determined phases, and again the phase extension process is carried through until it stops. Then a next symbol is assigned and so on until enough of the strong reflections have been signed (80%). This process can also be combined with a preliminary convergence process. As a rule it ends with more symbols than the methods described in the preceding section because, if symbolic phase indications are not identical, the extension process stops, and a new symbol will have to be chosen. This is the way one intuitively follows when a list of phase relationships is available and one works it out by hand; it will be used in this manner in the practical classes.

Figures of Merit

After the extension process a vast number of reflections have been phased symbolically, however, to calculate a Fourier summation numerical values for all phases are necessary. In order to determine numerical values for the symbols a variety of figures of merit (FOM's) can be used. In general figures of merit can be based on any quantity which can be expected to have an extreme value for the set of correct signs, e.g. on the internal consistency of the Σ_2 relation and tangent formula, on Sayre's equation, on Harker Kasper inequalities, on many invariants and seminvariants, and on electron density considerations. The FOM-values are evaluated by assigning numerical values to the symbols.

In Symbolic Addition, all figures of merit can be calculated in relatively small amounts of computing time. The general tactics is to calculate first the symbolic phase sums of the relations entering in a particular FOM, and then combine all weights of identical phase sums, thus reducing the number of relations to be summed. Finally, the summation is carried out for specified numerical values for the symbols. This gives FOM values for all specified solutions. In noncentrosymmetric cases trial values of either 0, 90, 180 and 270°, or 45, 135, 225 and 315° are used. In general this leads to 4^n values for the FOM in which n is the number of unknown symbols. Of course, in centrosymmetric cases only 2^n values of a FOM have to be calculated.

In the remaining part of this section some FOM's will be dealt with in more detail, whilst others will be mentioned only. The section ends with an overview when to use which FOM.

The Σ_2-consistency FOM can be derived using the properties of the arithmetic mean for relation 4 (Schenk, 1972):

$$\Sigma_2 FOM = \sum_H \sum_K E_3 (-\varphi_H + \varphi_K + \varphi_{H-K})^2 \tag{9}$$

$$\text{with} \quad -\pi < (-\varphi_H + \varphi_K + \varphi_{-H+K}) < \pi$$

which in the centrosymmetric case is equivalent to

$$\Sigma_2 FOM = \sum_H \sum_K E_3 (1 - S_H S_K S_{H-K}) \tag{10}$$

Similarly, from the exponential expression 6, the consistency FOM for the 'tangent' refinement can be obtained being

$$TanFOM = \sum_H \sum_K E_3 \sin^2 (\tfrac{1}{2}(-\varphi_H + \varphi_K + \varphi_{H-K})) \tag{11}$$

In the expressions 9, 10 and 11 the reflections H all belong to the asymmetric part of reciprocal space, but K and H–K may be found in the entire reciprocal space. Since the phases in the asymmetric part have to be determined, all phases in 9, 10 and 11 outside this part are replaced by phases inside by application of the space–group symmetry rules. E.g. expression 9 then changes into

$$\Sigma_2 FOM = \sum_H \sum_K E_3 (-\varphi_H* + S_K \varphi_K* + S_{H-K} \varphi_{(H-K)}* + c) \tag{12}$$

The starred indices belong to the asymmetric part, $S = \pm 1$ follows from symmetry rules, and c is a phase shift occurring in non–symmorphic space groups, i.e. space groups with translational symmetry operations (Schenk, 1972).

From expression 12 it is easily seen that for instance in P$\underline{1}$ (for typographical reasons the bar is not above 1 and instead 1 is underlined) where all φ=0 or π; $S_K = S_{H-K} = +1$; c=0 the trivial solution with all φ=0 gives the minimum value of $\Sigma_2 FOM$. Also the values for other solutions may be less reliable, so that $\Sigma_2 FOM$ is unreliable for selecting the correct Σ_2–solution in all centrosymmetric symmorphic space groups (space groups without translational symmetry) with c=0. Another example: in P1 ($S_K = \pm 1$, $S_{H-K} = \pm 1$, c=0) the most consistent solution with $\Sigma_2 FOM$=0 is again the one with all phases equal to 0. In general in the noncentric symmorphic space groups (c=0) the use of $\Sigma_2 FOM$ often leads to centrosymmetric solutions. In non–symmorphic polar space groups such as P2$_1$ refinement often results in centrosymmetric phases also, although here the centric projection reflections may have correct phases. In the other space groups (P2$_1$/c, P2$_1$2$_1$2$_1$) the smallest value of $\Sigma_2 FOM$ may indicate directly the set of correct phases.

It is emphasized here that all FOM's based on the consistency of either the Σ_2 relation or the tangent refinement have the same properties. This implies that:
1. for symmorphic spacegroups these FOM's are unreliable for selecting the correct phase set.
2. in many noncentrosymmetric spacegroups extra information is necessary to prevent a starting set from refining to a centrosymmetric solution.

In Symbolic–addition procedures the calculation of the consistency index $\Sigma_2 FOM$ is rather time consuming. A criterion Q (Schenk, 1971) is closely related to $\Sigma_2 FOM$:

$$Q = \sum_H \sum_i \sum_j E_{3i} E_{3j} | \varphi_{K_i} + \varphi_{(H-K)_i} - \varphi_{K_j} - \varphi_{(H-K)_j} | \tag{13}$$

and enables a much faster screening because most terms will have a phase sum of zero while many of the remaining ones have identical symbolic–phase sums and can be taken together. This leads to

$$Q = \sum_1 W_1 |\Delta\varphi_1| \qquad\qquad (14)$$

in which 1 ranges over a relative small number of terms. Since the $\Delta\varphi_1$ are expressions in the symbolic phases a, b, etc. the FOM Q is a function of them. Some of the terms in Q are relatively weak and can be neglected, which speeds up the calculation further. The phase function (Riche, 1973) is related to the Q Figure of Merit.

A large number of FOM's have been developed, most of them being dealt with in other chapters of this proceedings, for instance the Harker–Kasper criterion and the negative quartet criterion, which both solve the problems in symmorphic space groups, the residual R–factor (Karle and Karle, 1966), the ψ_0 check (Cochran & Douglas, 1957) and others.

In this chapter we will deal with two other FOM's: the Σ_1 criterion and the projection criterion (Schenk, 1971b). The latter is very helpful for phasing in noncentric spacegroups with a reasonable number of restricted phases, e.g. the h01 in space group $P2_1$. All symbolic–phase indications of these restricted reflections are collected, and they can be set equal to the restricted value modulo π. Then the projection FOM is defined as

$$PROFOM = \sum_H \sum_K E_3 \{ (\varphi_K + \varphi_{H-K}) - r_H \}$$

where H are the restricted reflections, and r_H their restricted values (e.g. in $P2_1$ $r_H = 0$ for the h01 reflections). The symbolic phases $\varphi_K + \varphi_{H-K}$ are summed, and for identical symbol combinations and r_H–values the weights E_3 are summed. This leads to:

$$PROFOM = \sum_1 W_1 (\alpha_1 - r_1)$$

in which α_1 is the phase expressed in the symbols a, b, etc. It is clear that values for the symbols are found modulo π only, so that PROFOM can never solve the phase problem without help of another FOM, such as the Σ_2FOM.

Finally one more criterion will be mentioned here, the Σ_1FOM (Overbeek & Schenk, 1976). It is well known that Σ_1 relations don't give very reliable results when used for the phasing of reflections and lead to problems when used in starting set procedures. A FOM based on Σ_1 relationships, however, is very effective in discriminating between the Σ_2–solutions.

In the following table recommendations are given which FOM should preferably be used in which space groups:

spacegroup	FOM
centrosymmetric symmorphic (e.g. P1)	HKC, zero check, S_1FOM, Negative Quartet FOM, <u>never</u> Σ_2FOM
centrosymmetric others (e.g. $P2_1/c$)	Σ_2FOM, optionally others
noncentrosymmetric symmorphic (e.g. P1)	enantiomorph specific FOM's <u>never</u> Σ_2FOM
noncentrosymmetric polar (e.g. $P2_1$)	enantiomorph specific FOM's PROFOM, Σ_1FOM, <u>never</u> Σ_2FOM
noncentrosymmetric others (e.g. $P2_12_12_1$)	Σ_2FOM, optionally others

It will lead to stronger Direct Methods procedures when for a particular space group the automatic programs use the proper FOM's only and neglect the unreliable ones.

Refinement of Symbol Values

All FOM's are calculated for a limited number of numerical values, and a minimum FOM-value therefore does not automatically imply that such a value is the lowest possible. This minimum can be approached by application of a non-linear least-squares procedure with the numerical values for the symbols belonging to the lowest FOM-values as starting values (Schenk, 1971c). In case the FOM is defined as:

$$FOM = \sum_i W_i \alpha_i$$

in which α_i stands for the symbolic relation, the least-squares procedure minimizes the quantity

$$LSFOM = \sum_i W_i \alpha_i^2$$

The resulting parameters for the symbols are better approximations for the symbols than those given by the FOM calculation.

Practical Exercises and Additional Reading

For additional reading the 1966 paper of Karle and Karle is strongly recommended. However, it is most instructive when the reader tries to solve the phase problem by hand given a list of relationships or triplets. Most programs like SIMPEL have an option by means of which tables of relations can be printed. Then given the information given in this chapter and in the chapter about the choice of origin and enantiomorph by Krabbendam it is possible to work out a Symbolic Addition by hand. An exercise like this will enable the reader to develop skill to redirect the phasing process in those cases where initially Direct Methods fail to solve a structure. In the authors labaratory a programmed text and Computer Assisted Learning module (PC based) have been developed, which is presented in this ASI in the poster session. This teaches students the basis and use of Symbolic Addition and applies it to a two dimensional structure (Wang et al., 1990).

References

Cochran, W. and Douglas, A.S. (1957). Proc.Roy.Soc.A. <u>243</u>, 281.
Cochran, W. and Woolfson, M.M. (1955). Acta Cryst. <u>8</u>, 1.
Germain, G. and Woolfson, M.M. (1968). Acta Cryst. <u>B24</u>, 91.
Germain, G., Main, P. and Woolfson, M.M. (1970). Acta Cryst. <u>B26</u>, 274.
Gilles, J. (1948). Acta Cryst. <u>1</u>, 174.
Harker, D. and Kasper, J.S. (1948). Acta Cryst. <u>1</u>, 70.
Karle, I.L. and Karle, J. (1963). Acta Cryst. <u>16</u>, 969.
Karle, J. and Karle, I.L. (1966). Acta Cryst. <u>21</u>, 849.
Overbeek, A.R. (1985). Description of SIMPEL, Crystallography, Amsterdam.
Overbeek, A.R. and Schenk, H. (1976). Proc. Kon. Ned. Akad. Wet. <u>B79</u>, 341.
Overbeek, A.R. and Schenk, H. (1978) in: Computing in Crystallography, ed. by
    H. Schenk, R. Olthof, H. van Koningsveld and G.C. Bassi, Delft University Press,
    Delft, p. 108.
Peschar, R. (1991), this proceedings
Riche, C. (1973), Acta Cryst. <u>A29</u>, 133.

Rumanova, I.M. (1954). Dokl. Acad. Nauk. SSSR 98, 399.

Schenk, H. (1971). Acta Cryst. B27, 2037.

Schenk, H. (1971b). Acta Cryst. B27, 2039.

Schenk, H. (1971c). Acta Cryst. B27, 2040.

Schenk, H. (1972. Acta Cryst. A28, 412.

Schenk, H. (1973). Acta Cryst. A29, 503.

Schenk, H. and Kiers, C.T. (1983). Simpel83, a Program System for Direct Methods. In: Crystallographic Computing 3, edited by Sheldrick, G.M., Kruger, C. and Goddard, R., Oxford: Clarendon Press.

Shi, J–Q., and Schenk, H. (1989). Proc. Kon. Ned. Akad. Wet. B91, 237.

Wang, Y.F., Molhoek, P., and Schenk, H. (1990), this proceedings.

Zachariasen, W.H. (1952), Acta Cryst. 5, 68.

TANGENT FORMULA, TANGENT REFINEMENT AND MULTISOLUTION METHODS

Michael M. Woolfson

Physics Department
University of York
York YO1 5DD, U.K.

THE TANGENT FORMULA

A single three-phase relationship, cast in the form

$$\phi(\mathbf{h}) \approx \phi(\mathbf{k}) + \phi(\mathbf{h-k}) \tag{1}$$

is capable of giving an estimate of the value of $\phi(\mathbf{h})$ if the values of the phases on the right-hand side are known. The question now arises of finding the overall estimate if there are several pairs of known phases, the estimate from each of which might well be different.

The answer to this important problem was given by Karle and Hauptman (1956) when they introduced the tangent formula

$$\tan\{\phi(\mathbf{h})\} = \frac{\sum_{\mathbf{k}} |E(\mathbf{k})\ E(\mathbf{h-k})|\ \sin\{\phi(\mathbf{k}) + \phi(\mathbf{h-k})\}}{\sum_{\mathbf{k}} |E(\mathbf{k})\ E(\mathbf{h-k})|\ \cos\{\phi(\mathbf{k}) + \phi(\mathbf{h-k})\}} . \tag{2}$$

With the various phase values inserted on the right-hand side an estimate for $\tan\{\phi(\mathbf{h})\}$ is found. Note that there is no phase ambiguity in deducing $\phi(\mathbf{h})$ from equation (2). The signs of the sine and cosine of $\phi(\mathbf{h})$ must be the same as those of the top and bottom respectively of equation (2). Thus if the top and bottom of the right-hand side are equal in magnitude and both positive then $\phi(\mathbf{h})$ equals $45°$ and not $225°$. The latter value would result if the top and bottom had the same magnitude but were both negative.

Direct Methods of Solving Crystal Stuctures
Edited by H. Schenk, Plenum Press, New York, 1991

There is an advantage theoretically in multipying the top and bottom of equation (2) by $|E(\mathbf{h})|\sigma_3/\sigma_2^{3/2}$ (see section on the Cochran distribution for the meaning of σ_n and of $\kappa(\mathbf{h},\mathbf{k})$, which follows). The tangent formula then becomes

$$\tan\{\phi(\mathbf{h})\} = \frac{\sum_{\mathbf{k}} \kappa(\mathbf{h},\mathbf{k})\ \sin\{\phi(\mathbf{k}) + \phi(\mathbf{h}-\mathbf{k})\}}{\sum_{\mathbf{k}} \kappa(\mathbf{h},\mathbf{k})\ \cos\{\phi(\mathbf{k}) + \phi(\mathbf{h}-\mathbf{k})\}} = \frac{B(\mathbf{h})}{A(\mathbf{h})} \ . \quad (3)$$

It was with a combination of the phasing formulae (1) and (3) that Karle and Karle (1964) solved the first non-centrosymmetric structure with direct methods.

DERIVING THE TANGENT FORMULA

The tangent formula can be derived from several different approaches and we shall look at two of them here. In the section on the Cochran distribution, from the imaginary part of Sayre's equation it was found that

$$\sum_{\mathbf{k}} \kappa(\mathbf{h},\mathbf{k})\ \sin\{\phi(\mathbf{h}) - \phi(\mathbf{k}) - \phi(\mathbf{h}-\mathbf{k})\} = 0 \quad . \quad (4)$$

The individual terms in the summation form a large population of numbers whose average value is zero. Normal statistical sampling theory then tells us that the expectation value of the sum of any number of these terms is also zero; while equation (4) is, in principle exactly valid for a complete data set it has a statistical validity if only some terms are used on the right-hand side. Equation (4) may now be rewritten

$$\sin\{\phi(\mathbf{h})\} \sum_{\mathbf{k}} \kappa(\mathbf{h},\mathbf{k})\ \cos\{\phi(\mathbf{k}) + \phi(\mathbf{h}-\mathbf{k})\}$$
$$- \cos\{\phi(\mathbf{h})\} \sum_{\mathbf{k}} \kappa(\mathbf{h},\mathbf{k})\ \sin\{\phi(\mathbf{h}) + \phi(\mathbf{h}-\mathbf{k})\} = 0 \ . \quad (5)$$

which can be rearranged to give equation (3).

Of course it could be argued that Sayre's equation only

applies to equal-atom structures which would invalidate the use of Sayre's equation as a route for deriving the tangent formula if all atoms were not equal. However, it turns out that the tangent formula is not rigorously valid for the non-equal-atom case. If it *was* rigorously valid then it would be so for any number of terms in the summation including the case when all the observed reflexions are included. For the non-equal atom case the equation which replaces Sayre's equation is

$$E(h)^{sq} = \frac{1}{V} \sum_{k} E(k) \; E(h-k) \tag{6}$$

where $E(h)^{sq}$ is the structure factor corresponding to the squared E-map density. The ratio of the imaginary part to the real part of equation (6) appears on the right-hand side of the tangent formula as it is structured in equation (2) but from equation (6) we can see that this ratio actually gives the phase of the $E(h)^{sq}$ - which is not the same as that of $E(h)$. It must be said that this is a somewhat pedantic point; the tangent formula is a very powerful tool in direct methods and, unless the heavy-atom content of the structure is extreme, gives good results.

There is another way of deriving the tangent formula based on the Cochran distribution. This uses the result that if there are several *independently derived* estimates of the probability distribution for a quantity x, all of equal reliability, then the best estimate for a combined probability density is

$$P(x)_{combined} = C \prod_{i=1}^{m} P_i(x) \tag{7}$$

where the $P_i(x)$ are the independent probability distributions and C is a scaling constant to give normalization.

When $\phi(k)$ and $\phi(h-k)$ are known then the Cochran distribution can be thought of as giving an estimated probability density for $\phi(h)$, based just on three magnitudes,

$$P\{\phi(h)\} = \frac{1}{2\pi I_0\{\kappa(h,k)\}} \exp\left[\kappa(h,k) \; \cos\{\phi(h)-\phi(k)-\phi(h-k)\}\right]. \tag{8}$$

Regarding the distributions for different **k** as independent then they contribute to the combined distribution

$$P\{\phi(\mathbf{h})\} = C \exp\left[\sum_{i=1}^{m} \kappa(\mathbf{h},\mathbf{k}_i) \cos\{\phi(\mathbf{h})-\phi(\mathbf{k}_i)-\phi(\mathbf{h}-\mathbf{k}_i)\}\right]. \qquad (9)$$

The peak of this distribution will occur for the value of $\phi(\mathbf{h})$ which makes the summation in equation (9) a maximum which occurs when

$$\frac{\partial}{\partial\,\phi(\mathbf{h})} \sum_{i=1}^{m} \kappa(\mathbf{h},\mathbf{k}_i) \cos\{\phi(\mathbf{h})-\phi(\mathbf{k}_i)-\phi(\mathbf{h}-\mathbf{k}_i)\} = 0. \qquad (10)$$

It is easily verified that this leads to equation (4) and hence to the tangent formula. Through the Cochran distribution the question of whether or not the structure contains equal atoms does not arise but then it is doubtful whether the individual distributions, each derived from three magnitudes with one in common, can be regarded as independent.

THE VARIANCE OF A TANGENT-FORMULA ESTIMATE

Intuitively it is clear that an estimate for $\phi(\mathbf{h})$ based on several three-phase relationships, especially if the separate estimates are all similar, should be much more reliable, as manifested by having a smaller variance, than that from a single relationship. It is possible to find this variance through equation (9). With A(h) and B(h) as defined in equation (3) it can be shown that equation (9) becomes

$$P\{\phi(\mathbf{h})\} = C \exp\left[A(\mathbf{h}) \cos\{\phi(\mathbf{h})\} + B(\mathbf{h}) \sin\{\phi(\mathbf{h})\}\right]. \qquad (11)$$

Writing the tangent formula estimate for $\phi(\mathbf{h})$ as $\phi(\mathbf{h})_{est}$ and also

$$\alpha(\mathbf{h})^2 = A(\mathbf{h})^2 + B(\mathbf{h})^2 \qquad (12)$$

then equation (11) becomes

$$P\{\phi(h)\} = C \exp\left[\alpha(h) \cos\{\phi(h) - \phi(h)_{est}\}\right] \quad . \quad (13)$$

This shows that the peak of the distribution is at $\phi(h)_{est}$ and that the distribution is a Cochran distribution with $\alpha(h)$ playing the role of $\kappa(h,k)$. Thus to find the variance of a tangent-formula estimate we calculate A(h) and B(h) from the tangent formula, then use equation (12) to find $\alpha(h)$ and finally look up a table or graph of $V(\kappa)$ against κ treating $\alpha(h)$ as though it was a κ for a single relationship.

Where several relationships give similar indications for a given phase then the tangent-formula will give a large value of α. We can see this illustrated in Figure 1.

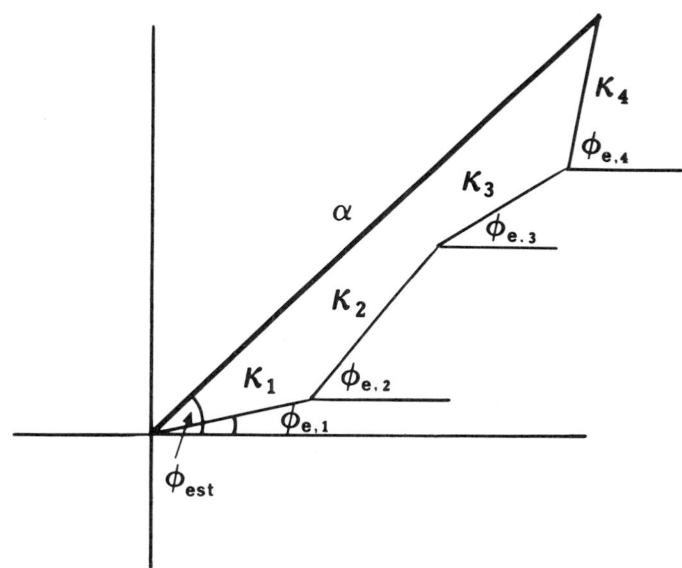

Fig.1   Individual estimates ϕ_{e1}, ϕ_{e2}, etc. with strengths indicated by κ_1, κ_2 etc. combining to give an overall estimate ϕ_{est} with strength α.

Although in the process of phase determination a large value of α may be found one should be cautious in interpreting what it means. The estimated variance will only be valid if the phases on the right-hand side of the tangent formula are all completely correct - which is hardly ever true in practice. Nevertheless large values of α, indicating consistency in phase indications, are usually a welcome sign that the phasing may be developing favourably.

PHASE REFINEMENT WITH THE TANGENT FORMULA

In most direct methods applications phase estimates are developed one after another in a chain-like process by means of the tangent formula. Some of the early phase estimates will have been based on quite a small body of previously-estimated phases and will not have had the benefit of phase relationships involving phases not yet determined. Once phase estimates have been found for all the reflexions in the system then a process of phase refinement is begun which seeks to modify the phases to the greatest consistency with the tangent formula.

Phase refinement consists of an iterative process in which the reflexions, taken in some order,have their phase estimates updated by putting estimated phases of all other reflexions on the right-hand side of the tangent formula. After a few cycles of such a process the phases have usually converged to self consistency and refinement is complete.

A small-scale example shows how the process works. We take a system of ten reflexions, all of equal normalized structure amplitude, with phases represented by ϕ_1 to ϕ_{10} and connected by the listed phase relationships. In phase relationships negative signs or constant phases sometimes occur because the phases to be determined are confined to one asymmetric unit of reciprocal space. For some space groups, equivalent reflexions involved in relationships may then introduce negative signs or constants such as π.

$$\phi_1 + \phi_2 - \phi_3 \approx 0 \qquad \phi_1 - \phi_4 - \phi_5 \approx 0 \qquad \phi_1 + \phi_5 + \phi_7 \approx 0$$
$$\phi_1 - \phi_6 + \phi_8 \approx 0 \qquad \phi_2 + \phi_3 - \phi_9 \approx 0 \qquad \phi_2 + \phi_4 - \phi_8 \approx 0$$
$$\phi_2 + \phi_9 - \phi_{10} \approx 0 \qquad \phi_3 + \phi_4 - \phi_6 \approx 0 \qquad \phi_3 + \phi_7 + \phi_8 \approx 0$$
$$\phi_4 + \phi_5 - \phi_{10} \approx 0 \qquad \phi_4 - \phi_6 + \phi_9 \approx 0 \qquad \phi_5 - \phi_6 + \phi_7 \approx 0$$
$$\phi_5 - \phi_7 + \phi_{10} \approx 0 \qquad \phi_7 - \phi_8 - \phi_{10} \approx 0 \qquad \phi_8 - \phi_9 - \phi_{10} \approx 0$$

In using these relationships it is easier to recast them in a form in which, for each reflexion, we can see the pairs of phases that contribute to a phase estimate. Such lists, which also include the κ weighting for each relationship, are normally called Σ_2 lists following a notation first introduced by Karle and Hauptman (1953). In our example we are going to assume that two phases are known explicitely, because of origin definition or in some other way, so we take $\phi_2 = 180°$ and $\phi_8 = 90°$. The Σ_2 list for the remaining reflexions is:

ϕ_1	ϕ_3	ϕ_4	ϕ_5
$\phi_3 - \phi_2$	$\phi_1 + \phi_2$	$\phi_1 - \phi_5$	$\phi_1 - \phi_4$
$\phi_4 + \phi_5$	$\phi_9 - \phi_2$	$\phi_8 - \phi_2$	$-\phi_1 - \phi_7$
$-\phi_5 - \phi_7$	$\phi_6 - \phi_4$	$\phi_6 - \phi_3$	$\phi_{10} - \phi_4$
$\phi_6 - \phi_8$	$-\phi_7 - \phi_8$	$\phi_{10} - \phi_5$	$\phi_6 - \phi_7$
		$\phi_6 - \phi_9$	$\phi_7 - \phi_{10}$

ϕ_6	ϕ_7	ϕ_9	ϕ_{10}
$\phi_1 + \phi_8$	$-\phi_1 - \phi_5$	$\phi_2 + \phi_3$	$\phi_2 + \phi_9$
$\phi_3 + \phi_4$	$-\phi_3 - \phi_8$	$\phi_{10} - \phi_2$	$\phi_4 + \phi_5$
$\phi_4 + \phi_9$	$\phi_6 - \phi_5$	$\phi_6 - \phi_4$	$\phi_7 - \phi_5$
$\phi_5 + \phi_7$	$\phi_5 + \phi_{10}$	$\phi_8 - \phi_{10}$	$\phi_7 - \phi_8$
	$\phi_8 + \phi_{10}$		$\phi_8 - \phi_9$

Given initial estimates for the ϕ's which are not fixed it is possible with a simple computer programme to refine initial phase estimates. With initial values

$$\phi_1 = 30°, \quad \phi_3 = 160°, \quad \phi_4 = 40°, \quad \phi_5 = 20°, \quad \phi_6 = 200°,$$
$$\phi_7 = 160°, \quad \phi_9 = 300° \quad \text{and} \quad \phi_{10} = 150°$$

then a few cycles of tangent formula refinement gives values:

$$\phi_1 = 78°, \quad \phi_3 = 189°, \quad \phi_4 = 322°, \quad \phi_5 = 100°, \quad \phi_6 = 211°,$$
$$\phi_7 = 151°, \quad \phi_9 = 313° \quad \text{and} \quad \phi_{10} = 87°.$$

It will be seen that the mean phase shift due to the refinement process is $42°$ but insertion of the original and refined phase values into the set of equations will show that the refined values satisfies them much better.

THE BASIS OF MULTISOLUTION METHODS

The important breakthrough in direct methods, which made non-centrosymmetric structures amenable to solution, was the symbolic - addition method as demonstrated by Karle and Karle (1964).   However, a problem with symbolic addition is that it is not possible to combine phase indications if they involve different symbols.   Thus if a phase is indicated as a+b by one phase relationship and c+d by another then it is not permitted to take the combined indication as an average (a+b+c+d)/2.   This is because of the " 2π ambiguity " in defining phases.   Thus with a = b = $\pi/2$ and c = d = $-\pi/2$ the two individual indications are the same, since $\pi \equiv -\pi$, but the averaged indication would be 0 which is as wrong as it possibly could be.   In the early stages of phase determination there are usually very few phase indications available and it is therefore desirable that all can be used in a positive way.   For that reason Germain and Woolfson (1968) proposed the use of explicit phase values, rather than symbols, and this idea gave rise to the programme system MULTAN and other similar techniques.   In the normal multiple-solution methods a starting set of reflexions is found, as for symbolic addition, but instead of symbols being used to indicate phases several different starting sets of explicit phases are assigned to the reflexions and, for each set, phase development and refinement is carried with the tangent formula, or some modification of it.   This gives rise to a multisolution process and then, with several phase sets available for a large number of reflexions, comes the problem of selecting the correct, or more plausible, phase sets and deriving the structure solution.   This is achieved by the use of Figures of Merit (FOM's) whch will be described later.

Multisolution methods are of many different varieties which are in a constant state of development and where new strategies are introduced from time to time.   Here we shall just consider the basic characteristics of the MULTAN kind of method without going into the fine details of implementation.

Calculation of |E| values

The complete data set is read into the programme and |E|
values are calculated automatically by one of the methods
described elsewhere.   If information about molecular
geometry is known then this can be read in and used to
improve the estimates of the normalized structure factors.

Selection of a set of large reflexions

The programme is now required to select a number of
reflexions in terms of which the structure may be solved - in
the sense that if their phases were known then an
interpretable electron-density map could be calculated.
Over the years experience has given an empirical set of rules
for selecting this number; the York procedure is as follows:-

        Number = 4 × Number of atoms + 100
    If system is triclinic then
        Number = Number + 100
    If system is monoclinic then
        Number = Number + 50
    Number = 1.1 × Number
    If Number exceeds capacity of programme then
        Number = Capacity of programme.

The reflexions chosen are those with the largest |E|
values and they are allocated code numbers, 1,2,3 etc. for
subsequent identification in the programme.

For one of the FOM's, by which it is hoped to recognize
a correct phase set, there is required a number (usually 50
to 100) of reflexions of very small (ideally zero) magnitude
and these are also selected at this stage.

Generating the Σ_2 list

By the techniques which have already been described a
Σ_2 list is generated for the set of reflexions with large |E|
values. In addition there are also found the pairs of

contributors to the small |E| reflexions i.e. the combination
of reflexion code numbers, and constant angle where
appropriate, which is equivalent to E(k) E(h-k) for an E(h)
of small magnitude.

Selection of a starting set

The algorithm for choosing the starting set is the
"convergence procedure" described by Germain, Main and
Woolfson (1970). The flow diagram of the algorithm is shown
in Figure 2.

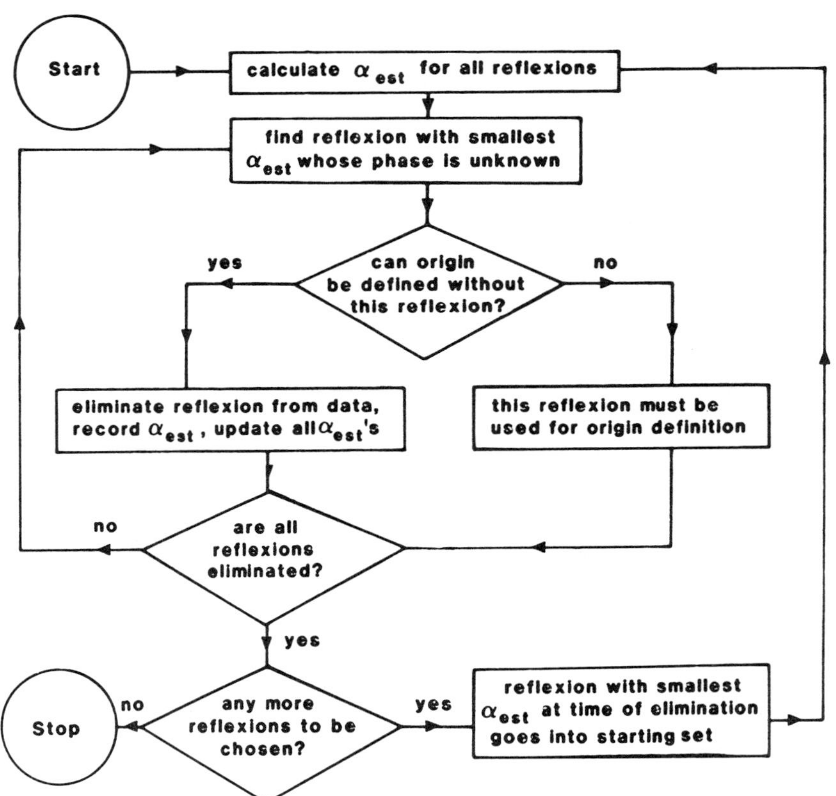

Fig. 2 The convergence algorithm.

The quantity α_{est} which appears in Figure 2 is the value
of α (see section on the Cochran distribution) which would be
expected on the basis of theory from the contributors to $\phi(h)$
under consideration.   No knowledge of phases is required to

estimate α_{est} which is found from

$$\alpha_{est}^{\,2} = \sum_{k} \kappa(h,k)^2 + \sum_{\substack{k \ l \\ k \neq l}} \kappa(h,k)\ \kappa(h,l)\ \frac{I_1\{\kappa(h,k)\}}{I_0\{\kappa(h,k)\}}\ \frac{I_1\{\kappa(h,l)\}}{I_0\{\kappa(h,l)\}} \quad (14)$$

At each stage the convergence algorithm identifies that reflexion which is least strongly linked to the remaining reflexions in the sense that it has the smallest value of α_{est}. If without this reflexion origin and enantiomorph definition would not be possible with the remaining reflexions then this reflexion is used for origin and enantiomorph definition. Another case that can arise is when α_{est} is zero, which means that this reflexion cannot be determined from the phases of the remaining reflexions. In this event then the reflexion becomes a member of the starting set whose phases are given specific values at the beginning of the phasing procedure. Otherwise the reflexion is eliminated, the values of α_{est} are updated and the process is continued.

At the end of the convergence procedure there is obtained a number of reflexions sufficient to fix the origin and enantiomorph and a number of other reflexions which can be given values to create different starting points for phase determination through the tangent formula. The strength of convergence is that it ensures, as far as possible, that the initial phases will strongly develop in the initial stages through strong and reliable phase relationships.

Allocation of initial phases

In early versions of multisolution methods, in particular MULTAN, starting set phases were permuted to give different starting points. A general phase in the starting set was given values $\pi/4$, $3\pi/4$, $5\pi/4$ and $7\pi/4$ (called quadrant permuation) while a reflexion whose phase could take one of two values, $(0,\pi)$ or $\pm\pi/2$, for example, would be given each of those values. Thus in in a particular case a starting set could have four reflexions for which specific values could be given to fix the origin and enantiomorph together with   two general   and two special-phase-value reflexions which would give 4 × 4 × 2 × 2 = 64 different starting phase sets .

Nowadays, where phase permutation is done at all, it is done differently using a concept called magic integers (White and Woolfson, 1975).   They   showed   that a sequence of n integers, m_1, m_2, ..., m_n , could be found such that n phases could be approximately represented by a single variable x by

$$\phi_i = m_i x \qquad \mathrm{mod}(2\pi) \qquad . \qquad (15)$$

For any values of the phases some value of x in the range 0 to 2π will approximately represent them.   To take an example the four magic integers, 5, 7, 8 and 9 could be taken to represent four general phases.   The values of x are taken as $\pi/32$ to $63\pi/32$ in steps of $\pi/16$, i.e. there are 32 values of x.   If the four values of phase are 0.96π, 0.16π, 1.77π and 1.32π respectively then for x = $47\pi/32$ the values of $m_i x$ mod (2π) are 1.34π, 0.28π, 1.75π and 1.22π respectively corresponding to discrepancies (in degrees) of 68°, 22°, 3° and 18° respectively.   To use quadrant permutation for four general phases would require 4^4 = 256 permutations; the 32 sets of phases would, in general entail having a larger error for the set closest to the true values but there is a saving of a factor of eight in the computational effort required.   Main (1977) gave efficient sets of magic integers for different numbers of phases to be represented, as shown below:

n	sequence	number of sets	rms error(°)
2	2 3	12	26
3	3 4 5	20	29
4	5 7 8 9	32	37
5	8 11 13 14 15	50	42
6	13 18 21 23 24 25	80	47
7	21 29 34 37 39 40 41	128	48
8	34 47 55 60 63 65 66 67	206	49

For quadrant permutation it would require $4^8 = 65,536$ starting points for n = 8 although the rms error for the closest one would be $26°$. It turns out that the errors given by magic-integer permutation are acceptable and the reduction in the number of starting points enables much larger starting sets to be used.

Phase extension and refinement

It was found quite early that phase development was improved by using a weighted form of the tangent formula

$$\phi(h) \quad = \quad phase \quad of\left\{ \sum_k w(k)\ w(h-k)\ E(k)\ E(h-k) \right\} \qquad (16)$$

where w(h) is the weight associated with $\phi(h)$. A very simple, but quite effective, weight for many structures is

$$w(h) = minimum\ of\ (0.2\ \alpha(h)\ ,\ 1.0) \qquad (17)$$

but with this weighting scheme the weights quickly go to unity and sometimes phases go towards values which make the the phase relationships work too well. Another weighting scheme, due to Hull and Irwin (1978), is

$$w(h) = minimum\ of\ [0.2\ \alpha(h),\ 1.0, \{\alpha(h)_{est} + 5\}/\ \alpha(h)] \quad .(18)$$

This function, shown in Fig.3, has the effect of reducing the weight if the value of $\alpha(h)$ goes substantially above the theoretically estimated value, $\alpha_{est}(h)$.

The process of phase extension and refinement is carried out in a very controlled way in MULTAN. In each cycle of phase development or refinement a new phase is accepted only if the associated weight is larger than some cut-off value. Thus the first new phases accepted are those most reliably indicated but with each cycle the cut-off level is reduced until eventually all new phase indications are accepted.

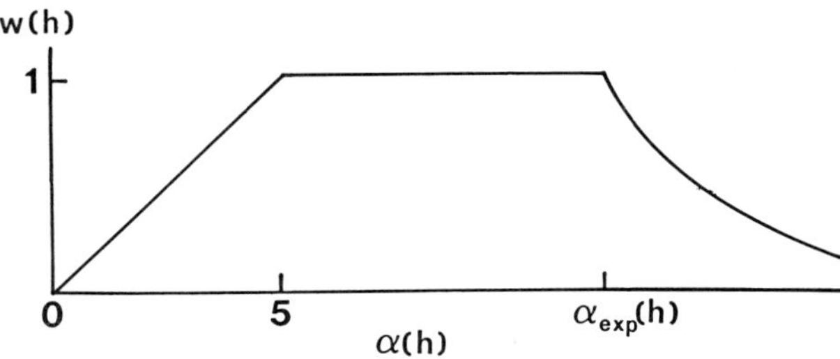

Fig.3. The Hull-Irwin weighting scheme

Figures_of_merit

When a large number of sets of phases have been developed by MULTAN it is necessary to rank them according to some figures of merit.   Those used are

$$\text{ABSFOM} = \frac{\sum_h \{\alpha(h) - \alpha_r(h)\}}{\sum_h \{\alpha_{est}(h) - \alpha_r(h)\}} \qquad (19)$$

which measures the extent to which the phase relationships hold and ranges in value from zero, for random phases where $\alpha(h)$ equals the random value, $\alpha_r(h)$, to unity where the α's equal their expected values.

$$\psi_0 = \frac{\sum_h |\sum_k E(k) \ E(h-k)|}{\sum_h \left[\sum_k |E(k) \ E(h-k)|^2\right]^{1/2}} \qquad (20)$$

where the summations contain the large $|E|$s whose phases are to be determined and the $E(h)$ have small magnitudes.   For correct phases the expected value of ψ_0 is unity.

$$R_\alpha = \frac{\sum_h |\alpha(h) - \alpha_{est}(h)|}{\sum_h \alpha_{est}(h)} \qquad (21)$$

which is a residual between the actual and estimated α's.

At the end of the phasing process it is possible to combine the three figures of merit into a single combined figure of merit

$$
\text{CFOM} = w_1 \frac{\text{ABSFOM} - (\text{ABSFOM})_{min}}{(\text{ABSFOM})_{max} - (\text{ABSFOM})_{min}} + w_2 \frac{(\psi_0)_{max} - \psi_0}{(\psi_0)_{max} - (\psi_0)_{min}}
$$

$$
+ w_3 \frac{(R_\alpha)_{max} - R_\alpha}{(R_\alpha)_{max} - (R_\alpha)_{min}} \tag{22}
$$

The weights w_1, w_2, w_3 are usually set at 0.6, 1.2 and 1.2 respectively which gives values of CFOM in the range 0 to 3. The most likely correct sets of phases are those with the highest values of CFOM.

Interpretation of E-maps

The final stage of MULTAN is to compute E-maps for plausible phase sets and by a peak-search process to interpret them as possible structures. Coherent clusters of peaks are analysed according to the rules of chemistry and projections are output on line-printer paper for visual examination and interpretation.

REFERENCES

Germain, G. and Woolfson, M.M., 1968, On the application of phase relationships to complex structures. *Acta Cryst.*, B**24**, 191-196.

Germain, G., Main, P. and Woolfson, M.M., 1970, On the application of phase relationships to complex structures. II Getting a good start. *Acta Cryst.*, B**26**, 274-285.

Hauptman, H. and Karle, J., 1953, Solution of the phase problem. I The centrosymmetric crystal., A.C.A. Monograph No.3, Polycrystal Book Service, Pittsburgh.

Hull, S.E. and Irwin, M.J., 1978, On the application of phase relationships to complex structures. XIV The additional use of statistical information in tangent-formula refinement. *Acta Cryst.*,**A34**, 863-870.

Karle, J. and Hauptman, H., 1956, A theory of phase determination for the four types of non-centrosymmetric space groups 1P222, 2P22, $3P_12$, $3P_22$. *Acta Cryst.*, **9**, 635-651.

Karle, I. L. and Karle, J., 1964, The application of the symbolic addition method to the structure of L-arginine dihydrate. *Acta Cryst.,*, **17**, 835-841.

Main, P., 1977, On the application of phase relationships to complex structures. XI A theory of magic integers. *Acta Cryst.*, **A33**, 750-757.

White, P.S. and Woolfson, M.M., 1975, On the application of phase relationships to complex structures. VII Magic integers. *Acta Cryst.*, **A31**, 53-56.

QUARTETS AND OTHER PHASE RELATIONSHIPS

H. Schenk

Laboratory for Crystallography, University of Amsterdam
Nieuwe Achtergracht 166, 1018 WV Amsterdam
The Netherlands

Introduction

In earlier chapters it was shown that the phase sum

$$\varphi_3 = \varphi_H + \varphi_K + \varphi_{-H-K} \tag{1}$$

is independent of the choice of origin. In direct methods quantities which are independent of the choice of origin are called structure invariants. When it is possible to estimate the phase sum of an invariant, this sum is generally called a phase relationship. Although in principle a phase relationship is always a structure invariant, it is sensible to make the distinction between all invariants and those, for which it is possible to estimate phase sums. Of course, for direct methods the latter are the only important invariants, because reliable phase relationships are the basis of all existing phasing methods, whereas for any structure there are millions more of useless structure invariants.

The purpose of this chapter is to explore in an intuitive way the nature of structure invariants and their space group dependent counterparts, the seminvariants, and also under what conditions they may prove to be useful phase relationships. In particular positive and negative quartets are dealt with in more detail.

The quartet relations

The triplet relation, although a two-dimensional phase relation, is very successful in solving three-dimensional crystal structures. However, it may be more appropriate to try to solve structures with three-dimensional phase relationships, the so-called quartet relations. A quartet relation is given by

$$\varphi_4 = \varphi_H + \varphi_K + \varphi_L + \varphi_{-H-K-L} = q \tag{2}$$

Expression (2) describes a three-dimensional relation when three linearly independent reflections H, K and L are involved. Quartets with $q \approx 0$ have been known since 1953 when they appeared as Σ_5-relations in the monograph of Hauptman and Karle. Independently Simerska (1956) derived them from a generalisation of the Sayre–Hughes equation for products of three reflections in stead of two. In both cases it was shown that the probability that $q = 0$ was larger for larger values of

$$E_4 = N^{-1} |E_H E_K E_L E_{-H-K-L}| \tag{3}$$

Direct Methods of Solving Crystal Stuctures
Edited by H. Schenk, Plenum Press, New York, 1991

or with other words the probability distribution of φ_4 is sharper for larger values of E_4. Simerska also tested the reliability of the quartet relations in a sample structure and suggested that they could be used for phasing in the same way as the triplet relation.

However, quartets were not used until 1973 when it was shown (Schenk) that quartets (2) with a probability distribution based on E_4 have inferior properties for actual phase determinations compared to those of the triplet phase relation (1) with its probability distribution based on the triple product

$$E_3 = N^{-1/2} |E_H E_K E_{-H-K}| \qquad (4)$$

But from that time they could be used profitably because in the same paper it was also shown (Schenk, 1973) that using E_4 and the $|E|$'s of three more structure factors, the cross-term magnitudes $|E_{H+K}|$, $|E_{H+L}|$ and $|E_{K+L}|$, the quartets change into very useful relationships. A simple explanation of the seven-magnitude positive (q≈0) and negative (q≈π) quartet relations can be based again on electrondensity considerations, following a similar reasoning as used in the first chapter of this proceedings.

It is recalled here that when a large intensity I_{hkl} is observed, the atoms will lie near the set of planes with indices hkl, successive planes of which are separated by d_{hkl} from each other. Or in other words a strong reflection implies concentration of the electron density in the set of planes defined by hkl. On the other hand, if I_{hkl} is small, than the atoms are not systematically lying in these planes, but randomly distributed.

<u>The positive quartet relation</u>

The positive quarted relation is formulated as:

$$\varphi H + \varphi K + \varphi L + \varphi_{-H-K-L} \simeq 0 \qquad (5)$$

for large E_4. Analogous to the treatment of the triplet relation now three reflections H, K and L are combined. In case the three reflections are all strong, the electron density must be found in the three sets of planes as indicated in Fig. 1. As a result the electron density will be found near the points of intersection of the three planes. These likely areas of electron density are depicted for only two successive planes from each set in Fig. 2. Then for a fourth strong reflection −H−K−L it is much more likely that its plane of maximum electron density will run through the points of intersection (Fig. 3) than that it will clear these points. In case an arbitrary origin O is defined in Fig. 3 the quartet relation (5) follows as straightforward as the triplet relation as is shown in Fig. 4 (Schenk, 1981). Again the quartet relation is independent of the position of the origin and it is thus a structure invariant (origin invariant). This quartet relation, however, is not as strong as the triplet relation because of the factor N^{-1} in E_4 whilst in E_3 a term $N^{-1/2}$ appears only, compare the expressions 3 and 4.

In Table 1 it is shown that the number of useful quartets of this kind is only a fraction of the number of useful triplet relations. However, the reliability can be

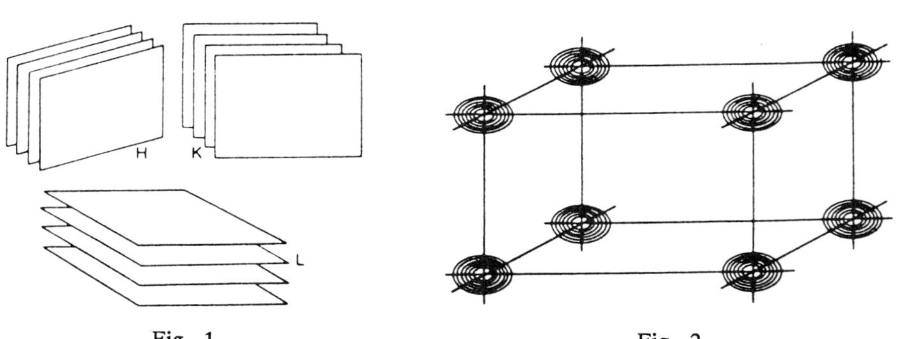

Fig. 1   Fig. 2

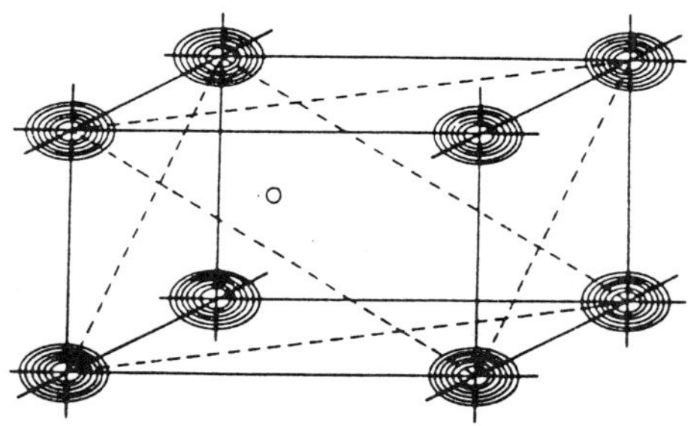

Fig. 3

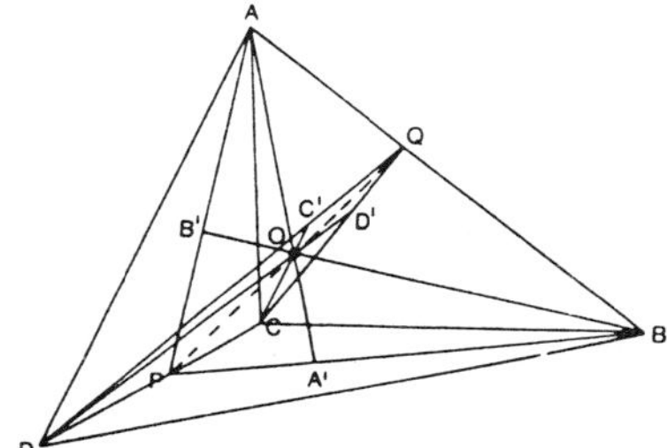

Fig. 4 Given
AO/AA'+BO/BB'+PO/PQ=2 and
CO/CC'+DO/DD'+QP/PQ=2.
Then the sum
AO/AA'+BO/BB'+CO/CC'+DO/DD'=
=4−PO/PQ−QO/PQ=3. Thus
$\varphi_H + \varphi_K + \varphi_L + \varphi_{-H-K-L} = 0 \pmod 1$

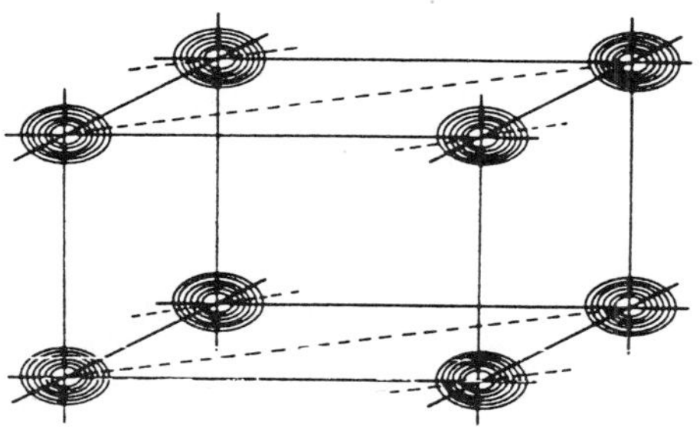

Fig. 5

improved by combining the quartet with an identical one constructed from two triplets:

$$\left. \begin{array}{l} \varphi_H + \varphi_K + \varphi_{-H-K} \simeq 0 \\[2mm] \varphi_L + \varphi_{-H-K-L} + \varphi_{H+K} \simeq 0 \end{array} \right\} \qquad \varphi_H + \varphi_K + \varphi_L + \varphi_{-H-K-L} = 0 \qquad (6)$$

which holds for large E_4 and large $|E_{H+K}|$ (Schenk, 1973). This can be understood by drawing the H+K reflection in Fig. 2, as indicated in Fig.5. When reflection H+K has also a large intensity then this strongly indicates that the electron density will be found near the intersection of H and K. Thus a large $|E_{H+K}|$ is an additional indication that quartet (5) holds. Similar situations can be sketched for the other two cross terms K+L and H+L. So, in conclusion, a large value of E_4 and large $|E_{H+K}|$, $|E_{H+L}|$ and $|E_{K+L}|$ are four different indications that the positive quartet relation (4) is likely to be true. As a result positive quartets are controlled by the magnitude of 7 structure factors, which all have to be large.

In Table 2 for the $P2_1/c$ structure the number of quartets based on seven

Table 1. Triplets and quartets for a $P2_1/c$ structure.

Triplets			Quartets		
E_3	number relations	% correct	E_4	number relations	% correct
3.0	49	100	3.0	–	
2.5	105	100	2.5	–	
2.0	265	100	2.0	7	100
1.5	771	99.4	1.5	74	100
1.0	2782	96.8	1.0	837	96.1

Table 2. Triplets and seven-magnitude quartets for a P2₁/c structure.

E_3	Triplets		$E_4{}^*$	7-m Quartets	
	number relations	% correct		number relations	% correct
3.0	49	100	3.0	17	100
2.5	105	100	2.5	70	100
2.0	265	100	2.0	259	100
1.5	771	99.4	1.5	965	99.5
1.0	2782	96.8	1.0	5078	96.3

magnitudes is compared with the triplets. Now the number of useful quartets has been improved considerably and moreover, this has not affected the reliability of them as is shown in the column marked with % correct. This implies that in actual phase determinations triplets and quartets can be used simultaneously and this will result in a more easy phase propagation and structure solution.

The negative quartet relation

Quartet relations for which the sum of the four phases equals π:

$$\varphi_H + \varphi_K + \varphi_L + \varphi_{-H-K-L} \simeq \pi \tag{7}$$

are referred to as negative quartets and such relations exist also for reasonably strong intensities for the relections H, K, L and −H−K−L.

The planes of maximum density for the four reflections involved in relation (7), taking account of the phase shift π, are indicated in Fig. 6.
It can be seen that for all indicated positions three out of four planes intersect. If electron density is located at these points the resulting unitary structure factors of H, K, L and −H−K−L will be 0.5, because three atoms lie in the planes and one lies halfway between (a unitary structure factor for an equal atom structure is equal to $U=E/(N)^{1/2}$). Thus for a negative quartet relation the reflections H, K, L and −H−K−L will in general not be found amongst the very strongest, but still they will be strong. But what will be the intensity of reflection H+K in case the electron density is located near the marked points of Fig. 6? From Fig. 7 it can easily be seen that H+K will have a small $|E|$ magnitude: equal numbers of points of electron density concentration lie on the

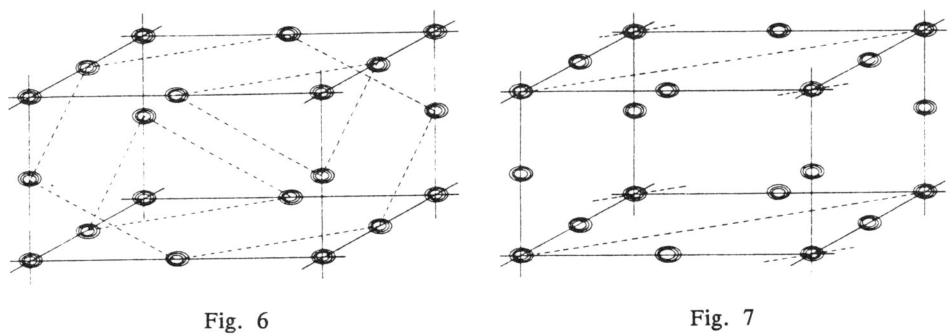

Fig. 6   Fig. 7

H+K−planes and halfway in between. As the same holds for the other cross terms it can be stated that the negative quartet relation (7) is likely to be true for reasonably large values of E_4 and small $|E_{H+K}|$, $|E_{H+L}|$ and $|E_{K+L}|$.

Negative quartets are not so numerous as positive quartets and triplets. Usually they

comprise a small fraction of all relationships available, and as such they are in general not very important for the process of phase propagation. However, in a vast number of space groups they may play a crucial role when used in the Negative Quartet Criterion, a figure of merit to select the correct solution in multisolution methods or to determine the numerical values of the symbols in symbolic addition methods.

<u>Estimating phase sums of other invariants and seminvariants</u>

Estimating the phase sums of structure invariants and seminvariants is the most important step in direct methods. The reason for this lies in the fact that knowledge of the actual values of phase sums will enable the precise calculation of the phases of the reflections. Consquently, the better the invarants are being estimated the more powerful the direct method procedures will be.

Reliable estimates of phase sums have become available for invariants and seminvariants other than the triplet phase relation φ_3. As described above the first results in this field were based on two observations, one that the estimation of the quartet invariant (4) is not only dependent on E_4 but also on the cross terms (Schenk, 1973a) and another that for special two-dimensional quartets the phase sum is equal to π (Schenk and De Jong, 1973; Schenk, 1973b). For both types of invariants useful applications were worked out at the same time.

This work has been the starting point for a more systematic approach to the search for those structure factors upon which the structure (sem)- invariants most sensitively depend, ultimately resulting in the Neighborhood Concept of Hauptman (1977, 1978) and the Representation Theory of Giacovazzo (1977). Both theories are based on three principles: a) the cross-term principle, b) the multipole or identity principle, and c) the extension principle. In the representation theory the space group symmetry is included in an explicit and systematic way, whereas this is done in a more implicit way in the neighborhood concept.

<u>Cross terms</u>

As already mentioned cross terms were used for the first time by Schenk (1973a,b) and Schenk & e Jong (1973) in order to estimate phase sums $|\varphi_4|$ of quartets. The example of quartets shows that cross terms originate from the fact, that higher order invariants can be obtained by summing suitable lower order invariants. In general it can be stated that for any invariant

$$\varphi_n = \varphi_H + \varphi_K + \varphi_L + \varphi_M + \ldots\ldots \tag{8}$$

its value φ_n is controlled by the magnitudes of the main reflections H, K, L, M.... and those of the cross reflections H+K, H+L, ... K+L, ... H+K+L, .. . Examples for quintets, sextets, etc., can be easily worked out. The number of cross terms is fast increasing; whereas quartets have three cross terms only quintets have already 10 of them and sextets 25.

<u>Identities</u>

The identity principle is most easily explained on the basis of quadrupoles of triplets, implicitly introduced by Sayre (1952), and further explored by e.g. Vaughan (1956), Karle and Hauptman (1957), De Vries (1963) and Viterbo (1976). A quadrupole is an identity among four triplet phase sums:

$$\left.\begin{array}{l} \varphi_H + \varphi_K + \varphi_{-H-K} = \varphi_3 \\ \varphi_P + \varphi_{-H} + \varphi_{H-P} = b \\ \varphi_{-P} + \varphi_{-K} + \varphi_{K+P} = c \\ \varphi_{H+K} + \varphi_{P-H} + \varphi_{-K-P} = d \end{array}\right\} \quad \text{with } \varphi_3 + b + c + d = 0 \tag{9}$$

From expression 9 it follows that the value of φ_3 is also dependent on b, c, and d. Thus in case all reflections involved, H,K,−H−K,P,H−P and K+P, are strong and b,c, and d are likely to be near zero on the basis of their own triple products, the identity will support $|\varphi_3|=0$. The vector P is not defined and in order to get full support for $|\varphi_3|$ P must range over all reciprocal vectors. Thus where a triplet phase sum is primarily controlled by the $|E|$−values of H, K and −H−K, it is in second approximation dependent on the $|E|$−values of P, H−P and K+P where P ranges over all reciprocal space.

In the same way identities (or multipoles) can be set up for any invariant. Another example is the triple of quartets:

$$\left. \begin{array}{l} \varphi_H + \varphi_K + \varphi_L + \varphi_{-H-K-L} = \varphi_4 \\ \varphi_{-H} + \varphi_{-K} + \varphi_P + \varphi_{H+K+P} = b \\ \varphi_{-L} + \varphi_{H+K+L} + \varphi_{-P} + \varphi_{-H-K+P} = c \end{array} \right\} \quad \text{with } \varphi_4 + b + c = 0 \qquad (10)$$

From the cross−term principle we know already that $|\varphi_4|$ is controlled in first approximation by the $|E|$−values of the main reflections H, K, L and −H−K−L and in second approximation by those of the cross reflections H+K, H+L, and K+L. Now in third approximation expression 10 learns that the reflections P, H+K−P and the cross terms P−H, P−K, −P−L, and −P+H+K+L, where P ranges over all reciprocal space, can be added to the controlling reflections; this is the third neighbourhood introduced by Hauptman (1977).

Many other identities are possible and the results have in common that the $|E|$'s which are identified to be important for the calculation of the phase sum of an invariant contain a free reciprocal lattice vector P. This implies that P always ranges over a large number of reciprocal vectors and that calculations based on the multipole or identity principle are computer−time demanding.

Representations and extensions

The third principle to be dealt with, the representation principle, is used primarily to identify the $|E|$−values controlling the phase sums of structure seminvariants. Seminvariants are quantities which are invariant in a particular space group only, sometimes even in a particular setting. In general the representation principle is used as follows: a structure (sem)invariant is summed with a second (sem)invariant to form an invariant. This invariant can be estimated following the first two pinciples. Now, in case the added second (sem)invariant has a trivial value or is determined by symmetry alone, the first (sem)invariant is estimated at the same time. All structure factors used in this process are then identified as being phase−sum controlling reflections.
Although this principle was used implicitly from as early as 1948 (Harker & Kasper, 1948, Hauptman & Karle, 1953, etc.), it is used explicitly by Giacovazzo (1977) and Hauptman (1978) under the names representation and embedding (extension) respectively.

The most simple example of the principle is the Σ_1−relation or the one phase seminvariant φ_{2H} in P$\underline{1}$. By making the special triplet phase sum

$$\varphi_3{}' = \varphi_{2H} + \varphi_{-H} + \varphi_{-H} \qquad (11)$$

it is seen that for large values of $|E|$ and $|E_{2H}|$ the magnitude of $|\varphi_{2H}|$ is known because φ_3 is most probably zero and in P$\underline{1}$ the phase sum $\varphi_{-H} + \varphi_{-H}$ is trivially zero.
As a next example the 0g0 reflections in monoclinic space groups are chosen. These can be determined by means of a large number of invariants; for instance reflection 020 can be embedded in three phase invariants of the type

$$\varphi_3{}'' = \varphi_{020} + \varphi_{h-11} + \varphi_{-h-1-1} \qquad (12)$$

The seminvariant $\varphi_{h-11} + \varphi_{-h-1-1}$ again is determined by symmetry only and thus for large $|E|$-values the invariant φ_3'' is known and as a result φ_{020} is also known. Giacovazzo (1977, 1980) has tackled this principle very systematically.

Some more examples

At this point all information is available to identify all structurefactor magnitudes which control the phase sum of any structure invariant or seminvariant. The general receipe for invariants is to identify first the cross terms and if a next group of (less important) structure factors is needed the invariant is used in identities or in extensions. Seminvariants are first extended to invariants and then, by the cross-term principle, and if necessary by further embedding in identities, the important reflections are identified. A couple of additional examples is dealt with in the folwing paragraphs.

The following example seems rather trivial. Starting with the triplet relation the invariant $\varphi_p+\varphi_{-p}$, trivially equal to 0, is added and the resulting invariant is a quintet:

$$\varphi_3 = \varphi_H + \varphi_K + \varphi_{-H-K} + \varphi_P + \varphi_{-P} \tag{13}$$

This quintet can be estimated using the $|E|$-values of the main terms H, K, -H-K, P, and -P and those of the cross terms H+K, H, K, O, ±P+H, ±P+K and ±P-H-K. The first four obviously do not add information, but the other six are new. In fact -P+H and P+K can be recognized as terms also occurring in the quadrupoles and by taking other triplet sequences it can be shown that the six cross terms involving P give rise to six quadrupoles. Thus the rather trivial trick of adding $\varphi_p+\varphi_{-p}$ is closely related to the quadrupole or identity principle.

In the literature (notably the paper of Giacovazzo, 1977) many more examples can be found and it can easily be seen that they all are based on the three principles treated here. Of course one has to take care of the expected effectiveness of the respective reflections. This can be worked out in an example for quartets φ_4. The most important reflections, of course, are the four main terms H,K,L and -H-K-L. Then the next effective reflections are the cross terms H+K, H+L and K+L. The reflections P, H+K-P, P-H, P-K, -P-L and -P+H+K+L which follow from the multipole principle are third in order of importance (Hauptman, 1977). An alternative third group may consist of those reflections obtained by taking $\varphi_4 + \varphi_P + \varphi_{-P}$, namely reflection P and the sextet cross terms, H±P, K±P, L±P, -H-K-L±P, H+K±P and K+L±P (Giacovazzo, 1977).

By using the three principles it is also possible to predict roughly the effect of the $|E|$'s of the reflections on the value of the invariant. As an example again the quartet relation: we have seen that in case $|E_{H+K}|$, $|E_{H+L}|$ and $|E_{K+L}|$ are small, φ_4 is likely to be π. Now the third neighbourhood follows from the identity (10) and thus either b or c have to be π and the other 0. In case b $\simeq \pi$ it is expected that the cross terms P-H and P-K are small and the other terms -P-L and -P+H+K+L corresponding to c=0 are strong. Thus using the third neighbourhood of the quartet it can be stated that $|E|_H$, $|E|_K$, $|E|_L$ and $|E|_{-H-K-L}$ large, $|E_{H+K}|$, $|E_{H+L}|$ and $|E_{K+L}|$ small, $|E_p|$, $|E_{H+K-P}|$ large $|E_{P-H}|$ small and $|E_{-P-L}|$, $|E_{-P+H+K+L}|$ large, indicate that φ_4 is likely to be π.

References

De Vries, A (1963), Thesis Utrecht.
Giacovazzo, C. (1977). Acta Cryst. A33, 933.
Giacovazzo, C. (1980). Acta Cryst. A36, 362.
Harker, D. and Kasper, J.S., (1948). Acta Cryst. 1, 70.
Hauptman, H. (1977). Acta Cryst. A33, 553.
Hauptman, H. (1978). Acta Cryst. A34, 525.
Hauptman, H. and Karle, J. (1953). A.C.A. Monograph No. 3. Pittsburgh, Polycrystal.
Karle, J. and Hauptman, H. (1957). Acta Cryst. 10, 515.

Sayre, D. (1952). Acta Cryst. 5, 60.
Schenk, H. (1973a). Acta Cryst. A29, 77.
Schenk, H. (1973b). Acta Cryst. A29, 480.
Schenk, H. (1981). Acta Cryst. A37, 573.
Schenk, H. and De Jong, J.G.H. (1973). Acta Cryst. A29, 31.
Simerska, M. (1956). Czech. J. Phys. 6, 1-7.
Vaughan, P.A. (1956). Am. Cryst. Ass. Ann. Meeting, Frenchlick, Ind.
Viterbo, D. (1976). In "Direct Methods in Crystallography", edited by H.A. Hauptman, Buffalo, p. 117.

THE JOINT PROBABILITY DISTRIBUTION OF THREE STRUCTURE FACTORS LEADING TO A DISTRIBUTION FOR THE TRIPLET PHASE SUM

Rene´ Peschar

Laboratory for Crystallography, University of Amsterdam,
Nieuwe Achtergracht 166, 1018 WV Amsterdam, The Netherlands

INTRODUCTION

After a short explanation of the basic probabilistic terminology, in this paper a method will be discussed to derive a joint probability distribution of three structure factors via its characteristic function. Assuming an a priori probabilistic model of uniformly and independently distributed atoms, the calculations result in the well-known Cochran distribution for the triplet phase sum.

PROBABILITY DISTRIBUTION OF A SINGLE RANDOM VARIABLE

A continuous random variable X is a real-valued variable which may have certain values on a particular interval $(x,x+dx)$. The function describing these possible values is the probability density distribution $p(x)$ which yields the probability $p(x)dx$ that X lies in the interval $(x,x+dx)$. $p(x)$ has the following two properties :

$$p(x) \geq 0 \tag{1}$$

$$\int_{-\infty}^{\infty} p(x)dx = 1$$

If $p(x)$ is constant over some interval and zero elsewhere, X is said to be uniformly distributed in that interval. A function $g(X)$ of a random variable X is also a random variable. X itself is then called a primitive random variable. The mathematical expectation value $g(X)$ is defined as

$$<g(X)> = \int_{-\infty}^{\infty} g(x)p(x)dx \tag{2}$$

Particularly useful is the function $g(X)= C(u) = \exp(iuX)$. $C(u)$ is the Fourier-transform of the probability density distribution $p(x)$ if u is real-valued. $C(u)$ is called the characteristic function of the random variable X. $C(u)$ and $p(x)$ define each other uniquely,

Direct Methods of Solving Crystal Stuctures
Edited by H. Schenk, Plenum Press, New York, 1991

$$p(x) = (2\pi)^{-1} \int_{-\infty}^{\infty} \exp[-iux)]C(u)du \tag{3}$$

$$C(u) = \int_{-\infty}^{\infty} \exp[i(ux)]p(x)dx = <\exp[iux]> \tag{4}$$

JOINT PROBABILITY DISTRIBUTION OF n RANDOM VARIABLES.

The joint probability density distribution (from now on joint probability distribution) $p(x_1,..,x_n)$ of n (≥ 2) random variables $(X_1,..,X_n)$ can be defined analogously to the single variable case. $p(x_1,..,x_n)dx_1..dx_n$ is the probability that simultaneously X_1 belongs to the interval (x_1,x_1+dx_1), X_2 to the interval (x_2,x_2+dx_2) ,.., and X_n to the interval (x_n,x_n+dx_n). The characteristic function $C(u_1,..,u_n)$ is the n-dimensional Fourier-transform of the probability distribution $p(x_1,..,x_n)$,

$$p(x_1,...,x_n) = (2\pi)^{-n} \int_{-\infty}^{\infty} .. \int_{-\infty}^{\infty} \exp[-i(u_1x_1+..+u_nx_n)]C(u_1,...,u_n)du_1..du_n \tag{5}$$

$$C(u_1,...,u_n) = \int_{-\infty}^{\infty} .. \int_{-\infty}^{\infty} \exp[i(u_1x_1+..+u_nx_n)]p(x_1,...,x_n)dx_1..dx_n \tag{6}$$

$$= <\exp[i(u_1x_1+..+u_nx_n)]>$$

CONDITIONAL PROBABILITY DISTRIBUTION

From (5) the conditional probability distribution of a subset of k random variables can be derived given that the remaining n-k assume particular values a_i : $X_{k+1} = a_{k+1},..,X_n = a_n$,

$$p(x_1,...,x_k|x_{k+1},...,x_n) = \frac{p(x_1,...,x_k,x_{k+1},..,x_n)}{\int_{-\infty}^{\infty} .. \int_{-\infty}^{\infty} p(x_1,...,x_k,a_{k+1},...,a_n)dx_1..dx_k} \tag{7}$$

JOINT PROBABILITY DISTRIBUTION OF A SUM OF N INDEPENDENT RANDOM VARIABLES

An important property of the characteristic function is that it simplifies the joint probability distribution calculation for the sum of n independently distributed random variables. If $p(x_1)$ and $p(x_2)$ are the probability distributions for X_1 and X_2 respectively and $p(x_1,x_2)$ their joint probability distribution, then X_1 and X_2 are independently distributed if

$$p(x_1,x_2) = p(x_1).p(x_2) \tag{8}$$

If X_1 and X_2 are random variables, then any function of X_1 and X_2 is itself a random variable. Applying this to a sum $y= x_1+..+x_n$ of n independently distributed random variables, the characteristic function of

y can be shown to be the product of the characterictic functions of the
summands,

$$C(u) = <\exp[i(ux_1)]>..<\exp[i(ux_n)]>$$

$$= C_1(u)..C_n(u) \tag{9}$$

Consequently, the probability distribution of y can be expressed as a
single integration

$$p(y) = (2\pi)^{-1} \int_{-\infty}^{\infty} \exp[-iu(x_1+x_2+..+x_n)]C(u)du \tag{10}$$

CHANGE OF VARIABLES

If each of the n random variables x_i is a single-valued function h_i
of the random variables $y_1,..,y_n$, say $x_i = h_i(y_1,..,y_n)$ then,

$$p(x_1,..,x_n)dx_1..dx_n = p[h_1(y_1,..,y_n),..,h_n(y_1,..,y_n)]|J|dy_1..dy_n \tag{11}$$

J is the Jacobian of the transformation,

$$J = \frac{\delta(x_1,..,x_n)}{\delta(y_1,..,y_n)} \tag{12}$$

Frequently a transformation to polar coordinates is used, e.g. $x_1 = R\cos(\Phi)$
and $x_2 = R\sin(\Phi)$. Then

$$|J| = \begin{vmatrix} \partial x_1/\partial R & \partial x_1/\partial \Phi \\ \partial x_2/\partial R & \partial x_2/\partial \Phi \end{vmatrix} = R \tag{13}$$

so $p(x_1,x_2)dx_1dx_2 = p[x_1=R\cos(\Phi),x_2=R\sin(\Phi)]RdRd\Phi$

THE JOINT PROBABILITY DISTRIBUTION OF THREE STRUCTURE FACTORS

As an example we will calculate the joint probability distribution of
three structure factors in the space group P1 via its characteristic
function. It will be shown that the calculations leads to the Cochran
distribution for the triplet phase sum in terms of normalized structure
factor magnitudes. A general description including space group symmetry
has been given elsewhere (Peschar & Schenk, 1987).
The joint probability distribution has to be defined first before the
corresponding characteristic function can be evaluated.
The real and imaginary parts A_H and B_H of the structure factors F_H are an
obvious choice as random variables. This means that the A_H and B_H must be
inserted in (5) and (6). For convenience, the carrying variables v and w
are used in connection with A and B respectively.
Using the short-hand notations A_ν, B_ν and F_ν for A_{H_ν}, B_{H_ν} and F_{H_ν}, the
joint probability distribution of the real and imaginary parts of three
structure factors is the six-fold integral

$$P(A_1,A_2,A_3,B_1,B_2,B_3) = (2\pi)^{-6} \int_{-\infty}^{\infty}..\int_{-\infty}^{\infty} \exp\left[-i\left\{ \sum_{\nu=1}^{3} (A_\nu v_\nu + B_\nu w_\nu) \right\} \right].$$

$$\cdot\ C(v_1,v_2,v_3,w_1,w_2,w_3)\,dv_1dv_2dv_3dw_1dw_2dw_3 \qquad (14)$$

while $C(v_1,..,w_3)$ can be expressed in view of (6) as

$$C(v_1,v_2,v_3,w_1,w_2,w_3) = \left\langle \exp\left[+i\left\{ \sum_{\nu=1}^{3}(A_\nu v_\nu + B_\nu w_\nu) \right\}\right]\right\rangle \qquad (15)$$

The average in (15) being over the A_ν and B_ν.
It is convenient to change to polar coordinates,

$$v_\nu = \rho_\nu\cos(\theta_\nu) \quad\text{and}\quad w_\nu = \rho_\nu\sin(\theta_\nu) \quad\text{for } \nu = 1,2,3$$

and
$$\qquad (16)$$

$$A_\nu = R_\nu\cos(\Phi_\nu) \quad\text{and}\quad B\nu = R_\nu\sin(\Phi_\nu) \quad\text{for } \nu = 1,2,3$$

The latter transformation changes (14) into the more readily appliciable joint probability distribution of the magnitudes R_ν and phases Φ_ν of the three structure factors. Referring to paragraph 4., $p(A,B)dAdB$ changes into $p(R,\Phi)RdRd\Phi$ and $C(v,w)dvdw$ into $C(\theta,\rho)\rho d\theta d\rho$. Moreover $A_\nu v_\nu + B_\nu w_\nu$ becomes $\rho_\nu R_\nu\cos(\theta_\nu-\Phi_\nu)$ after employing (16). As a result, (14) is changed into (e.g. Karle & Hauptman, 1958)

$$P(R_1,R_2,R_3,\Phi_1,\Phi_2,\Phi_3) = R_1R_2R_3(2\pi)^{-6}\cdot$$

$$\int_0^\infty\int_0^\infty\int_0^\infty\int_0^{2\pi}\int_0^{2\pi}\int_0^{2\pi} \rho_1\rho_2\rho_3\exp\left[-i\left\{ \sum_{\nu=1}^{3}\left[\rho_\nu R_\nu\cos(\theta_\nu-\Phi_\nu)\right]\right\}\right]\cdot$$

$$C(\rho_1,\rho_2,\rho_3,\theta_1,\theta_2,\theta_3)\,d\theta_1d\theta_2d\theta_3d\rho_1d\rho_2d\rho_3 \qquad (17)$$

In the characteristic function the explicit structure factor expression is introduced. In P1 A_ν and B_ν are

$$A_\nu = \sum_{j=1}^{N} f_{j\nu}\cos(2\pi\vec{H}_\nu\cdot\vec{r}_j)$$

$$\qquad (18)$$

$$B_\nu = \sum_{j=1}^{N} f_{j\nu}\sin(2\pi\vec{H}_\nu\cdot\vec{r}_j)$$

with the atomic scattering factor $f_{j\nu} = f_{j\nu}^0 \exp\left[-B[\sin(\theta)/\lambda]^2\right]$, $f_{j\nu}^0$ the zero-angle scattering factor and B the overall temperature factor. It is assumed that $f_{j\nu}$ is real-valued.

After inserting (16) (first transformation) and (18) (15) becomes

$$C = \left\langle\exp\left[i\left\{ \sum_{j=1}^{N}\sum_{\nu=1}^{3}\rho_\nu f_{j\nu}\cos(2\pi\vec{H}_\nu\cdot\vec{r}_j-\theta_\nu)\right\}\right]\right\rangle \qquad (19)$$

PRIMITIVE RANDOM VARIABLES

The average in (19) is over the A_ν and B_ν or, equivalently, the R_ν and Φ_ν. These random variables are themselves functions of both the Miller indices and the atomic coordinates. Consequently, the average in (19) may also be evaluated with either choice for the primitive random variables. However, the underlying probabilistic models are different,

i. Miller indices as the primitive random variables

In this model the crystal structure is regarded to be fixed (though unknown) while the indices are variable. $P(\Phi_1,..,R_3)d\Phi_1..dR_3$ is the probability that a particular combination of phase values and magnitudes occurs for the fixed unknown structure and varying indices values. A main disadvantage of this model is that it is virtually impossible to incorporate structural information.

ii. Atomic coordinates as primitive random variables

The crystal structure is regarded to be variable and the reciprocal vectors are kept fixed. Now $P(\Phi_1,..,R_3)d\Phi_1..dR_3$ is the probability that a particular combination of phase values and magnitudes occurs for fixed indices values but for any structure. The atomic coordinates are variable so the particular crystal structure under investigation is simply one of the numerous number of structures possible which have the same phase and magnitude combinations. In view of the variability of the coordinates, the use of a priori structural information should be possible in a logical way.
In order to evaluate (19) an additional assumption must be made concerning the distribution of the atoms. If nothing is known about the structure, except the contents of the unit cell, no preference can be stated concerning the atomic positions. This leads to the concept of uniformly and independently distributed atoms,

$$p(r_j) = 1 \quad \text{for} \quad j = 1,..,N \quad \text{and} \quad r_j \text{ varying over unit cell} \tag{20}$$

The strength of this model is that no feasible structure is excluded. Hence, the probabilistic expressions should be generally appliciable. However, all (chemically) unfeasible and unwanted structures are included in this model as well. Therefore it may be suspected that this general model may not be strong enough to solve all structures. Obviously, introduction of non-uniform structural information may modify the final distribution in a helpful way (Heinerman, 1977).

THE EVALUATION OF THE CHARACTERISTIC FUNCTION

Since the r_j are independent, C is the product of N averages. Expressing C as $\exp(\ln(C))$ leads to a sum of N logarithms (the logarithm of a product being the sum of logarithms) so only a single average needs to be calculated. The characteristic function can be expressed,

$$C = \exp\left[\sum_{j=1}^{N}\left\{\mathrm{Ln}<\exp\left[i\left\{\sum_{\nu=1}^{3}\rho_\nu f_{j\nu}\cos(2\pi\vec{H}_\nu.\vec{r}_j - \Theta_\nu)\right\}\right]>_{\vec{r}_j}\right\}\right] \tag{21}$$

From expression (21) the following steps can be inferred :
i. The calculation of the average. This can be done by a Taylorseries expansion of the exponential expression inside the average brackets followed by averaging over r_j term by term. This is called the moments calculation and the result is a series expansion in the integration variables ρ_ν and Θ_ν with the moments as coefficients.
ii. Taking the logarithm of the series expansion resulting from step 1. This leads to a second series in the integration variables ρ_ν and Θ_ν though with different coefficients, the cumulants, which are expressable

in the already calculated moments via the series-expansion of the Ln. This proces is called moments-cumulants transformation.
iii. Adding the cumulant terms for the N independent primitive random variables which results in the characteristic function.
 Finally, the characteristic function must be Fourier-transformed in order to get the joint probability distribution aimed for. These four steps will be discussed now in some detail.

THE CALCULATION OF THE MOMENTS

After the Taylor-series expansion of the exponential argument of the logarithm, the characteristic function can be cast into the form,

$$C = \exp\left[\sum_{j=1}^{N} \mathrm{Ln}\left\{ \sum_{nmax=0}^{\infty} U_{nmax}\left[m^{\beta_1,\ldots,\beta_3}_{\alpha_1,\ldots,\alpha_3} \right] \right\} \right] \tag{22}$$

U_{nmax} is a six-fold summation involving the integer summation indices $\alpha_1, \alpha_2, \alpha_3, \beta_1, \beta_2$ and β_3, all running from 0 to nmax, such that their sum equals to nmax,

$$U_{nmax} = \sum^{nmax} \ldots \sum^{nmax} \prod_{\nu=1}^{3} \left\{ \frac{(i\rho_\nu)^{\alpha_\nu+\beta_\nu}}{2^{\alpha_\nu+\beta_\nu}\,\alpha_\nu!\beta_\nu!} \cdot \exp[\,i\theta_\nu(\beta_\nu-\alpha_\nu)\,] \right\} \cdot$$

$$\alpha_1,\alpha_2,\alpha_3,\beta_1,\beta_2,\beta_3 \; = \; 0$$
$$\alpha_1 + \alpha_2 + \alpha_3 + \beta_1 + \beta_2 + \beta_3 \; = \; nmax$$

$$\cdot \; m^{\beta_1,\beta_2,\beta_3}_{\alpha_1,\alpha_2,\alpha_3} \tag{23}$$

with

$$m^{\beta_1,\beta_2,\beta_3}_{\alpha_1,\alpha_2,\alpha_3} = \left[\prod_{\nu=1}^{3} f_{j\nu}^{\alpha_\nu+\beta_\nu} \right] \cdot \left\langle \exp\left[2\pi i\left\{ \sum_{\nu=1}^{3} (\alpha_\nu - \beta_\nu)H_\nu \right\} \cdot \vec{r}_j \right] \right\rangle_{\vec{r}_j} \tag{24}$$

In this arrangement U_{nmax} contains only terms of order $N^{-(nmax/2-1)}$.
At this point the probabilistic a priori model should be invoked. Employing (20) the average in (24) can be evaluated as an integral which results in a delta-function expression,

$$\int \exp\left[2\pi i\left\{ \sum_{\nu=1}^{3} (\alpha_\nu - \beta_\nu)\vec{H}_\nu \right\} \cdot \vec{r}_j \right] p(\vec{r}_j)\,d\vec{r}_j = \delta\left\{ \sum_{\nu=1}^{3} (\alpha_\nu - \beta_\nu)\vec{H}_\nu \right\} \tag{25}$$

Hence, only when the restrictive condition,

$$\sum_{\nu=1}^{3} (\alpha_\nu - \beta_\nu)\vec{H}_\nu = \vec{0} \tag{26}$$

is fulfilled, a non-zero contribution to the moments exists
The first contributing terms are :

116

i. $nmax = 0$: $\alpha_\nu = \beta_\nu = 0$ for $\nu = 1,2$ and 3. Hence $U_0 = 1$

ii. $nmax = 2$: $\alpha_1 = \beta_1 = 1$ and $\alpha_2 = \beta_2 = \alpha_3 = \beta_3 = 0$
$\alpha_2 = \beta_2 = 1$ and $\alpha_1 = \beta_1 = \alpha_3 = \beta_3 = 0$
$\alpha_3 = \beta_3 = 1$ and $\alpha_1 = \beta_1 = \alpha_2 = \beta_2 = 0$

If H_1, H_2 and H_3 do not fulfill any particular requirements no other non-zero moments can be found. However, if $H_3 = - H_1 - H_2$, i.e. if a triplet relation exist amongst the three reflections, additional contributing moments will be found for $nmax \geq 3$. In this paper we will consider only contributions for which $nmax = 3$:

iii. $nmax = 3$: $\alpha_1 = \alpha_2 = \alpha_3 = 1$ and $\beta_1 = \beta_2 = \beta_3 = 0$
$\alpha_1 = \alpha_2 = \alpha_3 = 0$ and $\beta_1 = \beta_2 = \beta_3 = 1$

THE CALCULATION OF THE CUMULANTS

The next step is the calculation of the logarithm of the series-expansion in (22). In order to retain the same functional form the new arguments in the U_{nmax} series, the cumulants, are expressed in terms of the moments by means of the series-expansion of the logarithm. This procedure, the moments-cumulants transformation, can be expressed as (note that $U_0 = 1$),

$$\sum_{nmax=1}^{\infty} U_{nmax}\left[k\begin{matrix}\beta_1,\ldots,\beta_3\\\alpha_1,\ldots,\alpha_3\end{matrix}\right] = Ln\left\{1+ \sum_{nmax=1}^{\infty} U_{nmax}\left[m\begin{matrix}\beta_1,\ldots,\beta_3\\\alpha_1,\ldots,\alpha_3\end{matrix}\right]\right\} \tag{27}$$

Upon expansion of both sides of (27), identical series must appear. In this paper we consider only terms upto order $N^{-0.5}$. No terms exist with $nmax = 1$ so $Ln(1+x) \approx x$ with x the terms for which $nmax=2$ or 3. In other words, upto this expansion the cumulants for which $nmax=2$ or $nmax=3$ are equal to the moments.

THE CHARACTERISTIC FUNCTION

The last step to obtain the characteristic function consists of the summation of the cumulants contributions of the N different primitive random variables for each of the contributing summation index combinations which leads to

$$C = \exp\left[\sum_{nmax=2}^{3} U_{nmax}\left[Z\begin{matrix}\beta_1,\ldots,\beta_3\\\alpha_1,\ldots,\alpha_3\end{matrix} K\begin{matrix}\beta_1,\ldots,\beta_3\\\alpha_1,\ldots,\alpha_3\end{matrix}\right]\right] \tag{28}$$

with

$$Z\begin{matrix}\beta_1,\ldots,\beta_3\\\alpha_1,\ldots,\alpha_3\end{matrix} = \sum_{j=1}^{N}\left[\left\{\prod_{\nu=1}^{3} f_{j\nu}^{\alpha_\nu+\beta_\nu}\right\}\right] \tag{29}$$

The cumulants K defined in (28) equal the number of times condition (25) is fulfilled for the particular $(\alpha_1,\ldots,\beta_3)$ combination. For the cumulants for which $nmax=2$ and $nmax =3$ (29) can be expressed alternatively,

$$Z_{\nu\nu} = \sum_{j=1}^{N} f_{j\nu}^{2} \qquad \text{for } \nu = 1,2,3$$

and (29')

$$Z_{123} = \sum_{j=1}^{N} f_{j1} f_{j2} f_{j3}$$

As a result the characteristic function correct up to $O(N^{-0.5})$ is

$$C = \exp\left[-\sum_{\nu=1}^{3} Z_{\nu\nu}\frac{\rho_{\nu}^2}{4} - iZ_{123}\frac{\rho_1\rho_2\rho_3}{4} \cos(\theta_1 + \theta_2 + \theta_3) \right] \tag{30}$$

After inserting (30) in (17), the final task is to calculate the six integrations. In this respect it is convenient to introduce the following variable transformation and definitions,

$$\rho_{\nu} = t_{\nu} Z_{\nu\nu}^{-\frac{1}{2}} \quad \text{for } \nu = 1,2,3$$

$$|E_{\nu}| = R_{\nu} Z_{\nu\nu}^{-\frac{1}{2}} \quad \text{for } \nu = 1,2,3$$

and (31)

$$\sigma_3 = Z_{123} \cdot [Z_{11}Z_{22}Z_{33}]^{-\frac{1}{2}}$$

The joint probability expression becomes now

$$P(R_1,R_2,R_3,\Phi_1,\Phi_2,\Phi_3) = |E_1E_2E_3|(2\pi)^{-6}[Z_{11}Z_{22}Z_{33}]^{-\frac{1}{2}} \cdot$$

$$\int_0^\infty\int_0^\infty\int_0^\infty\int_0^{2\pi}\int_0^{2\pi}\int_0^{2\pi} t_1t_2t_3 \ \exp\left[-i\sum_{\nu=1}^{3} t_{\nu}|E_{\nu}|\cos(\Theta_{\nu}-\Phi_{\nu}) + \right.$$

$$\left. -\sum_{\nu=1}^{3}\frac{t_{\nu}^2}{4} - i\sigma_3\frac{t_1t_2t_3}{4}\cos(\theta_1+\theta_2+\theta_3) \right] d\theta_1 d\theta_2 d\theta_3 dt_1 dt_2 dt_3 \tag{32}$$

THE FOURIER-TRANSFORMATION OF THE CHARACTERISTIC FUNCTION

For the evaluation of (32) the following trigonometric expressions are employed (Hauptman, 1971). For real c_k

$$\sum_{k} c_k\cos(\Phi+\alpha_k) = Z\cos(\Phi+\xi)$$

with (33)

$$Z^2 = \sum_{k,l}\sum c_kc_l\cos(\alpha_k-\alpha_l) \quad \text{and} \quad Z\exp[i\xi] = \sum_{k} c_k\exp[i\alpha_k]$$

In addition, the following integral is used (Giacovazzo, 1980;E.16),

$$\int_0^\infty\int_0^{2\pi} \exp\left[-p^2x^2 - iaxcos(\Phi)\right] x\,dx\,d\Phi = \pi p^{-2}\exp\left[-\frac{a^2}{4p^2}\right] \tag{34}$$

118

Collect now in (32) terms depending on θ_1 and neglect the terms of $O(N^{-1})$ and larger,

$$- it_1|E_1|\cos(\theta_1-\Phi_1) \quad -i\sigma_3\frac{t_1t_2t_3}{4}\cos(\theta_1+\theta_2+\theta_3) \quad =$$

$$- it_1X_1\cos(\theta_1+\xi_1) \quad \text{with} \quad X_1^2 \simeq |E_1|^2+ \sigma_3\frac{|E_1|t_2t_3}{2}\cos(\Phi_1+\theta_2+\theta_3)$$

Employing (34) to the t_1,θ_1 integrations gives then $2\pi\exp[-X_1^2]$

The same procedure is followed for t_2,θ_2 and t_3,θ_3 with

$$X_2^2 \simeq |E_2|^2- i\sigma_3|E_1E_2|t_3\cos(\Phi_1+\Phi_2+\theta_3)$$

and

$$X_3^2 \simeq |E_3|^2- 2\sigma_3|E_1E_2E_3|\cos(\Phi_1+\Phi_2+\Phi_3)$$

respectively. The joint probability distribution of the three magnitudes and the three phases becomes now

$$P(R_1,R_2,R_3,\Phi_1,\Phi_2,\Phi_3) \simeq |E_1E_2E_3|\pi^{-3}[Z_{11}Z_{22}Z_{33}]^{-\frac{1}{2}}.$$

$$\exp\left\{-|E_1|^2- |E_2|^2- |E_3|^3+ 2\sigma_3|E_1E_2E_3|\cos(\Phi_1+\Phi_2+\Phi_3) \right\} \tag{35}$$

It is interesting to note that from this probabilistic approach it follows that the normalized structure factors $|E|$ are the final parameters instead of the structure factors itself.

Integrating (35) over Φ_1, Φ_2 and Φ_3 under the condition $\Psi_3 = \Phi_1+\Phi_2+\Phi_3$ gives the joint probability distribution of the triplet phase sum Ψ_3 and the magnitudes R_1, R_2 and R_3, $P(\Psi_3,R_1,R_2,R_3)$. From this distribution readily the conditional probability distribution of Ψ_3 given the three magnitudes is obtained.

$$P(\Psi_3|R_1,R_2,R_3) \simeq \left[2\pi I_0(2\sigma_3|E_1E_2E_3|)\right]^{-1}\exp\left[2\sigma_3|E_1E_2E_3\cos(\Psi_3)\right] \tag{36}$$

in which the expansion of the exponential in modified Bessel functions I_n has been employed

$$\exp[x\cos(\Phi)] = \sum_{n=-\infty}^{\infty} I_n(x)\exp[in\Phi] \tag{37}$$

REFERENCES

Cochran, W. (1955). Acta Cryst. __8__, 473-477.
Giacovazzo, G. (1980). "Direct Methods in Cystallography". Academic Press. London.
Hauptman, H. (1971). Z. Kristallogr. __134__, 28-43.
Heinerman, J.J.L. (1977). Acta Cryst. __A33__, 100-106.
Karle, J. & Hauptman, H. (1958). Acta Cryst. __11__, 264-269.
Peschar, R. & Schenk, H. (1987) Acta Cryst. __A43__, 751-763.

SOME APPLICATIONS OF PROBABILITY THEORY IN DIRECT METHODS

Suzanne Fortier and Ian R. Castleden

Dept. of Chemistry, Queen's University
Kingston, Canada K7L 3N6

INTRODUCTION

It may seem somewhat surprising at first that probability theory has played such an important role in the solution of the crystal structure determination problem. Crystals, which are usually defined in terms of the long range ordering they display, might appear to be poor samples on which to apply a mathematical model devised to deal with random experiments. It is the realization that the periodically repeating motif could itself be depicted as consisting of atoms randomly distributed (Wilson, 1949) that allowed the phase problem to be phrased and solved in a probabilistic framework. Since then, there have been numerous applications of probability theory to the problem of crystal structure determination. They have essentially provided a solution for the case of small molecule crystal structures.

Various probabilistic approaches have been used in crystallographic applications. Each approach was essentially introduced to answer a very specific question. In this paper we will concentrate on applications of probability theory in direct methods based on three specific approaches: 1) the use of the Central Limit Theorem 2) the method of joint probability distributions and 3) Bayesian statistics. We will examine both the type of questions addressed by these approaches and the solutions they offer.

There are many excellent textbooks on basic probability theory, for example Cramér (1962) and Papoulis (1984), that can be used for the necessary background material and definitions. The article published recently by the International Union of Crystallography Subcommittee on Statistical Descriptors (Schwarzenbach et al, 1989) is also highly recommended. Finally, there are several books on direct methods and crystallographic computing which present a thorough description of the theory and applications of the probabilistic approach in direct methods. We note in particular the books of Hauptman (1972) and Giacovazzo (1980).

1) THE USE OF THE CENTRAL LIMIT THEOREM

It was initially questions about the diffraction intensities that prompted the first efforts to use probability theory in crystallography. In particular, the question of how absolute intensity data could be

Direct Methods of Solving Crystal Stuctures
Edited by H. Schenk, Plenum Press, New York, 1991

obtained from the relative data was at the center of these early investigations. This work marked the beginning of a new field in crystallography, that of crystallographic statistics, and laid the foundation of the direct methods. Many important questions were then formulated and answered, although the initial question concerning the translation of relative data into absolute data still cannot be answered in an exact way.

Provided that the unit cell contains a sufficiently large number of atoms with no atom or small group of atoms dominating the scattering, the asymmetric unit of a crystal structure can be depicted as consisting of a collection of atoms which are randomly distributed. The atomic fractional coordinates can thus be regarded as random variables and, in turn, the structure factors, as functions of random variables, can themselves be regarded as random variables. It is this realization that allowed Wilson (1949) to invoke the Central Limit Theorem and derive probability distribution functions for diffraction intensities.

The Central Limit Theorem states that the sum of a sufficiently large number of independent random variables tends to the normal distribution, no matter what the distributions for the individual variables are. In other words, the probability distribution $f(x)$ of the sum of random variables

$$x = x_1 + x_2 + \ldots\ldots + x_n \text{ is} \tag{1.1}$$

$$f(x) \simeq \frac{1}{\sigma (2\pi)^{1/2}} \exp -(x - m)^2 / 2\sigma^2 \tag{1.2}$$

$$\text{where } m = \sum_{j=1}^{n} m_j = \text{ sum of the mean values} \tag{1.3}$$

$$\text{and } \sigma^2 = \sum_{j=1}^{n} \sigma_j^2 = \text{ sum of the variances} \tag{1.4}$$

In applying this theorem to the structure factors, Wilson (1949) considered separately the distributions for the real part (ξ_j), and the imaginary part (η_j) of the structure factors. Combining the two led to the intensity distribution. For example in space group P1, we have:

$$\text{for the real part, } \xi_j = f_j \cos (2\Pi \, \vec{s}.\vec{r}_j) \tag{1.5}$$

$$\text{where } \vec{s} = h\vec{a}^* + k\vec{b}^* + l\vec{c}^* \tag{1.6}$$

$$\vec{r}_j = x_j\vec{a} + y_j\vec{b} + z_j\vec{c} \tag{1.7}$$

The mean, m, $= \Sigma \, m_j = 0$ assuming that $2\Pi \, \vec{s}.\vec{r}_j$ is uniformly distributed

$$\text{over the trigonometric circle} \tag{1.8}$$

The variance, $\sigma^2 = \sum_{j=1}^{n} \sigma_j^2 = 1/2 \sum_{j=1}^{n} f_j^2 \tag{1.9}$

$$\text{since } \sigma_j^2 = \langle \, (\, \xi_j - \langle \, \xi_j \, \rangle \,)^2 \rangle \qquad (1.10)$$

$$= \langle \, \xi_j^2 \, \rangle = f_j^2 \, \langle \, \cos 2\Pi \, \vec{s}.\vec{r}_j \, \rangle \qquad (1.11)$$

$$= 1/2 \, f_j^2 \qquad (1.12)$$

$$\text{and } P(A) \, dA = \frac{1}{\left(\Pi \sum_{j=1}^{n} f_j^2 \right)^{1/2}} \exp \, (-A^2 / \sum_{j=1}^{n} f_j^2) \, dA \qquad (1.13)$$

$$\text{For the imaginary part, } \eta_j = f_j \sin (\, 2\Pi \, \vec{s}.\vec{r}_j \,) \qquad (1.14)$$

The mean, m, = 0 because the sin function is an odd function $\qquad (1.15)$

$$\sigma^2 = 1/2 \sum_{j=1}^{n} f_j^2 \qquad (1.16)$$

$$\text{and } P(B) dB = \frac{1}{\left(\Pi \sum_{j=1}^{n} f_j^2 \right)^{1/2}} \exp \, (-B^2 / \sum f_j^2) \, dB \qquad (1.17)$$

$$P(A,B)dAdB = P(A)P(B)dAdB = \frac{1}{\Pi \sum_{j=1}^{n} f_j^2} \exp \, (-(A^2 + B^2) / \sum_{j=1}^{n} f_j^2) \, dAdB \qquad (1.18)$$

$$\text{from which one obtains: } P(I)dI = \frac{1}{\sum f_j^2} \exp \, (-I / \sum_{j=1}^{n} f_j^2) \, dI \qquad (1.19)$$

$$\text{and } \langle I \rangle = \int_{0}^{\infty} I \, P(I) \, dI = \sum f_j^2 \qquad (1.20)$$

The importance of this result is of course well known. It is at the basis of the Wilson plot (Wilson, 1942), the most commonly used procedure for obtaining initial estimates of the scale and overall thermal factor which allows the conversion, at least in an approximate way, of the relative intensity data into absolute values. This step is crucial in crystal structure determination exercises, both of small molecules and macromolecules.

In his 1949 paper, Wilson also introduced the first tools for distinguishing between centrosymmetric and non-centrosymmetric space groups on the basis of the diffraction intensities. For example, he showed that:

$$\text{For centrosymmetric crystals, } \langle |F| \rangle \simeq 1/2 \, (\Pi \sum f_j^2)^{1/2} \qquad (1.21)$$

$$\text{for non-centrosymmetric ones, } \langle |F| \rangle \simeq (2 \sum f_j^2 / \Pi)^{1/2} \, . \qquad (1.22)$$

Subsequently, the Central Limit Theorem was used for the derivation of phase determining formulae, for example by Woolfson (1954) and Cochran and Woolfson (1955) in the derivation of the sign determining formula:

$$P_+ (U_h) = 1/2 + 1/2 \tanh (N |U_h| \sum_h U_h U_{h-h'}) \tag{1.23}$$

and by Heinerman, Krabbendam and Kroon (1977) who proposed a method which makes use of the Central Limit Theorem for the simple and rapid derivation of the phase invariant joint probability distributions.

2) THE METHOD OF JOINT PROBABILITY DISTRIBUTIONS

With the introduction of the method of joint probability distributions, the main questions under consideration were those concerning the unknown phases, as stated by Hauptman and Karle (1953b): "Thus the concept of the joint or compound probability distribution forms the basis for a direct attack on the phase problem". The importance of the approach, again according to Hauptman and Karle (1953a), is that "The joint distribution not only leads to probability distributions for both the magnitude and the phase of a structure factor, but also permits one to take into account an *a priori* knowledge of observed X-ray intensities". This *a priori* knowledge comes from the assumption, as in Wilson's work, that the atomic coordinates are uniformly and independently distributed in the asymmetric unit of the unit cell. The joint probability distributions are then updated by the values of the measured intensities. This leads to *a posteriori* probability distributions and allows the unknown phases to be estimated via conditional probability distributions.

In general, the joint probability distribution (j.p.d.) of n random variables $x_1 + x_2 + \ldots\ldots + x_n$ is given by :

$$P (x_1, x_2, \ldots x_n) = \frac{1}{(2\Pi)^n} \int_0^\infty \ldots \int_0^\infty \exp [-i (u_1 x_1 + u_2 x_2 + \ldots + u_n x_n)]$$

$$\times C (u_1, u_2, \ldots u_n) \, du_1 du_2 \ldots du_n \tag{2.1}$$

where $C(u_1, u_2 \ldots u_n)$ is the characteristic function of the random variables $x_1, x_2 \ldots x_n$ and,

$$C(u_1, u_2 \ldots u_n) \simeq < \exp [i(u_1 x_1 + u_2 x_2 + \ldots u_n x_n)]>. \tag{2.2}$$

Thus the joint probability distribution is obtained through the Fourier transform of its characteristic function.

In direct methods applications, the random variables are selected from the pool of structure factor magnitudes and phases. These can be viewed as random variables since they are themselves functions of random variables, according to the *a priori* model used. In some derivations, for example Giacovazzo (1976), the atomic positions are assumed to be the primitive random variables, while in others, for example Hauptman (1975a), it is the reciprocal vectors that are taken as the primitive random variables. The two approaches, however, often lead to nearly identical results, as pointed out by several authors (Giacovazzo, 1980; Hauptman and Karle, 1953a; Heinerman, 1977).

Most of the joint probability distributions used in current direct methods are joint probability distributions formed from a small number of structure factors. The selection of structure factors in joint probability distribution derivations has followed two main guidelines.

The first comes from the theory of invariants and semi-invariants which suggests that the set of structure factor phases in the joint probability distributions be independent of the choice of origin or permissible origins. The second comes from the neighborhood principle (Hauptman, 1975b) or representation theory (Giacovazzo, 1977) which, based on the observation that reliable invariant or semi-invariant estimates can be obtained from an appropriately selected small subset of structure factor magnitudes, offers guidelines for the selection of these subsets.

Most of the initial joint probability distributions included structure factors belonging to only one data set. More recently, a large number of joint probability distributions of structure factors gathered from more than one data set have been derived, particularly following the efforts to extend direct methods applications to macromolecules. For example, distributions for the case of single isomorphous replacement, SIR, (Hauptman, 1982a), anomalous scattering, SAS, (Hauptman, 1982b; Giacovazzo, 1983a) and partial/complete structure (Giacovazzo, 1983b) have been presented. It has also been shown recently that all current sources of phase information can be incorporated in a general joint probability distribution framework (Bricogne, 1988) and indeed that the the two-phase and three-phase invariant distributions presented for the SIR, SAS, and partial/complete structure cases are isomorphous (Fortier and Nigam, 1989).

The derivation of joint probability distributions was initially a very tedious and time-consuming task. Nowadays, such a derivation is greatly facilitated by the availability of general joint probability distributions which can be used for any number of phases and can be specialized for any space groups (Peschar and Schenk, 1987; Castleden, 1987). Furthermore, it has been shown that a general algorithm can be implemented for rapid, computer-aided derivations of joint probability distributions (Peschar and Schenk, 1987). Also, efforts to obtain exact joint probabilty distributions have recently been successful for cases involving a small number of terms which have found applications in the area of intensity statistics (Shmueli, Weiss, Kiefer and Wilson, 1984; Shmueli and Weiss, 1987).

Despite these important theoretical advances, the method of joint probability distribution has not proven capable yet of solving macromolecular crystal structures or, for that matter, of solving routinely small molecule crystal structures containing more than 100 or so independent non-hydrogen atoms. One of the most severe limitations of this approach, as it is presently used, comes from the underlying random atom distribution model. Indeed, the phase estimates are found to deteriorate greatly as the actual structure moves further and further away from this model.

3) BAYESIAN STATISTICS

The impetus for rephrasing the phase problem in a Bayesian statistical framework comes from the realization that, for structures of the complexity of macromolecules, the problem of crystal structure determination will require for its solution a scheme that allows the incorporation and updating of all possible sources of phase information (Bricogne, 1988).

Bayes' Theorem states that the probability of B_r given A, denoted by $P(B_r | A)$ can be calculated by :

$$P(B_r|A) = \frac{P(A|B_r) \times P(B_r)}{\sum\limits_{i=1}^{k} P(A|B_i) \times P(B_i)} \tag{3.1}$$

Thus the Bayesian approach combines sampling information with other available prior information. $P(B_r|A)$ is often referred to as the posterior distribution and $P(B_r)$ as the prior distribution.

An alternative formulation for Bayes' Theorem is:

$$O(B_r|A) = LS \times O(B_r) \tag{3.2}$$

where O, which stands for " odds", is defined by :

$$O(X) = \frac{P(X)}{1 - P(X)} \tag{3.3}$$

and

$$P(X) = \frac{O(X)}{1 + O(X)} \tag{3.4}$$

and LS, the likelihood ratio, is calculated as:

$$LS = \frac{P(A|B_r)}{P(A|\neg B_r)} \tag{3.5}$$

where $P(A|\neg B_r)$ is the probability of A given not B_r. $\tag{3.6}$

There had been few crystallographic applications of Bayesian statistics until the work of French and Wilson (1978) on the treatment of negative intensity data and the work of Oatley and French (1982) on profile fitting techniques for diffractometer data. Bricogne (1988) has recently proposed a full Bayesian inference scheme for crystal structure determination which offers the promise of being able to tackle macromolecular problems.

CONCLUDING REMARKS

Probability theory has made very important contributions to crystallography and, in particular, to the solution of the phase problem. Nevertheless, many problems still remain untractable. As yet, the direct methods have been unable to solve problems of the complexity of macromolecular structures. It is very encouraging, though, to witness the recent vigor and rigor with which such problems have been addressed.

REFERENCES

Bricogne, G., 1988, A Bayesian Statistical Theory of the Phase Problem. I. A Multichannel Maximum-Entropy Formalism for Constructing Generalized Joint Probability Distributions of Structure Factors, Acta Cryst., A44:517.

Castleden. I.R., 1987, A Joint Probability Distribution of Invariants for all Space Groups, Acta Cryst., A43:384.

Cochran, W. and Woolfson, M.M., 1955, The Theory of Sign Relations Between Structure Factors, Acta Cryst., 8:1.

Cramér, H., 1962, "The Elements of Probability Theory and some of its Applications", John Wiley & Sons, New York.

Fortier, S. and Nigam, G.D., 1989, On the Probabilistic Theory of Isomorphous Data Sets: General Joint Distributions for the SIR, SAS, and Partial/Complete Structure Cases, Acta Cryst., A45:247.

French, S. and Wilson, K., 1978, On the Treatment of Negative Intensity Observations, Acta Cryst., A34:517.

Giacovazzo, C., 1976, A Probabilistic Theory of the Cosine Invariant cos $(\phi_h + \phi_k + \phi_l - \phi_{h+k+l})$, Acta Cryst., A32:91.

Giacovazzo, C., 1977, A General Approach to Phase Relationships: The Method of Representations, Acta Cryst., A33:933.

Giacovazzo, C., 1980, "Direct Methods in Crystallography", Academic Press, London.

Giacovazzo, C., 1983a, The Estimation of Two-Phase and Three-Phase Invariants in P1 when Anomalous Scatterers are Present, Acta Cryst., A39:585.

Giacovazzo, C., 1983b, From a Partial to the Complete Crystal Structure, Acta Cryst., A39:685.

Hauptman, H.A., 1972, "Crystal Structure Determination. The Role of the Cosine Semi-Invariants", Plenum Press, New York.

Hauptman, H., 1975a, A Joint Probability Distribution of Seven Structure Factors, Acta Cryst., A31:671.

Hauptman, H., 1975b, A New Method in the Probabilistic Theory of the Structure Invariants, Acta Cryst., A31:680.

Hauptman, H., 1982a, On Integrating the Techniques of Direct Methods and Isomorphour Replacement: I. The Theoretical Basis, Acta Cryst., A38:289.

Hauptman, H., 1982b, On Integrating the Techniques of Direct Methods with Anomalous Dispersion. I. The Theoretical Basis, Acta Cryst., A38:632.

Hauptman, H. and Karle, J., 1953a, "Solution of the Phase Problem. I. The Centrosymmetric Crystal", A.C.A. Monograph No. 3, Polycrystal Book Service, Brooklyn.

Hauptman, H. and Karle, J., 1953b, The Probability Distribution of the Magnitude of a Structure Factor. II. The Non-Centrosymmetric Crystal, Acta Cryst., 6:136.

Heinerman, J.J.L., 1977, Some Contributions to the Theory of Triplet and Quartet Structure Invariants, in "Direct Methods in Crystallography", H. Hauptman, ed., Proceedings of the 1976 Intercongress Symposium.

Heinerman, J.J.L., Krabbendam, H. and Kroon, J., 1977, The von Mises Distribution of the Phase of a Structure Invariant, Acta Cryst., A33:873.

Oatley, S. and French, S., 1982, A Profile-Fitting Method for the Analysis of Diffractometer Intensity Data, Acta Cryst., A38:537.

Papoulis, A., 1984, "Probability, Random Variables, and Stochastic Processes", McGraw-Hill Book Company, New York.

Peschar, R. and Schenk, H., Computer-Aided Derivation of Theoretical Joint Probability Distributions of Normalized Structure Factors, Acta Cryst.

Schwarzenbach, D., Abrahams, S.C., Flack, H.D., Gonschorek, W., Hahn, T., Huml, K., Marsh, R.E., Prince, E., Robertson, B.E., Rollett, J.S. and Wilson, A.J.C., 1989, Statistical Descriptors in Crystallography, Acta Cryst., A45:63.

Shmueli, U., Weiss, G.H., Keifer, J.E. and Wilson, A.J.C., 1984, Exact Random-Walk Models in Crystallographic Statistics. I. Space Groups P1 and P$\bar{1}$, Acta Cryst., A40:651.

Shmueli, U. and Weiss, G.H., 1987, Exact Random-Walk Models in Crystallographic Statistics. III. Distributions of $|E|$ for Space Groups of Low Symmetry, Acta Cryst., A43:93.

Wilson, A.J.C., 1942, Determination of Absolute from Relative X-Ray Intensity Data, Nature, 150:152.

Wilson, A.J.C., 1949, The Probability Distribution of X-Ray Intensities, Acta Cryst., 2:318.

Woolfson, M.M., 1954, The Statistical Theory of Sign Relationships, Acta Cryst., 7:61.

THE NEIGHBORHOOD PRINCIPLE AND THE EXTENSION CONCEPT

Herbert A. Hauptman

Medical Foundation of Buffalo, Inc.
73 High Street
Buffalo, N.Y. 14203 U.S.A.

ABSTRACT

Three major themes are developed:

(i) The fundamental principle of direct methods: The structure
seminvariants link the observed magnitudes $|E|$ with the desired phases ϕ
of the normalized structure factors E. Specifically, for fixed
enantiomorph, the observed magnitudes $|E|$ determine, in general, unique
values for all the structure seminvariants; the latter, in turn, as
certain well defined linear combinations of the phases, lead to unique
values for the individual phases ϕ.

(ii) The neighborhood principle: For fixed enantiomorph, the
value of any structure seminvariant T is primarily determined, in
favorable cases, by the values of one or more small sets of magnitudes
$|E|$, the neighborhoods of T, and is relatively insensitive to the great
bulk of remaining magnitudes. The conditional probability distribution
of T, given the magnitudes in any of its neighborhoods, yields an
estimate for T which is particularly good in the favorable case that the
variance of the distribution happens to be small.

(iii) The extension concept: By embedding the structure
seminvariant T and its symmetry related variants in suitable structure
invariants Q, one obtains the extensions Q of the seminvariant T. In
this way the probabilistic theory of the structure seminvariants is
reduced to that of the structure invariants, which is well developed.

1. INTRODUCTION

1.1 The Phase Problem

The intensities of a sufficient number of x-ray diffraction maxima
determine a crystal structure. The available intensities usually exceed
the number of parameters needed to describe the structure. From these
intensities a set of numbers $|E_H|$ can be derived, one corresponding to
each intensity. However, the elucidation of the crystal structure
requires also a knowledge of the complex numbers $E_H = |E_H|\exp(i\phi_H)$, the
normalized structure factors, of which only the magnitudes $|E_H|$ can be
determined from the experiment. Thus a "phase" ϕ_H, unobtainable from
the diffraction experiment, must be assigned to each $|E_H|$, and the

problem of determining the phases when only the magnitudes $|E_H|$ are
known is called "the phase problem". Owing to the known atomicity of
crystal structures and the redundancy of observed magnitudes $|E_H|$, the
phase problem is solvable in principle.

1.2 The Structure Invariants

The values of the individual phases are determined by the crystal
structure and the choice of origin. However, there always exist certain
linear combinations of the phases whose values are determined by the
structure alone and are independent of the choice of origin. These
linear combinations of the phases are called the structure invariants.

1.3 The Structure Seminvariants

For all space groups other than P1 the origin may not be chosen
arbitrarily if one is to exploit fully the space group symmetries.
Those linear combinations of the phases whose values are uniquely
determined by the crystal structure and are independent of the choice of
permissible origin are known as the structure seminvariants. Thus the
collection of structure invariants is a subset of the collection of
structure seminvariants. In the space group P1 the two classes
coincide.

1.4 The Fundamental Principle of Direct Methods

For fixed enantiomorph, the observed magnitudes $|E|$ determine, in
general, unique values for all the structure seminvariants. The latter
in turn, as certain well-defined linear combinations of the phases, lead
unambiguously to unique values for the individual phases. Thus the
structure seminvariants serve to link the known magnitudes $|E|$ with the
desired phases ϕ (the fundamental principle of direct methods). By the
term "direct methods" is meant that class of methods which exploits
relationships among the structure factors in order to go directly from
the observed magnitudes $|E|$ to the needed phases ϕ.

1.5 The Neighborhood Principle

For fixed enantiomorph, the value of any structure seminvariant T
is primarily determined, in favorable cases, by the values of one or
more small sets of observed magnitudes $|E|$, the neighborhoods of T, and
is relatively insensitive to the values of the great bulk of remaining
magnitudes (the neighborhood principle). The conditional probability
distribution of T, given the magnitudes in any of its neighborhoods,
yields an estimate for T that is particularly good in the favorable case
that the variance of the distribution happens to be small.

1.6 The Extension Concept

By embedding the structure seminvariant T and its symmetry related
variants in suitable structure invariants Q one obtains the extensions Q
of the seminvariant T. Owing to the space group-dependent relationships
among the phases the value of T is simply related to the values of its
extensions. In this way the probabilistic theory of the structure
seminvariants is reduced to that of the structure invariants, which is
well developed. In particular, the neighborhoods of T are defined in
terms of the neighborhoods of its extensions.

2. THE PHASE PROBLEM

The atomic arrangement in the unit cell of a crystal, i.e., the
crystal structure, is determined once the intensities of a sufficient
number of x-ray diffraction maxima have been measured. The number of

these intensities usually exceeds by far the number of parameters required to describe the structure. From these intensities a set of numbers $|E_\mathbf{H}|$ can be derived, one corresponding to each intensity. However, the elucidation of the crystal structure requires also a knowledge of the complex numbers $E_\mathbf{H} = |E_\mathbf{H}|\exp(i\phi_\mathbf{H})$, of which only the magnitudes $|E_\mathbf{H}|$ can be determined from experiment. Thus a "phase" $\phi_\mathbf{H}$ must be assigned to each $|E_\mathbf{H}|$, and the problem of determining the phases when only the magnitudes $|E_\mathbf{H}|$ are known is called "the phase problem." Owing to the known atomicity of crystal structures and the redundancy of observed magnitudes $|E_\mathbf{H}|$, the phase problem is solvable in principle.

It is the redundancy of the system of equations relating the magnitudes $|E|$ with the desired phases ϕ, as well as errors in the observed $|E|$'s, which makes possible, even indispensable, the use of probabilistic techniques in the solution of the phase problem.

3. THE NORMALIZED STRUCTURE FACTORS E

The relationship between the (complex) normalized structure factors E and the crystal structure is given by the pair of equations

$$E_\mathbf{H} = |E_\mathbf{H}|\exp(i\phi_\mathbf{H})$$

$$= \frac{1}{\sigma_2^{1/2}} \sum_{j=1}^{N} Z_j \exp(2\pi i \mathbf{H}\cdot\mathbf{r}_j), \tag{1}$$

$$\left\langle E_\mathbf{H}\exp(-2\pi i \mathbf{H}\cdot\mathbf{r}) \right\rangle_\mathbf{H} = \left.\begin{array}{ll} \dfrac{Z_j}{\sigma_2^{1/2}} & \text{if } \mathbf{r} = \mathbf{r}_j \\[2em] = 0 & \text{if } \mathbf{r} \neq \mathbf{r}_j \end{array}\right\} \tag{2}$$

where $\mathbf{H}$ is an arbitrary reciprocal lattice vector, Z_j is the atomic number and $\mathbf{r}_j$ is the position vector of the atom labeled j, N is the number of atoms in the unit cell, and

$$\sigma_2 = \sum_{j=1}^{N} Z_j^2 . \tag{3}$$

Clearly (2) shows that the crystal structure (i.e., the position vectors $\mathbf{r}_j$) is determined in terms of the normalized structure factors $E_\mathbf{H}$. However, it turns out that although the magnitudes $|E_\mathbf{H}|$ may be determined, at least approximately, from experiment, the phases $\phi_\mathbf{H}$, which are also needed if the crystal structure is to be found via (2), cannot be determined experimentally. Nevertheless, because the number of equations (1) usually exceeds by far the number of unknowns $\mathbf{r}_j$, the available data, i.e., the known $|E_\mathbf{H}|$'s, are in general more than sufficient to determine crystal structures uniquely. In fact, from (1) one naturally formulates the problem as the determination of the N position vectors $\mathbf{r}_j$ which minimize the weighted sum of squares

$$\sum_\mathbf{H} w_j \left[|E_\mathbf{H}| - \frac{1}{\sigma_2^{1/2}} \left| \sum_{j=1}^{N} Z_j \exp(2\pi i \mathbf{H}\cdot\mathbf{r}_j) \right| \right]^2 , \tag{4}$$

in which the sum is taken over all reciprocal lattice vectors $\mathbf{H}$ for which magnitudes $|E_\mathbf{H}|$ are available, and the $W_\mathbf{H}$'s are a suitably chosen set of weights. This formulation clearly calls for a probabilistic approach. In practice the problem of finding the global minimum of (4) is too intractable to be solved ab initio. Instead the unknown position vectors $\mathbf{r}_j$ are eliminated from the system (1) to yield relationships among the $|E_\mathbf{H}|$'s having probabilistic validity. These in turn lead to approximate values of the unknown phases $\phi_\mathbf{H}$ which can then be used in (2) to determine a trial structure, i.e., approximate values of the unknowns $\mathbf{r}_j$. Employing standard iterative techniques (4), or something similar, is then used to obtain refined values for the $\mathbf{r}_j$. The techniques which employ the phases to determine crystal structures are known as direct methods, since the phases $\phi_\mathbf{H}$ are determined directly from the observed magnitudes $|E_\mathbf{H}|$ [rather than from a presumed known structure via (1)].

4.   THE STRUCTURE INVARIANTS

Equation (2) implies that the normalized structure factors $E_\mathbf{H}$ determine the crystal structure. However (1) does not imply that, conversely, the crystal structure determines the values of the normalized structure factors $E_\mathbf{H}$ since the position vectors $\mathbf{r}_j$ depend not only on the structure but on the choice of origin as well. It turns out nevertheless that the magnitudes $|E_\mathbf{H}|$ of the normalized structure factors are in fact uniquely determined by the crystal structure and are independent of the choice of origin, but that the values of the phases $\phi_\mathbf{H}$ depend also on the choice of origin. Although the values of the individual phases depend on the structure and the choice of origin, there exist certain linear combinations of the phases, the so-called structure invariants, whose values are determined by the structure alone and are independent of the choice of origin.

The most important structure invariants are the linear combinations of three phases (triplets):

$$\phi_\mathbf{H} + \phi_\mathbf{K} + \phi_\mathbf{L}, \tag{5}$$

where

$$\mathbf{H} + \mathbf{K} + \mathbf{L} = 0; \tag{6}$$

and the linear combination of four phases (quartets):

$$\phi_\mathbf{H} + \phi_\mathbf{K} + \phi_\mathbf{L} + \phi_\mathbf{M}, \tag{7}$$

where

$$\mathbf{H} + \mathbf{K} + \mathbf{L} + \mathbf{M} = 0; \tag{8}$$

et cetera.

5.   THE FUNDAMENTAL PRINCIPLE OF DIRECT METHODS

Two structures related by reflection through a point are said to be enantiomorphs of each other. The x-ray diffraction experiment is not capable of distinguishing the enantiomorphs when they are distinct. For this reason the fundamental principle of direct methods is formulated as follows. For fixed enantiomorph the observed magnitudes $|E|$ determine, in general, unique values for all the structure invariants. The latter, as certain well defined linear combinations of the phases, lead in turn to unique values for the phases ϕ. In short, the structure invariants serve to link the observed magnitudes $E_\mathbf{H}$ with the desired phases ϕ (the fundamental principle of direct methods).

6. THE NEIGHBORHOOD PRINCIPLE

It has been seen that for fixed enantiomorph the values of the observed magnitudes $|E|$ determine the values of all the structure invariants. A major recent insight is that, for fixed enantiomorph, the value of any structure invariant T is primarily determined, in favorable cases, by the values of one or more small sets of magnitudes $|E|$, the neighborhoods of T, and is relatively insensitive to the values of the great bulk of remaining magnitudes $|E|$ (the neighborhood principle). The conditional probability distribution of T, assuming as known the magnitudes $|E|$ in any of its neighborhoods, yields an estimate for T which is particularly good in the favorable case that the variance of the distribution happens to be small.

7. THE NEIGHBORHOODS OF THE STRUCTURE INVARIANTS

The first neighborhood of the triplet (5), where (6) holds, consists of the three magnitudes

$$|E_H|, \quad |E_K|, \quad |E_L|. \tag{9}$$

The first neighborhood of the quartet (7) where (8) holds, consists of the four magnitudes

$$|E_H|, \quad |E_K|, \quad |E_L|, \quad |E_M|. \tag{10}$$

The second neighborhood of the quartet consists of the four magnitudes (10) in the first neighborhood plus the three additional magnitudes

$$|E_{H+K}|, \quad |E_{K+L}|, \quad |E_{L+H}|, \tag{11}$$

i.e., seven magnitudes $|E|$ in all.

8. CONDITIONAL PROBABILITY DISTRIBUTIONS OF THE STRUCTURE INVARIANTS

8.1 Triplets

Suppose that a crystal structure consisting of N atoms per unit cell is fixed. Denote by W the collection of all reciprocal lattice vectors H and by ϕ_H the phase of the normalized structure factor E_H. Assume also that R_1, R_2, and R_3 are fixed nonnegative numbers. Suppose finally that the primitive random variable (vector) is the ordered triple (H, K, L) of reciprocal vectors H, K, L which is assumed to be uniformly distributed over the subset of the threefold Cartesian product $W \times W \times W$ defined by

$$|E_H| = R_1, \qquad |E_K| = R_2, \qquad |E_L| = R_3 \tag{12}$$

and

$$H + K + L = 0. \tag{13}$$

Then the structure invariant

$$\phi_3 = \phi_H + \phi_K + \phi_L \tag{14}$$

is a function of the primitive random variables H, K, L and therefore is itself a random variable. Denote by

$$P_{1/3} = P(\Phi | R_1, R_2, R_3) \tag{15}$$

the conditional probability distribution of ϕ_3, given the three magnitudes (12), the first neighborhood of ϕ_3. Then

$$P_{1/3} \simeq \frac{1}{2\pi I_0(A)} \exp(A \cos \Phi), \qquad (16)$$

where I_0 is the modified Bessel function, A is defined by

$$A = \frac{2\sigma_3}{\sigma_2^{3/2}} R_1 R_2 R_3, \qquad (17)$$

and

$$\sigma_n = \sum_{j=1}^{N} Z_j^n . \qquad (18)$$

Graphs of the distribution (16) for $A = 2.316$ and $A = 0.731$ are shown in Figures 1 and 2. Clearly this distribution has a unique maximum at $\Phi = 0$ in the interval $(-\pi, \pi)$ so that the most probable value of ϕ_3 is zero. The larger the value of A the smaller is the variance of the distribution and the more reliable is the estimate of ϕ_3, zero in this case.

8.2 Quartets

As before, suppose that a crystal structure consisting of N atoms per unit cell is fixed. Two distributions will be described. The first, in strict analogy with the preceding section ("Triplets"), is the conditional probability distribution of the quartet, given the four magnitudes in its first neighborhood; the second is the conditional probability distribution of the quartet, assuming that the seven magnitudes in its second neighborhood are known.

The Four-Magnitude Distribution. Assume that R_1, R_2, R_3, and R_4 are fixed nonnegative numbers. Next suppose that the primitive random variable (vector) is the ordered quadruple $(\mathbf{H},\mathbf{K},\mathbf{L},\mathbf{M})$ of reciprocal vectors $\mathbf{H},\mathbf{K},\mathbf{L},\mathbf{M}$ which is assumed to be uniformly distributed over the subset of the fourfold Cartesian product $W \times W \times W \times W$ defined by

$$|E_{\mathbf{H}}| = R_1, \qquad |E_{\mathbf{K}}| = R_2,$$

$$\qquad (19)$$

$$|E_{\mathbf{L}}| = R_3, \qquad |E_{\mathbf{M}}| = R_4$$

and

$$\mathbf{H} + \mathbf{K} + \mathbf{L} + \mathbf{M} = 0. \qquad (20)$$

In view of (20), the linear function of four phases

$$\phi_4 = \phi_{\mathbf{H}} + \phi_{\mathbf{K}} + \phi_{\mathbf{L}} + \phi_{\mathbf{M}} \qquad (21)$$

is a structure invariant which, as a function of the primitive random variables $\mathbf{H},\mathbf{K},\mathbf{L},\mathbf{M}$, is itself a random variable. Denote by

$$P_{1/4} = P(\Phi|R_1,R_2,R_3,R_4) \qquad (22)$$

the conditional probability distribution of ϕ_4, given the four

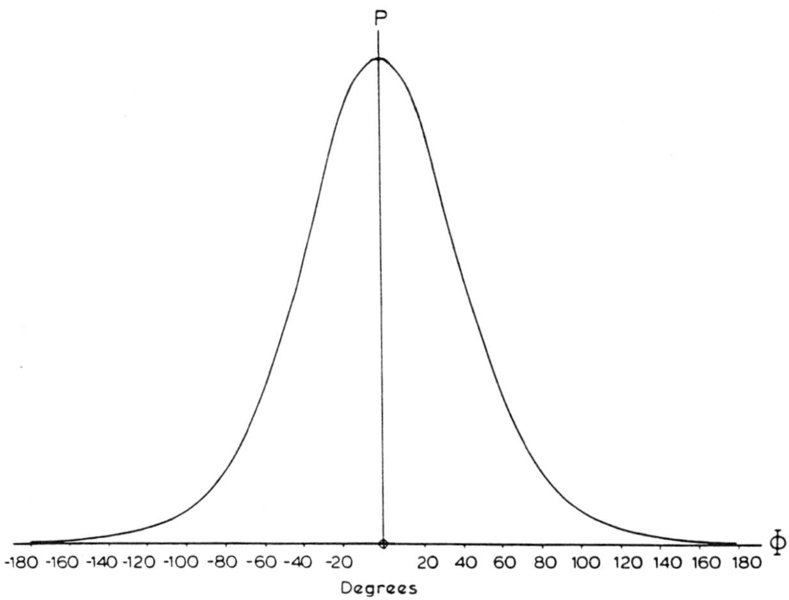

Figure 1.   The distribution $P_{1/3}$, (16), for A = 2.316.

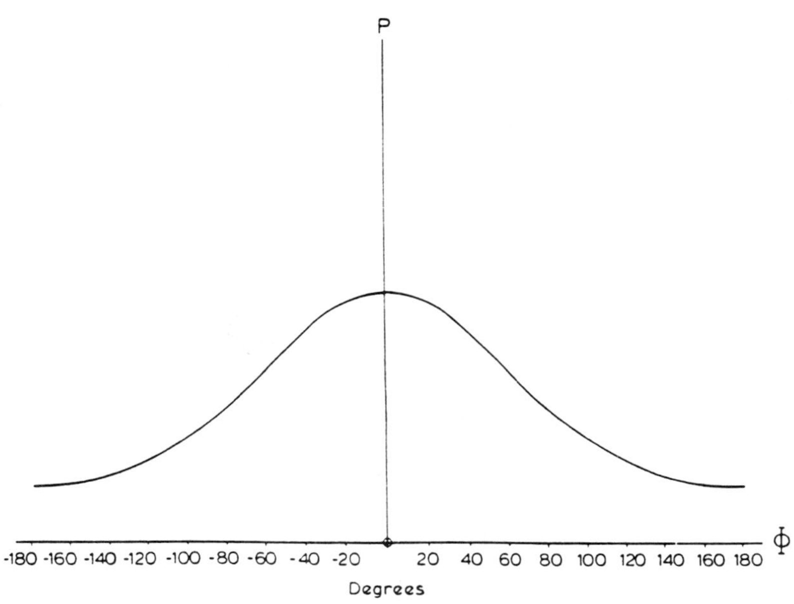

Figure 2.   The distribution $P_{1/3}$, (16), for A = 0.731.

magnitudes in its first neighborhood (19). Then

$$P_{1/4} \simeq \frac{1}{2\pi I_0(B)} \exp(B \cos \Phi),$$
(23)

where B is defined by

$$B = \frac{2\sigma_4}{\sigma_2^2} R_1 R_2 R_3 R_4$$
(24)

and σ_n by (18). Thus $P_{1/4}$ is identical with $P_{1/3}$, but B replaces A.
Hence similar remarks apply to $P_{1/4}$. In particular (23) always has a
unique maximum at $\Phi = 0$, so that the most probable value of the
structure invariant (21) is zero, and the larger the value of B the more
likely it is that $\phi_4 \simeq 0$. Since B values, of order 1/N, tend to be less
than A values, or order $1/N^{1/2}$, the estimate (zero) of the quartet (21)
is in general less reliable than the estimate (zero) of the triplet
(14). Hence (23) is no improvement over (16) and the goal of obtaining
a reliable nonzero estimate for a structure invariant is not realized by
(23). The decisive step in this direction is made in the next section.

    The Seven-Magnitude Distribution. If one assumes as known not only
the four magnitudes (19), but the additional three magnitudes $|E_{H+K}|$,
$|E_{K+L}|$, and $|E_{L+H}|$ then, in favorable cases, one obtains a more reliable
estimate for the quartet (21), and, furthermore, the estimate may lie
anywhere in the interval 0 to π.

    Assume that the seven nonnegative numbers R_1, R_2, R_3, R_4, R_{12}, R_{23},
R_{31} are fixed. Suppose next that the ordered quadruple of reciprocal
vectors (H,K,L,M) is a random variable which is uniformly distributed
over the subset of the fourfold Cartesian product W × W × W × W defined
by

$$|E_H| = R_1, \qquad |E_K| = R_2,$$
$$|E_L| = R_3, \qquad |E_M| = R_4;$$
(25)

$$|E_{H+K}| = R_{12}, \qquad |E_{K+L}| = R_{23},$$
$$|E_{L+H}| = R_{31};$$
(26)

and

$$H + K + L + M = 0.$$
(27)

Then the quartet (21) is a structure invariant which, as a function of
the primitive random variable (H,K,L,M), is itself a random variable.
Denote by

$$P_{1/7} = P(\Phi) |R_1, R_2, R_3, R_4, R_{12}, R_{23}, R_{31})$$
(28)

the conditional probability distribution of the quartet (21), given the
seven magnitudes in its second neighborhood, (25) and (26). The
explicit formula for $P_{1/7}$ has been found.

    Figures 3 to 5 show the distribution (28) (solid line ———) for
typical values of the seven parameters (25) and (26). For comparison
the distribution (23) (broken line - - -) is also shown. Since the
magnitudes $|E|$ have been obtained from a real structure, comparison with
the true value of the quartet is also possible. As already emphasized,

the distribution (23) always has a unique maximum at $\Phi = 0$. The
distribution (28), on the other hand, may have a maximum at $\Phi = 0$, or π,
or any value between these extremes, as shown by Figures 3 to 5.
Roughly speaking the maximum of (28) occurs at 0 or π according as the
three parameters R_{12}, R_{23}, R_{31} are all large or all small, respectively.
These figures also clearly show the improvement which may result when,
in addition to the four magnitudes (25), the three magnitudes (26) are
also assumed to be known. Finally, in the special case that

$$R_{12} \sim R_{23} \sim R_{31} \sim 0 \tag{29}$$

the distribution (28) reduces to

$$P_{1/7} \sim \frac{1}{L} \exp(-2B'\cos\Phi), \tag{30}$$

where

$$B' = \frac{1}{\sigma_2^3}(3\sigma_3^2 - \sigma_2\sigma_4)R_1R_2R_3R_4, \tag{31}$$

and L is a suitable normalizing parameter. Clearly (31) has a unique
maximum at $\Phi = \pi$ (Fig. 5).

9. THE STRUCTURE SEMINVARIANTS

For all space groups other than P1 the origin may not be chosen
arbitrarily if the simplification permitted by the space group
symmetries is to be realized. For example, if a crystal has a center of
symmetry it is natural to place the origin at such a center while if a
two-fold screw axis, but no other symmetry element is present, the
origin would normally be situated on this symmetry axis. In such cases
the permissible origins are greatly restricted and it is therefore

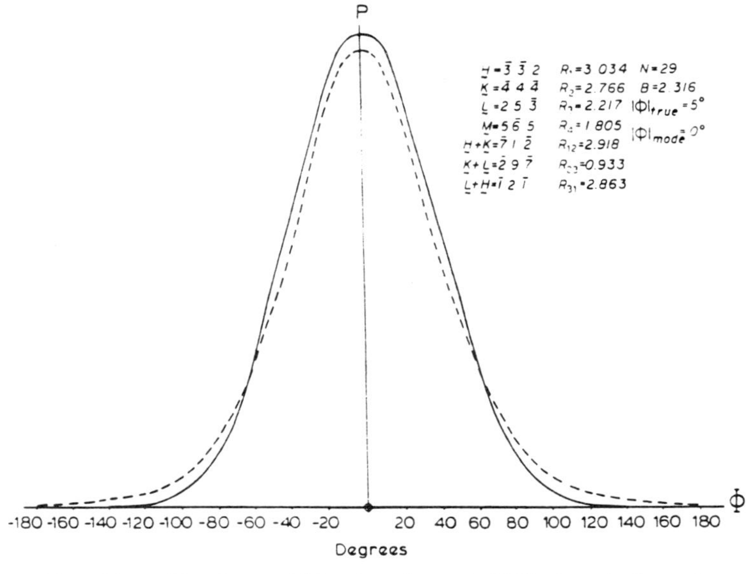

Figure 3. The distributions (28) (——) and (23) (---) for the values of
the seven parameters (25) and (26) shown. The mode of (28) is 0, of
(23) always 0.

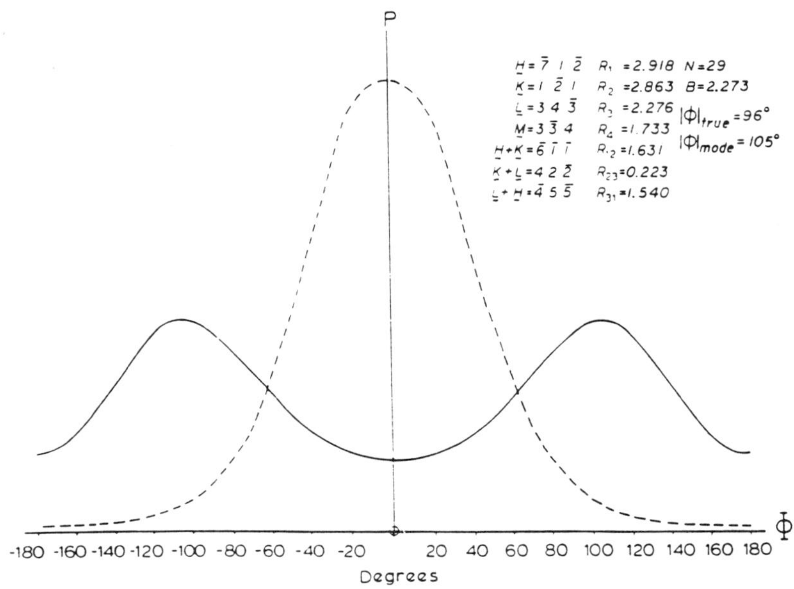

Figure 4.   The distributions (28) (——) and (23) (---) for the values of the seven parameters (25) and (26) shown.   The mode of (28) is 105°, of (23) always 0.

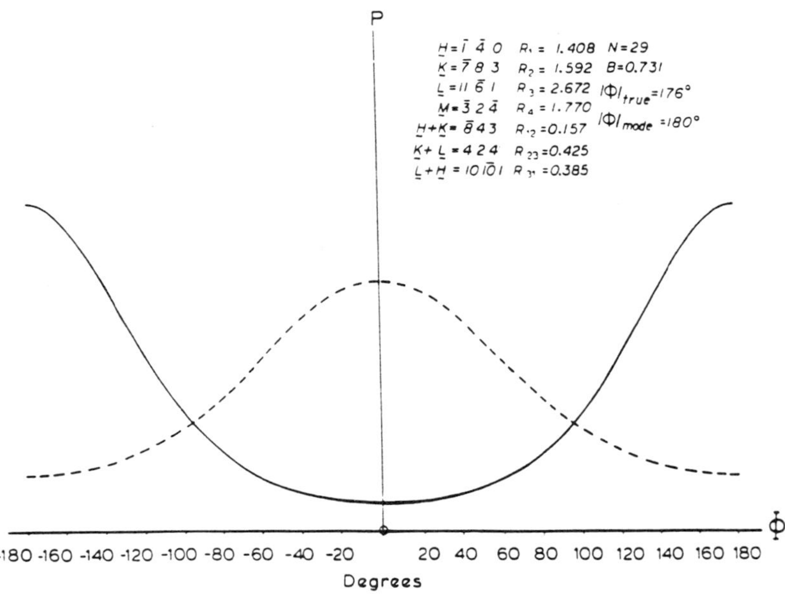

Figure 5.   The distributions (28) (——) and (23) (---) for the values of the seven parameters (25) and (26) shown.   The mode of (28) is 180°, of (23) always 0.

138

plausible to assume that many linear combinations of the phases will
remain unchanged in value when the origin is shifted only in the
restricted ways allowed by the space group symmetries. One is thus led
to the notion of the structure seminvariant, those linear combinations
of the phases whose values are independent of the choice of permissible
origin.

9.1 The Equivalence Concept

For any space group the coordinates of equivalent positions depend
upon the choice of origin. Hence the functional form for the geometric
structure factor also depends on the choice of origin. Two origins will
be said to be equivalent if they give rise to the same functional form
for the geometric structure factor. Alternatively, two points are
equivalent if they are geometrically situated in the same way with
respect to the symmetry elements. Thus, in the space group P1 all
points are equivalent; in $P\bar{1}$ all eight centers of symmetry are
equivalent to each other, but no other point is equivalent to any of
these eight; in $P2_1$ all points on any of the four two-fold screw axes
are equivalent to each other, but no other point is equivalent to a
point on a two-fold screw axis; in P4 all points on either of the two
four-fold axes are equivalent to each other, all points on either of the
two two-fold axes are equivalent to each other, but no point on a two-
fold axis is equivalent to any point on a four-fold axis. All points
equivalent to a given point are equivalent to each other and are said to
form an equivalence class.

9.2 The Primary Origin

Refer to "International Tables of X-Ray Crystallography", Vol. 1,
in order to define the primary origin for each space group.

9.3 The Permissible Origins

All points equivalent to the primary origin constitute the
permissible origins for each space group.

9.4 The Structure Seminvariants

The structure seminvariants are those linear combination of the
phases whose values are uniquely determined by the crystal structure
alone, no matter what the choice of permissible origin. Alternatively,
for a given functional form for the geometric structure factor, the
values of the structure seminvariants are determined by the structure
alone. A few examples will illustrate the method used to identify the
structure seminvariants in the different space groups.

If the origin of coordinates is shifted to a new point having
position vector $\mathbf{r}_0$ with respect to the old origin, then, from the
definition (equation (1)) of $E_{\mathbf{H}}$, it follows readily that the phase $\phi_{\mathbf{H}}$ of
the normalized structure factor $E_{\mathbf{H}}$ with respect to the old origin is
replaced by the new phase $\phi'_{\mathbf{H}}$ with respect to the new origin given by

$$\phi'_{\mathbf{H}} = \phi_{\mathbf{H}} - 2\pi\mathbf{H}\cdot\mathbf{r}_0. \tag{32}$$

9.4.1 Space Group $P\bar{1}$

Refer to "International Tables of X-Ray Crystallography", Vol. 1 to infer that there are eight permissible origins in $P\bar{1}$:

$$\left(000\right), \ \left(\tfrac{1}{2}\,00\right), \ \left(0\,\tfrac{1}{2}\,0\right), \ \left(00\,\tfrac{1}{2}\right),$$

$$\left(\tfrac{1}{2}\,\tfrac{1}{2}\,0\right), \ \left(\tfrac{1}{2}\,0\,\tfrac{1}{2}\right), \ \left(0\,\tfrac{1}{2}\,\tfrac{1}{2}\right), \ \left(\tfrac{1}{2}\,\tfrac{1}{2}\,\tfrac{1}{2}\right). \tag{33}$$

The one-phase structure seminvariants $\phi_{\mathbf{H}}$. From (32) it follows that $\phi'_{\mathbf{H}} = \phi_{\mathbf{H}}$ for all permissible origins $\mathbf{r}_0$ if and only if $\mathbf{H}\cdot\mathbf{r}_0$ is an integer, i.e. in view of (33), if and only if the components of $\mathbf{H}$ are even. Hence the single phase $\phi_{\mathbf{H}}$ is a structure seminvariant in $P\bar{1}$ if and only if the three components of $\mathbf{H}$ are even integers.

The two-phase structure seminvariants $\phi_{\mathbf{H}}+\phi_{\mathbf{K}}$. Follow the argument in the previous paragraph to infer that the linear combination of two phases $\phi_{\mathbf{H}}+\phi_{\mathbf{K}}$ in $P\bar{1}$ is a structure seminvariant if and only if the three components of $\mathbf{H}+\mathbf{K}$ are even integers.

9.4.2 Space Groups P2 and P2$_1$; Unique axis b

The permissible origins are

$$(0\ y\ 0), \ (0\ y\ \tfrac{1}{2}), \ (\tfrac{1}{2}\ y\ 0), \ (\tfrac{1}{2}\ y\ \tfrac{1}{2}), \tag{34}$$

where y is arbitrary. Then from (32) infer that the single phase ϕ_{hkl} is a structure seminvariant if and only if h and l are both even and $k = 0$; the linear combination of two phases

$$\phi_{h_1 k_1 l_1} + \phi_{h_2 k_2 l_2},$$

is a structure seminvariant if and only if h_1+h_2 and l_1+l_2 are both even and $k_1+k_2 = 0$; etc.

9.4.3 Space Group P4

The permissible origins are

$$(0\ 0\ z), \ (\tfrac{1}{2}\ \tfrac{1}{2}\ z), \tag{35}$$

where z is arbitrary. It follows from (32) that the single phase ϕ_{hkl} is a structure seminvariant if and only if $h+k$ is even and $l=0$; the linear combination of two phases

$$\phi_{h_1 k_1 l_1} + \phi_{h_2 k_2 l_2},$$

is a structure seminvariant if and only if $(h_1 + k_1) + (h_2 + k_2)$ is even and $l_1 + l_2 = 0$; etc.

10. THE TWO-PHASE STRUCTURE SEMINVARIANT T IN $P\bar{1}$

10.1 The Extensions of T

In view of § 9.4.1 the linear combination of two phases

$$T = \phi_{\mathbf{H}} + \phi_{\mathbf{K}} \tag{36}$$

is a structure seminvariant in the space group $P\bar{1}$ if and only if the three components of $\mathbf{H}+\mathbf{K}$ are even. In symbols

$$\mathbf{H} + \mathbf{K} \equiv 0(\mathrm{mod}\ \omega_s) \tag{37}$$

where ω_s, the seminvariant modulus in $P\bar{1}$, is defined by

$$\omega_s = (2,2,2). \tag{38}$$

In this case the three components of $\mathbf{H}-\mathbf{K}$ are also even so that the components of each of

$$h = \frac{1}{2} (H + K) \tag{39}$$

$$h_1 = \frac{1}{2} (H - K) \tag{40}$$

are integers and the linear combination of two phases

$$T_1 = \phi_H - \phi_K \tag{41}$$

is also a structure seminvariant. Furthermore, because of the space group dependent relationships among the phases,

$$T = T_1. \tag{42}$$

Embed the structure seminvariants T and T_1 in the respective quartets

$$Q = T + \phi_{\bar{h}} + \phi_{\bar{h}} = \phi_H + \phi_K + \phi_{\bar{h}} + \phi_{\bar{h}} \tag{43}$$

$$Q_1 = T_1 + \phi_{\bar{h}_1} + \phi_{\bar{h}_1} = \phi_H - \phi_K + \phi_{\bar{h}_1} + \phi_{\bar{h}_1} \tag{44}$$

so that, in view of (39) and (40), Q and Q_1 are structure invariants called the extensions of T. They are in fact special quartets since two of the four phases in each of them are equal. Again, because every phase in $P\bar{1}$ must be equal to 0 or π, it follows that

$$Q = Q_1 = T = T_1. \tag{45}$$

Hence it makes sense to define the neighborhoods of the structure seminvariant T in terms of the neighborhoods of its extensions Q and Q_1 which are well known (§ 7). In this way one reduces the probabilistic theory of the two-phase structure seminvariant T to that of the quartets.

10.2 The First Neighborhoods of the Extensions

Refer to (10) to infer that the first neighborhood of the special quartet Q consists of the three magnitudes

$$|E_H|, \quad |E_K|, \quad |E_h| \tag{46}$$

and the first neighborhood of the special quartet Q_1 of the three magnitudes

$$|E_H|, \quad |E_K|, \quad |E_{h_1}|. \tag{47}$$

10.3 The First Neighborhood of the Structure Seminvariant T

Refer to (46) and (47) in order to define the first neighborhood of the structure seminvariant T to consist of the four magnitudes

$$|E_H|, \quad |E_K|, \quad |E_h|, \quad |E_{h_1}|. \tag{48}$$

10.4 The Joint Probability Distribution of the Four Structure Factors E_H, E_K, E_h, E_{h_1} the Magnitudes of Which Constitute the First Neighborhood of T.

A crystal structure in $P\bar{1}$, consisting of N identical atoms in the unit cell, is supposed to be fixed. The twofold Cartesian product W × W of reciprocal space W is defined to be the collection of all ordered pairs (H,K) where H and K are reciprocal lattice vectors. The ordered pair (H,K) is assumed to be the primitive random variable uniformly distributed over the subset of $W \times W$ defined by (37). Then the four structure factors

$$E_{\mathbf{H}}, \ E_{\mathbf{K}}, \ E_h, \ E_{h_1}, \tag{49}$$

as functions of the primitive random variable $(\mathbf{H}, \mathbf{K})$, are themselves random variables. Denote by

$$P = P(S_1, \ S_2, \ s, \ s_1) \tag{50}$$

the joint probability distribution of the four structure factors (49) which, because the space group is $P\bar{1}$, are all real. Then, following methods which are by now standard, one finds

$$P = \frac{1}{2\pi^2} \exp \left\{ -\frac{1}{2} \left(S_1^{\ 2} + S_2^{\ 2} + s^2 + s_1^{\ 2} \right) + \frac{1}{N^{1/2}} \times \right.$$

$$\left. \left(S_1 + S_2 \right) s s_1 - \frac{1}{2N} S_1 S_2 \left(s^2 + s_1^{\ 2} \right) \right\} \left\{ 1 + 0 \left(\frac{1}{N} \right) \right\} \tag{51}$$

where $0(1/N)$ denotes terms of order $1/N$ or higher in which terms of order $1/N$ contain only even powers of S_1, S_2, s, s_1. Observe that the variables S_1, S_2, s, s_1 of (51) range over all real values from $-\infty$ to $+\infty$. The distribution (51) leads directly to the joint conditional probability distribution of the pair $\phi_{\mathbf{H}}$, $\phi_{\mathbf{K}}$, given the magnitudes of the four structure factors (49), as shown next.

10.5 <u>The Joint Conditional Probability Distribution of the Pair of Phases</u> $\phi_{\mathbf{H}}$, $\phi_{\mathbf{K}}$, <u>Given the Four Magnitudes</u> $\left|E_{\mathbf{H}}\right|$, $\left|E_{\mathbf{K}}\right|$, $\left|E_h\right|$, $\left|E_{h_1}\right|$

Suppose again that a crystal structure consisting of N identical atoms per unit cell in the space group $P\bar{1}$ is fixed and the four non-negative numbers R_1, R_2, r, r_1 are also specified. Assume now that the ordered pair $(\mathbf{H}, \mathbf{K})$ of reciprocal lattice vectors is a random variable which is uniformly distributed over the subset of the Cartesian product $W \times W$ defined by

$$\left|E_{\mathbf{H}}\right| = R_1, \ \left|E_{\mathbf{K}}\right| = R_2, \ \left|E_h\right| = r, \ \left|E_{h_1}\right| = r_1 \tag{52}$$

and (37). Then $\phi_{\mathbf{H}}$ and $\phi_{\mathbf{K}}$, the phases of the normalized structure factors $E_{\mathbf{H}}$ and $E_{\mathbf{K}}$, as functions of the primitive random variables $\mathbf{H}$ and $\mathbf{K}$, are themselves random variables. Denote by

$$P(\Phi_1, \ \Phi_2 | R_1, \ R_2, \ r, \ r_1) \tag{53}$$

the joint conditional probability distribution of the two phases $\phi_{\mathbf{H}}$, $\phi_{\mathbf{K}}$, given (52) and (37). Then $P(\Phi_1, \ \Phi_2 | R_1, \ R_2, \ r, \ r_1)$ is found from (51) by fixing the magnitudes of S_1, S_2, s, s_1 in accordance with the scheme

$$\left|S_1\right| = R_1, \ \left|S_2\right| = R_2, \ \left|s\right| = r, \ \left|s_1\right| = r_1 \tag{54}$$

i.e.

$$S_1 = R_1 \cos \Phi_1, \ S_2 = R_2 \cos \Phi_2, \ s = r \cos \phi, \ s_1 = r_1 \cos \phi_1, \tag{55}$$

where ϕ and ϕ_1 are the variables associated with the phases ϕ_h and ϕ_{h_1}, respectively, summing with respect to s and s_1 over their two possible signs (+ and -) or, equivalently, summing with respect to ϕ, ϕ_1 over their two possible values (0 and π), and multiplying the result by a suitable normalizing factor. Carrying out these summations one finally obtains, correct up to and including terms of order $1/N$ [since $0(1/N)$ of (51) consists of all terms of order $1/N$ or higher in which the terms of order 1/N contain only even powers of S_1, S_2, s, s_1].

$$P(\Phi_1, \ \Phi_2 \,|\, R_1, \ R_2, \ r, \ r_1) \; \simeq \; \frac{1}{K} \ \exp \left\{ - \frac{1}{2N} \ R_1 R_2 \ \cos\left(\Phi_1 + \Phi_2\right)\left(r^2 + r_1^2\right) \right\} \times$$

$$\cosh \left\{ \frac{rr_1}{N^{1/2}} \left(R_1 \ \cos \ \Phi_1 + R_2 \ \cos \ \Phi_2\right) \right\} \tag{56}$$

where K is a suitable normalizing constant, independent of Φ_1 and Φ_2. Although K is readily found by summing (56) over the four possible values of Φ_1, Φ_2 (each takes on the two values $0, \pi$) and setting the result equal to unity, the value of this normalizing parameter is not needed for the present purpose and is therefore not derived explicitly. Although (56) depends on the values of Φ_1 and Φ_2, it is readily confirmed that it actually is a function of $\Phi_1 + \Phi_2$ only, since $|\cos \Phi_1| = |\cos \Phi_2| = 1$ and the hyperbolic cosine is an even function of its argument. Hence (56) leads directly to the conditional distribution, give (52) and (37), of the sum $\phi = \phi_h + \phi_k$, as is shown next.

10.6   The Conditional Probability Distribution of the Structure Seminvariant $T = \phi_H + \phi_K$, Given the Four Magnitudes $|E_H|$, $|E_K|$, $|E_h|$, $|E_{h_1}|$.

Under the same hypotheses as in § 10.5, the structure seminvariant

$$T \; = \; \phi_H + \phi_K \tag{36}$$

is a random variable whose conditional probability distribution, given (52), $P(\Phi|R_1, R_2, r, r_1)$, is readily found from (56). Thus, correct up to and including terms of order 1/N, the major result of this section is given by

$$P(\Phi \,|\, R_1, \ R_2, \ r, \ r_1) \; \simeq \; \frac{1}{L} \ \exp \left\{ - \frac{1}{2N} \ (r^2 + r_1^2)\cos \ \Phi \right\} \times$$

$$\cosh \left\{ \frac{rr_1}{N^{1/2}} \left(R_1 + R_2 \ \cos \ \Phi\right) \right\} \tag{57}$$

where

$$L \; = \; \exp \left\{ - \frac{R_1 R_2}{2N} \left(r^2 + r_1^2\right) \right\} \ \cosh \left\{ \frac{r_1 r_2}{N^{1/2}} \left(R_1 + R_2\right) \right\}$$

$$+ \ \exp \left\{ \frac{R_1 R_2}{2N} \left(r^2 + r_1^2\right) \right\} \ \cosh \left\{ \frac{r_1 r_2}{N^{1/2}} \left(R_1 - R_2\right) \right\} . \tag{58}$$

Observe that, since Φ takes on only the two values $0, \pi$, the distribution (57) (like (56)) is discrete.

If one denotes by $P_+(P_-)$ the conditional probability, given (52) that

$$\phi \; = \; \phi_h + \phi_k \; = \; 0(\pi), \tag{59}$$

or that

$$\cos \phi = +1(-1) \tag{60}$$

or that

$$E_h E_k \text{ be positive (negative),} \tag{61}$$

then (57) is replaced by the more suggestive

$$P_{\pm} \simeq \frac{1}{M} \exp\left\{\mp \frac{1}{2N} R_1 R_2 \left(r^2 + r_1^2\right)\right\} \cosh\left\{\frac{1}{N^{1/2}} rr_1 \left(R_1 \pm R_2\right)\right\} \tag{62}$$

where upper (lower) signs go together and

$$M = \exp\left\{-\frac{1}{2N} R_1 R_2 \left(r^2 + r_1^2\right)\right\} \cosh\left\{\frac{1}{N^{1/2}} rr_1 \left(R_1 + R_2\right)\right\} +$$

$$\exp\left\{\frac{1}{2N} R_1 R_2 \left(r^2 + r_1^2\right)\right\} \cosh\left\{\frac{1}{N^{1/2}} rr_1 \left(R_1 - R_2\right)\right\}. \tag{63}$$

Equation (62) shows that if r and r_1 are both large then $P_+ > \frac{1}{2}$ and T is
probably equal to 0. If, on the other hand, one of r, r_1 is large and
the other small then $P_+ < \frac{1}{2}$ and T is probably equal to π, as shown next.

10.7 The case that $rr_1 \simeq 0$. In the case that

$$\left|E_h E_{h_1}\right| = rr_1 \simeq 0 \tag{64}$$

Eqs. (62) and (63) reduce to

$$P_{\pm} \simeq \frac{1}{M} \exp\left\{\mp \frac{1}{2N} R_1 R_2 \left(r^2 + r_1^2\right)\right\}, \tag{65}$$

$$M = 2 \cosh\left\{\frac{1}{2N} R_1 R_2 \left(r^2 + r_1^2\right)\right\} \tag{66}$$

so that, in this special case, $P_+ < \frac{1}{2}$, T is probably equal to π, and the

larger the value of $\frac{1}{2N} R_1 R_2 \left(r^2 + r_1^2\right)$ the more likely it is that $T = \pi$.
The reader should confirm that this result could have been anticipated
by plausible reasoning based on the quartet extensions of T.

   Although this section has been restricted to the two phase
structure seminvariant T in P$\overline{1}$ and to the equal atom case, it should be
stressed that the method used is perfectly general and may be employed
to define the neighborhoods of the structure seminvariants in all the
space groups and to derive the related conditional probability
distributions assuming as known magnitudes |E| in appropriate
neighborhoods of the structure seminvariants.

SUMMARY

Major emphasis has been placed on the neighborhood principle. The conditional probability distribution of a structure seminvariant T, given the magnitudes $|E|$ in any of its neighborhoods, yields a reliable estimate for T in the favorable case that the variance of the distribution happens to be small. Since the structure seminvariants are the essential link between magnitudes $|E|$ and phases ϕ, probabilistic methods are seen to play the central role in the solution of the phase problem. Owing to limitations of space, only the simplest cases have been treated in this article. However a much larger class of distributions is presently available. Not only have these distributions already proven to be of great value in the applications, particularly for complex structures when data sets have been limited in number and quality, but preliminary calculations strongly suggest that the distributions of the higher order structure seminvariants, particularly if one takes into account whatever elements of symmetry may be present, will play an important role in future applications.

THE METHOD OF REPRESENTATIONS OF STRUCTURE SEMINVARIANTS

I: THE GENERAL BACKGROUND

C. Giacovazzo

Dipartimento Geomineralogico
Campus Universitario
70124 BARI - Italia

1 - SYMBOLS AND ABBREVIATIONS

m: order of the space group ($\equiv$ number of symmetry operators C_s)

$C_s \equiv (R_s, T_s)$: sth symmetry operator. R_s is its rotational part, T_s its translational part

F_h: structure factor with vectorial index $\mathbf{h}$

ε_h: statistical Wilson's coefficient

$E_h = R_h \exp(i\phi_h)$: normalized structure factor

s.i.: structure invariant

s.s.: structure seminvariant

2 - INTRODUCTION

The solution of a crystal structure by Direct Methods usually requires a minimal amount of prior information, like:
a) the atomic nature of the electron density
b) a relatively large set of diffraction data
c) the chemical content of the unit cell
d) the space group symmetry.

Additional prior information of various kind may also be exploited like:
e) the geometry of the molecule, a well positioned or well oriented molecular fragment, the nature of a possible pseudo-translational symmetry, interatomic vectors, sets of diffraction data from isomorphous structures, etc.. The basic information quoted at the points a)-d) is used by Direct Methods in order to create numerous phase relationships among structure factors (triplet-, quartet-, quintet-, ... s.i.'s and 1-,2-,3-, ... phase s.s.'s). Their reliability is usually calculated via conditional distribution functions such as

$$P(\Phi|R_1, \ldots, R_n) \equiv P(\Phi|\{R\}).$$

The set $\{R\}$ giving information on Φ is not unique: in principle several sets could exist each of them giving rise to a different dependence of Φ on the selected magnitudes. The formulation of the "nested neighbourhoods principle" (Hauptman, 1976) first fixed the idea of defining a sequence of sets of reflections each contained within the succeeding one and having the property that the cosine invariant or seminvariant may be estimated via the magnitudes constituting any neighbourhood.

A more general theory (the so called "method of representations of structure invariants or seminvariants") has been independently described by Giacovazzo (1977, 1980). It is able to arrange, for any s.i. or s.s., the set of reflections in a sequence of subsets, each contained in the succeeding one, whose order is that of the expected effectiveness (in the statistical sense) for the estimation of Φ. A summary of such a theory is the main aim of this paper. How the additional information quoted at point e) may be exploited by Direct Methods in the framework of the theory of representations will be described in paper II.

3 - THE REPRESENTATIONS OF A STRUCTURE INVARIANT

We give the basic definitions of the method and essential information on its applications:

3.1. - <u>The first representation of a structure invariant</u>

Let

$$\Phi = \phi_{\mathbf{h}_1} + \phi_{\mathbf{h}_2} + \ldots + \phi_{\mathbf{h}_n} \qquad (\mathbf{h}_1 + \mathbf{h}_2 + \ldots + \mathbf{h}_n = 0)$$

be a s.i.. If the symmetry is higher than triclinic a number of symmetry operators may be found in favourable cases such that one or more s.i.'s

$$\psi_1 = \phi_{\mathbf{h}_1 R_s} + \phi_{\mathbf{h}_2 R_t} + \ldots + \phi_{\mathbf{h}_n R_\nu}$$

may be constructed, where R_s, R_t, $\ldots$, R_ν do not all coincide with identity. The collection of the distinct s.i.'s ψ_1 obtained when R_s, R_t, $\ldots$, R_ν vary over the set of the m rotation matrices of the space group is defined to be the <u>first</u> <u>representation</u> of Φ and will be denoted by $\{\psi\}_1$. The <u>first</u> <u>phasing</u> <u>shell</u> of ψ is given by the collection $\{B\}_1$ of magnitudes which are basis magnitudes (i.e., $R_{\mathbf{h}_1}$, $R_{\mathbf{h}_2}$, $\ldots$, $R_{\mathbf{h}_n}$) or cross magnitudes of at least one s.i. ψ_1. The set of cross magnitudes is defined by the cross vectors

$$m_1 \mathbf{h}_1 R_s + m_2 \mathbf{h}_2 R_t + \ldots + m_n \mathbf{h}_n R_\nu \qquad (m_p = 0, 1)$$

all of which are linear combinations of the basis vectors.

3.2 - <u>The upper representations of a s.i.</u>

For any ψ_1 belonging to $\{\psi\}_1$ let us construct the s.i.'s

$$\psi_2 = \psi_1 + \phi_{\mathbf{k}} - \phi_{\mathbf{k}} \tag{1}$$

where $\mathbf{k}$ is a free vector. The collection of the s.i.'s ψ_2 obtained when ψ_1 varies in the $\{\psi\}_1$ set and $\mathbf{k}$ is free is denoted by $\{\psi\}_2$ and is the second representation of Φ.

The second phasing shell $\{B\}_2$ of Φ is the collection of magnitudes which are basis or cross magnitudes of at least one ψ_2. Likewise

$$\{\psi\}_3 = \{\psi_2 + \phi_1 - \phi_1\}$$

where 1 is a free vector is the third representation of Φ. The procedure and the definition are recursive.

In this paper we will be interested, among the various s.i.'s, only to representations of triplet and quartet invariants. The reader will easily find algebraic rules holding for upper order invariants in a recursive way.

3.3 – The first representation of a triplet invariant

The representations theory makes full use of the symmetry and is the natural background for the application of the method of joint probability distribution functions. The role of the symmetry in the estimation of triplet phases has been studied by various authors (Giacovazzo, 1974a, b; Pontenagel and Krabbendam, 1983; Han and Langs, 1988; Burla and Giacovazzo, 1991). According to Burla and Giacovazzo (1991) we make three examples to show the role of the representation theory and its interaction with space group symmetry:

i) in $P3_1 [(x,y,z); (\bar{y},x-y,z+1/3); (-x+y,\bar{x},z+2/3)]$ the first representation of the triplet

$$\Phi = \phi_{30\bar{3}} + \phi_{\bar{3}31} + \phi_{0\bar{3}2}$$

contains, besides $\psi_1 = \Phi$, also

$$\psi'_1 = \phi_{30\bar{3}} + \phi_{0\bar{3}1} + \phi_{\bar{3}32} = \phi_{30\bar{3}} + \phi_{(\bar{3}31)R_3} + \phi_{(0\bar{3}2)R_2} = \Phi - 2\pi/3.$$

ψ_1 and ψ'_1 are in some way inconsistent if both of them are expected, according to Cochran (1955), to be close to zero.

ii) In $P4_1 2_1 2$ the triplet

$$\Phi = \phi_{1,-1,-19} + \phi_{1,1,12} + \phi_{-2,0,7}$$

is expected to be close to zero is no special care of symmetry is taken. However the values of $\phi_{1,-1,-19}$, $\phi_{1,1,12}$ and $\phi_{-2,0,7}$ are symmetry restricted to $(0,\pi)$, $(0,\pi)$, $(-\pi/4,3\pi/4)$ respectively, so that the triplet phase Φ is itself restricted to $-\pi/4$ or to $3\pi/4$. Such an inconsistency between Cochran's and symmetry rules is again due to the fact that the first representation of Φ contains besides $\psi_1 \equiv \psi$, also

$$\psi'_1 = \phi_{1,-1,-19} + \phi_{-1,-1,12} + \phi_{027} = \Phi + \pi/2$$

which is a triplet symmetry related to but inconsistent with Φ.

iii) In $P2_1 2_1 2_1$ the triplets

$$\psi_1 = \Phi = \phi_{h_1,k_1,0} + \phi_{-h_1,0,1_1} + \phi_{0,-k_1,-1_1}$$

and

$$\psi'_1 = \phi_{h_1,k_1,0} + \phi_{-h_1,0,-1_1} + \phi_{0,-k_1,1_1} = \Phi + \pi(h_1+k_1+1_1)$$

constitute the first representation of Φ. For odd values of $(h_1+k_1+1_1)$ both ψ_1 and ψ'_1 are symmetry restricted to $\pm\pi/2$: furthermore they differ by π, so violating Cochran's rule.

If probabilistic methods for phase estimation are applied to this algebraic background it may be seen that Cochran's parameter $G = 2R_{h_1}R_{h_2}R_{h_3}/\sqrt{N}$ has to be replaced by the complex parameter

$$G' = G \frac{1}{(\varepsilon_{h_1}\varepsilon_{h_2}\varepsilon_{h_3})^{1/2}} \sum_j \exp(+2\pi i\Delta_j) \qquad (2)$$

where the summation goes over the set of equivalent triplets constituting the first representation of Φ, $-\Delta_j$ is the phase shift of the j-th triplet with respect to the representative triplet Φ. The value Δ_f defined by

$$\Delta_f = (\sum_j \sin 2\pi\Delta_j)/(\sum_j \cos 2\pi\Delta_j) \qquad (3)$$

is the expected value of the phase Φ. In accordance with (2) and (3) the Φ's of the examples i) and ii) are expected to be equal to $\pi/3$ and $-\pi/4$ respectively, Φ of example iii) has a vanishing reliability.

3.4 – The second representation of a triplet invariant

According to (1) the second representation of Φ is the collection

$$\{\psi\}_2 = \left\{\psi_1 + \phi_{kR_i} - \phi_{kR_i}\right\} \qquad i = 1,2,\ldots,m$$

where ψ_1 is any of the triplets constituting the first representation of Φ_3 and $\mathbf{k}$ is a free vector. Each special quintet depends depends (in addition to the basis magnitudes R_h, R_h, R_h, R_k) on the 6m cross magnitudes

$$R_{h_1+kR_1}, \; R_{h_1-kR_1}, \; R_{h_2+kR_1}, \; R_{h_2-kR_1}, \; R_{h_3+kR_1}, \; R_{h_3-kR_1}$$
$$R_{h_1+kR_2}, \; R_{h_1-kR_2}, \; R_{h_2+kR_2}, \; R_{h_2-kR_2}, \; R_{h_3+kR_1}, \; R_{h_3-kR_2}$$
$$\cdots\cdots\cdots\cdots\cdots\cdots\cdots\cdots\cdots\cdots\cdots\cdots\cdots\cdots$$
$$R_{h_1+kR_m}, \; R_{h_1-kR_m}, \; R_{h_2+kR_m}, \; R_{h_2-kR_m}, \; R_{h_3+kR_m}, \; R_{h_3-kR_m}$$

In this case we may write

$$\{B\}_2 = \{R_{h_1},R_{h_2},R_{h_3},R_k,R_{h_1\pm kR_i},R_{h_2\pm kR_i},R_{h_3\pm kR_i}, \; i = 1,2,\ldots,m\}$$

The conditional probability $P(\Phi|\{B\}_2)$ now provides an estimate for Φ_3 in any space group. The probabilistic formula is quoted in literature as P_{10} formula (Cascarano et al., 1984).

3.5 - The first representation of a quartet invariant

Given the quartet

$$\Phi = \phi_{h_1} + \phi_{h_2} + \phi_{h_3} + \phi_{h_4} \qquad\qquad (h_1+h_2+h_3+h_4 = 0)$$

we look for the possible supplementary quartets

$$\psi = \phi_{h_1 R_s} + \phi_{h_2 R_t} + \phi_{h_3 R_u} + \phi_{h_4 R_v} \qquad\qquad (h_1 R_s + h_2 R_t + h_3 R_u + h_4 R_v = 0).$$

The collection of such quartets is the first representation $\{\psi\}_1$ of the quartet Φ. For each favourable set R_s, R_t, R_u, R_v the cross vectors

$$m_1 h_1 R_s + m_2 h_2 R_t + m_3 h_3 R_u + m_4 h_4 R_v \qquad\qquad (m_i = 0, 1)$$

may be calculated: together with the basis vectors h_1, h_2, h_3, h_4 they will give rise to the first phasing shell $\{B\}_1$.

The reader will easily find that in the space group Pmmm the first phasing shell of the quartet

$$\Phi = \phi_{123} + \phi_{234} + \phi_{5\bar{6}\bar{2}} + \phi_{\bar{8}1\bar{5}}$$

is

$$\{B\}_1 = \{R_{123}, R_{234}, R_{5\bar{6}\bar{2}}, R_{\bar{8}1\bar{5}}, R_{357}, R_{6\bar{4}1}, R_{7\bar{3}2}\}.$$

The conditional distribution $P(\Phi|\{B\}_1)$ in P1 and P$\bar{1}$ has been independently calculated by Hauptman (1975a, b) and Giacovazzo (1975, 1976a). The conclusive result by Giacovazzo is

$$P(\Phi|\{B\}_1) \cong [2\pi I_o(G)]^{-1} \exp(G \cos\Phi)$$

where $G = C(\varepsilon_5 + \varepsilon_6 + \varepsilon_7 + 1)/(1 + Q)$,
$\quad C = R_1 R_2 R_3 R_4$,
$\quad Q = [(\varepsilon_1\varepsilon_2 + \varepsilon_3\varepsilon_4)\varepsilon_5 + (\varepsilon_1\varepsilon_3 + \varepsilon_2\varepsilon_4)\varepsilon_6 + (\varepsilon_1\varepsilon_4 + \varepsilon_2\varepsilon_3)\varepsilon_7]/2N$,
$\quad \varepsilon = R^2 - 1$.

Negative value of G (and therefore negative estimated quartets) will occur for small values of the cross terms R_5, R_6, R_7.

A general theory of quartets, exploiting their complete first representation and therefore valid in all the space groups, was given by Giacovazzo (1976b). The joint probability distribution functions can now take advantage, in estimating Φ, of the prior information on more than 7 magnitudes. For example, if the representation theory is applied in Pmmm to the quartet

$$\Phi = \phi_{123} + \phi_{\bar{1}5\bar{3}} + \phi_{\bar{1}\bar{5}8} + \phi_{1\bar{2}\bar{8}}$$

then 16 cross-vectors arise according to Table 1, and the first phasing shell of Φ will contain 20 magnitudes.

Table 1.   List of   quartets symmetry   equivalent to
$\Phi = \Phi_1$ in the class mmm

Basis vectors	Cross-vectors
$(\underline{1},2,3)(\overline{1},5,\overline{3})(\overline{1},\overline{5},8)(1,\underline{\overline{2}},\overline{8})$	$(0,7,0)(\underline{0},\overline{3},11)(\overline{2},0,5)$
$(\overline{1},2,3)(1,5,\overline{3})(\overline{1},\overline{5},8)(1,\underline{\overline{2}},\overline{8})$	$(0,7,0)(\overline{2},\overline{3},11)\underline{(0},0,5)$
$(\underline{1},2,\overline{3})(\overline{1},5,3)(\overline{1},\overline{5},8)(1,\underline{\overline{2}},\overline{8})$	$(0,7,0)(\underline{0},\overline{3},5)(\overline{2},0,11)$
$(\overline{1},2,\overline{3})(1,5,3)(\overline{1},\overline{5},8)(1,\underline{\overline{2}},\overline{8})$	$(0,7,0)(\overline{2},\overline{3},5)(0,0,11)$
$(\overline{1},2,3)(\overline{1},5,\overline{3})(1,\overline{5},8)(1,\underline{\overline{2}},\overline{8})$	$(\overline{2},7,0)(0,3.11)(0,0,5)$
$(1,2,3)(\overline{1},\underline{5},\overline{3})(\overline{1},5,8)(1,\underline{\overline{2}},\overline{8})$	$(0,\overline{3},0)(\underline{0},7.11)(\overline{2},0,5)$
$(\overline{1},2,3)(\overline{1},\overline{5},\overline{3})(\overline{1},5,8)(1,\underline{\overline{2}},\overline{8})$	$(\underline{0},3,0)(\overline{2},\underline{7},11)(0,0,5)$
$(\overline{1},2,\overline{3})(\overline{1},5,3)(1,\overline{5},8)(1,\underline{\overline{2}},\overline{8})$	$(2,7,0)(0,\overline{3},5)(\underline{0},0,11)$
$(1,2,\overline{3})(\overline{1},\overline{5},3)(\overline{1},5,8)(1,\underline{\overline{2}},\overline{8})$	$(0,\overline{3}.0)(\underline{0},7,5)(\overline{2},0,11)$
$(\overline{1},2,\overline{3})(1,\overline{5},3)(\overline{1},5,8)(1,\underline{\overline{2}},\overline{8})$	$(\underline{0},\overline{3},0)(\overline{2},7,5)(0,0,11)$
$(\overline{1},2,3)(\overline{1},\overline{5},\overline{3})(1,5,8)(1,\underline{2},\overline{8})$	$(\overline{2},\overline{3},0)(0,7,11)(0,0,5)$

<u>3.6 - Upper representations of a quartet invariant</u>

The formulas for the upper representations are recursive: reader will readily apply definitions given in the section 3.2.

4 - THE FIRST REPRESENTATION OF A STRUCTURE SEMINVARIANT

Let us consider the combination of phases

$$\Phi = \phi_{u_1} + \phi_{u_2} + \dots + \phi_{u_n} \tag{4}$$

If at least one phase ϕ_h and two symmetry operators C_p and C_q exist such that

$$\psi_1 = \phi_{u_1 R_s} + \phi_{u_2 R_t} + \dots + \phi_{u_n R_\nu} + \phi_{h_1 R_p} - \phi_{h_1 R_q} \tag{5}$$

is a s.i., then Φ is a s.s. of first rank. If any of the ψ_1's is estimated, also Φ is estimated: indeed

$$\psi_1 = \Phi + \Delta$$

where Δ is a phase shift arising because of translation symmetry. The set $\{\psi\}_1$ is called the first representation of Φ.

Suppose now that Φ is a s.s. for which (5) cannot be stated: then two phases ϕ_{h_1} and ϕ_{h_2} and four symmetry operators C_p, C_q, C_i, C_j exist in principle such that

$$\psi_1 = \phi_{h_1 R_s} + \phi_{h_2 R_t} + \dots + \phi_{u_n R_\nu} + \phi_{h_1 R_p} - \phi_{h_1 R_q} + \phi_{h_2 R_i} - \phi_{h_2 R_j} \tag{6}$$

is a s.i.. Then Φ is a s.s. of second rank. In this case the first representation of Φ is the collection of s.i.'s $\{\psi\}$ defined by (6).

For both first and second rank s.s.'s the first phasing shell $\{B\}_1$ is the collection of the R magnitudes which are basis or cross-magnitudes of at least one $\psi_1 \in \{\psi\}_1$. From now on we will be mostly interested to s.s.'s of first rank. The reader is referred to Giacovazzo (1980) for information on s.s.'s of second rank.

The effective use of the s.s.'s in direct procedures for phase solution requires that: a) the linear combination of phases which are s.s.'s are identified; b) the rank of these s.s.'s is recognized; c) the first phasing shell is readily recognized.

The operations for a) do not present any difficulty. In fact, the algebraic conditions to be satisfied were clearly stated by Hauptman and Karle (1953, 1956, 1959) and, for centred cell, by Giacovazzo (1974c). The operations for b) present major difficulties. The following theorem is basic:

"If Φ as given by (4) is a s.s. of first rank then there are at least n vectors $\mathbf{h}_1$, $\mathbf{h}_2$, ..., $\mathbf{h}_n$ and n matrices R_α R_β, R_γ, ..., R_ν such that the system

$$\mathbf{h}_1 - \mathbf{h}_2 R_\beta = \mathbf{u}_1 \qquad\qquad (7.1)$$

$$\mathbf{h}_2 - \mathbf{h}_3 R_\gamma = \mathbf{u}_2 \qquad\qquad (7.2)$$

$$\mathbf{h}_3 - \mathbf{h}_4 R_\delta = \mathbf{u}_3 \qquad\qquad (7.3)$$

$$\cdots\cdots\cdots\cdots \qquad\qquad \cdots\cdots$$

$$\mathbf{h}_{n-1} - \mathbf{h}_n R_\nu = \mathbf{u}_{n-1} \qquad\qquad (7.n-1)$$

$$\mathbf{h}_n - \mathbf{h}_1 R_\alpha = \mathbf{u}_n \qquad\qquad (7.n)$$

holds. Viceversa, if eqs. (7) are satisfied, then (4) is a s.s. of the first rank."

The above theorem ensures that the more general expression for the s.s.'s of first rank are:

1) $\Phi = \phi_{\mathbf{u}} = \phi_{\mathbf{h}(I-R_\alpha)}$ for 1-phase s.s.'s

2) $\Phi = \phi_{\mathbf{u}_1} + \phi_{\mathbf{u}_2} = \phi_{\mathbf{h}_1-\mathbf{h}_2 R_\beta} + \phi_{\mathbf{h}_2-\mathbf{h}_1 R_\alpha}$ for 2-phase s.s.'s

3) $\Phi = \phi_{\mathbf{u}_1} + \phi_{\mathbf{u}_2} + \phi_{\mathbf{u}_3} = \phi_{\mathbf{h}_1-\mathbf{h}_2 R_\beta} + \phi_{\mathbf{h}_2-\mathbf{h}_3 R_\gamma} + \phi_{\mathbf{h}_3-\mathbf{h}_1 R_\alpha}$ for 3-phase s.s.'s

$$\cdots\cdots\cdots\cdots\cdots\cdots\cdots\cdots\cdots\cdots\cdots\cdots\cdots\cdots$$

This result ensures that, whatever the space group may be and for any s.s. of first rank Φ, the set of vectors $\{U\} = (\mathbf{u}_1, \mathbf{u}_2, ..., \mathbf{u}_n)$ can be expressed in terms of a fundamental set $\{\mathbf{h}\} = (\mathbf{h}_1, \mathbf{h}_2, ..., \mathbf{h}_n)$. In order to obtain $\{\mathbf{h}\}$ by $\{U\}$ we multiply (7.1) by I, (7.2) by R_β, (7.3) by $R_\gamma R_\beta$, ..., (7,n) by $R_\nu ... R_\gamma R_\beta$ and sum all of them. Then we find

$$\mathbf{h}_1(I - R_\alpha R_\nu \cdots R_\gamma R_\beta) = \mathbf{u}_1 + \mathbf{u}_2 R_\beta + \mathbf{u}_3 R_\gamma R_\beta + \cdots + \mathbf{u}_n R_\nu \cdots R_\gamma R_\beta. \qquad (8.1)$$

In a similar way one can obtain

$$\mathbf{h}_2(I - R_\beta R_\alpha R_\nu \cdots R_\delta R_\gamma) = \mathbf{u}_2 + \mathbf{u}_3 R_\gamma + \mathbf{u}_4 R_\delta R_\gamma + \cdots + \mathbf{u}_n R_\nu \cdots R_\delta R_\gamma$$

$$\cdots\cdots\cdots\cdots\cdots\cdots\cdots\cdots\cdots\cdots\cdots\cdots\cdots\cdots\cdots\cdots\cdots\cdots$$

$$+ \mathbf{u}_1 R_\alpha R_\nu \cdots R_\delta R_\gamma, \qquad\qquad (8.2)$$

$$\mathbf{h}_n(I - R_\nu \cdots R_\delta R_\gamma R_\beta R_\alpha) = \mathbf{u}_n + \mathbf{u}_1 R_\alpha + \mathbf{u}_2 R_\beta R_\alpha + \cdots + \mathbf{u}_{n-1} R_{\nu-1} \cdots R_\delta R_\gamma R_\beta R_\alpha. \qquad (8.n)$$

Equations (8) yield the set $\{h\}$ from the set $\{U\}$. However, $(I-R_i\ldots R_j)$ may be a singular matrix. Therefore, depending on the values of u_1, u_2, ..., u_n, R_α, R_β, ..., R_ν, the set $(h_1, h_2, \ldots h_n)$ may not be uniquely determined by (8). Indeed, one or more sets $(h_1, h_2, \ldots, h_n)$ may exist which satisfy (8). The more general way of dealing with this problem is to introduce the concept of the reflexive generalized inverse of a matrix (see, for example, Ben-Israel and Greville, 1974).

Definition. If A is an m x n matrix, an n x m matrix A^* is said to be a reflexive generalized inverse of A provided $AA^*A = A$ and $A^*AA^* = A^*$.

Property. A system of linear equations

$$Ax = b \tag{9}$$

has a solution if and only if

$$AA^*b = b. \tag{10}$$

Furthermore, if (9) has a solution then

$$x = A^*b + (I-A^*A)z, \tag{11}$$

where z is an arbitrary vector.

From a formal point of view, (11) may be easily applied to (8): i.e. for (8.1),

$$\begin{cases} A = (I-R_\beta R_\gamma \ldots R_\nu R_\alpha), \\ b = u_1 + u_2 R_\beta +\ldots u_n R_\nu \ldots R_\gamma R_\beta. \\ x = h_i \end{cases}$$

Furthermore, (8) suggests, for a given s.s. of the first rank, Φ, the expression of the s.i. $\psi_1 \in \{\psi\}_1$:

$$\phi_{u_1} +\phi_{u_2 R_\beta} +\phi_{u_3 R_\gamma R_\beta} +\ldots+\phi_{u_n R_\nu \cdots R_\gamma R_\beta} -\phi_{h_1} +\phi_{h_1 R_\alpha R_\nu \cdots R_\gamma R_\beta}, \tag{12.1}$$

$$\phi_{u_2} +\phi_{u_3 R_\gamma} +\phi_{u_4 R_\delta R_\gamma} +\ldots+\phi_{u_1 R_\alpha R_\nu \cdots R_\delta R_\gamma} -\phi_{h_2} +\phi_{h_2 R_\beta R_\alpha R_\nu \cdots R_\delta R_\gamma}, \tag{12.2}$$

$$\ldots\ldots\ldots\ldots\ldots\ldots\ldots\ldots\ldots\ldots\ldots\ldots$$

$$\phi_{u_n} +\phi_{u_1 R_\alpha} +\phi_{u_2 R_\beta R_\alpha} +\ldots+\phi_{u_{n-1} \ldots R_\delta R_\gamma R_\beta R_\alpha} -\phi_{h_n} +\phi_{h_n R_\nu \ldots R_\delta R_\gamma R_\beta R_\alpha} \tag{12.n}$$

It is now a trivial task to write down the magnitudes in the first phasing shell of Φ: one only needs to write down the set of magnitudes which are basis or cross magnitudes of at least one s.i. (12).

5 - UPPER REPRESENTATIONS OF A STRUCTURE SEMINVARIANT

Once the s.i.'s $\psi_1 \in \{\psi\}_1$ have been determined, the second representation of a s.s.'s is obtained by the first one by means of the same rules defined for s.i.'s.

REFERENCES

Ben-Israel, A., and Greville, T. N. E., 1974, "Generalized Inverse: Theory
    and Applications," John Wiley, New York.
Burla, M. C., and Giacovazzo, C., 1991, Submitted to Acta Cryst.
Cascarano, G., Giacovazzo, C., Camalli, M., Spagna, R., Burla, M. C., Nunzi,
    A., and Polidori, G., 1984, Acta Cryst., A40:278-283.
Cochran, W., 1955, Acta Cryst., 8:473-478.
Han, F., and Langs, D. A., 1988, Acta Cryst., A44:563-566.
Hauptman, H., 1975a, Acta Cryst., A31:671-679.
Hauptman, H., 1975b, Acta Cryst., A31:680-687.
Hauptman, H., 1976, Acta Cryst., A32:934-940.
Hauptman, H., and Karle, J., 1953, "Solution of the Phase Problem. I. The
    Centrosymmetric Crystal," A.C.A. Monogr. No. 3, Polycrystal Book
    Service, Pittsburgh.
Hauptman, H., and Karle, J., 1956, Acta Cryst., 9:45-55.
Hauptman, H., and Karle, J., 1959, Acta Cryst., 12:93-97.
Giacovazzo, C., 1974a, Acta Cryst., A30:626-630.
Giacovazzo, C., 1974b, Acta Cryst., A30:631-634.
Giacovazzo, C., 1974c, Acta Cryst., A30:390-395.
Giacovazzo, C., 1975, Acta Cryst., A31:252-259.
Giacovazzo, C., 1976a, Acta Cryst., A32:100-104.
Giacovazzo, C., 1976b, Acta Cryst., A32:958-966.
Giacovazzo, C., 1977, Acta Cryst., A33:933-944.
Giacovazzo, C., 1980, Acta Cryst., A36:367-372.
Pontenagel, W. M. G. F., and Krabbendam, H., 1983, Acta Cryst., A39:333-340.

INTRODUCTION TO MAXIMUM ENTROPY

Gérard Bricogne*

MRC Laboratory of Molecular Biology
Hills Road
Cambridge CB2 2QH, U.K.

1. OVERVIEW

1.1. Preamble

The Maximum Entropy Method (MEM) is not a completely different approach from Conventional Direct Methods (CDM) : it is an *improvement* of the latter at the level of the techniques of Analytic Probability Theory on which they are based.

Its first attraction is that it uses almost exclusively the "high-level language" of classical Probability Theory and Statistics with a minimum of *ad-hoc* jargon, and relegates to a subsidiary rôle the algebraic *minutiae* of phases invariants etc... which make Direct Methods seem so very esoteric to outsiders: if the MEM approach proves successful, these should ultimately be "phased out". Its main attraction, however, is that it is in principle much more powerful than CDM (although a lot of work still has to be done to turn this potential into a tangible reality), especially in the macromolecular field.

According to Michael Woolfson's own paradoxical criterion for assessing the potential of a new method, there must be *something* in this novel approach since he wrote (Woolfson, 1987): 'entropy maximization is adding nothing new to the crystallographic scene, and, since it involves a great deal of effort, perhaps nothing useful'.

It is my purpose in this first talk to help the reader form his or her own judgement on this matter.

---

* Permanent address : LURE, Bâtiment 209d, 91405 Orsay, France.

Direct Methods of Solving Crystal Stuctures
Edited by H. Schenk, Plenum Press, New York, 1991

1.2. The Phase Problem

The Phase Problem *per se*, i.e. posed for an arbitrary function, is mathematically indeterminate, but not all combinations of phases are equally acceptable when solving a crystal structure : a chemist can tell the difference between good phases and bad phases (the former give "fried eggs", the latter "scrambled eggs").

This criterion of *chemical validity* is very hard to translate into a tractable mathematical property of the structure factors. Therefore we will have to *weaken it*

      – *enough* for the mathematics to become tractable ;

      – *not too much* to retain some power of discrimination.

Note that any such method can only act as a *sieve*, since it will accept non-chemical solutions : an extra step of interpretation and validation (as e.g. in MULTAN) will always be needed.

1.3. Atomicity and the Random Atom Model

The property of *atomicity* seems to provide such a felicitous compromise. This word has two distinct meanings in CDM :

      – as used by Sayre, it results in an *exact analytical* relation between the structure factors of *every* structure ;

      – as used here, it gives rise to *statistical properties* which are obeyed only when viewed over an *ensemble* of structures.

It is perhaps surprising that the assumption that a crystal structure is made up of atoms gives rise to statistical constraints on the F's *even if we do not know where the atoms are*, i.e. even if they are assumed to be *randomly* placed, *independently* of each other.

Note that we have no need to invoke Hauptman's rationale that CDM work because the solution of the phase problem is highly overdetermined by the available data : indeed, the MEM works well *at any resolution*.

1.4. The basic idea

Direct phase determination solely from moduli data should be feasible by means of the following procedure (Hauptman & Karle, 1953) :

      (1) Derive $\mathcal{P}(F_{\mathbf{h}_1}, F_{\mathbf{h}_2}, \ldots, F_{\mathbf{h}_n})$, the joint probability distribution (jpd) of the structure factors, denoted $\mathcal{P}(\mathbf{F})$ in the sequel, from the random atom model .

      (2) Put $\left|F_{\mathbf{h}_j}\right| = \left|F_{\mathbf{h}_j}\right|^{\text{obs}}$ in each argument of $\mathcal{P}$.

      (3) Get $p\left(\varphi_{\mathbf{h}_1}, \varphi_{\mathbf{h}_2}, \ldots, \varphi_{\mathbf{h}_n} \middle| \left|F_{\mathbf{h}_1}\right|, \ldots, \left|F_{\mathbf{h}_n}\right|\right)$, the conditional probability distribution (cpd) of the phases given the observed values of the moduli, as (up to normalisation)

$$\mathcal{P}\left(\left|F_{\mathbf{h}_1}\right| e^{i\varphi \mathbf{h}}1, \ldots, \left|F_{\mathbf{h}_n}\right| e^{i\varphi \mathbf{h}}n\right) .$$

Thus, given the observed moduli, we get indications that certain combinations of phases are more probable than others, which is just what we want .

This sounds straightforward enough, but isn't !

1.5. The basic shortcoming

Conventional Direct Methods obtain $\mathcal{P}(\mathbf{F})$ for uniformly distributed random atoms (by methods to be examined later) as a *series expansion* which is most accurate for *small deviations* from its centre of mass $\langle \mathbf{F} \rangle = \mathbf{0}$, i.e. for small moduli only ; and yet these distributions are always used after substituting the *largest* moduli available

The problem cannot be cured by summing more terms, because the series in question is an *asymptotic series*.

To make matters worse, CDM have never (except for very recent work) made use of $\mathcal{P}(\mathbf{F})$ as a whole, but have instead considered *individual terms* of the series, and then have spent a lot of effort trying to put the series back together again by questionable methods (with problems of non-independence).

It is this author's opinion that this latter feature of CDM, i.e. the viewpoint that one wants to "estimate phase invariants/seminvariants" and then "combine" these estimates, constitutes a *perversion* of the initial probabilistic idea rather than its fruition.

Of course CDM have been extremely successful in the field of small molecules, and this *critique* must be understood as a prelude to their extension to bigger structures, and not misconstrued as gratuitous *criticism*.

1.6. Consequences

The analysis just given has two consequences :

(1) There is *no such thing* for practical purposes as a unique functional expression for $\mathcal{P}(\mathbf{F})$ (which had been naively assumed) into which observed moduli can be substituted to give the cpd of phases.

(2) We need an approximation method *other* than that used in CDM (called the Edgeworth series), since it must be able to work for *large deviations* from the expectation $\langle \mathbf{F} \rangle$ of $\mathbf{F}$.

1.7. The Maximum Entropy Method

Two essential departures from CDM are required in any method which proposes to deal with the problem of large deviations.

(1) Different approximations to $\mathcal{P}$ must be used in different regions of $\mathbf{F}$-space, i.e.

for different phase assumptions made on large moduli. These are most conveniently indexed by a *phasing tree* which describes the hierarchical dependence between all the phase assumptions made by means of a tree structure.

(2) Each approximation labelled by a node of the phasing tree is calculated using the *saddlepoint method* rather than the Edgeworth series. It is at this stage that the notion of *entropy* is encountered, and that it is required to *maximise* it.

Therefore the buzzword **maximum entropy** should not make us lose sight of the fact that what we are building is *an improvement of CDM in their own terms* : there are no new magic principles, no new justifications required; just more powerful mathematics.

1.8. A Tree–Directed Multisolution Strategy

The precise object attached to each node v of the phasing tree consists of :

- a *basis set* $H_v = \{ h_1^v , \dots , h_{n_v}^v \}$ of strong reflexions ;

- a set of *trial phase values* $\Phi_v = \{ \varphi_1^v , \dots , \varphi_{n_v}^v \}$ for them ;

- an *approximation* to $\mathcal{P}(F)$ near $F = F_{H_v}^*$ allowing the accurate computation of

conditional probabilities $\mathcal{P}\left(F_K \middle| F_{H_v} = F_{H_v}^* \right)$, where K is a set of unique

non-origin reflexions disjoint from H .

The use of a tree, borrowed from AI (see e.g. Nilsson, 1971), invites an analogy with *game-playing programs* which, although it can be dangerously distracting, has the merit of drawing attention to the need for *two types of criteria* to assist phase determination :

- one to measure the strength of the current position ; this will turn out to be *entropy* ;
- one to measure the ability of the current position to lead to strong positions after more moves have occurred, i.e. to provide a "look-ahead" capability ; this will turn out to be *likelihood* .

1.9. Calculation of Joint Probabilities

It is convenient to switch to unitary structure factors, which for equal point atoms of unit weight are given by $U_h = F_h / F_0$.

The *saddlepoint approximation* $\mathcal{P}^{SP}(U_H^*)$ to $\mathcal{P}(U_H^*)$ for N such atoms distributed in the unit cell with density $m(\mathbf{x})$ is given by :

$$\mathcal{P}^{SP}(U_H^*) = \frac{e^{NS}}{\left[\det (2\pi Q) \right]^{\frac{1}{2}}} \tag{1.1}$$

where

$$S = S_m(q^{ME}) \tag{1.2}$$

with
$$S_m(q) \;=\; -\int q(\mathbf{x})\;\log\!\left[\frac{q(\mathbf{x})}{m(\mathbf{x})}\right]\, d^3\mathbf{x} \tag{1.3}$$

and where q^{ME} maximises $S_m(q)$ under the constraints that its Fourier coefficients for the basis set reflexions should have the value U_H^*. The matrix $\mathbf{Q}$ can be computed from the full set of Fourier coefficients of q^{ME} by structure factor algebra. The method of calculation of q^{ME} for a given U_H^* will be described later.

Note that this expression can be evaluated numerically *even if H contains hundreds, thousands, ..., hundreds of thousands of reflexions*. This would be totally out of the question with the Edgeworth series.

1.10. Maximum Entropy Extrapolation

The process of forming the maximum-entropy distribution q^{ME} from the basis-set data U_H^* *extrapolates* Fourier coefficients U_k^{ME} for $\mathbf{k} \notin H$.

This extrapolation allows *phase extension* (see Bricogne, 1984, §7.3) in circumstances where CDM – including tangent formulae – give nothing or diverge.

However, starting from poor basis-set phases would produce worse, and probably useless, extrapolated phases : we need to be able to *refine* phases before taking their extrapolates seriously. This is where we need the look-ahead capability provided by *likelihood*.

The idea is to try and consult the unphased moduli outside the basis set to choose or refine the basis set phases we have chosen in U_H^*. Since this is also the purpose for which Hauptman's neighbourhoods (Hauptman, 1980) and Giacovazzo's phasing shells (Giacovazzo, 1980) were defined, it will be no surprise to find that they are somehow present implicitly in this likelihood criterion.

1.11. Calculation of Conditional Probabilities

Let K be a set of non-basis reflexions ($H \cap K = \varnothing$) and let U_K be the corresponding vector of structure factors. Then the conditional distribution of U_K given that $U_H = U_H^*$ is :

$$\mathcal{P}^{SP}\left(U_K \,\big|\, U_H = U_H^*\right) \;\propto\; \exp\left\{-\tfrac{1}{2} N \left(U_K - U_K^{ME}\right)^T Q_{KK}^{-1} \left(U_K - U_K^{ME}\right)\right\} \tag{1.4}$$

i.e. it is a multivariate Gaussian centered at U_K^{ME} (not at $\mathbf{0}_K$) whose covariance matrix Q_{KK} can be computed from the Fourier coefficients of q^{ME} (including the extrapolates) by structure factor algebra.

This conditional distribution is in general strongly *multimodal*, i.e. has many local maxima, making the process of phase extension intrinsically ambiguous ("branching problem"). Fortunately, the phasing tree constitutes an ideal book-keeping device to handle this ambiguity.

The distribution one would derive from the *null hypothesis* $U_H = 0$ would be the usual Gaussian centered at the origin :

$$\mathcal{P}\left(U_K \middle| \; U_H = 0 \right) \propto \exp\left\{ -\tfrac{1}{2} N \; U_K^T \varepsilon_{KK}^{-1} U_K \right\} \tag{1.5}$$

where ε_{KK} is the diagonal matrix of statistical weights (the Wilson distribution).

The mechanism by which these constantly updated conditional distributions confer an advantage over CDM is identical to the basis of the "card-counting" strategy at the game of blackjack (Thorpe, 1966).

1.12. Likelihood and Likelihood Ratio

The two distributions just derived under two different hypotheses on U_H cannot be easily compared because they involve the phases φ_k of non-basis reflexions. However we can *integrate these phases out* to obtain the *conditional probability distributions* not of the structure factors U_K but *of their moduli* $\left| U_K \right|$ (formulae will be given later).

We may then compare the conditional probabilities assigned to the *observed* values of moduli under each of the two hypotheses, the better assumption being that which assigns the *highest* probability to them. We may use this quantity, called the *likelihood* of the hypothesis, as a figure of merit.

The comparison is best carried out using the *likelihood ratio* :

$$\frac{\Lambda(U_H^*)}{\Lambda(0)} = \frac{\mathcal{P}\left(\left| U_K \right|^{obs} \middle| \; U_H = U_H^* \right)}{\mathcal{P}\left(\left| U_K \right|^{obs} \middle| \; U_H = 0 \right)} \tag{1.6}$$

or its logarithm, the *log-likelihood gain*. This is our "look-ahead criterion" which plays the rôle of the heuristic function in AI. We can also invoke Bayes's theorem and calculate the *a posteriori* probability of U_H^* by

$$\mathcal{P}^{post}\left(U_H^* \right) \propto \mathcal{P}^{SP}\left(U_H^* \right) \times \Lambda(U_H^*) \tag{1.7}$$

and *refine* the basis-set phases in U_H^* by maximising this quantity with respect to them.

1.13. Comparison with Conventional Direct Methods

The likelihood contains in an implicit form the entire theory of cross-terms, of neighbourhoods (Hauptman, 1980) and of representations (Giacovazzo, 1980). The definition of the neighbourhoods of the basis set is very natural (no need for a "neighbourhood principle") : they consist of those non-basis reflexions where ME extrapolation is the strongest, and they can be ranked according to the power of $\dfrac{1}{\sqrt{N}}$ to which that mean strength is proportional. Note that we define in this way the neighbourhoods of the entire basis set, not only those of individual invariants; the latter are generated *numerically* in the course of entropy maximisation and of integration over the non-basis phases, and never have to be handled explicitly.

As a simple example, the Hauptman quartet formula (Hauptman, 1975) can be derived in a few lines for small U values, and can be made intuitively obvious (Bricogne, 1984, §4.2.2 ;

Bricogne, 1988b, p.70-72) . For large U values, however, only the general likelihood is quantitatively correct.

According to a fundamental theorem of Neyman & Pearson (1933), likelihood is a very special figure of merit as an aid to making optimal decisions. In the context of Direct Methods this implies that, although many quantities can be concocted which will be *useful* for selecting good phase sets and refining them (the York school seems to be an inexhaustible source of these), *likelihood is the best possible.* Thus the optimality of likelihood is not an advertising slogan, it is a classical theorem of Statistics.

2. THEORETICAL BASIS OF THE CURRENT IMPLEMENTATION

2.1. Joint distributions of structure factors

Joint probability distributions (jpd's for short) of structure factors are calculated by means of the saddlepoint method according to equations (1.1) to (1.3), but neglecting the effect of the determinant. The maximum-entropy distributions q^{ME} required by this method are constructed numerically by exponential modelling, coupled with a very robust plane-search which often simplifies to a line-search (Bricogne & Gilmore, 1990, §2).

2.2. Conditional distributions of structure factors

The maximum-entropy distribution of atoms q^{ME} constructed to evaluate $\mathcal{P}(\, U_H^*\,)$ according to (1.1) provides a means of approximating the conditional distribution of any vector U_K with K disjoint from H as a multivariate Gaussian (1.4) centered at U_K^{ME} and whose covariance matrix Q_{KK} is essentially the Toeplitz matrix associated to q^{ME} (Bricogne, 1984, §4.2.1). The Wilson distribution (Wilson, 1949, 1950) would be $\mathcal{P}^{SP}(\, U_K \mid U_H = 0\,)$, corresponding to the "null hypothesis" of a uniform distribution of atoms $q^{ME}(x) = \dfrac{1}{V}$.

The quadratic form in the exponent of (1.4) may be rewritten (Bricogne, 1984, § 4.2.1) by using Parseval's theorem, to give another approximation in terms of maps in real space :

$$\mathcal{P}^{SP}(\, U_K \mid U_H = U_H^*\,) \;\; \propto \;\; \exp \left\{ -\frac{1}{2}\; N \int_V \frac{[\delta q(x)]^2}{q^{ME}(x)}\; d^3x \right\} \tag{2.1}$$

where $\delta q(x)$ is the Fourier synthesis calculated from the coefficients $U_K - U_K^{ME}$ (with symmetry expansion according to the space group of the structure under study). This new expression has the following very simple interpretation in real space : the role of the non-uniform distribution q^{ME} is to create a *differential cost* for the addition of new features at different locations in the unit cell, making it less costly (in terms of loss of conditional probability) to introduce new detail where q^{ME} is large than where it is small.

In this work we will use a cruder approximation than $\mathcal{P}^{SP}$ to the cpd of U_K , called the *diagonal approximation* in the sequel, which consists in using as a covariance matrix the diagonal matrix ε_{KK} of statistical weights. Thus we retain the first-moment information U_K^{ME} generated by the non-uniformity of q^{ME}, but discard the off-diagonal second-moment information associated to that non-uniformity. This decouples the various U_k's so that the resulting $\mathcal{P}_{diag}^{SP}$ may be written as :

$$\mathcal{P}_{diag}^{SP} \, (\, U_K \mid U_H = U_H^* \,) \;\; = \;\; \prod_{k \in K} \; \mathcal{P}_{diag}^{SP} \, (\, U_k \mid U_H = U_H^* \,) \qquad (2.2),$$

each factor being a one- or two-dimensional Gaussian offset from the origin by U_k^{ME} .

The diagonal approximation has the serious disadvantage that it is *unimodal*, so that it is unable to represent the branching behaviour which introduces ambiguities in the process of phase extension. This scrambling of the multimodal structure of $\mathcal{P}^{SP}$ is however less serious when the phases are integrated out to form conditional distributions of moduli and likelihood functions. The advantage of the diagonal approximation is that it yields relatively simple explicit expressions for these various distributions (see Bricogne & Gilmore (1990) for the derivations).

(i) For **k** be an acentric reflexion, write

$$U_k \; = \; A_k + i \, B_k \; = \; |U_k| \, e^{i \varphi_k} \, ,$$

and let

$$dA_k \, dB_k \; = \; |U_k| \, d|U_k| \, d\varphi_k$$

be the plane measure in the corresponding copy of the complex plane. The conditional probability distribution $\mathcal{P}_{diag}^{SP}$ is then :

$$\mathcal{P}_{diag}^{SP} \, (\, U_k \mid U_H = U_H^* \,) \, dA_k \, dB_k \;\; = \;\; \frac{N}{\pi \varepsilon_k} \, \exp \left[-\frac{N}{\varepsilon_k} \, |U_k - U_k^{ME}|^2 \right] \, dA_k \, dB_k$$

$$= \;\; \frac{2N}{\varepsilon_k} \, |U_k| \, \exp \left\{ -\frac{N}{\varepsilon_k} \, [\, |U_k|^2 + |U_k^{ME}|^2 \,] \right\}$$

$$\times \;\; \exp \left[\frac{2N}{\varepsilon_k} \, |U_k| \, |U_k^{ME}| \cos(\varphi_k - \varphi_k^{ME}) \right] \, d|U_k| \, \frac{d\varphi_k}{2\pi} \qquad (2.3).$$

The conditional distribution of the phase φ_k when $|U_k|$ is known is therefore :

$$P \, (\, \varphi_k \mid U_H = U_H^*, \, |U_k| = |U_k|^{obs} \,) \;\; = \;\; \frac{1}{2\pi I_0(X_k)} \, \exp \left[X_k \cos(\varphi_k - \varphi_k^{ME}) \right] \qquad (2.4)$$

where

$$X_k \; = \; \frac{2N}{\varepsilon_k} \, |U_k|^{obs} \, |U_k^{ME}| \qquad (2.5).$$

In spite of all the approximations made, this expression already incorporates the use of the triple-phase relationship, as shown in Bricogne (1984), equations 3.23 and 4.15.

(ii) For **k** a centric reflexion, the phase φ_k has only two possible values, π apart, say ω_k and $\omega_k + \pi$; the quantity $\cos(\varphi_k - \omega_k)$ is a mere sign, which will be denoted s_k . If dU_k denotes the line measure along the corresponding real line in the complex plane, then the equivalent of the polar coordinate relations is

$$dU_k \;=\; d|U_k| \, \tfrac{1}{2}[\,\delta(s_k-1) + \delta(s_k+1)\,]$$

where the normalised sum of δ-functions (a discrete measure representing a choice of sign) is the centric equivalent of $\dfrac{d\varphi_k}{2\pi}$. It follows that

$$\mathcal{P}^{SP}_{diag}(\,U_k \mid U_H = U_H^*\,)\, dU_k \;=\; \left(\frac{N}{2\pi\varepsilon_k}\right)^{\tfrac{1}{2}} \exp\left[-\frac{N}{2\varepsilon_k}\,|U_k - U_k^{ME}|^2\right] dU_k$$

$$=\; \left(\frac{2N}{\pi\varepsilon_k}\right)^{\tfrac{1}{2}} \exp\left\{-\frac{N}{2\varepsilon_k}\,[\,|U_k|^2 + |U_k^{ME}|^2\,]\right\}$$

$$\times\; \exp\left[\frac{N}{\varepsilon_k}\,|U_k|\,|U_k^{ME}|\, s_k\, s_k^{ME}\right]\, d|U_k|\,\tfrac{1}{2}[\,\delta(s_k-1) + \delta(s_k+1)\,] \qquad (2.6),$$

giving for the conditional distribution of the sign s_k when $|U_k|$ is known :

$$P(\,s_k \mid U_H = U_H^*,\, |U_k| = |U_k|^{obs}\,) \;=\; \frac{1}{2\cosh(X_k)}\,\exp\left[X_k\, s_k\, s_k^{ME}\right] \qquad (2.7)$$

where

$$X_k \;=\; \frac{N}{\varepsilon_k}\,|U_k|^{obs}\,|U_k^{ME}| \qquad (2.8).$$

A straightforward adaptation to the centric case of eqns. 3.23 and 4.15 in Bricogne (1984) shows that this approximation, although crude, incorporates the use of the hyperbolic-tangent formula of Cochran & Woolfson (1955).

<u>2.3. Conditional distributions of moduli</u>

The calculations carried out so far convey the (correct) impression that entropy maximisation has a useful built-in ability to extrapolate phase information. Indeed, even in the case of the small protein crambin, very significant phase extension was shown to take place in conditions where direct methods are powerless (Bricogne, 1984, §7.3) . But in this case, the starting phases to 3Å resolution were exact, and the question immediately arose of knowing whether in a real situation, errors in the starting phases would not be amplified so as to make the extrapolated phases worthless. In other words maximum-entropy phase extension looks like an inherently divergent process, capable of some extrapolation from good starting phases, but perhaps incapable of refining bad starting phases prior to their extrapolation. This section will be devoted to showing that this is not the case : phase refinement *is* possible through the use of *likelihood functions* introduced in Bricogne (1984), §4.2.2.

The basic observation is the following : if we integrate $\mathcal{P}^{SP}(\,U_K \mid U_H = U_H^*\,)$ with respect to the (unobservable) phases in U_K, we obtain a conditional marginal distribution for the (observable) moduli $\mathcal{P}^{SP}(\,|U_K| \mid U_H = U_H^*\,)$ *which differs from the standard Wilson distribution of moduli*, the latter being $\mathcal{P}^{SP}(\,|U_K| \mid U_H = 0\,)$.

This consideration is of fundamental importance : through the use of the saddlepoint approximation, phase choices for the "basis" reflexions in H induce a *deformation* of the conditional marginal distribution of the moduli in K away from the usual (Wilson) distribution. In other words, the maximum-entropy extrapolation which relates U_H^* to U_K^{ME} acts as a *transducer*, converting hypotheses about phase values in U_H^* (which cannot be tested directly from measured intensities) into hypotheses about a change in the statistical distribution of the moduli $|U_K|$ (which *can* be tested).

This transduction effect persists (although somewhat weakened) even if we use the diagonal approximation $\mathcal{P}_{diag}^{SP}$, because of the origin offset U_K^{ME}. Explicit expressions for this level of approximation are easily obtained by integrating out phases or signs.

For an acentric reflexion $\mathbf{k}$ the above derivations immediately yield :

$$\mathcal{P}_{diag}^{SP} (|U_\mathbf{k}| \mid U_H = U_H^*) \; d|U_\mathbf{k}|$$

$$= \frac{2N}{\varepsilon_\mathbf{k}} |U_\mathbf{k}| \exp\left\{ -\frac{N}{\varepsilon_\mathbf{k}} [|U_\mathbf{k}|^2 + |U_\mathbf{k}^{ME}|^2] \right\} I_0\left(\frac{2N}{\varepsilon_\mathbf{k}} |U_\mathbf{k}| |U_\mathbf{k}^{ME}| \right) \; d|U_\mathbf{k}| \qquad (2.9)$$

(a distribution first derived by Rice (1944, 1945) in another context), while for a centric reflexion

$$\mathcal{P}_{diag}^{SP} (|U_\mathbf{k}| \mid U_H = U_H^*) \; d|U_\mathbf{k}|$$

$$= \left(\frac{2N}{\pi\varepsilon_\mathbf{k}} \right)^{\frac{1}{2}} \exp\left\{ -\frac{N}{2\varepsilon_\mathbf{k}} [|U_\mathbf{k}|^2 + |U_\mathbf{k}^{ME}|^2] \right\} \cosh\left(\frac{N}{\varepsilon_\mathbf{k}} |U_\mathbf{k}| |U_\mathbf{k}^{ME}| \right) \; d|U_\mathbf{k}| \qquad (2.10).$$

2.4. Likelihood functions, likelihood ratios

For the purpose of testing hypotheses, we use the customary *likelihood* criterion where the likelihood of a hypothesis is defined as the probability it assigned to the actual result of an observation before that observation was performed. In the case at hand, we therefore define the likelihood Λ of the phase choices in U_H^* as the conditional probability of the observed values $|U_K|^{obs}$:

$$\Lambda (U_H^* \mid |U_K| = |U_K|^{obs}) = \mathcal{P}^{SP}(|U_K|^{obs} \mid U_H = U_H^*) \qquad (2.11).$$

In practice, it is convenient to normalise this quantity with respect to the null-hypothesis (H_0) that the distribution of atoms is uniform (i.e. that $U_H = 0$) and to use the *likelihood ratio*

$$\frac{\Lambda(U_H^* \mid |U_K| = |U_K|^{obs})}{\Lambda(0 \mid |U_K| = |U_K|^{obs})} = \frac{\mathcal{P}^{SP}(|U_K|^{obs} \mid U_H = U_H^*)}{\mathcal{P}^{SP}(|U_K|^{obs} \mid U_H = 0)} \qquad (2.12)$$

or its logarithm, the *log-likelihood gain*

$$L(U_H^* \mid |U_K| = |U_K|^{obs}) - L(0 \mid |U_K| = |U_K|^{obs}) \qquad (2.13),$$

where $L = \log \Lambda$. For brevity, we will often simply write e.g. $\Lambda (U_H^*)$ or $\Lambda (0)$ when the observation results used in forming the likelihood (here, $|U_K| = |U_K|^{obs}$) are unambiguously defined by the context.

According to Neyman & Pearson (1933), the likelihood ratio just defined constitutes the most powerful statistical criterion for identifying the best set of phases for U_H^* on the basis of the information $|U_K| = |U_K|^{obs}$. Concretely, this criterion measures the extent to which the observed values of the yet unphased moduli $|U_K|^{obs}$ are made more probable by the assumption that $U_H = U_H^*$ than by the assumption that $U_H = 0$.

Explicit expressions for likelihoods are easily obtained in the diagonal approximation from the cpd's of moduli derived above. For $\mathbf{k}$ acentric,

$$\Lambda\,(\,U_H^*\,\mid\,|U_{\mathbf{k}}| = |U_{\mathbf{k}}|^{obs}\,)\;=$$

$$\frac{2N}{\varepsilon_{\mathbf{k}}}\,|U_{\mathbf{k}}|^{obs}\,\exp\Big\{-\frac{N}{\varepsilon_{\mathbf{k}}}\,[\,(|U_{\mathbf{k}}|^{obs})^2 + |U_{\mathbf{k}}^{ME}|^2\,]\Big\}\;\times\;I_0\Big(\frac{2N}{\varepsilon_{\mathbf{k}}}\,|U_{\mathbf{k}}|^{obs}\,|U_{\mathbf{k}}^{ME}|\Big) \qquad (2.14)$$

while

$$\Lambda(\,0\;\mid\,|U_{\mathbf{k}}| = |U_{\mathbf{k}}|^{obs}\,)\;=\;\frac{2N}{\varepsilon_{\mathbf{k}}}\,|U_{\mathbf{k}}|^{obs}\,\exp\Big\{-\frac{N}{\varepsilon_{\mathbf{k}}}\,(|U_{\mathbf{k}}|^{obs})^2\Big\} \qquad (2.15)$$

hence

$$\frac{\Lambda(\,U_H^*\,\mid\,|U_{\mathbf{k}}| = |U_{\mathbf{k}}|^{obs}\,)}{\Lambda(\,0\;\mid\,|U_{\mathbf{k}}| = |U_{\mathbf{k}}|^{obs}\,)}\;=\;\exp\Big\{-\frac{N}{\varepsilon_{\mathbf{k}}}\,|U_{\mathbf{k}}^{ME}|^2\Big\}\;I_0\Big(\frac{2N}{\varepsilon_{\mathbf{k}}}\,|U_{\mathbf{k}}|^{obs}\,|U_{\mathbf{k}}^{ME}|\Big) \qquad (2.16)$$

so that the global log-likelihood criterion from all the acentric moduli in K reads :

$$L(U_H^*) - L(\,0\,)\;=\;\sum_{\substack{\mathbf{k}\in K\\ \mathbf{k}\ \text{acentric}}}\Big[\,\log I_0\Big(\frac{2N}{\varepsilon_{\mathbf{k}}}\,|U_{\mathbf{k}}|^{obs}\,|U_{\mathbf{k}}^{ME}|\Big) - \frac{N}{\varepsilon_{\mathbf{k}}}\,|U_{\mathbf{k}}^{ME}|^2\,\Big] \qquad (2.17).$$

Similarly, for $\mathbf{k}$ centric,

$$\Lambda\,(\,U_H^*\,\mid\,|U_{\mathbf{k}}| = |U_{\mathbf{k}}|^{obs}\,)\;=$$

$$\Big(\frac{2N}{\pi\varepsilon_{\mathbf{k}}}\Big)^{\frac{1}{2}}\,\exp\Big\{-\frac{N}{2\varepsilon_{\mathbf{k}}}\,[\,(|U_{\mathbf{k}}|^{obs})^2 + |U_{\mathbf{k}}^{ME}|^2\,]\Big\}\;\times\;\cosh\Big(\frac{N}{\varepsilon_{\mathbf{k}}}\,|U_{\mathbf{k}}|^{obs}\,|U_{\mathbf{k}}^{ME}|\Big) \qquad (2.18)$$

while

$$\Lambda(\,0\;\mid\,|U_{\mathbf{k}}| = |U_{\mathbf{k}}|^{obs}\,)\;=\;\Big(\frac{2N}{\pi\varepsilon_{\mathbf{k}}}\Big)^{\frac{1}{2}}\,\exp\Big\{-\frac{N}{2\varepsilon_{\mathbf{k}}}\,(|U_{\mathbf{k}}|^{obs})^2\Big\} \qquad (2.19)$$

hence

$$\frac{\Lambda(\,U_H^*\,\mid\,|U_{\mathbf{k}}| = |U_{\mathbf{k}}|^{obs})}{\Lambda(\,0\;\mid\,|U_{\mathbf{k}}| = |U_{\mathbf{k}}|^{obs})}\;=\;\exp\Big\{-\frac{N}{2\varepsilon_{\mathbf{k}}}\,|U_{\mathbf{k}}^{ME}|^2\Big\}\;\cosh\Big(\frac{N}{\varepsilon_{\mathbf{k}}}\,|U_{\mathbf{k}}|^{obs}\,|U_{\mathbf{k}}^{ME}|\Big) \qquad (2.20)$$

so that the global log-likelihood criterion from all the centric moduli in K reads :

$$L(U_H^*) - L(\,0\,)\;=\;\sum_{\substack{\mathbf{k}\in K\\ \mathbf{k}\ \text{centric}}}\Big[\,\log\cosh\Big(\frac{N}{\varepsilon_{\mathbf{k}}}\,|U_{\mathbf{k}}|^{obs}\,|U_{\mathbf{k}}^{ME}|\Big) - \frac{N}{2\varepsilon_{\mathbf{k}}}\,|U_{\mathbf{k}}^{ME}|^2\,\Big] \qquad (2.21).$$

Because the diagonal approximation has been used, these likelihood functions are sensitive *only to the extrapolated moduli* $|U_K^{ME}|$, and not to the associated phases. An exact expression for the likelihood, incorporating the full covariance matrix Q_{KK} (eq. (1.4)), has been derived and will be described elsewhere ; it is sensitive to the extrapolated phases as well. Other forms of this likelihood function, suitable for using diffraction intensities from fibres or powders rather than from single crystals, will also be published shortly.

<u>1.5. Phase refinement by Bayes's theorem</u>

The a priori probability $\mathcal{P}^{SP}(\mathbf{U}_H^*)$ may be combined with the likelihood $\Lambda(\mathbf{U}_H^* \mid |\mathbf{U}_K| = |\mathbf{U}_K|^{obs})$ by means of Bayes's theorem (see e.g. Lindley, 1965), yielding the *a posteriori* probability

$$\mathcal{P}_{post}(\mathbf{U}_H^* \mid |\mathbf{U}_K| = |\mathbf{U}_K|^{obs}) \;\; \propto \;\; \mathcal{P}^{SP}(\mathbf{U}_H^*) \times \Lambda(\mathbf{U}_H^* \mid |\mathbf{U}_K| = |\mathbf{U}_K|^{obs}) \qquad (2.22)$$

whose maximisation with respect to the phases in $\mathbf{U}_H^*$ provides a procedure for phase refinement (Bricogne, 1988a, §0.6 ; Bricogne, 1988b, p 68-72). It was shown in Bricogne (1984), §4.2.2, and Bricogne (1988b), p. 70-72, that this approach is an enhancement of the use of a "quartet figure of merit" which can be implemented without any explicit manipulation of phase invariants.

Using the approximation $\mathcal{P}^{SP}(\mathbf{U}_H^*) \approx \exp\left(N\,\mathcal{S}_m(q^{ME})\right)$, it is clear that Bayes's theorem leads to considering the compound criterion

$$N\,\mathcal{S}_m(q^{ME}) + L(\mathbf{U}_H^* \mid |\mathbf{U}_K| = |\mathbf{U}_K|^{obs}) \qquad (2.23)$$

as the best discriminator between phase sets for $\mathbf{U}_H^*$ on the basis of the information contained in the observed moduli associated to H and K.

In practice, L is much more sensitive to $\mathbf{U}_H^*$ than $\mathcal{S}_m$, especially at the early stages of phase determination. As a result, phase refinement is often carried out by maximisation of likelihood alone.

<u>1.6. Centroid extrapolated structure factors</u>

Once the trial phases assigned to the moduli $|\mathbf{U}_H^*|$ have been refined by likelihood maximisation against the unphased moduli $|\mathbf{U}_K|^{obs}$, the phase dependence of the conditional probability distributions $\mathcal{P}^{SP}_{diag}(\mathbf{U}_k \mid \mathbf{U}_H = \mathbf{U}_H^*)$, which is integrated out in order to form the likelihood function, may be preserved and used to calculate the centroid value $\langle \mathbf{U}_k \rangle$ of each structure factor $\mathbf{U}_k$. These centroid values are analogous to the "best" structure factor values defined by Blow & Crick (1959) in another context, and have the property that they minimise the mean-square error in the corresponding maps caused by the residual phase uncertainties.

Using the expressions derived in §2.2 for the conditional distribution of the phase φ_k , we get

(i) for k acentric,

$$\langle \mathbf{U}_k \rangle \;=\; |\mathbf{U}_k|^{obs} \frac{I_1(X_k)}{I_0(X_k)}\, e^{\,i\varphi_k^{ME}} \quad \text{with} \quad X_k = \frac{2N}{\varepsilon_k}|\mathbf{U}_k|^{obs}\,|\mathbf{U}_k^{ME}| \qquad (2.24);$$

(ii) for k centric,

$$\langle \mathbf{U}_k \rangle \;=\; |\mathbf{U}_k|^{obs} \tanh(X_k)\, e^{\,i\varphi_k^{ME}} \quad \text{with} \quad X_k = \frac{N}{\varepsilon_k}|\mathbf{U}_k|^{obs}\,|\mathbf{U}_k^{ME}| \qquad (2.25).$$

Maps calculated from this extended set of structure factors will already show enhanced resolution without any new phase choices having been made.

3. STRATEGY AND IMPLEMENTATION

3.1. Survey

The practical task of constantly updating the distribution of atoms $q(\mathbf{x})$ to the maximum-entropy distribution compatible with the phase choices made at each stage requires a book-keeping scheme capable of recording these various phase choices and the parentage relations between them – in other words, of representing all the phasing paths one may wish to explore. For this purpose it is convenient to use a *multisolution tree* (Bricogne, 1984, §2.4, §8.1 ; Bricogne, 1988b, p. 73-75) .

3.1.1. Growth of the multisolution tree.

Let the nodes of this multisolution tree be labelled by the index ν , let H_ν denote the set of "basis" reflexions (i.e. of reflexions for which trial phase values are being treated as free parameters) attached to that node, and let K_ν denote the second neighbourhood of H_ν . The overall strategy then goes as follows (Bricogne, 1984, §8.1). The root node (labelled $\nu = 1$) of the tree consists of the set H_1 of origin-fixing and enantiomorph-defining reflexions. The subsequent growth of the tree is governed by the following sequence of computations at each node ν :

(a_ν) update the prior distribution of atoms $q(\mathbf{x})$ to the maximum-entropy distribution $q_\nu^{ME}(\mathbf{x})$ compatible with the phase choices (giving rise to structure factor values $U_{H_\nu}^*$) attached to that node ;

(b_ν) construct the conditional distribution $\mathcal{P}_{diag}^{SP}\ (\ U_{K_\nu}\ |\ U_{H_\nu} = U_{H_\nu}^*\ )$ of the yet unphased structure factors in K_ν ;

(c_ν) construct the likelihood function $\Lambda\ (\ U_{H_\nu}^*\ |\ |U_{K_\nu}| = |U_{K_\nu}|^{obs}\ )$ by integrating the conditional distribution over the unknown phases and substituting the observed values of the moduli in K_ν ;

(d_ν) refine the basis phases in $U_{H_\nu}^*$ by maximizing Λ with respect to these basis phases ;

(e_ν) solve the local branching problem for (i.e. identify the local maxima of) the conditional distribution with respect to a subset of unknown phases when the values of the corresponding moduli are introduced ;

(f_ν) expand the current node by creating a branch leading to a new tip node for each of these maxima ; if the conditional distribution is too flat for (e_v) to be meaningful, just do some phase permutation ; in both instances the current node v generates a certain number m_v of daughter nodes $(v^{(1)}, v^{(2)}, \ldots, v^{(m_v)})$; each such node has an enlarged basis set $H_{v^{(1)}}, H_{v^{(2)}}, \ldots, H_{v^{(m_v)}}$ with associated trial structure factor values $U^*_{H_{v^{(1)}}}, U^*_{H_{v^{(2)}}}, \ldots, U^*_{H_{v^{(m_v)}}}$, and process (a) may be applied to each of them.

3.1.2. Optimal scheduling of tree growth

The growth of the tree must be supervised so as to maximize the chance of finding the correct set of phases without having to develop too many of its spurious branches. For this purpose, we may use two criteria at each node v :

(i)   the loss of entropy $\mathcal{S}_m(q_v^{ME})$ in going from a uniform distribution of atoms $m(x) = \dfrac{1}{V}$ to the current non-uniform $q_v^{ME}(x)$; according to Shannon (Shannon & Weaver, 1949), this measures the shrinkage of the population of "reasonably probable structures" when the constraints $U^*_{H_v}$ are enforced ;

(ii)   the likelihood $\Lambda (U^*_{H_v} \mid |U_{K_v}| = |U_{K_v}|^{obs})$, which measures the degree to which the phase choices made in $U^*_{H_v}$ are able to anticipate correctly, through their maximum-entropy extrapolation, some of the information present in the yet inactive constraints, and hence measures the "chances of survival" of the current node in the face of these forthcoming constraints.

Intuitively, the entropy of a node measures the "girth" of the branch directly leading to it from the root node, while its likelihood is an estimate of the sum of the girths of the branches bifurcating away from that node once the values of the unphased moduli in its second neighbourhood are taken into account. One would thus expect likelihood (which looks "ahead" at the degree of pre-established harmony between its predictions and the actual observations for the yet unphased data) and entropy (which looks "back" at the cost of accommodating the current phase assumptions) to be complementary measures of the worth of a trial phase set.

It is this very analysis which is embodied in the criterion (2.23) defined in § 2.5 on the basis of Bayes's theorem. This combined criterion can therefore be used as a score to determine the priority with which each node will be considered for further growth in step (f), thus affording an optimal scheduling of the growth of the multisolution tree.

3.2. Principles of operation

We will now examine in detail the procedures used in each of the above steps. Programming and practical considerations are dealt with in Gilmore, Bricogne & Bannister (1990).

3.2.0. Origin and enantiomorph definition.

The direct methods program MITHRIL (Gilmore, 1984 ; Gilmore & Brown, 1988) is used to select those reflexions which optimally define the origin and enantiomorph. We have however found that the present phasing strategy is much more effective if an extra resolution criterion is used (together with the usual $|E|$ magnitude criterion) to choose these reflexions: if relatively strong reflexions are present at low resolution, even though they may not have the strongest $|E|$ values, they should be incorporated preferentially into the initial basis set H_1. A plausible explanation of this behaviour is that the conditional distributions of the phases of strong high-resolution reflexions are less multimodal (hence make the branching problem more tractable) if the non-uniformity in $q(\mathbf{x})$ has been created by low-resolution constraints rather than by other high-resolution terms.

3.2.1. Updating to a maximum-entropy distribution.

Step (a) is carried out by the method described in §2 of Bricogne & Gilmore (1990), involving bicubic modelling and plane search. This algorithm is extremely robust and capable of dealing with U values close to unity.

3.2.2. Conditional distributions.

At step (b) , the conditional distributions of the structure factors in K_V (in the diagonal approximation) are calculated by the explicit formulae given in §2.2 . Centroid values for these structure factors may be calculated as described in §2.6 and used to calculate a map for visual inspection.

3.2.3. Likelihood functions, Σ refinement

Step (c) uses the expressions (2.14) through (2.21) derived in §2.4 for the likelihood functions in the diagonal approximation. The second neighbourhood K_V of the current basis set H_V is generated and the value of two "annealing parameters" Σ_a and Σ_c are refined by a Newton method. A weighted average of $\dfrac{1}{\Sigma_c}$ and $\dfrac{1}{2\Sigma_a}$ is subsequently used as a substitute for the number of atoms N in all formulae involving N (see §3.3).

Gilmore, Bricogne & Bannister (1990) have found the likelihood criterion to be of immense power in ranking trial phase sets, even at very early stages when none of the existing figures of merit used in traditional direct methods would give any useful indications. As phasing progresses, the correct node rapidly emerges above the others, its likelihood being orders of magnitudes greater.

3.2.4. Phase refinement.

Step (d) is based on the principle outlined in §2.5 and is carried out by the algorithm described in §2.5 of Bricogne & Gilmore (1990).

This new method of phase refinement has proved extremely effective, even with very small basis sets which would in no way lend themselves to a tangent refinement.

3.2.5. Survey of branching behaviour

As explained in §1.2.4, the diagonal approximation destroys the multimodality of conditional distributions, and thus the expressions set up at step (b) are unsuitable to carry out step (e). For the latter purpose, we use the real-space expression (1.4) of the full conditional distribution for the structure factors.

In order to survey this cpd with respect to as many new phases as possible, we use the magic-integer technique (White & Woolfson, 1975 ; Main, 1977) to encode several phases into fewer numerical symbols. The integral $\int_V \dfrac{[\delta q(\mathbf{x})]^2}{q^{ME}(\mathbf{x})}\, d^3\mathbf{x}$ is calculated for each point of a grid of simultaneous values of these symbols, and its minima (corresponding to the maxima of the cpd) are located. Absence of contrast at this stage is usually a sign that the current basis set is a bad choice, and that backtracking to another node of the tree is advisable. If no better node can be identified, simple-minded phase permutation is used to create the daughter nodes, whose ranking must then await the evaluation of the likelihood criterion at the next round.

This method of surveying the multimodality of the cpd has proved an effective and reliable one, and an essential aid in controlling the phasing process.

3.2.6. Expansion of the basis set

At steps (e) and (f), it is of vital importance to exercise great care in choosing the set of new reflexions with respect to whose phases the branching behaviour will be surveyed. Choosing those reflexions where the strongest ME extrapolation takes place always leads to an impasse, undoubtedly because the phasing process becomes over-consistent on a strongly-coupled subset of the data, and is then unable to incorporate any of the information present in the rest of the observations (this is a well-known phenomenon in tangent refinement). We have found that the correct strategy for expanding the basis set consists in choosing those strong reflexions for which no strong extrapolation has yet taken place, i.e. whose strength is maximally unexpected to the current statistical model. For equal degrees of unexpected strength, the preferential choice of low-resolution *vs.* high-resolution reflexions is helpful, for the same reason as explained in §1.2.2 and §3.2.0 .

3.3. Use and meaning of the annealing parameter Σ

The use of an empirical and refinable parameter $\dfrac{1}{\Sigma}$ (calculated as the weighted average of $\dfrac{1}{\Sigma_c}$ and $\dfrac{1}{2\Sigma_a}$) instead of the number of atoms N deserves further comments. It will be recalled that, for unitary structure factors in the equi-atom case, the quantity Σ appearing in Wilson's statistics (the sum of the squared form factors over the contents of the unit cell) becomes $\dfrac{1}{N}$ (see Bricogne (1984), § 4.2.2(1)). As defined in §2.3.2 and used in §2.4 of Bricogne & Gilmore (1990), Σ measures the statistical dispersion of the distribution of the observed moduli , taking into account the modulation of the latter by ME-extrapolation from the basis set of reflexions (§2.3) .

172

If $q(\mathbf{x})$ is the uniform distribution, Σ is related to the dispersion of the unphased $|E|$ values around their root-mean-square value of 1.0 . If the atoms in the structure at hand are not truly statistically independent at the working resolution but occur as "globs" (Harker, 1953), then the refined value of Σ will be *greater* than $\frac{1}{N}$, being roughly the inverse of the number of such globs rather than of the number of atoms. This phenomenon has actually been observed in the course of the direct phasing of contrast variation data from a protein crystal at low resolution, and the size of the globs thus indicated correlates well with the existence of a sharp maximum in the radial intensity distribution at about $10.5 \, \text{Å}$ resolution (Carter, Crumley, Coleman, Hage & Bricogne, 1990). The quantity $\frac{1}{\Sigma}$ is thus N_{eff}, the "effective N" defined in I, §8.3, and this observation corroborates the conclusion reached there that the potential strength of statistical phase relations should be gauged from the value of N_{eff} rather than of N . Our practice of using Σ rather than N guarantees that this effect is automatically taken into account.

At the other extreme, when a great deal of correct phase information has accumulated at certain nodes, the ME-extrapolation becomes more and more exact, as the Toeplitz determinants approach zero (Goedkoop, 1950). In these circumstances the parameter Σ refines to values much *smaller* than $\frac{1}{N}$, but as the U values are not renormalised the effect is to make the conditional distributions much sharper, and phase extension much stronger. We have repeatedly observed this "critical behaviour" of Σ , which is accompanied by soaring values of the log-likelihood of the correct node, to be an indication that the phasing process has reached successful completion.

The nature and behaviour of Σ is best rationalised by viewing it as a *mean thermal energy per reflexion*, i.e. as the parameter kT, in the Simulated Annealing protocol of Kirkpatrick, Gelatt & Vecchi (1983). With this interpretation, the conditional distributions (2.4) and (2.7) for the phases are formally identical to Boltzmann distributions for a dipole in a field under the identification : dipole $= U_{\mathbf{k}}$, field $= U_{\mathbf{k}}^{ME}$; the factor of 2 in (2.5) follows from the fact that the energy (i.e. the variance) of an acentric reflexion is evenly split between 2 degrees of freedom. This justifies in the centric case, and shows how to extend to the acentric case, the heuristic use of "Phase Annealing" reported by George Sheldrick in this Volume. The analogy between phasing and magnetic phase transitions, especially those in spin glasses, will be pursued further elsewhere.

4. SUMMARY

We have examined in detail a previously proposed multisolution strategy for direct phase determination by combined maximisation of entropy and of likelihood, as well as the theoretical results on which its implementation is based. Increasingly specific phase assumptions are made as part of a tree-directed search. For each such assumption, saddlepoint approximations to joint and conditional probabilities of structure factors are calculated by means of maximum-entropy distributions of atoms. A likelihood criterion is constructed for testing phase assumptions in the light of the observed distribution of the yet unphased moduli. Different trial phase sets are ranked according to a criterion combining entropy and likelihood according to Bayes's theorem, and the phases are refined by optimising this criterion. Finally, the ambiguity inherent in the process of phase extension (the "branching problem") is explicitly surveyed by evaluation of a real-space expression which has a simple interpretation in terms of a differential cost for the creation of new detail, and the search tree is expanded accordingly. Practical applications to small molecules are described by Gilmore, Bricogne & Bannister (1990).

ACKNOWLEDGEMENTS

The collaboration between the writer and Dr. C.J. Gilmore was initiated at L.U.R.E. (Orsay, France) thanks to a Senior Ciba-Geigy Fellowship (to CJG); subsequent work in Glasgow was supported by grants from the SERC (to CJG). GB acknowledges support from CNRS (France) and partial support from Trinity College (Cambridge, U.K.) during the course of this work.

REFERENCES

BLOW, D.M. & CRICK, F.H.C. (1959). *Acta Cryst.* A12, 794-802.

BRICOGNE, G. (1984). *Acta Cryst.* A40, 410-445.

BRICOGNE, G. (1988a). *Acta Cryst.* A44, 517-545.

BRICOGNE, G. (1988b). In *Crystallographic Computing 4* , edited by N.W. ISAACS & M.R. TAYLOR, pp. 60-79. New York : Oxford Univ. Press.

BRICOGNE, G. & GILMORE, C.J. (1990). *Acta Cryst.* A46, 284-297.

CARTER, C.W.Jr., CRUMLEY, K.V., COLEMAN, D.E., HAGE, F. & BRICOGNE, G. (1990). *Acta Cryst.* A46, 57-68.

COCHRAN, W. & WOOLFSON, M.M. (1955). *Acta Cryst.* 8, 1-12.

GIACOVAZZO, C. (1980). *Direct Methods in Crystallography.* London : Academic Press.

GILMORE, C.J. (1984). *J. Appl. Cryst.* 17, 42-46.

GILMORE, C.J., BRICOGNE, G. & BANNISTER, C. (1990). *Acta Cryst.* A46, 297-308.

GILMORE, C.J. & BROWN, S.R. (1988). *J. Appl. Cryst.* 22, 571-572.

GOEDKOOP, J.A. (1950). *Acta Cryst.* 3, 374-378.

HARKER, D. (1953). *Acta Cryst.* 6, 731-736.

HAUPTMAN, H. (1975). *Acta Cryst.* A30, 472-476.

HAUPTMAN, H. (1980). In *Theory and Practice of Direct Methods in Crystallography,* edited by M.C.F. LADD & R.A. PALMER, pp. 151-197. New York : Plenum.

HAUPTMAN, H. & KARLE, J. (1953). *The Solution of the Phase Problem. I. The Centrosymmetric Crystal. Am. Crystalogr. Assoc. Monogr. No. 3.* Pittsburgh : Polycrystal Book Service.

KIRKPATRICK, S., GELATT, C.D.Jr. & VECCHI, M.P. (1983). *Science* **220**, 671-680.

LINDLEY, D.V. (1965). *Introduction to Probability and Statistics from a Bayesian Viewpoint*, Vols. 1 and 2. Cambridge Univ. Press.

MAIN, P. (1977). *Acta Cryst.* **A33**, 750-757.

NEYMAN, J. & PEARSON, E. (1933). *Phil. Trans. Roy. Soc.* **A231**, 289-337.

NILSSON, N.J. (1971). *Problem-Solving Methods in Artificial Intelligence.* New York : McGraw-Hill.

RICE, S.O. (1944, 1945). *Bell System Tech. J.* **23**, 283-332 (parts I and II) ; **24**, 46-156 (parts III and IV). Reprinted in *Selected Papers on Noise and Stochastic Processes* (1954), edited by N. WAX, pp. 133-294. New York : Dover Publications.

SHANNON, C.E. & WEAVER, W. (1949). *The Mathematical Theory of Communication.* Urbana : Univ. of Illinois Press.

THORPE, E.O. (1966). *Beat the Dealer.* New York : Vintage Books.

WHITE, P.S. & WOOLFSON, M.M. (1975). *Acta Cryst.* **A31**, 53-56.

WILSON, A.J.C. (1949). *Acta Cryst.* **2**, 318-321.

WILSON, A.J.C. (1950). *Acta Cryst.* **3**, 258-261.

WOOLFSON, M.M. (1987). *Acta Cryst.* **A43**, 593-612.

DIRECT METHODS PACKAGES IN 1990

C.J. Gilmore

Department of Chemistry, University of Glasgow
Glasgow G12 8QQ, Scotland

1.0 INTRODUCTION

Direct methods packages are the interface between theory and the user of that theory. It is safe to say that without easy-to-use programs direct methods would still be an esoteric, little used branch of crystallography rather than the most important technique available for solving the structures of 'small' molecules. The packages which will be discussed here are (in alphabetical order):

(i) MITHRIL (Gilmore, 1984, Gilmore & Brown 1989)

(ii) MULTAN88: developed in its current form by Debaerdemaeker, Germain, Main, Refaat, Tate and Woolfson. MULTAN was the first direct methods package assembled for use by non-experts, and it has had a revolutionary impact on crystallography. It is still the most widely used suite of programs.

(iii) SHELXS-86: SHELX in its many forms is the most widely used crystallographic package for structure solution and refinement. I will concentrate here on the direct methods routines in SHELXS-86, but with some comment on new developments in the SHELXS-90 system.

(iv) SIMPEL: In its current form produced by Peschar and Schenk, and based on symbolic addition to produce a starting set.

(iv) SIR88: The principal authors are Cascarano, Giacovazzo, Burla, Polidori, Camalli, Spagna and Viterbo. MULTAN-like in its construction it contains novel features based around structure seminvariants.

(v) XTAL: This is a vast program package edited by Hall and Stewart designed for the structure solution and refinement of molecular structures both small and large. The direct methods routines in version 2.6 will be discussed.

(vi) MICE: This is quite unlike any other program suite in that it uses entropy maximisation and likelihood to solve crystal structures (Bricogne & Gilmore 1990, Gilmore, Bricogne & Bannister 1990, Bannister, Bricogne & Gilmore 1989) It contains a MITHRIL interface.

Two other programs which specialise in modulated structures, pseudo-symmetry and situations where only small fragments can be located are:
(i) DIRDIF (Buerskens *et al.* 1990)
(ii) SAPI (Fan Hai-fu *et al.* 1983, Fan Hai-fu et al. 1990)
Both these packages are described elsewhere in this book, and I shall concentrate on the situations where pseudo-symmetry is not a critical problem.

Direct Methods of Solving Crystal Stuctures
Edited by H. Schenk, Plenum Press, New York, 1991

A survey of this type should not be wholly uncritical, and there is one comment to make at the very outset: in general, direct methods calculations are very undemanding for modern computers. Given the proliferation of networked workstations of great power offering colour graphics, windowing etc., direct methods programs do not seem to exploit this environment; they still seem routed in the past of monolithic mainframes, even MITHRIL which was designed with interactive use in mind.

Conventional direct methods follows a sequence of calculations that is well defined:

(i) Normalisation: going from $|F_h|$ to $|E_h|$.
(ii) The generation of structure invariants and seminvariants.
(iii) The assembly of a starting set.
(iv) Phase extension and refinement.
(v) Production of Fourier maps and their interpretation.
(vi) Recycling of fragments.

Let us look at each of these calculations in turn, and examine the similarities and differences which exist in the above direct methods packages. (This section needs to be read in conjunction with my chapter on 'Solving Difficult Structures')

2.0 NORMALISATION

In many ways, normalisation is the key to a successful program package for two reasons:

(i) The whole process of normalisation is critical, and largely automatic, and hence can be ignored by the user without realising the consequences.
(ii) It is the interface with other programs and packages, and the user's own data.

Ideally, all packages should provide at least two methods of normalisation, the ability to input E-magnitudes from other programs as well as checking for duplicate reflections and systematic absences plus editing facilities and overrides on calculated scale and temperature factors. SHELXS has an elegant single re-normalisation formula:

$$E_{new} = \exp(8\,\pi\,\delta U\,\sin^2\vartheta/\lambda^2)/(E_{old}^{-4} + E_{max}^{-4})^{1/4}$$

where δU is a temperature factor shift (positive or negative) and E_{max} is the maximum permitted E-magnitude. MITHRIL provides similar features plus a set of filters that are described by Gilmore & Brown, 1988. XTAL enables the user to control scale, temperature factor and $\sin\vartheta/\lambda$ cut-off; SIR has only the ability to define one's own temperature factor - the remaining packages are greatly deficient in this respect. Some packages also provide only one method of normalisation, and only MITHRIL and SIR allow the direct input of E's from other sources. No program allows an anisotropic temperature factor.

The handling of weak and missing reflections is important. Although modern practice in data collection usually presents a full data set of measured reflections to direct methods programs, this is not always the case, and it is often useful to have alternative strategies for dealing with weak data sets.

XTAL has a limited Bayesian approach based on work by French and Wilson (French and Wilson 1978) or the use of $0.5|F_{min}|$ for missing reflections in user-defined shells of $\sin\vartheta/\lambda$. MITHRIL uses the method of Vickovic and Viterbo (Vickovic & Viterbo, 1979) to fill in missing reflections.

MITHRIL is unique in providing a full Bayesian method of normalisation as an alternative to K-curve or Wilson methods. One's prior prejudice for B can be input, and sensible standard deviations, $\sigma(|E_{\underline{h}}|)$, are generated. XTAL also estimates these σ's but the values obtained are too large. XTAL does, however, provide some interesting alternative handling of the Wilson plot. Using these estimates of $\sigma(|E_{\underline{h}}|)$ is another problem, but they can be useful in removing triplets where the three E's invoved all have exceptionally large variances.

Index rescaling is an important, if sometimes contentious, issue. All programs offer parity group rescaling, and all but MITHRIL allow one to define additional groups for rescaling which can be very useful in the case of pseudosymmetry. In a similar way, all but SHELX allow one to input groups of random or known orientation; such information is used both in the Wilson plot and in the generation of structure invariants (See Section 2.2)(Main, 1975) Such normalised structure factors may not necessarily be more reliable than their simpler counterparts, but they can provide alternative phasing paths through the direct methods process.

Both MICE and SIMPEL do not (yet) provide a normalisation program, but take their E's from other sources. MICE is tightly linked to the normalisation module of MITHRIL.

At this point we have normalised structure factors; we now generate the invariants and seminvariants.

3.0 INVARIANTS AND SEMINVARIANTS

3.1 Three-Phase Invariants

Now nearly 40 years old, the triple phase structure invariant (triplet) is still the dominant phasing equation in direct methods, and forms the backbone of every program suite.   Define Φ_3 as:

$$\Phi_3 = \varphi_{\underline{h}} + \varphi_{\underline{k}} + \varphi_{\underline{1}} \qquad \text{where} \qquad \underline{h} + \underline{k} + \underline{1} = 0$$

Ideally, triplets should be given a statistical weight which reflects the space-group phase restrictions imposed on the triplet contributors. These are of minor importance, but handled correctly by most packages.

Let us examine options which estimate $\cos \Phi_3$ rather than assuming that it equals unity:

(i) MITHRIL offers the L.E. equations method (Hauptman 1985, Gilmore & Hauptman 1985) and the MDKS formula (Hauptman, 1972). Both of these are used only as a weight for $\varkappa$ in convergence mapping and tangent refinement, where:

$$\varkappa = 2|E_{\underline{h}} E_{\underline{k}} E_{\underline{1}}|/\sqrt{N} \qquad N = \text{No. of atoms in the unit cell.}$$

(ii) SIR offers the P_{10} formula (Cascarano, Giacovazzo, Camalli, Burla, Nunzi & Polidori 1984). $\varkappa$ is now modified *via:*

$$\varkappa' = \varkappa \, (1 + Q)$$

where Q is obtained from P_{10} and $\varkappa'$ is used to re-rank the invariants which are actively used in the phasing process. Figures of merit are constructed from

In the author's experience with MDKS and L.E., attempts to estimated three-phase invariants in this way are both time consuming (although this is of limited importance today) and unreliable, but the methods have a useful perurbative effect, so the options can be valuable. However, the P_{10} formula, exploiting as it does the crystallographic symmetry, appears to be much more useful. Equally useful is the use of group data as defined by Main (1975), in which Φ_3 is estimated in the prescence of fragments of known orientation and optionally position. MITHRIL, MULTAN and XTAL use this formalism although its success rate is variable.

3.2 Four-Phase Invariants

Quartets are a newer addition to the invariants family, although they were known and discussed in papers in the 1950's. Schenk and Hauptman introduced the concepts of negative and enantiomorph sensitive quartets, and these, especially the former, have found their way in to most packages. Let:

$$4 \quad \underline{h} \quad \underline{k} \quad \underline{l} \quad \underline{m}$$

(i) Negative quartets having $\Phi_4 \approx \pi$ are the most useful, being independent of triplets. These are used actively as an option in MITHRIL, XTAL and SHELX, as well as finding use as a figure of merit. Strangely, MULTAN uses them not at all except in the SAYTAN (See Section 5.3)
(ii) Positive quartets with $\Phi_4 \approx 0$ are more problematical being derivable from triplets, and so highly correlated with them. MITHRIL calculates such relationships if requested and attempts to cope with correlations in a simple way based on cross term magnitudes. XTAL also permits the user to generate these relationships and use them, but no correction for correlation seems to be available.
(iii) Enantiomorph sensitive quartets having $\cos \Phi_4 \approx 0$ are also calculated, at user request, but they are very unreliable even as a figure of merit.

Problems also arise with missing cross-terms because they are outside the sphere of measurement. MITHRIL and XTAL allow missing cross-terms, which are given magnitudes of unity. MITHRIL also permits the use of the P_{13} formula (Hauptman 1977) which gives increasing quartet accuracy at the cost of computer time.

Quintets are calculated in MITHRIL, but they were inserted as an experiment, and very little use has been made of them. Their slow computation time, their correlation with quartets and their $N^{-3/2}$ dependence makes them an unattractive proposition.

3.3 Seminvariants

The one-phase seminvariant (Σ_1) finds its way into most programs although there are differences in approach. By far the most elaborate is to be found in SIR where representation theory comes into its own, and suitable phases are estimated by both first and second representation, and used in phasing. MITHRIL and MULTAN by way of contrast extract their seminvariant information from special triplets, which is a less powerful way of doing things. SHELX also calculates the Σ_1 reflections for use as a figure of merit, but does not use such reflections in the starting set unless specifically given them *via* the **PHAS** command. It is quite possible to use SIR to estimate the one-phase ssi's and then input them to SHELX.

The two-phase ssi's are found in SIR and SIMPEL, although MITHRIL can use them if they are explicitly loaded at convergence mapping time. SIR calculates them *via* their first representation, whereas SIMPEL obtains them from quartets in which they are embedded. SIMPEL also can extract three-phase seminvariants from quintets in a similar way.

In total contrast to all this, MICE generates no ssi's or indeed any phase relationships of any sort, the entropy maximisation procedure uses them *without* their explicit generation.

It is interesting to see how the whole subject of phase relationships is more complex in its implementation than one might otherwise expect, and this complexity of choice will continue through the problems associated with both selecting the starting set and phase refinement and expansion.

4.0 SELECTING A STARTING SET

Convergence mapping (Germain, Main & Woolfson 1970) was a revolutionary algorithm which had a great impact on direct methods allowing as it did for a sensible choice of starting reflections that defined the origin and enantiomorph as well as selecting reflections for phase permutation without user intervention. It is still widely used, although often supplemented by divergence mapping, and forms an important part of MITHRIL, MULTAN, SIMPEL, SIR and XTAL. Only SHELX does not use it at all (although MULTAN and MITHRIL have this option) but has a random phasing procedure built into it in such a way that the definition of origin and enantiomorph are deemed irrelevant. This, of course, allows one to write very compact FORTRAN code, which has always been a hallmark of the Sheldrick approach to package development.

MITHRIL, SIR and XTAL also provide a random phasing procedure as an alternative, although they use the convergence mapping results to select the reflections which will be subject to random phasing. SIR has a unique option in which the phases are subject to random permutations generated *via* a perturbation of a number of triplet phases by fitting their von Mises distribution (Burla, Giacovazzo & Polidori 1987). This can be used to generate a large number of phase sets which can keep a workstation gainfully employed over the weekend.

In general, the phase permutation process involves magic integers, but XTAL allows one to use a permutation sequence of one's own including quadrant permutation, and MITHRIL allows some control over the magic integer sequence used, which in general makes the search in phase space a coarser one.

Everything is now ready for phase extension and refinement.

Although these are separate concepts, they are inextricably bound in conventional direct methods and will be considered together.

5.0 PHASE EXTENSION and REFINEMENT

5.1 Symbolic Addition

Although it was one of the earliest phasing methods, symbolic addition has found little favour with direct methods program writers, partly because it is difficult to program. SIMPEL provides the only elaborate symbolic addition routines, and is unique in this respect. It also utilises the principles of dynamic programming to create an optimum phasing path (Peschar & Schenk 1987). MITHRIL in its latest formulation provides symbolic addition for the centrosymmetric case only, where it can be useful in heavy atom structures. Symbolic addition has a benefit of not needing an origin to be defined fully, or indeed at all, until a later time, and this is exploited by SIMPEL. It also tends to be much faster than its tangent formula equivalent.

5.2 The Tangent Formula

The tangent formula comes in a wide variety of forms; these involve permutations of the following design choices:

(i) Types of weighting scheme.
(ii) Choice of invariants and seminvariants.
(iii) Phase permutation and/or random phasing.
(iv) The use of early figures of merit.

Let us look at each program package in turn:

MITHRIL: Two types of weighting - standard MULTAN weights or Hull-Irwin weights (Hull & Irwin 1978). The latter are very useful in cases of pseudosymmetry or in the presence of heavy atoms although there is the danger of oscillation of weights in going from cycle to cycle. Triplets, quartets and quintets are used plus any seminvariant phase relationships up to the order five which may have been generated elsewhere. Both phase permutation and random phasing methods are available with two early figures of merit, which tend to be little used. There is, however, an early
stop mechanism whereby a solution with outstanding figures of merit may stop the tangent refinement process.

MULTAN: Again two types of weight generally with triplets only (but see Section 5.3), and the choice of random or phase permutation methods and the early stop mechanism. As a default only a subset of reflections is phased initially when using random phasing methods.

SIMPEL: Weighted tangent refinement using triplets and quartets. Early figures of merit and stop options are not relevant here because of the initial phasing that has been carried out already by an extensive symbolic addition.

SHELXS: Triplets and negative quartets are used on a reflection subset for reasons of computational efficiency. A filter , equivalent to an early figure of merit, is applied to select the best 10% of these subset phase permutations which are generated by random phasing methods. This filter, F, is based on the criterion:

$$F = R_\alpha + \max[0, (0.25 + NQEST)]$$

where NQEST is the negative figure of merit and R_α a triplet consistency criterion. The sets thus filtered are subject to full tangent refinement; the user has considerable control over the selection of suitable filtered sets *via* a SUBS command.

In SHELX-90 the methods of simulated annealing (Kirkpatrick, Gelatt & Vecchi 1983) are applied in the form of phase annealing (Sheldrick, 1990). Given the recent successes of this optimisation method in other fields, it offers some exciting prospects in solving large structures.

SIR: A tangent formula using triplets and the most reliable two-phase seminvariants is used; apart from standard weights, there is a scheme in which the α parameter is forced to match the theoretical ones (Burla, Cascarano, Giacovazzo, Nunzi & Polidori 1987) The parameter α has its usual definition:

$$\alpha = (T_{\underline{h}}^2 + B_{\underline{h}}^2)^{1/2}$$

where $T_{\underline{h}}$ and $B_{\underline{h}}$ are the numerator and denominator of the tangent formula. There are no early figures of merit or stop options.

XTAL: The GENTAN routines carry out phase extension and refinement. Any mixture of quartets and triplets is permitted. There are three weighting schemes:
(i) Probabilistic weights similar to those used in MULTAN,
(ii) Hull-Irwin weights.
(iii) Modified Hull-Irwin weights in which the weighting scheme is a function of $\alpha_{\underline{h}}/\langle\alpha_{\underline{h}}\rangle$ rather than $|\alpha_{\underline{h}}/\langle\alpha_{\underline{h}}\rangle|^2$. The weights here are more heavily damped than in (ii) but not as sensitive as (i) to variations in phase agreement. An early stop option is implemented with override. There is more control available over the tangent formula than in any other package, but it takes considerable experience to use this.

5.3 Alternatives to the Tangent Formula

MULTAN is quite unique here in offering two quite distinct phasing routines:
(i) A Sayre equation tangent formula derived by least-squares minimisation of the residual of the system of Sayre equations (Debaerdemaeker, Tate & Woolfson 1985, 1988) Quartets, both positive and negative, can appear here in an active form.
(ii) A related X-Y parameter shift refinement procedure.
Both of these methods can be used with random phasing procedures or phase permutation. Woolfson *et al.* are currently testing these methods on small protein structures.

MITHRIL also offers two alternatives to tangent refinement:
(i) The YZARC procedure of Baggio *et al.* (Baggio, Woolfson, Declercq & Germain 1978; Declercq, Germain & Woolfson 1979) which treats the triplets and quartets as a set of simultaneous linear equations which its solves by conventional matrix algebra. An option exists to bypass the convergence map in these circumstances.
(ii) MAGEX (White & Woolfson 1975; Declercq, Germain & Woolfson 1975) which is a form of automated symbolic addition, although it does use the tangent refinement for its final stages of phase extension.

MICE takes a sarting set (called the 'basis set') and generates new phases by a process of extrapolation *via* entropy maximisation. This starting set can come from a convergence map or a technique based on optimal third neighbourhood extension.

6.0 FIGURES OF MERIT

This is where the packages appear at their most individualistic. An ideal figure of merit (FOM) involves using phase relationships which have not been actively employed in phasing so that it is independent of the phasing process. However, this can be difficult in practice: if the phase relationships are good enough for a FOM, why not use them actively to aid phasing? The standard FOM's in current use are based on:

(i) Triplet consistency $-R_\alpha$, ABSFOM, R_{Karle} etc.
(ii) Quartet consistency, usually negative ones (NQEST) but not necessarily.
(iii) Special triplets Ψ_0.
(iv) Structure seminvariants Σ_1 and two-phase seminvariants.

The way in which the required phase relationships are selected can be important. In summary the packages we have been considering use the following FOM's

MITHRIL: ABSFOM, R_α, Ψ_0, NQEST and a negative quintet figure of merit. A current experiment with likelihood type functions culled from maximum entropy research is also proving successful.

MULTAN: ABSFOM, R_α, Ψ_0, NQEST.
SHELXS: R_α with suitable weights, and NQUAL which involves triplet and negative quartet consistency.
SIMPEL: There are seven used:
(i) Two based on positive quartets.
(ii) A negative quartet criterion.
(iii) Three based on 1-, 2-, and 3-phase seminvariants.
(iv) Projection criteria which check the consistency of phase relationships in which one or more have their phases restricted by space-group symmetry.
SIR: The figures of merit are based on fitting all the invariants and seminvariants that have been generated. (Cascarano, Giacovazzo & Viterbo 1987) These include the use of:
(i) 1- and 2-phase seminvariants.
(ii) NQEST.
(iii) Enantiomorph sensitive triplets and quartets (as a warning only)
(iv) Positive and negative triplets.
(v) Modified forms of Ψ_0 , ABSFOM and R_α.
XTAL: RFOM (the inverse of ABSFOM), Ψ_0, R_α and a form of NQEST.

In all cases the individual FOM's are combined to give a single figure of merit on which the solutions are sorted before being presented in a suitable order to the Fourier routines with their associated peak searches.

7.0 FINISHING THE JOB AND A FEW REMARKS ON THE FUTURE

The programs nearly all contain facilities for recycling fragments with or without the presence of pseudosymmetry. For structures that are difficult to complete for various reasons the Buerskens DIRDIFF program is often used. SHELXS has a facility for automatically recycling fragments coupled with crystallographic least-squares, which can be a painless way of interpreting E-maps which contain only small features that seem to be correct MITHRIL has a new interface to the FRODO program of Alwyn Jones much used by protein crystallographers. This allows E-maps to be investigated for fragments by eye, which can be useful in situations where the peaks are poorly resolved or difficult to interpret automatically.

What of the future? I think that the next generation of packages should take on board:

(i) Windowing and interactive graphics to provide a more efficient human interface.
(ii) AI methods for interpreting maps.
(iii) Better diagnostics for difficult structures.

ACKNOWLEDGEMENTS

Discussing other author's packages is not a way to make friends and influence people. I am, however, grateful to all the authors of the programs discussed here for their assistance in providing me with manuals and other information.

REFERENCES

Baggio, R. Woolfson, M.M., Declercq, J-P and Germain G. (1978) Acta Cryst. **A34**, 883-892.

Bannister, C., Bricogne, G and Gilmore, C.J. In *Maximum Entropy and Bayesian Methods* Ed. J. Skilling, Kluwer (1989) 225-232.

Bricogne, G. and Gilmore, C.J. (1990) Acta Cryst. **A46**, 284-297.

Burla, M.C., Cascanaro, G., Giacovazzo, C., Nunzi, A. and Polidori, G. (1987) **A43**, 370-374.

Burla, M.C., Giacovazzo, C. and Polidori, G. (1987) Acta Cryst. **A43**, 797-802.

Buerskens, P.T., Bosman, W.P., Doesburg, H.M., Gould, R.O., van der Hark, E.M., Prick, P.A.J., Noordick, J.H., Buerskend, G., Parthasarathi, V., Bruins-Slot, H.J., Haltiwanger, R.C., Strumpel, M.K., Smits, J.M.M., Garcia-Granda, S., Smykalla, C., Behm, H.J.J., Schafer, G. and Admiraal, G. (1990) *The DIRDIF90 program system*. University of Nijmegen. In press.

Cascanaro, G., Giacovazzo, C., Camalli, M., Burla, M.C., Nunzi, A. and Polidori, G. (1984) Acta Cryst. **A40,** 278-283.

Cascanaro, G., Giacovazzo, C. and Viterbo, D. (1987) Acta Cryst. **A43,** 22-29.

Declercq, J-P., Germain, G. and Woolfson, M.M. (1975) Acta Cryst. **A31,** 367-372.

Declercq, J-P., Germain, G. and Woolfson, M.M. (1979) Acta Cryst. **A35,** 622-626.

Debaerdemaeker, T., Tate, C. and Woolfson (1985) Acta Cryst. **A41,** 286-290.

Debaerdemaeker, T., Tate, C. and Woolfson (1988) Acta Cryst. **A44,** 353-357.

French, S. and Wilson, K. (1978) Acta Cryst. **A34**, 517-525.

Fan Hai-fu, Yao Jia-xing, Main, P. and Woolfson, M.M. (1983) Acta Cryst. **A39**, 566-569.

Fan Hai-fu, Qian Jin-zi, Zheng Chao-de, Gu Yuan-xin Ke Heng Min and Huang Sheng-hua (1990) Acta Cryst. **A46,** In press.

Germain, G., Main, P. and Woolfson, M.M. (1970) Acta Cryst. **26**, 274-285.

Gilmore, C.J. (1984) J. Appl. Cryst. **17**, 42-46.

Gilmore, C.J. and Brown, S.R. (1988) Acta Cryst. **A44**, 1018-1021.

Gilmore, C.J. and Brown, S.R. (1989) J. Appl. Cryst. **21**, 571-572.

Gilmore, C.J., Bricogne, G and Bannister, C. Acta Cryst. (1990) **A46** 297-308.

Gilmore, C.J. and Hauptman, H.A. (1985) **A41,** 457-462.

Hauptman, H.A. *Crystal Structure Analysis: The Role of the Cosine Seminvariants* Plenum Press (1972) .

Hauptman, H.A. (1977) Acta Cryst. **A33**, 556-568.

Hauptman, H.A. (1985) <u>Acta Cryst.</u> **A41,** 452-457.
Hull, S.E. and Irwin, M.J. (1978) <u>Acta Cryst.</u> **A34,** 863-870.
Kirkpatrick, S, Gelatt, C.D. and Vecchi, M.P. (1983) <u>Science</u> **220,** 671-680.
Main, P. in *Crystallographic Computing Techniques* Ed. F.R. Ahmed, Munskgaard (1976) 97-105.
Peschar, P and Schenk, H. (1987) <u>Acta Cryst.</u> **A43,** 751-763.
Sheldrick, G.M., (1990) <u>Acta Cryst.</u> **A46,** In press.
Viscovic, V. and Viterbo, D. <u>Acta Cryst.</u> (1979) **A35,** 500-501.
White, P.S. and Woolfson, M.M. (1975) <u>Acta Cryst.</u> **A31,** 53-55.

PATTERSON METHODS

Patrick Tollin

Department of Applied Physics and Electronic
and Manufacturing Engineering
University of Dundee
Dundee, Scotland, DD1 4HN

INTRODUCTION

The solution of a crystal structure consists in combining information about the magnitudes of the $|F|^2$'s (obtained from experimental observation of a diffraction experiment) and some other information about the structure to obtain the positions of the atoms within the crystal unit cell. Other facts about the structure, such as thermal motion etc., may be obtained and with restricted data it is true that atomic resolution may not be achieved, but stripped of all else, this statement expresses the fundamental process of structure determination. The important point is that some extra information however meagre has to be added to the diffraction data.

Direct methods, as usually understood, form the subject of most of this school, and use very general pieces of additional information expressed as the positive nature of the electron density or the equality of ρ and ρ^2, or the presence of equal and resolved atoms or in some such way. Certainly at some stage an E-map, or its equivalent, will have to be interpreted. The inter- pretation will involve using the information that the structure contains atoms separated by distances between 1 and 3Å, and having acceptable angular relations.

Patterson (1934, 1935) realised that the information contained in the $|F|^2$'s could be re-expressed in the form of the function which now bears his name:

$$P(\underline{u}) = \sum_{\underline{h}} |F(\underline{h})|^2 \; \mathrm{Cos}\; 2\pi\underline{h}.\underline{u} = \int \rho(\underline{x})\rho(\underline{x}+\underline{u})d\underline{x}$$

and that this function contained information about the interatomic vectors within the crystal. The extra information which must be added to this expression of the $|F|^2$ information, to obtain the crystal structure may take a number of forms. It might be that the structure contains one or a few particularly heavy atoms, or that the structure contains identical sub units not related by the crystal structure, or that part of the structure has a known stereochemistry or such like.

Direct Methods of Solving Crystal Stuctures
Edited by H. Schenk, Plenum Press, New York, 1991

The purpose of this paper is to review the ways in which Patterson methods can be used to solve crystal structures, either alone or as a useful addition to direct methods and to describe one method, not necessarily the best, in more detail which uses stereochemical information for this purpose.

The book 'Patterson and the Patterson' published by the International Union of Crystallography to commemorate fifty years of the Patterson function gives an excellent overview of the history of the function and is referred to in this article.

HEAVY ATOM METHODS

Patterson recognised that since $|F(\underline{h})|^2$ could be written in the form

$$|F(\underline{h})|^2 = \sum_{i=1}^{N} f_i^2 + \sum_{i,j=1}^{N} f_i f_j \exp\left(2\pi i(\underline{x}_i - \underline{x}_j)\right)$$

the peaks in the Patterson function would have heights which were proportional to $Z_i Z_j$ the atomic numbers of the atoms at the ends of interatomic vectors $(\underline{x}_i - \underline{x}_j)$.

If a structure contains one, or a few, atoms of outstandingly large Z_H compared with the other atoms in the structure, then the Patterson function will contain large peaks of height Z_H^2 due to the vectors between these atoms. The coordinates of the heavy atoms can be deduced for the positions of these peaks and phases α_H can be calculated for a structure consisting of these atoms alone. Combining these phases with the observed structure amplitudes and calculating an electron density

$$\rho(\underline{r}) = \sum_{\underline{h}} |F(\underline{h})| e^{i\alpha_H} \exp{-2\pi i \underline{h}.\underline{r}}$$

will in favourable cases reveal the rest of the structure.

Very soon Harker (1936) realised that the vectors between atoms related by crystallographic symmetry would be on special planes and lines. Thus an atom at $x_1 y_1 z_1$ if repeated by a 2_1 axis would give an atom at $-x_1 \; \tfrac{1}{2}+y_1 z_1$ and hence a peak in the Patterson function at $2x_1 \; \tfrac{1}{2} \; 2z_1$. So all atoms related by this axis will give rise to peaks in the section $y = \tfrac{1}{2}$ in $P(\underline{u})$. Similarly a mirror plane perpendicular to y gives atoms at $x\,y\,z$ and $x-y\,z$ and a peak at $0\,2y\,0$. These Harker sections and Harker lines were used to solve many structures and are still important in the interpretation of Patterson functions containing heavy atoms.

In the isomorphous replacement method the information supplied is that two crystals have near identical structures but one contains in addition one or a few heavy atoms. Here the heavy atom vectors are not predominant enough to stand out from the light atom background. However since two sets $|F_H|^2$ and $|F_P|^2$, of data are available, the difference Patterson, using $|F_H|^2 - |F_P|^2$, or the ΔF^2 synthesis, using $||F_H| - |F_P||^2$ as coefficients will lead to the determination of the heavy atom positions.

SUPERPOSITION METHODS

Two problems arise in the interpretation of Harker sections when there is not one heavy atom. The first is that if in the first structure above two atoms are fortuitously separated by $\frac{1}{2}$ in y these will give a non-Harker peaks in the Harker section at $x_1 - x_2$ $\frac{1}{2}$ $z_1 - z_2$. More seriously atoms at $(x+\frac{1}{2}\ y\ z)$ $(x\ y\ z+\frac{1}{2})$ and $x'+\frac{1}{2}\ y\ z+\frac{1}{2})$ will all give rise to Harker peaks at $2x\ 0\ 2z$. So that attempt to solve the Harker section for many atoms will fail.

Buerger (1951) and others recognised that the Patterson function could be described as a set of images of the structure shifted to the ends of interatomic vectors. If one could identify a peak in the Patterson function corresponding to a single vector between two atoms and if one shifted the Patterson origin to the end of this vector and superimposed it on the original, coincidences of high Patterson density could show up the underlying structure. He proposed as image-seeking functions superpositions which used the product or the minimum of these coincidences. These ideas were extended by Ramachandran and Raman (1959) and an extensive discussion is given by Buerger (1959). Problems arise in the detection of single vector peaks so that multiple superpositions may be needed and in any case more than one is needed for a non-centrosymmetric crystal. The minimum function or some modification of it undoubtedly leads to the clearest result in real space. On the other hand, using the sum of the superposed Pattersons, the sumfunction, or the related α-synthesis of Ramachandran and Raman (1959) has the advantage of being able to be expressed in reciprocal space terms. A modern computer analysis of minimum superposition maps for the determination of crystal structures without any prior chemical information is described by Richardson and Jacobson (1987).

THE KNOWN FRAGMENT

If the sterochemistry of part of the structure is known, we know the relative coordinates $\underline{s}_i$ of the atoms of that part of the structure. The problem is to find the matrix [C] and the vector $\underline{R}_0$ such that $[C]\underline{s}_i + \underline{R}_0$ gives the position vectors of the atoms in the structure. In general the matrix [C] is a function of three angular parameters, $\theta_1 \theta_2 \theta_3$ say, and the vector $\underline{R}_0$ can be expressed in terms of three position parameters, $X_0 Y_0 Z_0$.

However, some separation of the determination of these parameters is always possible. At the very least the orientation and position determinations can be separated. This corresponds in terms of the Patterson function to examining intermolecular and intramolecular vectors separately.

Whatever Patterson function interpretation method is used, the orientation of the group is determined, in principle, by calculating the position vectors $[C]s_i$ for all possible values of the angles defining the orientation of the group. The range of possible angles which must be examined is fixed by the symmetry of the crystal and of the model. The Patterson function is then examined at the ends of the vectors $(\underline{r}_i - \underline{r}_j)$ for all i and j, to find the best fit with this model vector set.

Once the orientation of the group is determined, the true relative coordinates of the group, $r_i = [C]\underline{s}_i$, are known. The remaining problem is that of determining the translation vector t which correctly positions the group.

The vector $\underline{R}_O$ has components X_O, Y_O, Z_O. In all methods the procedure is essentially the same. If r_i are correct relative coordinates and T is some symmetry operator then the molecule is positioned at $(r_i + \underline{R}_O)$. There are, therefore, intermolecular vectors of the form $(r_i + \underline{R}_O - T(r_j + \underline{R}_O))$ for all i and j. The Patterson function is, therefore, examined at these positions, i.e. $P[(r_i + \underline{R}_O) - T(r_j + \underline{R}_O)]$ for all i and j, and for all possible values of $\underline{R}_O$.

Criteria of Fit

Any method of Patterson function interpretation, consists of finding the best fit of the model vector set with the Patterson function. For example, having chosen a particular orientation of model, the measure of fit of the vector set corresponding to this model could be the sum, product or minimum over the vector $r_i - r_j$;

$$\sigma(\theta_1 \theta_2 \theta_3) = \sum_{i\,j} P(r_i - r_j)$$

$$P(\theta_1 \theta_2 \theta_3) = \prod_{i\,j} P(r_i - r_j)$$

$$M(\theta_1 \theta_2 \theta_3) = \min_{i\,j} P(r_i - r_j)$$

An alternative description of these Patterson function interpretation techniques can often be given in terms of Patterson superposition techniques. In the expressions given above, the right hand sides represent operations on the corresponding superposition function. Since $P(r - r_j)$ represents the Patterson function with its origin shifted to r_j then the sum over j, $\sum_j P(r - r_j)$ is just the sum function. The second summation then corresponds to examining the sum function at the points r_i and using the sum of the values of sum function found at these points as the criterion of the best fit. In a similar way, the other two expressions can be though of in terms of the product and minimum functions.

Real Space Methods

It is impossible to review here all the variations on real space methods which have been used to solve this problem. Characteristically, they use the sum function for example in the program by Braun, Hornstra and Leenhouts (1969) or the minimum function, or a function related to it, for example Nordman (1980) and Simonov (1980). Program descriptions are given by Schilling et al (1981) and by Beurskens et al (1987). Egert (1983) and Egert and Sheldrick (1985) describe a method and program (PATSEE) which use a modified sum function which lies somewhere between the sum function and the minimum function for the rotation search, but use a modified direct method based on the three phase structure invariants to solve the translation problem. This is then refined by a Patterson Correlation. Doesburg and Beurskens (1983) also describe new translation functions which can follow orientation determination using the methods of Braun et al or Nordman or others. The 'Faltmolekül' method (Huber, 1970) is a real space method which uses as its criterion of fit the sum of the differences between the Patterson function and the convolution molecules.

A Reciprocal Space Method – PATMET

One of the features of modern Patterson function interpretation programs, such as those mentioned above, is to make them more user friendly and thus be accessible to crystallographers used to Direct Methods packages. This has also been the aim of the PATMET program (Wilson, 1989) to be described here. In fact, if the user is willing to follow the program he largely need not concern himself with the fact that the operations are occurring in reciprocal space. However for a clearer understanding the following notes may be useful.

The basic method used is the sum function transformed into reciprocal space (Wilson, 1988).

The $|F(\underline{h})|^2$ values are sharpened and these sharpened values are used as the data throughout.

Coordinates $\underline{s}_i$ of a model of the known fragment are supplied with respect to an arbitrary defined cell. This might be the crystal cell itself or the cell of another crystal which contains the known fragment.

The orientation of the model is found by the search for a maximum in the function $\sigma(\theta_1\theta_2\theta_3)$ (Tollin, 1976) where

$$\sigma(\theta_1\theta_2\theta_3) = \sum_{\underline{h}}|F(\underline{h})|^2\left[(\sum_i \cos 2\pi\underline{h}\underline{r}_i)^2 + (\sum_j \sin 2\pi\underline{h}.\underline{r}_i)^2\right]$$

and $\theta_1\theta_2\theta_3$ are the Eulerian angles. From these the rotation matrix [C] is calculated and applied to the $\underline{s}_i$. The range of orientations to be explored depends on the symmetry of the crystal and of the model (Tollin et al, 1966).

If the model is planar a further separation of variables is possible. Describing the orientation of the group now using spherical polar angles ($\theta\phi\psi$), the orientation of the planar group defined by the orientation of its plane normal ($\theta\phi$) can first be determined and then the azimuthal angle of the model in the plane ψ can be found separately.

The plane normal is found using the $I(\theta\phi)$ function defined by Tollin and Cochran (1964). A maximum in the function

$$I(\theta\phi) = \int P(\underline{r})t(\underline{r})d\underline{r}$$

$$\text{where} \quad t(r) \quad = \quad 1 \text{ on a disc radius R}$$
$$= \quad 0 \text{ elsewhere}$$

as θ and ϕ are varied indicates the normal to the plane.

The function is calculated in reciprocal space

$$I(\theta\phi) = \sum_{\underline{h}}|F(\underline{h})|^2\, 2\pi R^2\, \frac{J_1(2\pi RS)}{2\pi RS}$$

where S is the distance of the reciprocal lattice point $\underline{h}$ from the plane normal.

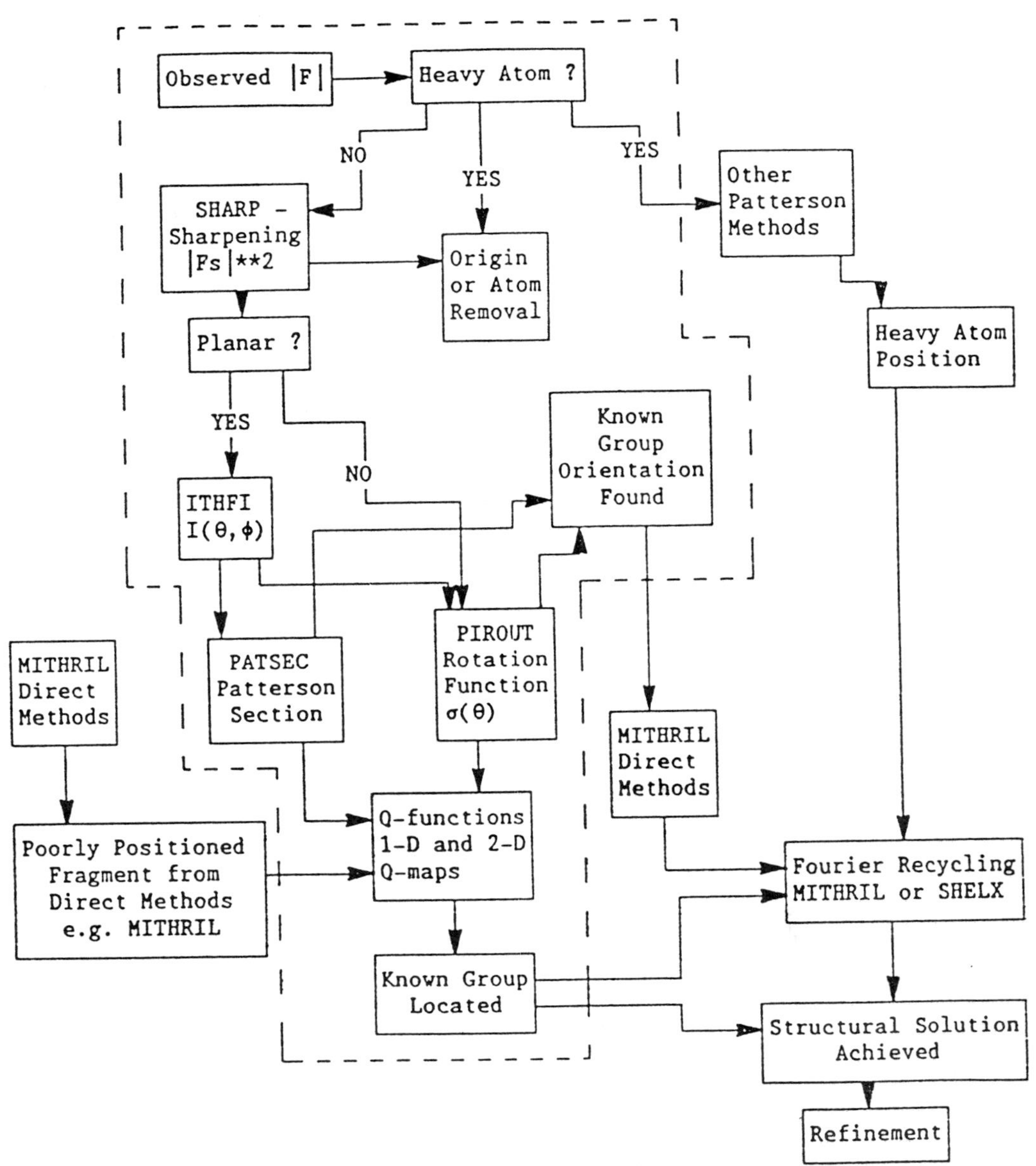

Figure 1. Basic flow diagram of PATMET

The angle ψ is found by calculating a line through the rotation function $\sigma(\theta_1 \theta_2 \theta_3)$ for which $0 < \theta_1 < 2\pi$, $\theta_2 = -\theta$, $\theta_3 = -\phi - \frac{\pi}{2}$. A maximum on this line defines ψ and hence the matrix $[C]$ is calculated and applied to the $\underline{s}_i$ to give correctly oriented coordinates $\underline{r}_i$. As a check, the program can output the section through the Patterson function corresponding to the orientation $(\theta\phi)$ for visual inspection.

The correctly oriented model now is used to calculate the Q-functions (Tollin, 1966) to determine the translation vector R_0 required. The Q functions can be written as

$$Q(R_0) = \sum_{ij} P(\underline{r}_i + \underline{R}_0 - T(\underline{r}_j + R_0))$$

or in terms of the sum function as

$$Q(R_0) = \sum_{j=1}^{n} S_n \left[T(\underline{r}_j + \underline{R}_0) - \underline{R}_0 \right]$$

and is calculated in reciprocal space as

$$Q(\underline{R}_0) = \sum_{\underline{h}} |F_P(h)|^2 \sum_{ij} \cos 2\pi \underline{h} \cdot [\underline{r}_i + \underline{R}_0 - T(r_j + \underline{R}_0)]$$

In fact the components $X_0 Y_0 Z_0$ of $\underline{R}_0$ are determined independently or in pairs with respect to the individual symmetry elements of the space group.

The derivation of these separate Q-functions is given by Tollin, 1966, for planes of symmetry and for two-fold axes. Wilson has recently (1990) generalised the results to cover all space groups and this is taken care of internally within the program. A maximum in $Q(R_0)$ defines the true translation vector. If the correctly oriented model given to the Q-function has pairs of atoms which would give rise to non-Harker peaks in the Harker section then false peaks will occur in the Q-map. However the program calculates and marks their position and also excludes all translation vectors which would give rise to unacceptably close contacts between molecules in the crystal.

The correctly oriented and positioned fragment may be sufficient to give adequate phases to allow Fourier refinement to proceed. If not a further option, the Partial Fragment Rotation Function can be used. If a further group of known stereochemistry is known to be attached to the now determined group but linked by a bond about which it can rotate, a one-dimensional search in the PFRF can now be undertaken (Wilson, 1988). Even when the confirmation is not known, if it can be related to one or two parameters than a modified PFRF can be used (Wilson, 1989).

The results from translation functions can sometimes be unsatisfactory and this can be improved by correlating different Q-functions either those due to different symmetry elements (Wilson and Tollin 1988) or by calculating them with data from different ranges of $\sin\theta$, an idea suggested by Karle (1972) (Wilson and Tollin, 1988).

PATTERSON INTERPRETATION AND DIRECT METHODS

Patterson methods may have advantages when the structure may pose problems for direct methods, for example in the presence of strong planar structures. More importantly, Patterson methods can be used where the data are very inaccurate or of low resolution or from such unlikely sources as powder diffraction. Wilson (1989) has examined in some detail the effect of applying random errors (of up to 100% on average) to experimental data, and using very small data sets. Wilson and Wadsworth (1990) have also considered the application to powder data as have Rius and Miraritlles (1987) who have used a technique similar to that described here.

Rather than as an alternative, Patterson methods can co-operate with Direct Methods. If the orientation of a fragment can be found then this information can be used to improve the phases and weights in the Σ_2 relations and generally improve the statistics (Main, 1976). Thus the orientation of a group found from a rotation function or $I(\theta\phi)$ calculation can be supplied to the Direct Methods program. The program described here is designed to interact with Direct Methods packages. It is sometimes the case that direct methods lead to correctly oriented fragments misplaced in the cell, here again Patterson methods can use these as a starting model for translation functions to resolve the difficulty (Karle 1972).

THE MOLECULAR REPLACEMENT METHOD

The subject of Patterson Methods cannot be left without some mention of the Molecular Replacement Method (Rossmann, 1972). The basic ideal is that if a structure is composed of subunits related by local symmetry elements which are not part of the group symmetry of the crystal, which can often occur in macromolecular structures such as protein crystals and viruses, there is redundant information in the diffraction data which can be used to solve the structure.

The first problem is to determine the non-crystallographic symmetry, the angular relation between the sub-units. The method is to rotate the Patterson of the structure on itself and look for coincidences either in real or reciprocal space. In the latter case, which is by far and away the most common, the function used is the Rotation Function of Rossmann and Blow (1962) or a derivative of it. A number of solutions of the translation problems have been suggested. One by Agarwal (1978) uses the power of modern computers and the FFT algorithm to calculate the R-factor as a function of position.

> After more than 50 years the Patterson function is alive
> and well and can still make significant contributions to
> structure determination.

REFERENCES

Agarwal, R.C., 1978, A new least-squares refinement technique based on the fast Fourier transform algorithm, _Acta. Cryst._, A34:791-809.

Beurskens, P.T., Beurskens, G., Strumpel, M. and Nordman, C.E. 1987, Automation of rotation fuctions in vector space, in: "Patterson and Pattersons", J.P. Glusker, B.K. Patterson and M. Rossi, eds., Oxford University Press, New York.

Buerger, M.J., 1951, A new approach to crystal structure analysis, Acta Cryst., 4:531-544.

Buerger, M.J., 1959, "Vector Space", John Wiley and Sons Inc. New York.

Doesburg, H.M. and Beurskens, P.T. 1983, Acta Cryst. A39: 368-376.

Egert, E., 1983, Patterson search - an alternative to direct methods, Acta Cryst. A39:936-940.

Egert E., and Sheldrick, G.M. 1985, Search for a fragment of known geometry by integrated Patterson and direct methods, Acta Cryst., A41:262-268.

Harker, D., 1936, The application of the three dimensional Patterson method and the crystal structures of proustite Ag_3AsS_3 and pyrargyrite Ag_3SbS_3, J. Chem. Phys., 4:381-390.

Main, P., 1976, Recent Developments in the MULTAN System. The Use of Molecular Structures, in "Crystallographic Computing Techniques", F.R. Ahmed, ed., Munksgaard, Copenhagen.

Nordman, C.E., 1980, Vector space search, in:"Computing and Crystallography" R. Diamond, S. Ramaseshan and K. Venkatesan, eds., The Indian Academy of Sciences, Bangalore.

Patterson, A.L., 1934, A Fourier series method for the determination of the components of interatomic distances in crystals, Phys. Rev. 46:372-376.

Patterson, A.L., 1935, A direct method for the determination of the components of interatomic distances in crystals, Z. Krist., A90:517-542.

Ramachandran, G.N., and Raman, S., 1959, Synthesis for the deconvolution of the Patterson function Part 1 General Principles, Acta. Cryst., 12:957-564.

Richardson, J.W., and Jacobson, R.A., 1987, Computer-aided analysis of multi-solution Patterson superpositions, in: "Patterson and the Pattersons", J.P. Glusker, E.K. Patterson and M. Rossi, eds., International Union of Crystallography, Oxford University Press.

Rius, J. and Miravitlles, C., 1987, An automated full-symmetry Patterson search method, J. Appl. Cryst., 20:261-264.

Rossmann, M.G., 1972, "The Molecular Replacement Method", Gordon and Breach, New York.

Rossmann, M.B., and Blow, D.M., 1962, The detection of sub-units within the crystallographic asymmetric unit, Acta. Cryst., 15:24-31.

Schilling, J.W., Hage, L.G., Strumpel, M., and Nordman, C.E., 1981, "Patterson Search Programs - Write up and User's Guide", The University of Michigan, Ann Arbor, Michigan.

Siminov, V.I., 1980, Automatic interpretation of the Patterson function, in:
"Computing and Crystallography", R. Diamond, S. Ramaseshan and
K. Venkatesan, eds., The Indian Academy of Sciences, Bangalore.

Tollin, P., 1976, Some Patterson interpretation techniques and their
application to biologically important structures, in:
"Crystallographic Computing Techniques", F.R. Ahmed, ed., Munksgaard,
Copenhagen.

Tollin, P., and Cochran, W., 1964, Patterson function interpretation for
structures containing planar groups, Acta Cryst., 17:1332.

Tollin, P., Main, P., and Rossman, M.G., 1966, The symmetry of the rotation
function, Acta. Cryst., 20:404 -407.

Wilson, C.C., 1988, Recent Advances in Reciprocal space Patterson Methods
and the PATMET program. S.E.R.C. Rutherford Appleton Laboratory,
Didcot, Oxford Publication. RAL-88-087.

Wilson, C.C. 1988, Increasing the size of search fragments for use in
Patterson method calculations – the partial-fragment rotation function,
Acta. Cryst., A44:478-481.

Wilson, C.C., 1989, Practical aspect of the Patterson methods program
PATMET, S.E.R.C. Rutherford Appleton Laboratory, Chilton, Didcot,
Oxford Publication, RAL-89-125.

Wilson, C.C., 1989, Determination of crystal structures for poor-quality
data using Patterson methods, Acta. Cryst., A45:833-839.

Wilson, C.C., 1990, Molecular confirmation determination in unknown crystal
structures by Patterson methods, J. Appl. Cryst. (in press).

Wilson, C.C., and Tollin, P., 1988, Improving the interpretation of
translation functions by a simple map-correlation procedure,
Acta. Cryst., A44:226-230.

Wilson, C.C., and Wadsworth,.J.W., 1990, Crystal structure determination
from low-resolution X-ray powder diffraction data, Acta. Cryst.,
A46 (in press).

OPTIMAL SYMBOLIC PHASE DETERMINATION

Rene´ Peschar and Henk Schenk

Laboratory for Crystallography, University of Amsterdam
Nieuwe Achtergracht 166, 1018 WV Amsterdam, The Netherlands

INTRODUCTION

In spite of various differences, direct-method programs have one
procedure in common. After normalizing the structure factors and
generating the phase relations, a convergence procedure (Germain, Main &
Woolfson, 1970) is invoked in order to get a (small) set of starting
reflections. After fixing the origin with a subset of the starting set
reflections, the remaining starting set reflections are assigned either
various numerical trial values or symbolic phase values. In a second step
all other phases are expressed successively in terms of the starting set
phases, taking into account the statistical weights of the newly phased
reflections. A drawback of this phasing technique may be the fixed
starting set. The statistical weights can not be incorporated in the
convergence procedure as they are calculated during and propagating
throughout the phasing process. Therefore, a fixed starting set does not
guarantee the most efficient starting set nor the best possible order of
phasing reflections. A more effective procedure, in which both the
starting set and the order of phasing reflections are flexible, can be
based on the dynamic programming technique (Bellman, 1957).

DYNAMIC PROGRAMMING

Dynamic programming is a technique for handling so-called multi-stage
decision processes in order to obtain an optimal sequence of decisions
with respect to a predefined criterion function. The subsequent optimal
decisions form together an optimal policy. Dynamic programming is applied
to a system T. A state of the system is denoted as T_j for $j = 0,..,N$. T_0 is
called the initial or zero state, T_N the final state. Each state T_j is
characterized by a small set of M parameters, the state variables.

$$\vec{t}_j = (t_{1j},...,t_{Mj}) \quad j = 0,..,N \tag{1}$$

The system starts in T_0 and changes in N steps, called stages, into T_N.
In stage j the system goes from state T_{j-1} to state T_j. The goal of
dynamic programming is to maximize a predefined criterion function R of
the state variables,

Direct Methods of Solving Crystal Stuctures
Edited by H. Schenk, Plenum Press, New York, 1991

$$R = \max[F_j(\vec{t}_j)] \qquad (2)$$

The state variables choice which optimizes the criterion function will be denoted by $\vec{t}_j^{\,o}$. Maximization of R is achieved by ensuring that only and all optimized criterion function value(s) from stage j are used as input to the optimization of the criterion function in stage j+1.
This leads to a functional recurrence relation,

$$F_j(\vec{t}_j^{\,o}) = \max[\ F(\vec{t}_j)\ \text{using}\ F_{j-1}(\vec{t}_{j-1}^{\,o})] \quad \text{for}\ j = 1,..,N \qquad (3)$$

The dynamic programming process for a function of M state variables consists of the following steps. In the first stage R is maximized for each of the state variables. This ensures that for the next stage all and only all optimal R values are present. In each of the subsequent stages R is maximized again for all stage variables using only the optimal R values from the previous stage. As a result, the N-stage process is divided into N distinct 1-stage processes. The latter are maximized separately. The only decisions made concern the maximization of R, ensuring a final unique optimum. Nevertheless, several equivalent optimal paths may lead to the same unique optimum. This reflects the Principle of Optimality (Bellman, 1957) which states that whatever the initial state and initial decision are, the remaining decisions must constitute an optimal policy with regard to the state resulting from the first decision.

<u>Exercise : Optimal path problem</u>

A sales-man living in city A wants to visit the cities, B, C, D, and E before finally returning again in A. An intercity travelling-time table (see table 1) is available. Determine the least time-consuming route.

Table 1. Inter-city time-table in minutes

	A	B	C	D	E
A	0	40	60	30	20
B	40	0	50	70	40
C	60	50	0	10	50
D	30	70	10	0	40
E	20	40	50	40	0

OPTIMAL PHASE DETERMINATION

The dynamic programming technique as applied to the phasing process, will be referred to as Dynamic Programming Phasing or DPP. It constitutes an optimal phasing technique with respect to a predefined criterion function and the a priori probabilistic assumptions. We shall first formulate the requirements for the phase determination process :
1. As many phases as possible are to be phased with a unique phase indication. If symbolic phase values are employed, phases may get different symbolic phase indications. These phases can not be used to determine other phases unless suitable assumptions are made.
2. The final set of phases should be as reliable as possible.
3. The starting set, consisting of reflections the phases of which must be chosen, should be as small as possible.

4. Neither the starting set reflections nor the order in which the phases should preferably be determined must be defined a priori.

The Dynamic-Programming approach

In case of the phase determination of M reflections a total of M! different phasing sequences exist. In addition, the problem of selecting the starting set reflections must be solved. It is quite clear that in general one cannot possibly handle all these possible sequences in order to arrive at the optimal one (e.g. for thirty reflections $30! \approx 2.65 \times 10^{32}$) by simply enumerating all possibilities. Dynamic programming however does provide the tool for a systematic phasing procedure, yielding a optimal phasing sequence and involving only a small fraction of the effort required for the enumerating approach.

The first thing to do is to select the state variables and the criterion function for this problem. The order in which the reflections must be phased should be kept variable. This suggests to take the phasing of a reflection as a state variable. It should be noted that this phasing may consist of either a determination on basis of other reflections or the selection as a new starting set reflection. In view of the requirement that the final phase set should be as reliable as possible, it is obvious that the criterion function should be a function of the statistical weights of the phased reflections. The goal of DPP to phase as many reflections as possible with as few starting set reflections as possible implies that the size of the starting set should be variable within certain practical limits. Obviously, the order in which the reflections are to be phased is variable since it will be calculated by the DPP process. This leads to the conclusion that the criterion function must be evaluated in every stage for each stage variable. The actual choice for R is of utmost importance since DPP aims at maximizing R only. As a result, the phasing will be optimal only with respect to R. Several functional forms are possible which incorporate the above requirements. An example is the criterion function W1, consisting of maximizing the average statistical weight of the phased reflections,

$$W1_i(Nst) = \frac{\sum\limits_{H} g_H}{j} \quad \text{maximal} \tag{4}$$

Obviously, W1 is a function of the number of starting set reflections Nst.

The execution of DPP

The execution of the DPP procedure will now be explained in more detail. In each stage j, $1 \leq j < M$ optimal j-sequences are to be selected (a sequence of n subsequently phased reflections will be called a n-sequence). This selection is executed by maximizing R for each reflection (state variable) using all optimal (j-1)-sequences which result from the previous stage. In T_0 no sequences are present yet : the procedure starts from scratch. In the first stage optimal 1-sequences are selected. Obviously, no phases can be determined yet since the 0-sequence contains no reflections. Consequently, any reflection i_1, $i_1=1,..,M$, will be selected as a starting-set reflection, resulting in M optimal 1-sequences. In the second stage optimal 2-sequences are to be found. If only triplets amongst three different reflections are considered, any reflection i_2 can be selected as a starting-set reflection in combination with any optimal 1-sequence reflection i_1 provided $i_2 \neq i_1$. Hence, after the second stage $M \times (M-1)$ optimal 2-sequences result. The first stage to be performed in practice is the third stage. Considering the maximization of

R for a particular reflection i, this reflection may form a triplet with
any of the optimal 2-sequences. Obviously, the triplet with the largest E_3
determines reflection i most reliably which leads directly to the optimal
3-sequence ending at reflection i. This optimal 3-sequence and its
associated weight are assigned to reflection i. In this way each
reflection i, i = 1,..,M, is provided with an optimal 3-sequence $S_i(o,3)$,
ending at reflection i, and an associated weight $w_i(o,3)$. In the next
stage, the maximization of R comes down to combining each of the M
reflections with each of the optimal 3-sequences, in which the reflection
is not present yet, forming 4-sequences. For each of the 4-sequences
ending at a particular reflection i the statistical weight of the
reflection i, $w_i(4)$, can be calculated and added to the weight of the
3-sequence j, the total weight being $W_i(4) = w_j(o,3) + w_i(4)$. Obviously,
the 4-sequence which yields the largest weight $W_i(4)$ maximizes the
criterion function so the optimal 4-sequence ending at this reflection is
obtained. In this way for each of the reflections 1,..,M an optimal
4-sequence $S_i(o,4)$, with an associated weight $W_i(o,4)$ can be calculated.
As a result, at the end of the fourth stage we are left with at least M
optimal 4-sequences, one for each i. If a reflection is combined with an
optimal sequence with which it does not form triplets, no weight can be
calculated for this reflection. As a result, it will be included as a
starting set reflection.

This procedure is repeated until finally the M th reflections are
combined with the optimal (M-1)-sequences, resulting in optimal
M-sequences, $S_i(o,M)$, with weights $W_i(o,M)$. Of these sequences, the one
with the largest $W_i(o,M)$ is the optimal M-sequence looked for. Some
additional remarks should be made concerning this technique. At each stage
more than one sequence, all having the same number of starting set phases,
may exist which optimize R for the same reflection. These equivalent
sequences should all included in the procedure though for practice reasons
there number may be limited. A difficulty in optimizing R may seem to
arise when comparing sequences, differing in their number of starting set
reflections, at the determination of the same reflection. Obviously,
sequences with a different average weight are acceptable provided the
averages increase with the number of starting set reflections present.
However, if a sequence has a lower average weight in spite of the more
starting set phases it will be rejected.

In the current implementation of DPP in the direct method program SIMPEL,
sequences containing up to six different number of starting set
reflections may be accepted when maximizing each reflection in any stage j

CALCULATION OF STATISTICAL WEIGTHS

The statistical weights to be employed in the phasing process are not
unique and may be defined in various ways (Germain, Main & Woolfson, 1971;
Hull & Irwin, 1978; Giacovazzo, 1980). Nevertheless, in general they will
derived from a joint probability distribution of a phase given a number of
triplets,

$$P(\Phi_H) = L^{-1}\exp\left[2\alpha_H\cos(-\Phi_H + \Xi_H)\right] \tag{5}$$

with

$$\alpha_H\exp(i\Xi_H) = \sum_{j=1}^{n} E_{3j}\exp[i(\Phi_{K_j} + \Phi_{H-K_j})] \tag{6}$$

and

$$\alpha_H^2 = \sum_{j,j'} E_{3j}E_{3j'}\cos[(\Phi_{K_j} + \Phi_{H-K_j}) - (\Phi_{K_j'} + \Phi_{H-K_j'})] \tag{7}$$

L^{-1} being a normalization constant and E_{3j} the reliability indicator of a triplet (Cochan, 1955; Cochran & Woolfson, 1955; Giacovazzo, 1974)

An example is the tangent formula (Karle & Hauptman, 1956; Karle & Karle, 1966) which can be obtained by equating to zero the derivative of expression (5) with respect to Φ_H. In practice weighted tangent expressions are used, including statistical weights w_{K_j} and w_{H-K_j} for the known phases

$$\tan(\Phi_H) = \frac{\sum_{j=1}^{n} E_{3j}w_{K_j}w_{H-K_j}\sin(\Phi_{K_j}+\Phi_{H-K_j})}{\sum_{j=1}^{n} E_{3j}w_{K_j}w_{H-K_j}\cos(\Phi_{K_j}+\Phi_{H-K_j})} \tag{8}$$

If symbolic phase values are employed (8) can not be used directly to phase reflections. If all symbolic phase parts are identical (A_H), the expectation value for Φ_H, $<\exp(i\Phi_H)>$, can be obtained in a standard way from (5) for both general and value restricted reflections and shown to be expressable as,

$$<\exp(i\Phi_H)> = g_H\exp[i(A_H+y_H)] \tag{9}$$

y_H is the numerical part of the symbolic phase indication A_H+y_H for Φ_H. The g_H lies between 0 and 1 and can be used as a weight reflecting the reliability of the symbolic phase indication. For a starting set reflection $g_H = 1$. If two or more different symbolic phase parts A_j are present, $<\exp(i\Phi_H)>$ can not be expressed as a single symbolic phase indication unless additional assumption are invoked (Peschar & Schenk, 1987).

REFERENCES

Bellman, R. (1957). Dynamic Programming. Princeton: Princeton
    University Press, New Jersey.
Cochran, W. (1955). Acta Cryst. <u>8</u>, 473-478.
Cochran, W. & Woolfson, M.M. (1955). Acta Cryst. <u>8</u>, 1-12.
Germain, G., Main, P. & Woolfson, M.M. (1970). Acta Cryst. <u>B26</u>, 274-285.
Germain, G., Main, P. & Woolfson, M.M. (1971). Acta Cryst. <u>A27</u>, 368-376.
Giacovazzo, C. (1974). Acta Cryst. <u>A30</u>, 631-634.
Giacovazzo, C. (1980). Acta Cryst. <u>A36</u>, 74-82.
Hull, S.E. & Irwin, M.J. (1978). Acta Cryst. <u>A34</u>, 863-870.
Karle J. & Hauptman, H. (1956). Acta Cryst. <u>9</u>, 635-651.
Karle, J. & Karle, I. (1966). Acta Cryst. <u>21</u>, 849-859.
Peschar, R. & Schenk, H. (1987). Acta Cryst. <u>A43</u>, 751-763.

RANDOM APPROACHES TO THE PHASE PROBLEM

Michael M. Woolfson

Physics Department
University of York
York YO1 5DD, U.K.

INTRODUCTION

A weakness of the original multisolution methods was that they relied on a very small base of initial phases so that some early phase indications were unreliable. If in the early stages there were a few phase relationships which held poorly then this could throw the phasing into confusion no matter what the starting phase set. Thus the pattern of phase relationships, for every starting phase set, could indicate that the phases of two particular reflexions differed by π when in fact they were almost equal. It was early realised that such difficulties could be obviated by having a much larger starting set and here we shall be examining various ways in which this goal was sought.

THE MAGIC-INTEGER CONCEPT

This idea was first described by White and Woolfson (1975). Let us consider n phases ϕ_j, $j = 1$ to n, expressed in cycles so that $0 \le \phi < 1$. We write

$$\phi_j = m_j x \qquad (\mathrm{mod}\ 1) \qquad\qquad (1)$$

where the m's are integers and $0 \le x < 1$. The magic integer

Direct Methods of Solving Crystal Stuctures
Edited by H. Schenk, Plenum Press, New York, 1991

approach is to say that for some value of x a near match will be found for equations (1).

For example with $\phi_1 = 0.3$, $\phi_2 = 0.2$, $\phi_3 = 0.7$
$$m_1 = 4 , \quad m_2 = 6, \quad m_3 = 7$$
then for $x = 0.538$ the values of $m_j x$ (mod 1) are 0.152, 0.228, 0.766 with errors (in degrees) of $53°$, $10°$ and $24°$.

Main (1977) has given the rules for generating efficient magic-integer sequences of any size and formulae for estimating the rms error.

INVARIANTS IN MAGIC-INTEGER FORM

For a given integer sequence a number of variables may be used so that the elements of the matrix

$$\begin{Bmatrix} 8 \\ 12 \\ 14 \\ 15 \end{Bmatrix} \quad (x \ y \ z)$$

may be used to generate twelve phases.

Since individual phases may be expressed in magic-integer form so triple-phase relationships may be similarly represented.   With

$$\phi_1 = 8x, \ \phi_2 = 14x, \ \phi_3 = 12y$$

the relationship

$$\phi_1 - \phi_2 + \phi_3 + 0.5 \ \approx \ 0 \ (\text{mod } 1) \tag{2}$$

appears in the form

$$-6x + 12y + 0.5 \ \approx \ 0 \ (\text{mod } 1) \tag{3}$$

and, in general any relationship appears as

$$Hx + Ky + Lz + b \ \approx \ 0 \ (\text{mod } 1) \tag{4}$$

where phases are expressed in cycles.   The correct values

would tend to satisfy most of the interconnecting relationships which means that for some values of x, y and z which give reasonable representations of the phases the relationships in form (4) should hold reasonably well. This leads us to look for peaks in the function

$$\psi(x,y,z) = \sum_r^M \kappa_r \cos\{2\pi(H_r x + K_r y + L_r z + b_r)\} \tag{5}$$

where the summation is over the M relationships linking the phases represented by magic integers plus any others with known phases. A peak in the ψ-map gives possible trial values for phases which can then be used as a starting point for a phase extension process.

THE PRIMARY-SECONDARY METHOD

It has been found possible to extend the magic-integer representation of phases by what is called the primary-secondary (P-S) method. A number of phases, called primaries, are represented directly by magic integers. Then strong relationships are found which link two primaries with some other reflexion (a secondary) which may thus be represented in magic-integer form. The ψ-map may then be calculated with all the relationships which link the primary and secondary reflexions. These ideas led to the programme MAGIC (Declercq, Germain and Woolfson, 1975) which was added to the MULTAN system. Later, with more improvements added, another magic-integer based procedure, MAGEX, appeared on the scene (Hull et al., 1981; Zhang and Woolfson, 1982).

The magic-integer-based methods were very often successful when the original MULTAN procedure failed. Their strength was that they enabled a large number of reflexions, and a corresponding large number of relationships, to be employed from the start and, most importantly, through the use of a ψ-map the relationships were used all together rather than via a chain process. Thus a few unreliable relationships would have little impact on the magic-integer methods as long as most of the relationships held reasonably

well. On the other hand one or two bad relationships early
in a chain process could have disasterous consequences that
subsequent steps could not correct.

LINEAR EQUATIONS

Woolfson (1977) suggested that, as an alternative to the
tangent formula, once phase estimates for all phases were
available, then refinement could be carried out with linear
equations and a least-squares process.

A triple-phase relationship, with κ weighting, can be
written

$$\kappa \, \phi_p \pm \kappa \, \phi_q \pm \kappa \, \phi_r \pm \kappa \, b \; \approx \; \kappa \, n \tag{6}$$

where phases are in cycles and n is some integer. If the
integers were known then the complete set of relationships
could be written in matrix form as

$$\mathbf{A} \, \mathbf{x} \; \approx \; \mathbf{c} \tag{7}$$

and, replacing $\approx$ by =,' there is a least-squares solution

$$\mathbf{x} = (\mathbf{A}^T\mathbf{A})^{-1} \, \mathbf{A}^T\mathbf{c} \tag{8}$$

If approximate phases are available then the nearest
integers may be found and equations (7) and (8) provide the
basis for the cyclic refinement of the phases. Tests show
that this method of phase refinement is always at least as
good as the tangent formula and usually much better.

PSEUDO-WEIGHTING

When the initial phases are of poor quality a weighting
scheme, in addition to κ weighting, is required. For
example, if with the current phase estimates the three-phase
invariant has the value 1.5 then one cannot choose a nearest
integer. It would be desirable to eliminate such a
relationship. If, on the other hand the value was, say,
1.45 then the nearest integer is 1 but it should clearly be

down-weighted.   If different weights were applied to the system of equations in every cycle then the refinement process would be very inefficient as it would be necessary to invert a different large matrix in every cycle.

An alternative approach, which gives a pseudo-weighting scheme, involves changing the right-hand side of the equations but keeps the matrix **A** fixed.   If the current value of the triple phase invariant was 1.5 then putting this value on the right-hand side would effectively eliminate the influence of the equation since it would give no tendency to modify the involved phases.   Investigation has shown that an effective pseudo-weighting scheme is obtained by writing the equations as

$$\kappa \ (\ \phi_p \ \pm \ \phi_q \ \pm \ \phi_r \ + b) = \kappa \ (n + 4\alpha^3) \tag{9}$$

where α is the deviation of the invariant from the nearest integer with the current phase estimates so that

$$-1/2 \ \leq \ \alpha \ < \ 1/2.$$

For an invariant value 1.5 then 1.5 would appear on the right-hand side, as required.   With invariant values of 1.9 and 1.1 the right-hand sides would appear as 1.936 and 1.064 respectively, closer to the nearest integers than the original invariant values.   Since the inversion of $\mathbf{A}^T\mathbf{A}$ needs only to be done once, quite large systems can be refined economically.

THE RADIUS OF CONVERGENCE

In examining the efficiency of any refinement process it is natural to examine its radius of convergence.   The way this was done was by imposing random errors, with various rms values, on to the correct phases to see whether the refinement process would give back a more-or-less correct phase set, which was take to be with a final rms error of less than $30°$.   The result of doing this for various trial structures was a surprise.   With one fairly difficult structure, 3-chloro-1,3,4-triphenyl-azetidin-2-one, starting with rms errors of $75°$ still led to correct phase sets six times with ten trials for a system with 70 phases. Increasing the initial rms phase error to $90°$ still gave four successful trials out of ten. The obvious thing to try was to start with random phases and again it was found that an appreciable fraction of trials were successful. The basis of *Random approaches to the phase problem*, the topic of this contribution, was established.

RANTAN

Following the discovery of the principle of using initially random phases a procedure called YZARC was devised. This used random phases for an initial set of up to 100 phases followed by linear equations refinement and then phase extension and further refinement by the tangent formula. At this stage in the development of direct methods the objective of starting with a large starting set had been taken quite a long way but it was taken even further by Yao (1981) with his introduction of the RANTAN procedure. In RANTAN initially-random phases are allocated to *all* the reflexions in the system which are then refined by the tangent formula. However, the success of the method depends on the controlled use of weights in the tangent refinement.

Weights are allocated to the initial phases as follows: origin-fixing phases, weight = 1.00; enantiomorph-fixing reflexions with special value, weight = 0.99; enantiomorph-fixing reflexions with general value, weight = 0.85; random phase, weight = 0.25. Any other 'known' phase, for example one estimated by a Σ_1 formula, is included with a weight 2P - 1, where P is the probability of the indicated special value.

This system of weights, and especially the weight for a random phase, has been determined empirically; the method is not too sensitive to the weights used as long as they have approximately the values suggested here. In the tangent refinement process in RANTAN the initial random phase is not changed until a phase estimate is obtained with a new weight

greater than 0.25. Then, and only then, is the phase allowed to vary and to follow its refinement path.

Experience has shown that sometimes one or other of the normal weighted or Hull-Irwin weighted tangent formalae is effective and the other is not. The former is faster to apply but if there is an enantiomorph-defining problem then Hull-Irwin weighting is preferred.

In RANTAN a large number of trial sets of random phases is refined and then subjected to the usual MULTAN figures of merit. For small structures ($\leq$ 100 independent atoms) one hundred trials or less usually suffices to obtain a solution. However, there have been applications, for small proteins, where one thousand trials have been used.

OTHER TECHNIQUES OF RANDOM PHASE REFINEMENT

Debaerdemaeker and Woolfson (1983) explored the use of seven different functions for refinement of initially-random phases by a parameter-shift method. Several of the functions involve the summations

$$X(\mathbf{h}) = \sum_{\mathbf{k}} |E(\mathbf{h})\ E(\mathbf{k})\ E(\mathbf{h-k})|\ \cos\{\phi(\mathbf{h}) - \phi(\mathbf{k}) - \phi(\mathbf{h-k})\} \qquad (10)$$

$$Y(\mathbf{h}) = \sum_{\mathbf{k}} |E(\mathbf{h})\ E(\mathbf{k})\ E(\mathbf{h-K})|\ \sin\{\phi(\mathbf{h}) - \phi(\mathbf{k}) - \phi(\mathbf{h-k})\} \qquad (11)$$

and the quantities $\eta\{\kappa(\mathbf{h,k})\} = I_1\{\kappa(\mathbf{h,k})\}/\ I_0\{\kappa(\mathbf{h,k})\}$,

$$C(\mathbf{h,k}) = \cos\{\phi(\mathbf{h}) - \phi(\mathbf{k}) - \phi(\mathbf{h-k})\}$$

$$\text{and} \quad S(\mathbf{h,k}) = \sin\{\phi(\mathbf{h}) - \phi(\mathbf{k}) - \phi(\mathbf{h-k})\} \qquad (12)$$

The functions are

$$\psi_A = \sum_{\mathbf{h}} [\{Z\ X(\mathbf{h}) - |E(\mathbf{h})|^2\}^2 + \{Z\ Y(\mathbf{h})\}^2] \qquad (13)$$

where $\quad Z = \sum_{\mathbf{h}} |E(\mathbf{h})|^2 / \sum_{\mathbf{h}}\sum_{\mathbf{k}} |E(\mathbf{h})\ E(\mathbf{k})\ E(\mathbf{h-k})|\ \eta\{\kappa(\mathbf{h,k})\}$;

$$\psi_B = \sum_{\mathbf{h}} [\{Z(\mathbf{h})\ X(\mathbf{h}) - |E(\mathbf{h})|^2\}^2 - \{Z(\mathbf{h})\ Y(\mathbf{h})\}^2] \qquad (14)$$

where $\quad Z(\mathbf{h}) = |E(\mathbf{h})|^2 / \sum_{\mathbf{k}} |E(\mathbf{h})\ E(\mathbf{k})\ E(\mathbf{h-k})|\ \eta\{\kappa(\mathbf{h,k})\}$;

$$\psi_C = \sum_h \{X(\mathbf{h}) - |Y(\mathbf{h})|\} \qquad ; \qquad (15)$$

$$\psi_D = \sum_h \{X(\mathbf{h}) - Y(\mathbf{h})\} \qquad ; \qquad (16)$$

$$\psi_E = \sum_h \sum_k \kappa(\mathbf{h},\mathbf{k})\,[C(\mathbf{h},\mathbf{k}) - \eta\{\kappa(\mathbf{h},\mathbf{k})\}]^2 \qquad ; \qquad (17)$$

$$\psi_F = \sum_h \sum_k \kappa(\mathbf{h},\mathbf{k})\ \eta\{\kappa(\mathbf{h},\mathbf{k})\}\ C(\mathbf{h},\mathbf{k}) \qquad (18)$$

and

$$\psi_G = \sum_h \sum_k \kappa(\mathbf{h},\mathbf{k})\left\{[C(\mathbf{h},\mathbf{k})-\eta\{\kappa(\mathbf{h},\mathbf{k})\}]^2 + \left|S(\mathbf{h},\mathbf{k})^2 - \frac{1}{2} - \frac{1}{2}\frac{I_2\{\kappa(\mathbf{h},\mathbf{k})\}}{I_1\{\kappa(\mathbf{h},\mathbf{k})\}}\right|\right\}$$

$$(19)$$

The function ψ_D arose because of a programming error during the investigation. The surprising result is that, despite the fact that it has no rational basis, the function is the most effective of all those which were examined. This function, incorporated in a procedure called XMY, is now a standard part of the York distributed package and has proved to be extremely useful (Debaerdemaeker and Woolfson, 1989).

SAYTAN

No account of random approaches to the phase problem would be complete without mention of the SAYTAN package, which has replaced MULTAN and is totally committed to the use of initially-random phase sets. At the heart of SAYTAN is the Sayre-equation tangent formula (Debaerdemaeker, Tate & Woolfson, 1985,1989). This has the characteristic that it tends to give phases for a set of reflexions with large |E| which satisfy Sayre's equation both for the reflexions with large |E| and also for a selected set of reflexions with

small (ideally zero) |E|. The equation is of the form

$$\phi(\mathbf{h}) = \text{phase of } \{\ t(\mathbf{h}) - 2\ K\ q(\mathbf{h})\ \} \qquad (20)$$

where

$$t(h) = \sum_{k} [1/g(h)+1/g(k)+1/g(h-k)] \; E(k) \; E(h-k) \qquad (21)$$

$$q(h) = \sum_{l} [1/g(l)^2] \sum_{k} E(k) \; E(l-k) \; E(h-l) \qquad (22)$$

$$K = \frac{\sum_{k} E(\bar{k}) \; t(k)}{\sum_{k} E(\bar{k}) \; q(k)} \qquad (23)$$

and g(h) is the scattering factor for atoms with squared electron density. In the summation for q(h) there are two kinds of term, those for which there is a cross term E(k) of large magnitude, referred to as a large quartet term and those for which E(k) is of small magnitude, called a small quartet term. The extra power of including a set of small |E|'s into the phasing procedure makes SAYTAN far more powerful than conventional MULTAN and SAYTAN is now the default method in the distributed York package. Running it in its fullest form with both large and small quartet terms takes about twice as long as MULTAN. However, there is a cheaper mode of running SAYTAN which uses only the small quartets and a Hull-Irwin weighting scheme for the triples term which simulates the behaviour of the large quartets.

There are many aspects of running SAYTAN, such as the use of weighting schemes, which are relevant to its effective use in various situations. For example, it has been found that having a non-unit weight for small quartet terms is advantageous although from the users point of view these matters are dealt with automatically by the programme.

REFERENCES

Debaedemaeker, T., Tate, C. and Woolfson, M. M., 1985, On the application of phase relationships to complex structures. XXVI The Sayre tangent formula. *Acta Cryst.*, A**41**, 286-290.

Debaerdemaeker, T., Tate, C. and Woolfson, M. M., 1988, On the application of phase relationships to complex structures. XXVI Developments of the Sayre-equation tangent formula. *Acta Cryst.*, **A44**, 353-357.

Debaerdemaeker, T. and Woolfson, M. M., 1983, On the application of phase relationships to complex structures. XXII Techniques for random phase refinement. *Acta Cryst.*, **A39**, 193-196.

Debaerdemaeker, T. and Woolfson, M. M., 1989, On the application of phase relationships to complex structures. XXVIII XMY as a random approach to the phase problem. *Acta Cryst.*, **A45**, 349-353.

Declercq, J. P., Germain, G., Main, P. and Woolfson, M. M., 1975, On the application of phase relationships to complex structures. VIII An extension of the magic integer approach. *Acta Cryst.*, **A31**, 367-372.

Hull, S. E., Viterbo, D., Woolfson, M.M. and Zhang Shao-hui, 1981, On the application of phase relationships to complex structures. XIX Magic integer representation of a large set of phases: the MAGEX procedure. *Acta Cryst.*, **A37**, 566-572.

Main, P., 1977, On the application of phase relationships to complex structures. XI A theory of magic integers. *Acta Cryst.*, **A33**, 750-757.

White, P. S. and Woolfson, M. M., 1975, On the application of phase relationships to complex structures. VII Magic integers. *Acta Cryst.*, **A31**, 53-56.

Woolfson, M. M., 1977, On the application of phase relationships to complex structures. X MAGLIN, a successor to MULTAN. *Acta Cryst.*, **A33**, 219-225.

Yao Jia-xing, 1981, On the application of phase relationships to complex structures. XVIII RANTAN - random MULTAN. *Acta Cryst.*, **A37**, 642-644.

Zhang Shao-hui and Woolfson, M. M., 1982, On the application of phase relationships to complex structures. XXI An extension of the MAGEX procedure. *Acta Cryst.*, A38, 683-685.

MAXIMUM ENTROPY AND THE SADDLEPOINT METHOD

Gérard Bricogne*

MRC Laboratory of Molecular Biology
Hills Road
Cambridge CB2 2QH, U.K.

INTRODUCTION

The purpose of this second talk is to examine in detail the nature of the relation between the traditional formulation of probabilistic direct methods, their maximum-entropy reformulation, and the saddlepoint method. Most of the material in this talk has already appeared in the literature (Bricogne, 1984, 1988a, 1988b) but it is hoped that the choice and presentation adopted here will make the main results stand out clearly.

1. ANALYTICAL METHODS OF PROBABILITY THEORY

This section is not intended as an introduction to probability theory – for which the reader is referred to Cramér (1946), Lindley (1965), Petrov (1975) or Bhattacharya & Ranga Rao (1976) – but only as a description of the analytical devices used in formulating and implementing direct methods of phase determination.

1.1. Convolution of probability densities

The addition of independent random variables or vectors leads to the convolution of their probability distributions : if X_1 and X_2 are two n-dimensional random vectors over the

---

* Permanent address : LURE, Bâtiment 209d, 91405 Orsay, France.

Direct Methods of Solving Crystal Stuctures
Edited by H. Schenk, Plenum Press, New York, 1991

"

reals, independently distributed with probability densities P_1 and P_2 respectively, then their sum $\mathbf{X} = \mathbf{X}_1 + \mathbf{X}_2$ has probability density $\mathcal{P}$ given by

$$\mathcal{P}(\mathbf{X}) = \int_{\mathbb{R}^n} P_1(\mathbf{X}_1)\, P_2(\mathbf{X}-\mathbf{X}_1)\, d^n\mathbf{X}_1 = \int_{\mathbb{R}^n} P_1(\mathbf{X}-\mathbf{X}_2)\, P_2(\mathbf{X}_2)\, d^n\mathbf{X}_2$$

i.e.
$$\mathcal{P} = P_1 * P_2 \quad . \tag{1.1}$$

This result can be extended, by means of the theory of distributions, to the case where P_1 and P_2 are singular measures and do not have a density with respect to the Lebesgue measure in $\mathbb{R}^n$.

1.2. Characteristic functions

This convolution can be turned into a simple multiplication by considering the Fourier transforms, called the *characteristic functions*, of P_1, P_2 and $\mathcal{P}$. The Fourier transformation used here is defined with a slightly different normalisation from the usual one in crystallography in that there is no factor of 2π in the exponent :

$$C(t) = \int_{\mathbb{R}^n} P(\mathbf{X})\, e^{\,i t \cdot \mathbf{X}}\, d^n\mathbf{X} \quad . \tag{1.2}$$

Then by the convolution theorem :
$$C(t) = C_1(t) \times C_2(t) \quad , \tag{1.3}$$

so that $P(\mathbf{X})$ may be evaluated by Fourier inversion of its characteristic function as :

$$\mathcal{P}(\mathbf{X}) = \frac{1}{(2\pi)^n} \int_{\mathbb{R}^n} C_1(t)\, C_2(t)\, e^{\,-i t \cdot \mathbf{X}}\, d^n t \tag{1.4}$$

(note the normalisation factor).

It follows from the differentiation theorem that the partial derivatives of the characteristic function $C(t)$ at $t = 0$ are related to the moments of a distribution P by the identities:

$$\mu_{r_1 r_2 \ldots r_n} \equiv \int_D P(\mathbf{X})\, X_1^{r_1}\, X_2^{r_2} \ldots X_n^{r_n}\, d^n\mathbf{X} \tag{1.5a}$$

$$= i^{-n}\, \left. \frac{\partial^{\,r_1 + \ldots + r_n} C}{\partial t_1^{r_1} \ldots \partial t_n^{r_n}} \right|_{t=0} \tag{1.5b}$$

for any n-tuple of non-negative integers $(r_1, r_2, \ldots, r_n)$.

1.3. Moment-generating functions

The above relation can be freed from powers of i by defining (at least formally) the *moment-generating function* :

216

$$M(t) = \int_{\mathbf{R}^n} P(X)\, e^{t \cdot X}\, d^n X \tag{1.6}$$

which is related to $C(t)$ by $C(t) = M(it)$ so that the inversion formula reads :

$$\mathcal{P}(X) = \frac{1}{(2\pi)^n} \int_{\mathbf{R}^n} M_1(it)\, M_2(it)\, e^{-it \cdot X}\, d^n t \quad . \tag{1.7}$$

The moment-generating function is well-defined, in particular, for any probability distribution with compact support, in which case it may be continued analytically from a function over $\mathbf{R}^n$ into an entire function of n *complex* variables by virtue of the Paley-Wiener theorem (Paley & Wiener, 1934 ; see also §5.1 of Bricogne (1984)). Its moment-generating properties are summed up in the following relations :

$$\mu_{r_1 r_2 \dots r_n} = \left. \frac{\partial^{r_1 + \dots + r_n} M}{\partial t_1^{r_1} \dots \partial t_n^{r_n}} \right|_{t=0} \quad . \tag{1.8}$$

1.4. Cumulant-generating functions

The multiplication of moment-generating functions may be further simplified into the addition of their logarithms :

$$\log \mathcal{M} = \log M_1 + \log M_2 \tag{1.9}$$

or equivalently of the coefficients of their Taylor series at $t=0$, *viz.* :

$$\kappa_{r_1 r_2 \dots r_n} = \left. \frac{\partial^{r_1 + \dots + r_n} (\log M)}{\partial t_1^{r_1} \dots \partial t_n^{r_n}} \right|_{t=0} \quad . \tag{1.10}$$

These coefficients are called *cumulants*, since they cumulate when the independent random vectors to which they belong are added, and $\log M$ is called the *cumulant-generating function*. The inversion formula for $\mathcal{P}$ then reads :

$$\mathcal{P}(X) = \frac{1}{(2\pi)^n} \int_{\mathbf{R}^n} \exp\left[\log M_1(it) + \log M_2(it) - it \cdot X \right] d^n t \quad . \tag{1.11}$$

1.5. Asymptotic expansions and limit theorems

Consider an n-dimensional random vector X of the form

$$X = X_1 + X_2 + \dots + X_N \tag{1.12}$$

where the N summands are independent n-dimensional random vectors identically distributed with probability density P . Then the distribution $\mathcal{P}$ of X may be written in closed form as a Fourier transform :

$$\mathcal{P}(\mathbf{X}) \;=\; \frac{1}{(2\pi)^n} \int_{\mathbb{R}^n} M^N(it)\, e^{-it\cdot\mathbf{X}}\, d^n t \qquad (1.13a)$$

$$=\; \frac{1}{(2\pi)^n} \int_{\mathbb{R}^n} \exp\left[\, N \log M(it) - it\cdot\mathbf{X} \,\right] d^n t \qquad (1.13b)$$

where

$$M(t) \;=\; \int_{\mathbb{R}^n} P(\mathbf{Y})\, e^{\,t\cdot\mathbf{Y}}\, d^n\mathbf{Y} \qquad (1.14)$$

is the moment generating function common to all the summands.

This is an exact expression for $\mathcal{P}$, which may be exploited analytically or numerically in certain favourable cases. Supposing for instance that P has compact support, then its characteristic function $M(it)$ can be sampled finely enough to accommodate the bandwidth of the support of $\mathcal{P}=P^{*N}$ (this sampling rate clearly depends on N) so that the above expression for $\mathcal{P}$ can be used for its numerical evaluation as the *discrete* Fourier transform of $M^N(it)$. This exact method is practical only for small values of the dimension n .

In all other cases some form of approximation must be used in the Fourier inversion of $M^N(it)$. For this purpose it is customary (Cramér, 1946 ; Bhattacharya & Ranga Rao, 1976) to expand the cumulant-generating function around $t = 0$ with respect to the carrying variables t :

$$\log\left[\, M^N(it) \,\right] \;=\; \sum_{\mathbf{r}\in N^n} \frac{N\kappa_{\mathbf{r}}}{\mathbf{r}!}\, (it)^{\mathbf{r}} \qquad (1.15)$$

where $\mathbf{r} = (r_1, r_2, \ldots, r_n)$ is an n-tuple of non-negative integers, $\mathbf{r}! = r_1!\, r_2!\ldots r_n!$, $t^{\mathbf{r}} = t_1^{\,r_1} \ldots t_n^{\,r_n}$, and for later use $|\mathbf{r}| = r_1 + r_2 + \ldots + r_n$. The first-order terms may be eliminated by recentering $\mathcal{P}$ around its vector of first-order cumulants

$$\langle\mathbf{X}\rangle \;=\; \sum_{j=1}^{N} \langle\mathbf{X}_j\rangle \qquad (1.16)$$

where $\langle\cdot\rangle$ denotes the mathematical expectation of a random vector. The second-order terms may be grouped separately from the terms of third or higher order to give :

$$M^N(it) \;=\; e^{-\frac{1}{2} N\, t^T Q\, t} \;\times\; \exp\left\{ \sum_{|\mathbf{r}|\geq 3} \frac{N\kappa_{\mathbf{r}}}{\mathbf{r}!}\, (it)^{\mathbf{r}} \right\} \qquad (1.17)$$

where $Q = \nabla\nabla^T (\log M)$ is the covariance matrix of the multivariate distribution P . Expanding the exponential gives rise to a series of terms of the form

$$e^{-\frac{1}{2} N\, t^T Q\, t} \;\times\; \text{monomial in } t_1, t_2, \ldots, t_n \qquad (1.18)$$

Each of these terms may now be subjected to Fourier inversion, giving rise to a product of Hermite functions of t with coefficients of involving the cumulants κ of P . Taking the

218

transformed terms in natural order gives an asymptotic expansion of P for large N called the *Gram-Charlier series* of $\mathcal{P}$, while grouping the terms according to increasing powers of $\frac{1}{\sqrt{N}}$ gives another asymptotic expansion called the *Edgeworth series* of $\mathcal{P}$. Both expansions comprise a leading Gaussian term which embodies the *Central Limit Theorem* :

$$\mathcal{P}(\mathbf{X}) \approx \frac{1}{\sqrt{\det (2\pi \mathbf{Q})}} e^{-\frac{1}{2}\mathbf{E}^T \mathbf{Q}^{-1}\mathbf{E}} \qquad \text{where} \quad \mathbf{E} = \frac{\mathbf{X} - \langle \mathbf{X}\rangle}{\sqrt{N}} \qquad . \qquad (1.19)$$

1.6. The saddlepoint approximation

A limitation of the Edgeworth series is that it gives an accurate estimate of $\mathcal{P}(\mathbf{X})$ only in the vicinity of $\mathbf{X} = \langle \mathbf{X}\rangle$, i.e. for small values of $\mathbf{E}$. These convergence difficulties are easily understood: one is substituting a *local* approximation to $\log M$ (*viz.* a Taylor series expansion valid near $\mathbf{t} = \mathbf{0}$) into an integral, whereas integration is a *global* process which consults values of $\log M$ far from $\mathbf{t} = \mathbf{0}$. More specifically, under the simplifying assumption made earlier that we are looking at a centered distribution (i.e. that $\langle \mathbf{X}\rangle = \mathbf{0}$), the exact expression (1.13a) of $\mathcal{P}(\mathbf{X})$ as the Fourier transform of $C^N(\mathbf{t})$ at $\mathbf{X}$ makes it intuitively clear that expanding $C(\mathbf{t})$ at $\mathbf{t} = \mathbf{0}$ (where it has its maximum modulus 1) will lead to a good approximation of the *integral* of $C^N(\mathbf{t})$, *viz* $\mathcal{P}(\mathbf{0})$, or of its low-order frequency components, *viz* $\mathcal{P}(\mathbf{X})$ for $\mathbf{X}$ near $\mathbf{0}$, but will fail to represent adequately the rapid oscillations of $C^N(\mathbf{t})$ which contribute to $\mathcal{P}(\mathbf{X})$ for $\mathbf{X}$ far from $\mathbf{0}$.

It is possible, however, to let the point $\mathbf{t}$ where $\log M$ is expanded as a Taylor series depend on the particular value $\mathbf{X}^*$ of $\mathbf{X}$ for which an accurate evaluation of $\mathcal{P}(\mathbf{X})$ is desired by means of (1.13b). By the previous reasoning, this will require identifying the point $\mathbf{t}$ where $C^N(\mathbf{t})$ oscillates as $e^{+i\mathbf{t}\cdot\mathbf{X}^*}$ and hence whose vicinity will contribute most to the integral representing $\mathcal{P}(\mathbf{X}^*)$. This is the essence of the *saddlepoint method* (Fowler, 1936 ; Khinchin 1949 ; Daniels, 1954 ; DeBruijn, 1970 ; Bleistein & Handelsman, 1986), which uses an analytical continuation of $M(\mathbf{t})$ from a function over $\mathbb{R}^n$ to a function over $\mathbb{C}^n$ (see e.g. §5.1 in Bricogne (1984)). Putting then $\mathbf{t} = \mathbf{s} - i\tau$, the $\mathbb{C}^n$-version of Cauchy's theorem (Hörmander, 1973) gives rise to the identity :

$$\mathcal{P}(\mathbf{X}^*) = \frac{e^{-\tau \cdot \mathbf{X}^*}}{(2\pi)^n} \int_{\mathbb{R}^n} \exp\left\{ N\left[\log M(\tau + i\mathbf{s}) - i\mathbf{s}\cdot\frac{\mathbf{X}^*}{N} \right] \right\} d^n\mathbf{s} \qquad (1.20)$$

for *any* $\tau \in \mathbf{R}^n$. By a convexity argument involving the positive-definiteness of covariance matrix $\mathbf{Q}$, there is a unique value of τ such that

$$\nabla \left(\log M \right)\Bigg|_{t=0-i\tau} = \frac{X^*}{N} \qquad . \qquad (1.21)$$

At the *saddlepoint* $t^* = 0 - i\tau$, the modulus of the integrand above is a maximum and its phase is stationary with respect to the integration variable s : as N tends to infinity, all contributions to the integral cancel because of rapid oscillation, except those coming from the immediate vicinity of t^* where there is no oscillation. A Taylor expansion of $\log M^N$ to second order with respect to s at t^* then gives :

$$\log M^N(\tau + is) \approx \log M^N(\tau) + i\,s\cdot X^* - \frac{N}{2}\left[s^T Q\, s \right] \qquad (1.22)$$

and hence :

$$\mathcal{P}(X^*) \approx e^{\log M^N(\tau) - \tau\cdot X^*} \frac{1}{(2\pi)^n} \int_{\mathbb{R}^n} e^{-\frac{1}{2} s^T Q\, s}\, d^n s \qquad .$$

The last integral is elementary and gives the "saddlepoint approximation":

$$\mathcal{P}^{SP}(X^*) = \frac{e^S}{\sqrt{\det(2\pi Q)}} \qquad (1.23)$$

where

$$S = \log M^N(\tau) - \tau \cdot X^* \qquad (1.24)$$

and where

$$Q = \nabla\nabla^T(\log M^N) = N\,Q \qquad . \qquad (1.25)$$

This approximation scheme amounts to using the "conjugate distribution" (Khinchin, 1949)

$$P_\tau(X_j) = P(X_j)\, \frac{e^{\tau\cdot X_j}}{M(\tau)} \qquad (1.26)$$

instead of the original distribution $P(X_j) = P_0(X_j)$ for the common distribution of all N random vectors X_j. The exponential modulation results from the analytic continuation of the characteristic (or moment-generating) function into $\mathbb{C}^n$, as in §5.2 of Bricogne (1984). The saddlepoint approximation $\mathcal{P}^{SP}$ is only the leading term of an asymptotic expansion (called the *saddlepoint expansion*) for $\mathcal{P}$, which is actually the Edgeworth expansion associated with P_τ^{*N} .

2. THE STATISTICAL THEORY OF PHASE DETERMINATION

The methods of Probability Theory just surveyed were applied to various problems formally similar to the crystallographic Phase Problem (e.g. the "Problem of the Random Walk" of Pearson (1905)) by Rayleigh (1880, 1899, 1905, 1918, 1919) and Kluyver (1906). They became the basis of the statistical theory of communication with the classic papers of Rice (1944,1945) .

The Gram-Charlier and Edgeworth series were introduced into crystallography by Bertaut (1955a, b, c ; 1956a) and by Klug (1958) respectively, who showed them to constitute the mathematical basis of numerous formulae derived by Hauptman & Karle (1953). The saddlepoint approximation was introduced by the present writer (Bricogne, 1984).

2.0. Definitions and conventions

Let H be set of unique non-origin reflexions $\mathbf{h}$ for a crystal with lattice Λ and space group G . Let H contain n_a acentric and n_c centric reflexions. Structure factor values attached to all reflexions in H will comprise $n = 2n_a + n_c$ real numbers. For $\mathbf{h}$ acentric, $\alpha_{\mathbf{h}}$ and $\beta_{\mathbf{h}}$ will be the real and imaginary parts of the complex structure factor ; for $\mathbf{h}$ centric, $\gamma_{\mathbf{h}}$ will be the real coordinate of the (possibly complex) structure factor measured along a real axis rotated by one of the two angles $\theta_{\mathbf{h}}$, π apart, to which the phase is restricted modulo 2π . These n real coordinates will be arranged as a column vector containing the acentric then the centric data, i.e. in the order :

$$\alpha_1 , \beta_1 , \alpha_2 , \beta_2 , \dots , \alpha_{n_a} , \beta_{n_a} , \gamma_1, \gamma_2, \dots, \gamma_{n_c} \qquad . \qquad (2.1)$$

2.1. Vectors of trigonometric structure factor expressions

Let $\xi(\mathbf{x})$ denote the vector of trigonometric structure factor expressions associated with $\mathbf{x} \in D$, where D denotes the asymmetric unit. These are defined as follows :

$$\alpha_{\mathbf{h}}(\mathbf{x}) + i\,\beta_{\mathbf{h}}(\mathbf{x}) = \Xi\,(\mathbf{h},\mathbf{x}) \qquad \text{for } \mathbf{h} \text{ acentric} \qquad (2.2a)$$

$$\gamma_{\mathbf{h}}(\mathbf{x}) = \exp(-i\theta_{\mathbf{h}})\,\Xi\,(\mathbf{h},\mathbf{x}) \qquad \text{for } \mathbf{h} \text{ centric} \qquad (2.2b)$$

with

$$\Xi\,(\mathbf{h},\mathbf{x}) = \frac{1}{|G_{\mathbf{x}}|} \sum_{g \in G} \exp \left\{ 2\pi i \mathbf{h} \cdot [S_g(\mathbf{x})] \right\} \qquad (2.3)$$

where we denote by $|G_{\mathbf{x}}|$ the number of elements in the isotropy subgroup of $\mathbf{x}$ ($|G_{\mathbf{x}}| > 1$ if $\mathbf{x}$ is a special position).

According to the convention above, the coordinates of $\xi(\mathbf{x})$ in $\mathbb{R}^n$ will be arranged in a column vector as follows :

$$\xi_{2r-1}(\mathbf{x}) = \alpha_{\mathbf{h}_r}(\mathbf{x}) \qquad \text{for} \qquad r = 1, \dots , n_a \qquad , \qquad (2.4a)$$

$$\xi_{2r}(\mathbf{x}) = \beta_{\mathbf{h}_r}(\mathbf{x}) \qquad \text{for} \qquad r = 1, \dots , n_a \qquad , \qquad (2.4b)$$

$$\xi_{n_a+r}(\mathbf{x}) = \gamma_{\mathbf{h}_r}(\mathbf{x}) \qquad \text{for} \qquad r = n_a+1, \dots , n_a+n_c \qquad . \qquad (2.4c)$$

Let position $\mathbf{x}$ in D now become a random vector with probability density $m(\mathbf{x})$. Then $\xi(\mathbf{x})$ becomes itself a random vector in $\mathbf{R}^n$, whose distribution $p(\xi)$ is the image of distribution $m(\mathbf{x})$ through the mapping $\mathbf{x} \to \xi(\mathbf{x})$ just defined. The locus of $\xi(\mathbf{x})$ in $\mathbf{R}^n$ is a compact algebraic manifold L (the multidimensional analogue of a Lissajous curve), so that p is a singular measure concentrated on that manifold, hence with compact support. The average with respect to p of any function Ω over $\mathbf{R}^n$ which is infinitely differentiable in a neighbourhood of L may be calculated as an average with respect to m over D by the "induction formula" :

$$\langle\, p \,,\, \Omega \,\rangle \;\; = \int_D m(\mathbf{x})\, \Omega\, [\xi(\mathbf{x})]\; d^3\mathbf{x} \qquad\qquad (2.5)$$

In particular, one can calculate the moment generating function M for distribution p as:

$$M(t) \;\;\equiv\;\; \langle\, p_\xi \,,\, e^{t \cdot \xi} \,\rangle \;\; = \int_D m(\mathbf{x})\; e^{t \cdot \xi(\mathbf{x})}\; d^3\mathbf{x} \qquad\qquad (2.6)$$

and hence calculate the moments μ (resp. cumulants κ) of p by differentiation of M (resp. $\log M$) at $t = 0$:

$$\mu_{r_1 r_2 \,\ldots\, r_n} \;\; \equiv \int_D m(\mathbf{x})\; \xi_1^{r_1}(\mathbf{x})\; \xi_2^{r_2}(\mathbf{x}) \ldots \xi_n^{r_n}(\mathbf{x})\; d^3\mathbf{x} \qquad\qquad (2.7a)$$

$$= \frac{\partial^{\,r_1+\ldots+r_n}\,(M)}{\partial t_1^{\,r_1} \ldots\, \partial t_n^{\,r_n}} \qquad\qquad (2.7b)$$

$$\kappa_{r_1 r_2 \,\ldots\, r_n} \;\; = \frac{\partial^{\,r_1+\ldots+r_n}\,(\log M)}{\partial t_1^{\,r_1} \ldots\, \partial t_n^{\,r_n}} \;\; . \qquad\qquad (2.7c)$$

The structure factor algebra for group G (Bertaut, 1955c, 1956b, 1959a,b ; Bertaut & Waser, 1957) then allows one to express products of ξ's in (2.7a) as linear combinations of other ξ's, and hence to calculate directly all moments and cumulants of distribution $p(\xi)$ as linear combinations of real and imaginary parts of Fourier coefficients of the prior distribution of atoms $m(\mathbf{x})$. This plays a key rôle in the use of non-uniform distributions of atoms.

<u>2.3. The joint probability distribution of structure factors</u>

In the random atom model of an equal-atom structure, N atoms are placed randomly, independently of each other, in the asymmetric unit D of the crystal with probability density $m(\mathbf{x})$. For point atoms of unit weight (scattering factors other than unity can easily be introduced, see Bricogne(1988a)), the vector $\mathbf{F}$ of structure factor values for reflexions $\mathbf{h} \in H$ may be written:

$$\mathbf{F} = \sum_{I=1}^{N} \xi^{[I]} \qquad (2.8)$$

where the N copies $\xi^{[I]}$ of random vector ξ are independent and have the same distribution $p(\xi)$.

The joint probability distribution $\mathcal{P}(\mathbf{F})$ is then (§1.1.5) :

$$\mathcal{P}(\mathbf{F}) = \frac{1}{(2\pi)^n} \int_{\mathbb{R}^n} \exp\left[N \log M(i\mathbf{t}) - i\mathbf{t} \cdot \mathbf{F} \right] d^n\mathbf{t} \qquad (2.9)$$

For low dimensionality n it is possible to carry out the Fourier transformation numerically after discretisation, provided the characteristic function $M(i\mathbf{t})$ is sampled sufficiently finely that no aliasing results from taking its Nth power (Barakat, 1974). This exact approach can also accommodate heterogeneity, and has been used first in the field of intensity statistics (Shmueli, Weiss, Kiefer & Wilson , 1984 ; Shmueli, Weiss & Kiefer, 1985; Shmueli & Weiss, 1987, 1988), then in the study of the Σ_1 and Σ_2 relations in triclinic space groups (Shmueli & Weiss, 1985, 1986). These authors have considered uniformly distributed atoms only, but the method could be extended to the construction of any jpd in any space group by using the generic expression for the mgf derived by Bricogne(1984) for non-uniformly distributed atoms. It is, however, limited to small values of n by the necessity to carry out n-dimensional FFT's on large arrays of sample values.

In all other cases, it is necessary to use one of the approximation schemes described in §1.5 and §1.6 .

2.4. Overcoming the limitations of the Edgeworth series

The asymptotic expansions of Gram-Charlier and Edgeworth have good convergence properties only if $F_\mathbf{h}$ lies in the vicinity of $\langle F_\mathbf{h} \rangle = N \, \overline{\mathcal{F}} [m] (\mathbf{h})$ for all $\mathbf{h} \in H$. Previous work on the jpd of structure factors has used for $m(\mathbf{x})$ a uniform distribution, so that $\langle F \rangle = 0$; as a result, the corresponding expansions are accurate only if all moduli $|F_\mathbf{h}|$ are small, in which case the jpd contains little phase information. As was first pointed in Bricogne (1984), conventional direct methods lost sight of the fact that the Edgeworth series is a functional expression suitable only for locally approximating $\mathcal{P}(\mathbf{F})$ near $\mathbf{F} = 0$ and have mistakenly assumed it to be the global functional form for that function "in the large".

This two observation shows that the traditional form of direct methods is a rather rudimentary implementation of the ideas of probability theory on which it is based. Fortunately, better analytical devices can be found which circumvent these difficulties; the reader is referred to Bricogne (1984) for a full exposition, which will be freely summarized here. The main idea is that, if the locus $\mathcal{T}$ (a high-dimensional torus) defined by the large moduli is too extended in structure factor space for a single asymptotic expansion of $\mathcal{P}(\mathbf{F})$ to be acccurate everywhere on it, then we should break $\mathcal{T}$ up into subregions, and construct a different local approximations to $\mathcal{P}(\mathbf{F})$ in each of them. Each of these subregions will consist of a "patch" of $\mathcal{T}$, surrounding a point $\mathbf{F}^{*} \neq \mathbf{0}$ located on $\mathcal{T}$. Such a point $\mathbf{F}^{*}$ is obtained by assigning "trial" phase values to the known moduli, but these trial values do not have to be viewed as "serious" assumptions concerning the true values of the phases : rather, they should be thought of as pointing to a patch of $\mathcal{T}$ and to a specialised asymptotic expansion of $\mathcal{P}(\mathbf{F})$ designed to be the most accurate approximation possible to $\mathcal{P}(\mathbf{F})$ on that patch. With a sufficiently rich collection of such constructs, $\mathcal{P}(\mathbf{F})$ can be accurately calculated anywhere on $\mathcal{T}$.

We are thus led inevitably to the notion of *recentering*. Recentering the usual Gram-Charlier or Edgeworth asymptotic expansion for $\mathcal{P}(\mathbf{F})$ away from $\mathbf{F} = \mathbf{0}$, by making trial phase assignments which define a point $\mathbf{F}^{*}$ on $\mathcal{T}$, is equivalent to using a non-uniform prior distribution of atoms $q(\mathbf{x}) \neq \frac{1}{V}$, reproducing $\mathbf{F}^{*}$ among its Fourier coefficients.

The latter constraint, however, leaves $q(\mathbf{x})$ highly indeterminate, but it turns out that there is a uniquely defined "best' choice for it : it is that distribution which has *maximum entropy* under these constraints, entropy being a measure of closeness to uniformity. This result can be justified in two ways (Bricogne (1984), §3 and §5) :

(1) The first way invokes Shannon's theory of information (Shannon & Weaver, 1949), according to which any decrease in the entropy of the distribution of atoms $q(\mathbf{x})$ tends to diminish the size of the statistical ensemble of structures which can be generated using $q(\mathbf{x})$; then it invokes Jaynes's heuristic principle (Jaynes, 1957) that one should rule out as few structures as possible on the basis of the limited information available in $\mathbf{F}^{*}$, to conclude that $q(\mathbf{x})$ should have maximum entropy (call it $q^{ME}(\mathbf{x})$).

(2) The second way relies on the saddlepoint approximation described in §1.6. The remarkable thing is that, without ever mentioning entropy nor invoking any semi-philosophical arguments, it yields an identical expression for the best choice of a non-uniform prior $q(\mathbf{x})$. The saddlepoint approximation to any conditional distribution conditioned on the knowledge of $\mathbf{F}^{*}$ is then the Gaussian approximation built from $q^{ME}(\mathbf{x})$.

These two approaches, and the equivalence between them, will now be described.

2.5. The maximum entropy method

2.5.1. The context of information theory

The standard scheme of direct methods uses as its starting point a source of random atomic positions. The picture of this process may be discretized by partitioning the asymmetric unit of the crystal into B equal boxes labelled 1 to B. Then the symbols used by the source are the labels of these boxes, in the sense that drawing symbol i is equivalent to placing an atom in box i . Sequences of such randomly produced abstract symbols are called messages ; in our case, a message of length N is an N-atom random structure.

Any such discrete stochastic process, considered as a discrete source of symbols, generates not only a set but an *ensemble* of messages, i.e. in Wiener's terms "a repertory of possible messages, and over this repertory a measure determining the probability of these messages" (Wiener, 1949). In our case, this probability measure reflects the fact that different random structures – and hence different sets of structure factors – will occur with different frequencies (or "statistical weights") in the ensemble.

2.5.2. Definition and meaning of entropy

An important numerical quantity associated with a discrete source is its *entropy per symbol* H , which gives a measure of the "amount of uncertainty" involved in the choice of a symbol. Suppose that successive symbols are independent and that symbol i has probability q_i. Then the general requirements that H should be a continuous function of the q_i, should increase with increasing uncertainty, and should be additive for independent sources of uncertainty, suffice to define H uniquely as :

$$H (q_1, \ldots , q_n) = - k \sum_{i=1}^{n} q_i \log q_i \qquad (2.10)$$

where k is an arbitrary positive constant (Shannon & Weaver (1949), Appendix 2) whose value depends on the unit of entropy chosen.

Two important theorems (Shannon & Weaver (1949), Appendix 3) provide a more intuitive grasp of the meaning and importance of entropy :

(1) H is approximately the logarithm of the reciprocal probability of a typical long sequence, divided by the number of symbols in the sequence ;

(2) H gives the rate of growth, with increasing sequence length, of the logarithm of the number of reasonably probable sequences, regardless of the precise meaning given to the criterion of being "reasonably probable".

The entropy H of a source is thus a direct measure of the strength of the restrictions placed on the permissible messages, greater restrictions leading to lower entropy. In our case, its maximum value $H_{max} = \log B$ is reached when all the symbols are equally probable, i.e. for a uniform prior distribution of the atoms. When the prior distribution is not uniform, the usage of the different symbols is biased away from this maximum freedom, and the entropy of the source is lower; by Shannon's theorem (2), the number of "reasonably probable" messages of a given length is decreased accordingly.

2.5.3. The maximum-entropy criterion

We are now in a position to address the problem posed in §2.4, namely the choice of a non-uniform prior distribution of atoms $q(\mathbf{x})$ from the assumed knowledge of a limited set of its Fourier coefficients.

By the two theorems of Shannon quoted above, the entropy H of the source of randon atomic positions defined by $q(\mathbf{x})$ affords a quantitative measure of the extent to which the range of random structures which can be generated with a reasonable probability has been narrowed down. Any reduction of the entropy of the source beyond that strictly necessary to accommodate the assumed knowledge will be reflected by a decrease in the number of reasonably probable random structures, and hence by an unjustified restriction to a subset of all possible structures consistent with the given data ; this is equivalent to imposing extra constraints, not warranted by the data.

If the assumed knowledge of some structure factors $\mathbf{F}^*$ is to be reflected in a non-uniform prior distribution of atoms, the previous reasoning leads uniquely to choosing for $q(\mathbf{x})$ the maximum-entropy distribution $q^{ME}(\mathbf{x})$ having the corresponding values $\mathbf{U}^*$ of the unitary structure factors for its Fourier coefficients, since the latter defines the source which reproduces the assumed data with minimum bias. This is a particular instance of the Maximum Entropy Principle of Jaynes (1957, 1968, 1983) which may be loosely phrased as follows : "Any prior probability assignment shall be the one with maximum entropy consistent with the available prior knowledge, so as to remain maximally non-committal with regard to missing information".

The quantity which measures most directly the strength of the restrictions introduced by the non-uniformity of $q(\mathbf{x})$ is not the source entropy $H(q)$ itself, but rather the difference $H(q) - H_{max}$ since the fraction of N-atom random structures which remain 'reasonably probable' in the ensemble of the corresponding source is $\exp\{N[H(q) - H_{max}]\}$. This difference may be written (using continuous rather than discrete distributions) :

$$H(q) - H_{max} \;=\; -\int_V q(\mathbf{x}) \, \log\left[\frac{q(\mathbf{x})}{m(\mathbf{x})}\right] \, d^3\mathbf{x} \qquad (2.11)$$

where $m(\mathbf{x}) = \frac{1}{V}$ is the uniform distribution which is such that $H(m) = H_{max} = \log V$.

The final form of the maximum-entropy criterion is thus that $q(\mathbf{x})$ should be chosen so as to maximise, under the constraints expressing the prior knowledge of some of its Fourier coefficients, its entropy

$$\mathcal{S}_m(q) \;=\; -\int_V q(\mathbf{x}) \, \log\left[\frac{q(\mathbf{x})}{m(\mathbf{x})}\right] \, d^3\mathbf{x} \qquad (2.12)$$

relative to the "prior prejudice" $m(\mathbf{x})$ which maximises H in the absence of such knowledge.

2.5.4. The maximum-entropy formalism

Jaynes (1957) solved the problem of explicitly determining such maximum-entropy distributions in the case of general linear constraints, using an analytical apparatus first exploited by Gibbs in statistical mechanics.

The maximum-entropy distribution $q^{ME}(\mathbf{x})$, under the prior prejudice $m(\mathbf{x})$, satisfying the linear constraint equations

$$C_j(q) \;\equiv\; \int_V q(\mathbf{x}) \, C_j(\mathbf{x}) \, d^3\mathbf{x} \;=\; c_j \qquad (j=1, 2, ..., M) \qquad (2.13)$$

226

where the $C_j(q)$ are linear *constraint functionals* defined by given *constraint functions* $C_j(\mathbf{x})$, and the c_j are given *constraint values*, is obtained by maximizing with respect to q the relative entropy defined by equation (2.12). An extra constraint is the normalization condition

$$C_0(q) \equiv \int_V q(\mathbf{x}) \; 1 \; d^3\mathbf{x} = 1 \tag{2.14}$$

to which it is convenient to give the label j=0, so that it can be handled together with the others by putting $C_0(\mathbf{x}) = 1$, $c_0 = 1$.

By a standard variational argument, in constant use in statistical mechanics, this constrained maximization is equivalent to the unconstrained maximization of the functional

$$S_m(q) + \sum_{j=0}^{M} \lambda_j \, C_j(q) \tag{2.15}$$

where the λ_j are Lagrange multipliers whose values may be determined from the constraints. This new variational problem is readily solved : if $q(\mathbf{x})$ is varied to $q(\mathbf{x}) + \delta q(\mathbf{x})$, the resulting variations in the functionals S_m and C_j will be :

$$\delta S_m = \int_V \left\{ -1 - \log\left[\frac{q(\mathbf{x})}{m(\mathbf{x})}\right] \right\} \delta q(\mathbf{x}) \, d^3\mathbf{x}$$

$$\delta C_j = \int_V \left\{ C_j(\mathbf{x}) \right\} \delta q(\mathbf{x}) \, d^3\mathbf{x} \tag{2.16}$$

respectively. If the variation of the functional (2.15) is to vanish for arbitrary variations $\delta q(\mathbf{x})$. the integrand in the expression for that variation from (2.16) must vanish identically. Therefore the maximum-entropy density distribution $q^{ME}(\mathbf{x})$ satisfies the relation :

$$-1 - \log\left[\frac{q(\mathbf{x})}{m(\mathbf{x})}\right] + \sum_{j=0}^{M} \lambda_j \, C_j(\mathbf{x}) = 0 \tag{2.17}$$

and hence :

$$q^{ME}(\mathbf{x}) = m(\mathbf{x}) \, \exp(\lambda_0 - 1) \, \exp\left[\sum_{j=1}^{M} \lambda_j \, C_j(\mathbf{x})\right] \tag{2.18}$$

It is convenient now to separate the multiplier λ_0 associated to the normalization constraint by putting :

$$\lambda_0 - 1 = -\log Z \tag{2.19}$$

where Z is a function of the other multipliers $\lambda_1, \ldots, \lambda_M$. The final expression for $q^{ME}(\mathbf{x})$ is thus :

$$q^{ME}(\mathbf{x}) = \frac{m(\mathbf{x})}{Z(\lambda_1, \ldots, \lambda_M)} \, \exp\left[\sum_{j=1}^{M} \lambda_j \, C_j(\mathbf{x})\right] \tag{ME1}$$

The values of Z and of $\lambda_1, \ldots, \lambda_M$ may now be determined by solving the initial constraint equations. The normalization condition demands that :

$$Z(\lambda_1, \ldots, \lambda_M) = \int_V m(\mathbf{x}) \, \exp\left[\sum_{j=1}^{M} \lambda_j \, C_j(\mathbf{x})\right] d^3\mathbf{x} \tag{ME2}$$

The generic constraint equations (2.13) determine $\lambda_1, \ldots, \lambda_M$ by the conditions that :

$$\int_V \frac{m(\mathbf{x})}{Z} \exp\left[\sum_{k=1}^{M} \lambda_k \, C_k(\mathbf{x}) \right] C_j(\mathbf{x}) \; d^3\mathbf{x} \;=\; c_j \qquad (2.20)$$

for $j = 1, 2, \ldots, M$. But, by Leibniz's rule of differentiation under the integral sign, these equations may be written in the compact form :

$$\frac{\partial (\log Z)}{\partial \lambda_j} \;=\; c_j \qquad (j = 1, 2, \ldots, M) \qquad \text{(ME3)}$$

Equations (ME1), (ME2) and (ME3) constitute the *maximum-entropy equations.*

The maximum value attained by the entropy is readily found :

$$\mathcal{S}_m(q^{ME}) \;=\; - \int_V q^{ME}(\mathbf{x}) \; \log\left[\frac{q^{ME}(\mathbf{x})}{m(\mathbf{x})} \right] d^3\mathbf{x}$$

$$=\; - \int_V q^{ME}(\mathbf{x}) \left[- \log Z + \sum_{j=1}^{M} \lambda_j \, C_j(\mathbf{x}) \right] d^3\mathbf{x}$$

i.e., using the constraint equations :

$$\mathcal{S}_m(q^{ME}) \;=\; \log Z - \sum_{j=1}^{M} \lambda_j \, c_j \qquad \qquad . \qquad (2.21)$$

The latter expression may be rewritten, by means of equations (ME3), as :

$$\mathcal{S}_m(q^{ME}) \;=\; \log Z - \sum_{j=1}^{M} \lambda_j \, \frac{\partial (\log Z)}{\partial \lambda_j} \qquad \qquad . \qquad (2.22)$$

which shows that, in their dependence on the λ's, the entropy and $\log Z$ are related by Legendre duality.

Jaynes's ME theory also gives an estimate for $\mathcal{P}(\mathbf{F}^*)$ as a consequence of Shannon's second theorem (§2.5.2) :

$$\mathcal{P}^{ME}(\mathbf{F}^*) \;\approx\; e^{\mathcal{S}} \qquad (2.23)$$

where

$$\mathcal{S} \;=\; \log Z^N - \lambda \cdot \mathbf{F}^* \;=\; N \, \mathcal{S}_m(q^{ME}) \qquad (2.24)$$

is the total entropy.

2.5.5. *The crystallographic maximum-entropy formalism*

It is possible to solve the maximum-entropy equations explicitly for the crystallographic case which has motivated this study, i.e. for the purpose of constructing $q^{ME}(\mathbf{x})$ from the knowledge of a set of trial structure-factor values. These derivations are given in §3.4 and §3.5 of Bricogne (1984). Extensive relations with the algebraic formalism of traditional direct methods are exhibited in §4, and connections with the theory of determinantal inequalities and with the maximum determinant rule of Tsoucaris (1970) are studied in §6, of the same paper, which the reader interested in these topics is invited to consult.

2.6. The saddlepoint method

The saddlepoint method (§1.6) constitutes an alternative approach to the problem of evaluating the joint probability $\mathcal{P}(\mathbf{F}^*)$ of structure factors when some of the moduli in $\mathbf{F}^*$ are large. As shown previously this approximation amounts to using the "conjugate distribution"

$$p_\tau(\xi) \;=\; p(\xi)\,\frac{e^{\tau\cdot\xi}}{M(\tau)} \qquad\qquad (2.25)$$

instead of the original distribution $p(\xi) = p_0(\xi)$ for the distribution of random vector ξ. This conjugate distribution p_τ is induced from the modified distribution of atoms

$$q_\tau(\mathbf{x}) \;=\; m(\mathbf{x})\,\frac{e^{\tau\cdot\xi(\mathbf{x})}}{M(\tau)} \qquad\qquad (\mathrm{SP1})$$

where, by the induction formula (2.5), $M(\tau)$ may be written as :

$$M(\tau) \;=\; \int_D m(\mathbf{x})\, e^{\tau\cdot\xi(\mathbf{x})}\, d^3\mathbf{x} \qquad\qquad (\mathrm{SP2})$$

and where τ is the unique solution of the saddlepoint equation :

$$\nabla_\tau\left(\log M^N\right) \;=\; \mathbf{F}^* \qquad\qquad (\mathrm{SP3})$$

The desired approximation is then :

$$\mathcal{P}^{\mathrm{SP}}(\mathbf{F}^*) \;=\; \frac{e^{\mathcal{S}}}{\sqrt{\det(2\pi\mathcal{Q})}} \qquad\qquad (2.26)$$

where

$$\mathcal{S} \;=\; \log M^N(\tau) - \tau\cdot\mathbf{F}^* \qquad\qquad (2.27)$$

and where

$$\mathcal{Q} \;=\; \nabla\nabla^T\left(\log M^N\right) \;=\; N\,\mathbf{Q} \qquad\qquad (2.28)$$

Note that the elements of the Hessian matrix $\mathbf{Q} = \nabla\nabla^T(\log M)$, which may be used to solve equation (SP3) for τ by Newton's method, are just the trigonometric second-order cumulants of distribution p, and hence can be calculated *via* structure factor algebra from the Fourier coefficients of $q_\tau(\mathbf{x})$. All the quantities involved in the expression for $\mathcal{P}^{\mathrm{SP}}(\mathbf{F}^*)$ are therefore effectively computable from the initial data $m(\mathbf{x})$ and $\mathbf{F}^*$.

2.7. Relation between the maximum entropy method and the saddlepoint method

Comparison of the three equations (SP1), (SP2) and (SP3) with (ME1), (ME2) and (ME3) in §2.5.4 shows complete identity, up to an obvious change in notation :
- the Z-function in (ME2) is the M-function in (SP2) ;
- the Lagrange multipliers in (ME3) are the coordinates of the saddlepoint in (SP3) .

There is thus complete equivalence between the maximum-entropy approach to the phase problem and the classical probabilistic approach by the method of joint distributions, provided the latter is enhanced by the adoption of the saddlepoint approximation.

This rather striking result is not without precedent. The analytic proof of the maximum-entropy principle given here is an exact parallel of the Darwin-Fowler formulation of Statistical Mechanics (Fowler, 1936), which also bypasses the established route - explicit combinatorial enumerations followed by an appeal to Stirling's formula - by means of a direct saddlepoint approximation to the thermodynamic partition function. Both results may ultimately be rationalised by recalling that Stirling's formula itself can be established by a saddlepoint approximation to Euler's Γ - function (de Bruijn, 1970).

The saddlepoint method is actually slightly superior to the maximum entropy method. Comparing (2.23) and (2.24) on the one hand with (2.26) and (2.27) on the other hand, we see that $\mathcal{P}^{ME}$ is almost identical to $\mathcal{P}^{SP}$, but lacks the denominator. The latter, which is the normalisation factor of a multivariate Gaussian with covariance matrix Q as given in (2.28), may easily be seen to arise from the extra logarithmic term in Stirling's formula :

$$\log(q!) \approx q \log q - q + \frac{1}{2} \log(2\pi q) \tag{2.29}$$

(see for instance Lebedev, 1972) beyond the first two terms which serve to define entropy, since by Szegö's theorem (Szegö, 1920)

$$\frac{1}{n} \log \det (2\pi \mathbf{Q}) \approx \int_V \log 2\pi q^{ME}(\mathbf{x}) \, d^3\mathbf{x} \qquad . \tag{2.30}$$

The relative effect of this extra normalisation factor depends on the ratio

$$\frac{n}{N} = \frac{\text{dimension of } \mathbf{F} \text{ over } \mathbf{R}}{\text{number of atoms}}$$

and hence becomes predominant at high resolution, where it tends to induce the *minimisation* rather than the previously advocated *maximisation*, of the determinant.

SUMMARY

The power of probabilistic direct methods of phase determination is limited by two factors :
    (1) the inability of the random atom model to represent the subtle correlations between atomic positions induced by the laws of chemistry ;
    (2) any extraneous inadequacy in implementing the random atom model itself.

This study, first presented in Bricogne (1984), has been directed at the second problem, and has shown that severe degradation of the potential of the random atom model has resulted from the time-honoured recourse to the Edgeworth series for approximating joint probability distributions of structure factors.

Two remedies, at first sight very different from each other, have been examined : the maximum-entropy method, and the saddlepoint approximation. They have been shown to be essentially equivalent. The reader may share the writer's relief at having replaced the invokation of an all-embracing epistemological argument (however sound and however profound) by the certainty of a purely analytical derivation, especially as the latter gives an even better result.

The above relation between entropy maximisation and the saddlepoint approximation is the basis of a Bayesian statistical approach to the Phase Problem (Bricogne, 1988a) where the assumptions under which joint distributions of structure factors are sought incorporate many new ingredients (such as molecular boundaries, isomorphous substitutions, known fragments, non-crystallographic symmetries, multiple crystal forms) besides trial phase choices for basis reflexions. The ME criterion intervenes in the construction of $q^{ME}(\mathbf{x})$ under these assumptions, and the distribution $q^{ME}(\mathbf{x})$ is a very useful computational intermediate in obtaining the approximate joint probability $\mathcal{P}^{SP}(\mathbf{F}^*)$ and the associated conditional distributions and likelihood functions.

REFERENCES

BARAKAT, R. (1974). *Opt. Acta* **21**, 903-921.

BERTAUT, E.F. (1955a). *Acta Cryst.* **8**, 537-543.

BERTAUT, E.F. (1955b). *Acta Cryst.* **8**, 544-548.

BERTAUT, E.F. (1955c). *Acta Cryst.* **8**, 823-832.

BERTAUT, E.F. (1956a). *Acta Cryst.* **9**, 322.

BERTAUT, E.F. (1956b). *Acta Cryst.* **9**, 322-323.

BERTAUT, E.F. (1956c). *Acta Cryst.* **9**, 769-770.

BERTAUT, E.F. (1959a). *Acta Cryst.* **12**, 541-549.

BERTAUT, E.F. (1959b). *Acta Cryst.* **12**, 570-574.

BERTAUT, E.F. & WASER, J. (1957). *Acta Cryst.* **10**, 606-607.

BHATTACHARYA, R.N. & RANGA RAO, R. (1976). *Normal Approximation and Asymptotic Expansions.* New York : John Wiley & Sons.

BLEISTEIN, N. & HANDELSMAN, R.A. (1986). *Asymptotic Expansions of Integrals.* New York : Dover Publications.

BRICOGNE, G. (1984). *Acta Cryst.* A**40**, 410-445.

BRICOGNE, G. (1988a). *Acta Cryst.* A**44**, 517-545.

BRUIJN, N.G. DE (1970). *Asymptotic Methods in Analysis.* Amsterdam : North-Holland.

CRAMÉR, H. (1946). *Mathematical Methods of Statistics.* Princeton Univ. Press.

DANIELS, H.E. (1954). *Ann. Math. Stat.* **25**, 631-650.

FOWLER, R.H. (1936). *Statistical Mechanics*, 2nd edition Cambridge Univ. Press.

HAUPTMAN, H. & KARLE, J. (1953). *The Solution of the Phase Problem. I. The Centrosym-metric Crystal. Am. Cryst. Assoc. Monogr. No. 3.* Pittsburgh : Polycrystal Book Service.

HÖRMANDER, L. (1973). *An Introduction to Complex Analysis in Several Variables*, 2nd edition. Amsterdam : North-Holland.

JAYNES, E.T. (1957). *Phys. Rev.* **106**, 620-630.

JAYNES, E.T. (1968). *IEEE Trans.* SSC-4, 227-241.

JAYNES, E.T. (1983). *Papers on Probability, Statistics and Statistical Physics.* Dordrecht : Reidel Publ. Co.

KHINCHIN, A.I. (1949). *Mathematical Foundations of Statistical Mechanics.* New York : Dover Publications.

KLUG, A. (1958). *Acta Cryst.* **11**, 515-543.

KLUYVER, J.C. (1906). *K.Ned. Akad. Wet. Proc.* **8**, 341-350.

LEBEDEV, N.N. (1972). *Special Functions and their Applications.* New York : Dover

LINDLEY, D.V. (1965). *Introduction to Probability and Statistics from a Bayesian Viewpoint,* Vols. 1 and 2. Cambridge Univ. Press.

PALEY, R.E.A.C. & WIENER, N. (1934). *Fourier Transforms in the Complex Domain.* Providence, R.I. : Amer. Math. Soc.

PEARSON, K. (1905). *Nature, Lond.* **72**, 294, 342.

PETROV, V.V. (1975). *Sums of Independent Random Variables.* Berlin : Springer-Verlag.

RAYLEIGH (J.W. STRUTT), Lord (1880). *Phil. Mag.* **10**, 73-78.

RAYLEIGH (J.W. STRUTT), Lord (1899). *Phil. Mag.* **47**, 246-251.

RAYLEIGH (J.W. STRUTT), Lord (1905). *Nature, Lond.* **72**, 318.

RAYLEIGH (J.W. STRUTT), Lord (1918). *Phil. Mag.* **36**, 429-449.

RAYLEIGH (J.W. STRUTT), Lord (1919). *Phil. Mag.* **37**, 321-347.

RICE, S.O. (1944, 1945). *Bell System Tech. J.* **23**, 283-332 (parts I and II) ; **24**, 46-156 (parts III and IV). Reprinted in *Selected Papers on Noise and Stochastic Processes* (1954), edited by N. WAX, pp. 133-294. New York : Dover Publications.

SHANNON, C.E. & WEAVER, W. (1949). *The Mathematical Theory of Communication.* Urbana : Univ. of Illinois Press.

SHMUELI, U.& WEISS, G.H. (1985). *Acta Cryst.* A**41**, 401-408.

SHMUELI, U.& WEISS, G.H. (1986). *Acta Cryst.* A**42**, 240-246.

SHMUELI, U.& WEISS, G.H. (1987). *Acta Cryst.* A**43**, 93-98.

SHMUELI, U.& WEISS, G.H. (1988). *Acta Cryst.* A**44**, 413-417.

SHMUELI, U., WEISS, G.H.& KIEFER, J.E. (1985). *Acta Cryst.* A**41**, 55-59.

SHMUELI, U., WEISS, G.H., KIEFER, J.E. & WILSON, A.J.C. (1984). *Acta Cryst.* A**40**, 651-660.

SZEGÖ, G. (1920). *Math. Z.* **6**, 167-202.

TSOUCARIS, G. (1970). *Acta Cryst.* A**26**, 492-499.

WIENER, N. (1949). *Extrapolation, Interpolation and Smoothing of Stationary Time Series.* Cambridge, Mass. : MIT Press.

THE METHOD OF REPRESENTATIONS OF STRUCTURE SEMINVARIANTS

II : SOME APPLICATIONS

C. Giacovazzo

Dipartimento Geomineralogico
Campus Universitario
70124 BARI - Italia

1 - INTRODUCTION

In the paper I it has been shown that reciprocal space may be arranged
in order to obtain accurate phase estimates. If some form of prior information is available its proper use can increase the rate of success of the
phasing process. We will show here that several types of information can be
exploited. Some of them are hidden in the diffraction data, others can arise
from additional sources. Special reference will be made in this paper to the
way in which the different types of prior information are exploited by the
program SIR88 (Burla et al., 1989a). We will deal with:
a) pseudotranslational symmetry
b) well located molecular fragment
c) well oriented but misplaced molecular fragment
d) Patterson information
e) attended and forbidden domains

2- PSEUDOTRANSLATIONAL SYMMETRY

Important contributions to this subject have been given by Fan Hai-fu
et al. (1983), Böhme (1982) and Gramlich (1984). We will shortly describe
the mathematical model used by Cascarano, Giacovazzo and Luic' (1985, 1987,
1988a, b). Let us suppose that a non-negligible amount of the electron
density $\rho(\mathbf{r})$, say $\rho_p(\mathbf{r})$, satisfies the condition

$$\rho_p(\mathbf{r} + \mathbf{u}) = \rho_p(\mathbf{r}).$$

Suppose also that in a space group of order m more than one independent
pseudotranslations are simultaneously present. $\mathbf{u}_i$ will denote the ith

independent pseudotranslation and n_i its index (the index of a pseudotrans-

lation is the smallest integer n for which $n\mathbf{u}$ is a lattice vector). Let p be
the overall number of atoms, symmetry equivalents included, whose positions
are related by the pseudotranslations $\mathbf{u}_i$, $i = 1, 2, \ldots$, and let t_p be the

number of independent atoms which generate the p atoms when the pseudotrans-
lation $\mathbf{u}_i$, $i = 1, 2, \ldots$ and the symmetry operators $\mathbf{C}_s = (\mathbf{R}_s, \mathbf{T}_s)$, $s = 1, 2,$

$\ldots, m$ are applied.

Direct Methods of Solving Crystal Stuctures
Edited by H. Schenk, Plenum Press, New York, 1991

For each atom in $\mathbf{r}_j$ suffering pseudotranslational symmetry, $mn_1n_2n_3$ equivalent atoms can be found at

$$C_s(\mathbf{r}_j + \nu_1\mathbf{u}_1 + \nu_2\mathbf{u}_2 + \nu_3\mathbf{u}_3 + \ldots),$$

where

$$1 \le s \le m, \qquad 0 \le \nu_i \le n_i - 1.$$

Then the structure factor equation may be defined as

$$F_\mathbf{h} = \sum_{j=1}^{t_p} f_j \sum_{\nu_1,\nu_2,\nu_3,\ldots,s} \exp[2\pi i \mathbf{h} C_s(\mathbf{r}_j + \nu_1\mathbf{u}_1 + \nu_2\mathbf{u}_2 + \nu_3\mathbf{u}_3 + \ldots)]$$

$$+ \sum_{j=t_p+1}^{t_p+t_q} f_j \sum_{s=1}^{m} \exp(2\pi i \mathbf{h} C_s \mathbf{r}_j). \tag{1}$$

Equation (1) may be rewritten as

$$F_\mathbf{h} = \sum_{j=1}^{t_p+t_q} f_j g_j, \tag{2}$$

where

$$g_j = \sum_{s=1}^{m} \frac{\sin(n_1\pi\mathbf{h}R_s\mathbf{u}_1)\sin(n_2\pi\mathbf{h}R_s\mathbf{u}_2)}{\sin(\pi\mathbf{h}R_s\mathbf{u}_1)\ \ \sin(\pi\mathbf{h}R_s\mathbf{u}_2)} \times \frac{\sin(n_3\pi\mathbf{h}R_s\mathbf{u}_3)}{\sin(\pi\mathbf{h}R_s\mathbf{u}_3)} \ldots$$

$$\times \exp\left[2\pi i \mathbf{h} C_s\left(\mathbf{r}_j + \frac{n_1-1}{2}\mathbf{u}_1 + \frac{n_2-1}{2}\mathbf{u}_2 + \frac{n_3-1}{2}\mathbf{u}_3 + \ldots\right)\right]$$

if $j \le t_p$, and,

if $j > t_p$

$$g_j = \sum_{s=1}^{m} \exp(2\pi i \mathbf{h} C_s \mathbf{r}_j).$$

From the above model one can calculate, for a given $\mathbf{h}$

$$\langle|F_\mathbf{h}|^2\rangle = \varepsilon_\mathbf{h}\left(\alpha_\mathbf{h}\Sigma_p + \Sigma_q\right) \tag{3}$$

where

$$\alpha_\mathbf{h} = (n_1n_2n_3\ldots)\gamma_\mathbf{h}/m.$$

$\gamma_\mathbf{h}$ is the number of times for which algebraic congruences

$$\mathbf{h}R_s\mathbf{u}_i \equiv 0(\text{mod } 1) \quad \text{for } i = 1,2,3,\ldots$$

are simultaneously satisfied when s varies from 1 to m. The fractional scattering powers Σ_q/Σ_N and $\Sigma_p/\Sigma_N = (1-\Sigma_q/\Sigma_N)$ are calculated according to

$$\frac{<|E'_h|^2>-\alpha_h}{1 - \alpha_h} = \frac{\Sigma_q}{\Sigma_N} \tag{4a}$$

or to

$$\frac{(<|E'_h|^2>-1)}{\alpha_h - 1} = \frac{\Sigma_p}{\Sigma_N} \tag{4b}$$

where the averages are taken over reflexions with a fixed value of α, and the E''s are normalized structure factors calculated in absence of any information on the pseudotranslational symmetry. F_h is defined to be a superstructure reflexion if $\gamma_h = 0$ (for them $<|E'_h|^2> = \Sigma_q/\Sigma_N$), otherwise F_h is a substructure reflexion. The maximum value of γ is m; consequently the maximum value of α is $n_1 n_2 n_3 \cdots$.

$$E_h = F_h / \left[\varepsilon_h\left(\alpha_h\Sigma_p+\Sigma_q\right)\right]^{1/2}.$$

The information so obtained on the nature of the pseudotranslational symmetry and on the percentage of electron density satisfying it is used for calculating the normalized structure factors and the distribution of the triplet phase $\Phi = \phi_h-\phi_k-\phi_{h-k}$. This last is found to be of von Mises type, with concentration parameter given by

$$G' = G \ Q$$

where G is the Cochran coefficient and Q is a function which depends on $\mathbf{h}$, $\mathbf{k}$, $\mathbf{h-k}$. Σ_p and Σ_q.

Real substructures often do not exactly comply with the above mathematical model (i.e., atoms related by pseudotranslational symmetry are not exactly located or are of different nature). A more general model has been introduced by Cascarano, Giacovazzo and Luic' (1988a, b) and implemented into SIR88. The program automatically looks for a pseudotranslational symmetry, estimates the percentage of electron density satisfying such a pseudosymmetry, modifies triplet estimates and applies tangent formula without any user intervention.

3 - FROM A PARTIAL TO A COMPLETE CRYSTAL STRUCTURE

Special procedures have been devised like :
a) tangent recycling methods (Karle, 1970; Hull and Irwin, 1978).
b) tangent recycling methods applied to difference structure factors (Beurskens et al., 1979).
c) joint probability distribution methods (Main, 1976; Heinerman, 1977).
Further insight into these methods has been given by Giacovazzo (1983), Camalli et al. (1985), Burla et al. (1989b). In this paper we will closely follow their point of view. The coordinates of the p atoms in known position were assumed to be fixed parameters of the probabilistic problem, while the other $q = N-p$ atomic positions define the primitive random variables. A basic result was the conditional joint probability distribution function

$$P \ (\ E_h \ | \ E_k, \ E_{h-k}, \ E_{ph}, \ E_{pk}, \ E_{ph-k} \ ) \tag{5}$$

from which the value of the ϕ_h may be derived. In a condensed form (5) leads to

$$E_h \cong E_{ph} + \frac{1}{\sqrt{q}} (E_k - E_{pk}) (E_{h-k} - E_{ph-k})$$

which generalizes Sayre equation.

An interesting question could be: "can the prior information on the location of the molecular fragment be integrated with the second representation formula for triplets?". As shown in paper I the P10 formula is based on the combined estimate of the special quintets

$$\{ \phi_h - \phi_k - \phi_{h-k} + \phi_l - \phi_l \}$$

where l is a free vector. Accordingly P10 exploits the larger subset of magnitudes

$$\{ R_h, R_k, R_{h-k}, R_l, R_{h\pm l}, R_{k\pm l}, R_{h-k\pm l} \}.$$

The canonical mathematical two-step process for gaining the probabilistic density function may be described as follows:

a) Calculate the joint probability distribution function

$$P (E_h, E_k, E_{h-k}, \{ E_l, E_{h\pm l}, E_{k\pm l}, E_{h-k\pm l} \},$$
$$| E_{ph}, E_{pk}, E_{ph-k}, \{ E_{ph}, \ldots, E_{ph-k\pm l} \},$$

b) Calculate the conditional distribution

$$P (\Phi | R_h, R_k, R_{h-k}, \{R_l, \ldots, R_{h-k\pm l}\}, \{E_{ph}, \ldots, E_{ph-k\pm l} \}).$$

Such a distribution is not easy to derive and probably too time consuming for computer applications. It may be approximated by

$$P (\Phi | \ldots) \simeq L \exp \left[S \cos (\Phi - \Phi_p) + K \cos \Phi \right]$$
$$= L \exp \left[\alpha \cos (\Phi - \Theta) \right] \tag{6}$$

where Θ is the most probable value of Φ, K is the (positive or negative) reliability parameter calculated according to the P10 formula, and S is the Sim's contribution, according to which $\Phi \cong \Phi_p$ with reliability parameter given by

$$D_1 (S) = D_1 (G_h) \; D_1 (G_k) \; D_1 (G_{h-k})$$

and $G = 2 R' R'_p$, where R' is a structure factor magnitude normalized on the scattering factors of the q-atoms. From (6)

$$\tan \Theta = (S \sin \Phi_p) / (S \cos \Phi_p + K) = A / B$$

Because of the prior information on the located fragment, Θ may lie anywhere between 0 and 2π; furthermore, the magnitudes of α depends on the magnitudes of S and K, and on the sign of K. If the ratio S/K is large (as, in a statistical sense, when a large molecular fragment is located) then $\Theta = \Phi_p$

as may be expected. If the ratio S/K is small (as, in a statistical sense, when only a small molecular fragment is located) then the P10 formula prevails: indeed Θ will be close to 0 or π according to the sign of K. Thus the full power of the P10 formula is preserved in (6).

4 – FINDING THE COMPLETE CRYSTAL STRUCTURE FROM A CORRECTLY ORIENTED ATOMIC GROUP

The position of a well oriented fragment with respect to the symmetry elements of the space group is usually achieved by:
a) translation functions in the vector space or in the reciprocal space (for shortness we do not quote literature).
b) special direct methods procedures (Karle, 1968; Bruins Slot and Beurskens, 1984; Main, 1976; Giacovazzo, 1988). We will closely follow here the point of view of the last author.

The application of translation search techniques is rather simplified if restrictions (in accordance with space group symmetry) are imposed on the admissible shifts. For example, shifts which leave invariant the algebraic form of the symmetry operators are superfluous: thus, thanks to the indeterminacy in choice of the origin, the coordinates of the molecular center of a fragment in P2 may be arbitrary restricted to

$$0 \le x < 1/2 \ , \qquad y = 0 \ , \qquad\qquad 0 \le z < 1/2 \ .$$

If suitable restrictions on the primitive random variable are imposed in the probabilistic methods, a non negligible benefit will arise for them. In this view, we first divide the crystal structure factor into two parts:

$$F_h = \sum_{i=1}^{p} g_i(h) + F_{qh} = F_{ph} + F_{qh}$$

where

$$g_i(h) = \sum_{j=1}^{n_i} f_j(h) \sum_{s=1}^{m} \exp(2\pi ih C_s r_j)$$

$$F_{qh} = \sum_{j=1}^{q} f_j(h) \sum_{s=1}^{m} \exp(2\pi ih C_s r_j)$$

n_i is the number of atoms in the ith molecular fragment, p is the number of fragment (symmetry independent) with unknown position and fixed orientation. According to the stated postulates one may also write

$$g_i(h) = \sum_{j=1}^{n_i} f_j(h) \sum_{s=1}^{m} \exp[2\pi ih C_s (u_j + t_i)]$$

where the u_j's are the trial atomic positions for the ith fragment and t_i is the shift to be applied to that fragment in order to translate atoms to the correct positions

$$u_j + t_i, \ \text{for} \ \ j = 1, \ \ldots, n_i.$$

A more useful form of $g_i(h)$ is

$$g_i(\mathbf{h}) = \sum_{s=1}^{m} g_{is}(\mathbf{h}) \exp\left(2\pi i \mathbf{h} R_s \mathbf{t}_i\right)$$

where

$$g_{is}(\mathbf{h}) = \sum_{j=1}^{n_i} f_j(\mathbf{h}) \exp\left(2\pi i \mathbf{h} C_s \mathbf{u}_j\right)$$

Factors $g_{is}(\mathbf{h})$ do not depend on $\mathbf{t}_i$ and may be calculated from prior information for any i and s. Thus the primitive random variables in our probab-
ilistic approach are the m shifts $\mathbf{t}_i$ and q atomic positions $\mathbf{r}_j$. Without loss
of generality, each primitive random variable may be now considered statist-
ically independent of the others, and uniformly distributed over the unit
cell. Such an assumption may be considered unfavourable in space groups for
which the allowed $\mathbf{t}_i$'s are arbitrary displacements along polar directions.

Owing to the indeterminacy in the choice of origin, one of the shifts $\mathbf{t}_i$ may
be restricted to a region that is smaller than the unit cell. As examples of
different dimensionality: a) in P2 we are free to specify the origin along
the diad axis by restricting $\mathbf{t}_1$ into the family of those vectors of type
[x 0 z]; b) in Pm the origin may be fixed by restricting $\mathbf{t}_1$ to the class of
vectors [0 y 0]; in P1 $\mathbf{t}_1$ may be restricted to the vector [0 0 0]. Once we
have restricted $\mathbf{t}_1$, no other restriction can be assigned to the others $\mathbf{t}_i$'s,
which may be assumed to be uniformly distributed over the unit cell. It has
been shown (Giacovazzo, 1988) that for $\mathbf{h}$ satisfying

$$\mathbf{h} \, \mathbf{t}_1 = 0$$

the conditional probability distribution of $\phi_\mathbf{h}$ may be calculated on the
assumption that $\mathbf{t}_1$ is a restricted random vector while $\mathbf{t}_i$, for $i = 2, \ldots,$
p and q positional vectors $\mathbf{r}_j$ are the primitive random variables, randomly
distributed over the unit cell. Again probability distribution functions can
be calculated for the combination of phases

$$\Phi = \phi_{\mathbf{h}_1} + \phi_{\mathbf{h}_2} + \ldots + \phi_{\mathbf{h}_n}$$

from which

$$(\mathbf{h}_1 + \mathbf{h}_2 + \ldots + \mathbf{h}_n)\mathbf{t}_1 = 0$$

Such phases Φ are not s.i.'s or s.s.'s.

Applications of the above theory have been presented by Cascarano,
Giacovazzo, Luic' and Vickovic' (1988).

5 - USE OF THE PATTERSON MAP

Direct methods and Patterson Methods are often used as independent
approaches to crystal structure solution. Their integration is possible.
Three different examples are:
a) The three-dimensional Patterson map is analyzed in order to obtain
estimates of triplet invariants (Hauptman and Karle, 1962; Hoppe, 1963; Main
and Woolfson, 1963; Allegra, 1979).
b) Orientation of the search model is determined by a Patterson rotation
search, while its position is found by maximizing the sum of the cosines of

a small number of strong translation sensitive triplet invariants (Egert and Sheldrick, 1985).
c) Harker sections of the Patterson map are calculated. Their Fourier transforms give information about moduli and phases of the 1-phase seminvariants of first rank (Ardito et al., 1985). In particular the calculated value of F_H is given by

$$F_H = \exp\ (2\pi i HT_p)\ F_{HS}[P(\mathbf{u})] \tag{7}$$

where F is the symbol of Fourier transform, $\mathbf{u}$ varies over the complete Harker section HS, and L is a suitable constant which takes into account the dimensionality of HS.

Eq. (7) is quite significant from a theoretical point of view: it apparently solves the phase problem owing to the fact that it does need any interpretation of the Patterson map. In the practice the application of (7) is limited by different factors:
1) Due to the finiteness of experimental data, the function P actually available is an imperfect representation of the true Patterson function.
2) For nearly equal-atom structures peaks in P are in general unresolved.
3) Occurrence in Harker section of non-Harker peaks. Points 1-3 suggest that relation may be successfully applied only to structures with few dominant heavy atoms. The approach could be extended to structures with medium weight atoms (S, Cl, Fe,...) provided a method is available for improving both positions and form of Harker peaks, and their discrimination from false Harker-peaks. A special least squares procedure is proposed (Cascarano, Giacovazzo, Luic', Pifferi and Spagna, 1987). The first applications show that 1-phase seminvariants of first rank may be reliably estimated for a large variety of crystal structures.

6 - ATTENDED AND FORBIDDEN DOMAINS

It may be asked now if the invariance and seminvariance concepts are the only ones which may be invoked in order to obtain phase relationships. In Section 4 we have already seen that, if some molecular fragments have been correctly oriented but have unknown position then phase relationships may be found which are not s.i. or s.s.'s. Such a case may be generalized by starting that phase relationships which are not s.i.'s or s.s.'s arise when suitable constraints are applied to some primitive random variables, or, in more general terms, when some primitive random variables are non-uniformly distributed over the unit cell. Extreme cases occur when some domains are forbidden (inaccessible domains) or are attended (some random variables are restricted to some domains).

The idea of forbidden or attended domains is not new in crystallography. It has been applied to structure-factor statistics (Wilson, 1964; Pradhan and Nigam, 1986; Wilson, 1987) and to determinantal methods (von Eller, 1962; Lajzerowicz and Lajezerowicz, 1986; Knossow et al., 1977). Prior information on inaccessible or attended domains may be exploited for the probabilistic estimation of phases which are or are not s.s.'s (Giacovazzo, 1989).

A large variety of forbidden domains may be imagined for macromolecular structures, where diffracting matter occupies a reduced percentage of the unit cell volume. Inaccessible domains may also be located around symmetry elements not compatible with molecule point symmetry. As a practical example let us consider the case of the structure CEPHA ($C_{18}H_{21}NO_3$, C2, Z = 8). The following phase values for seminvariants reflexions for which $|h + l| \leq 4$ can be found:

$$\emptyset_{200} = \pi; \quad \emptyset_{002} = \pi; \quad \emptyset_{004} = \pi$$

$$\emptyset_{400} = 0; \quad \emptyset_{202} = \pi; \quad \emptyset_{20\bar{2}} = \pi.$$

Inaccessible domains around binary axes suggest the phase value π for all of them.

As a practical case of attended domain, let us consider the crystal structure $CaNaN(CH_2CO_2)_3$ (BOBBY in code), space group $P2_13$, $Z = 3$. Ca, Na and N atoms must lie on ternary symmetry axes. A set of phases can be estimated by such prior crystallochemical information.

Consider also the crystal structure KCr_5Se_8 (Nguyen-Huy Dung et al., 1988), space group $B2/m$ with unique axis $\mathbf{c}$, $Z = 2$. The reflexion intensities for homologous series hkl, $l = 0,2,4$ were very similar: the same was true for hkl, $l = 1,3,5$. The heavy atoms were therefore expected to be located in the mirror planes (at $\mathbf{z} = 0$ and $\mathbf{z} = 1/2$). From the above information it has been possible to correctly evaluate the values of the reflexions

$$E_{002} = +2.416 \quad \text{and} \quad E_{004} = +2.741.$$

Furthermore 7968 pairs of reflexions satysfying the condition

$$H_1 + H_2 = (001)$$

have been found among the 208 reflexions with largest value of $|E|$. A large percentage of them are positive, as expected: probabilistic considerations, here not developed, will rank them in order of calculated reliability.

REFERENCES

Allegra, G., 1979, Acta Cryst., A35:213-220.

Ardito, G., Cascarano, G., Giacovazzo, C., and Luic', M., 1985, Z. Kristallogr., 172:25-34.

Beurskens, P. T., Prick, A. J., Doesburg, H. M., and Gould, R. O., 1979, Acta Cryst., A35:765-772.

Böhme, R., 1982, Acta Cryst., A38:316-326.

Bruins Slot, H. J., and Beurskens, P. T., 1984, Acta Cryst., A40:701-703.

Burla, M. C., Camalli, M., Cascarano, G., Giacovazzo, C., Polidori, G., Spagna, R., and Viterbo, D., 1989a, J. Appl. Cryst., 22:389-393.

Burla, M. C., Cascarano, G., Fares, V., Giacovazzo, C., Polidori, G., and Spagna, R., 1989b, Acta Cryst., A45:781-786.

Camalli, M., Giacovazzo, C., and Spagna, R., 1985, Acta Cryst., A41:605-613.

Cascarano, G., Giacovazzo, C., and Luic', M., 1985, Acta Cryst., A41:544-551

Cascarano, G., Giacovazzo, C., and Luic', M., 1987, Acta Cryst., A43:14-22.

Cascarano, G., Giacovazzo, C., and Luic', M., 1988a, Acta Cryst., A44:176-183.

Cascarano, G., Giacovazzo, C., Luic', M., Pifferi, A., and Spagna, R., 1987, Z. Kristallogr., 179:113-125.

Cascarano, G., Giacovazzo, C., Luic', M., and Vickovic', I., 1988, Z. Kristallogr., 185:448.

Egert, E., and Sheldrick, G. M., 1985, Acta Cryst., A41:262-268.

Eller, G. von, 1962, Acta Cryst., 15:590-595.

Fan Hai-fu, Yao Jia-xing, Main, P., and Woolfson, M. M., 1983, Acta Cryst., A39:566-569.

Giacovazzo, C., 1983, Acta Cryst., A39:685-692.

Giacovazzo, C., 1988, Acta Cryst., A44:294-300.

Giacovazzo, C., 1989, Acta Cryst., A45:150-157.

Gramlich, V., 1984, Acta Cryst., A40:610-616.
Hauptman, H., and Karle, J., 1962, Acta Cryst., 15:547-550.
Heinerman, J. J. L., 1977, Acta Cryst., A33:100-106.
Hoppe, W., 1963, Acta Cryst., 16:1056-1057.
Hull, S. E., and Irwin, M. J., 1978, Acta Cryst., A34:863-870.
Karle, J., 1968, Acta Cryst., B24:182-186.
Karle, J., 1970, in: "Crystallographic Computing," 155-164, F. R. Ahmed,
    ed., Munksgaard, Copenhagen.
Knossow, M., De Rango, C., Mauguen, Y., Sarrazin, M., and Tsoucaris, G.,
    1977, Acta Cryst., A33:119-125.
Lajzerowicz, J., and Lajzerowicz, J., 1966, Acta Cryst., 21:8-12.
Main, P., 1976, in: "Crystallographic Computing Techniques," 97-105, F. R.
    Ahmed, ed., Munksgaard, Copenhagen.
Main, P., and Woolfson, M. M., 1963, Acta Cryst., 16:1046-1051.
Nguyen-Huy Dung, Vo-van Tieu, Behm, H. J., and Beurskens, P. T., 1987, Acta
    Cryst., C43:2258-2260.
Pradhan, D., and Nigam, G. D., 1986, Z. Kristallogr., 176:271-281.
Wilson, A. J. C., 1964, Acta Cryst., 17:1591-1592.
Wilson, A. J. C., 1987, Acta Cryst., A43:250-252.

STRUCTURE DETERMINATION BY THE REPRESENTATION METHOD

Davide Viterbo

Dipartimento di Chimica
Universitá della Calabria
I-87030 Arcavacata di Rende (Cs), Italy

Phase determination procedures

The probabilistic formulae derived by means of the representation theory and the related algorithms for setting up structure invariants (i.s.) and structure seminvariants (s.s.) have been implemented in the SIR (SemInvariant Representation) program (Burla et Al., 1980a) and the most important features of the program will be described in this paragraph, while examples of its application to the solution of crystal structures will be illustrated in the next paragraph.

The program can be conveniently described in terms of the following steps:

<u>Normalization</u> - Following the procedure described in a previous lecture the values of the normalized structure factors are first calculated and output as a list of reflexions sorted in decreasing order of $|E|$.

The standard statistical analysis of the normalized amplitudes is performed; a more detailed statistical analysis (Cascarano, Giacovazzo & Luic, 1985, 1987, 1988a,b) is carried out in order to reveal the presence and type of pseudotranslational symmetry and, when needed, to renormalize the amplitudes accordingly.

The available <u>a-priori</u> information, such as the existence of pseudo-translational symmetry or the coordinates of a previously located fragment (Camalli, Giacovazzo & Spagna, 1985; Burla et Al., 1989b) may be input.

The subroutine SYM (Burzlaff & Hountas, 1982), allowing the derivation of all symmetry operators and of the seminvariance conditions directly form the space group symbol, is part of this section of the SIR program

<u>One and two-phase seminvariants</u> - The general algorithm, proposed by Cascarno and Giacovazzo (1983), for identifying structure seminvariants in all space groups, is employed in this section.

One-phase s.s.'s are estimated by means of their second representation (Giacovazzo, 1978; Burla et Al., 1980; Burla Giacovazzo & Polidori, 1980; Cascarano et Al., 1984a), while two-phase s.s.'s are estimated by their first representation (Giacovazzo et Al., 1979; Cascarano et Al., 1982)); more recently a more efficient algorithm for setting up and estimating two-phase s.s.'s (Burla, Giacovazzo & Polidori, 1989) has been implemented.

Direct Methods of Solving Crystal Stuctures
Edited by H. Schenk, Plenum Press, New York, 1991

All the estimated s.s.'s are stored; those with highest reliability will be used actively, while the others will be employed to compute two figures of merit (_fom_'s).

Triplet and quartet invariants - Following the procedure described in a previous lecture all triplets relating reflexions with $|E|$ greater than a given threshold, are generated and those with Cochran's G value

$$G_{hk} = 2\sigma_3\sigma_2^{-3/2}|E_hE_kE_{h-k}| \qquad (1)$$

above a chosen limit are stored.

At the same time, triplets relating two reflexions with large $|E|$ and one with $|E|$ close to zero (ψ_o triplets), are generated to be used later in computing several _fom_'s.

Negative quartets are generated by combining in pairs the ψ_o triplets with a common weak reflexion; those with all three cross-magnitudes smaller than a given threshold are estimated by means of their first representation (Busetta et Al., 1980), and will be used to compute a _fom_.

The estimates of triplets may be considerably improved by the use of their second representation (P10 formula) (Cascarano et Al., 1984b). The concentration parameter of the von Mises distribution is given by

$$G' = G (1 + R) \qquad (2)$$

where R is a function of all the magnitudes in the second phasing shell. The G' values are rescaled on the G values and the triplets are ranked in decreasing order of G'; the top relationships represent a better selection of triplets with value close to zero than obtained when ranking according to G. These triplets will be actively used in the phase determination process. R may also be negative and for some triplets a negative G' value may be obtained, indicating that their expected value is π. The reliability of the negative estimates is in general too small for an active use of negative triplets in the phase determination process, but a passive use to compute an effective _fom_ proved to be very useful. Besides, the capability of the P10 formula of sampling out most triplets with a value far from zero, allows their elimination from the phase determination process, which then becomes more stable.

Triplets and quartets estimated with G'≈0 are expected to have values around $\pm(\pi/2)$ and are used to compute enantiomorph sensitive functions.

When the _a-priori_ information concerning the existence of pseudo-translational symmetry (usually detected at the normalization step) is input to this section, then for each triplet G is suitably modified on the basis of the indices of the reflexions **h**, **k** and **h-k** and of the type of pseudotranslation (Cascarano, Giacovazzo & Luic, 1987).

Convergence/divergence procedure - An optimal starting set of phases can be defined by means of the convergence procedure (Germain, Main & Woolfson, 1970), which is an elimination process based on the values of

$$\langle\alpha_h\rangle = \sum_{j=1}^{r} G \frac{I_1(G)}{I_o(G)} = \sum_{j=1}^{r} G \, D_1(G) \qquad (3)$$

where $I_o(G)$ and $I_1(G)$ are modified Bessel functions of zero and first order and the summations are over all relations defining reflexion **h**.

In SIR, when one and two-phase s.s.'s are evaluated, the few selected one-phase s.s.'s are treated as known phases, while some tens of selected two-phase s.s.'s are used in the convergence process as phase relations

$$\phi_h \approx \pm \phi_k + S2 \qquad\qquad (4)$$

where S2 is the estimated value of the s.s. entering in (3) with a reliability parameter G_s.

Once the starting set has been defined by the convergence procedure, the best pathway for phase expansion is determined by a divergence procedure, which is an add-on process carried out starting from the starting set of phases.

A development of this step uses a new weighting procedure (Burla et Al., 1987); each reflexion is assigned a weight taking into account the accuracy (obtained as a function of $<\alpha_k>$ and $<\alpha_{h-k}>$), with which the phases contributing to it may be determined.

<u>Phase</u> <u>extension</u> <u>and</u> <u>refinement</u> - Starting from the limited set of phases defined in the previous step new phases are determined and refined using a weighted tangent formula. Two phase s.s.'s may also contribute to it. The tangent formula becomes then

$$\tan\phi_h = \frac{\Sigma_{k_t} W_t \sin(\phi_{k_t}+\phi_{h-k_t}) + \Sigma_{k_s} W_s \sin(\phi_{k_s}+S2)}{\Sigma_{k_t} W_t \sin(\phi_{k_t}+\phi_{h-k_t}) + \Sigma_{k_s} W_s \sin(\phi_{k_s}+S2)} = \frac{T_h}{B_h} \qquad (5)$$

where the summations over k_t are over all triplets contributing to ϕ_h, each with weight W_t, and those over k_s are over all two-phase s.s.'s involving ϕ_h, each with weight W_s. The starting set defined in the preceding step is usually formed by the origin (and enantiomorph) fixing reflexions, optionally by a few one-phase s.s.'s and by a number of other phases to be permuted; among these, the general phases are assigned magic integer values.

A new random procedure (Burla, Giacovazzo & Polidori, 1987) has been recently introduced in SIR. Triplets which are involved in the initial stages of phase determination, are assigned random values distributed according to a von Mises distribution with their G or G' as concentration parameters. The random value is then added as a phase-shift to each triplet in the tangent expansion and refinement.

<u>Figures</u> <u>of</u> <u>merit</u> - After expansion and refinement of each set of phases several <u>fom</u>'s are computed using all invariants and seminvariants estimated by means of the representation method. Their form and meaning is described in a paper by Cascarano, Giacovazzo & Viterbo (1987), together with an optimized way of combining all the computed functions to give a highly selective combined <u>fom</u>.

<u>E-maps</u> - The program for the calculation of the electron density maps also performs an automatic peak search and supplies a list of maxima sorted in decreasing order of height. This list is then analyzed in terms of distances and angles among the peaks; peaks related by a sensible chemical geometry are selected and labeled as a possible molecular fragment. Attempts have been recently made to establish an automatic link between the E-map output and the structure refinement programs (Cascarano, 1990); to this aim the map interpretation procedure has been improved and the peaks are also assigned an atomic symbol. For simple structures it becomes then possible to go from the diffraction intensities to the refined molecular picture without human intervention.

<u>Partial</u> <u>structure</u> <u>recycling</u> - It may happen that in the best E-map only a small molecular fragment can be recognized and in this case direct methods can be used to complete the structure. The first and simplest proce-

dure was proposed by Karle (1968) and is known as _Karle recycling_. The procedure for completing a partial structure at the basis of the DIRDIF program (Beurskens et Al., 1985) is described elsewhere.

In SIR a new probabilistic method (Camalli, Giacovazzo & Spagna, 1985; Burla et Al., 1989b) has been implemented. The structure factors for the known fragment are calculated and appropriate pseudo-normalized structure factors for the known and unknown part of the structure are obtained. These are then used in the new probabilistic formulae for the estimation of s.i.'s; on the basis of the estimated α's computed by these formulae, a convergence procedure will define the best starting set of phases, which will include both permuted phases and phases calculated from the partial structure. Phase expansion and refinement is performed by means of a new weighted tangent formula and two specific _fom_'s, related to R_α and ψ_o, but computed with the pseudo-normalized structure factors, are used to select the correct solution.

Examples of application of the sir program

We will first follow in detail the solution of the structure of the antibiotic _21-Acetoxy-11-(R)-Rifamicinol_, $C_{39}H_{49}NO_{13} \cdot CH_3OH \cdot H_2O$ (Cerrini et Al., 1988), which crystallizes in the space group $P2_1$ with $Z=2$. Its structural formula is

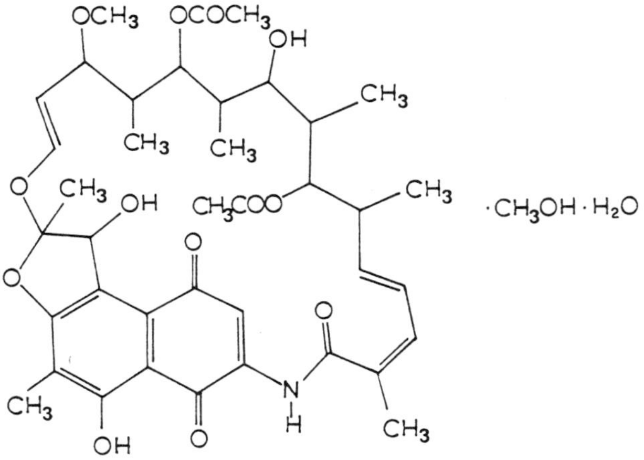

Normalized structure factors are computed by the normalization routine and 362 reflexions with $|E| > 1.723$ selected for phase determination; the non-centrosymmetric space group is confirmed by the statistical analysis an no pseudo-translational symmetry is detected.

None of the 17 one-phase s.s.'s estimated by means of the second representation has a sufficiently high _tanh_ argument to be accepted and these indications are only used to compute the related SS1FOM. 326 two-phase s.s.'s are estimated using their first representation; all of them are used to compute the corresponding SS2FOM, while 129 with $|G| > 0.6$ are actively used in the phase determination procedure.

Triplet invariants relating the 362 strongest reflexions are set up and estimated by means of their second representation For RIFOL 2280 positive triplets estimated with $G' \geq 0.6$ are actively used, while 141 negatively estimated triplets are employed to compute the corresponding NTREST _fom_. At the same time 1399 ψ_o triplets relating the 100 weakest reflexions with the considered 362 strongest reflexions are generated, to be used in the calculation of the related _fom_'s. Finally also 1009 nega-

246

tive quartets are estimated by their first representation and the most
reliable 500 are used to compute the related NQUEST _fom.

The convergence/divergence procedure is carried out using positive
triplets and two-phase s.s.'s. The derived starting set, shown in Table
1, includes three origin fixing reflexions and four general permuted
phases which are expressed in terms of magic integers, giving rise to 24
phase permutations (enantiomorph fixed by the magic integer procedure).

Phase expansion is carried out using the tangent formula (5).
Starting from the 8 reflexions in the staring set, the phase extension is
carried out following the order indicated by the divergence procedure. In
Table 2 the initial part of the divergence map of RIFOL is reported and
with its aid we can follow the phase determination process.

TABLE 1 - RIFOL: starting set of phases defined by the convergence procedure

Type	Code	h	k	l	E	phase
Origin	5	5	0	-18	3.57	0
	15	2	0	-1	2.93	0
	4	6	1	-18	3.85	0
Permuted	18	6	4	22	2.92	MI
	72	10	4	-5	2.40	MI
	149	3	1	-17	2.09	MI
	25	1	2	-16	2.79	MI
	17	8	4	-4	2.92	MI

Total number of permutations 24

TABLE 2 - RIFOL: divergence map illustrating the phase determination path
starting from the set of reflexions in Table 1.

Code	h	k	l	<α>				Contributors			
24	4	1	-17	8.41	4	-15	0	8.94			
99	8	1	-19	11.01	24	*	0	6.00	4	15	0	6.09
40	2	5	14	9.50	4	17	π	8.10	72	99	π	5.72
188	4	5	13	13.75	4	72	π	6.35	17	24	π	6.15
					15	40	0	5.78			
31	3	1	-1	7.80	-24	25	π	6.69	-5	99	0	3.71
13	3	2	-17	10.46	15	25	0	7.29	4	31	π	6.02
135	7	6	12	8.60	13	72	0	4.88	17	25	0	4.42
					31	188	0	3.92			
7	7	0	-19	6.72	5	15	0	7.26			
44	9	5	-5	11.47	7	40	0	7.26	5	188	0	5.28
37	3	4	14	10.22	-5	17	0	8.37	-7	72	0	6.40

Reflection 24 is the first to be determined by a single triplet relation; it will be used to define the following reflexions with its corresponding α value. The second reflexion no. 99 forms a two-phase s.s. with 24 (written as a triplet but with an asterisk instead of the third code number) and is related by a triplet to the starting set; each summation in (5) will have one term. The remaining reflexions are then determined by a similar chain process.

When a sufficiently large number (60-100) of phases has been determined, the program proceeds to their refinement. It is in fact possible to redetermine the initial phases with a greater number of contributors to the sums in (5); for instance the phase of reflexion 24 can now be determined using all relations involving 24 and two other reflexions among the 60-100 determined ones. The tangent refinement process is repeated until self-consistency and then the remaining phases are determined and refined. At this point different _fom_'s are computed.

Table 3 - RIFOL: list of the ten best sets of phases with their relatives figures of merit.

+++++++++++++ FIGURES OF MERIT +++++++++++++

SET	MABS	ALCOMB	PSCOMB	CPHASE	CFOM
1	1.093	0.920	0.925	0.885	0.911
2	1.090	0.843	0.921	0.878	0.885
3	1.016	0.388	0.683	0.479	0.554
4	1.056	0.457	0.624	0.440	0.539
5	1.032	0.348	0.601	0.495	0.517
6	1.052	0.422	0.554	0.431	0.502
7	1.051	0.432	0.538	0.435	0.501
8	1.049	0.401	0.566	0.431	0.500
9	1.044	0.383	0.567	0.431	0.496
10	1.054	0.429	0.533	0.426	0.495

The tangent procedure is repeated for each of the 24 permutations of the starting set phases. This step may become quite computer-time consuming when the number of phase permutations is increased. Table 3 gives the list of the 10 best sets output by SIR in decreasing order of the combined _fom_ CFOM. ALCOMB and PSCOMB are combinations of functions based on α and ψ_0 respectively. The first set with the highest CFOM yields the E-map shown in Figure 1, which may be interpreted in terms of the RIFOL structure as indicated by the connected circled maxima. Not all atoms are found in the map (solvent molecules plus eight atoms are missing, two to close the macrocycle and the others in the side chains), but the structure can be easily completed by standard Fourier methods. The final refined structure is shown in Figure 2.

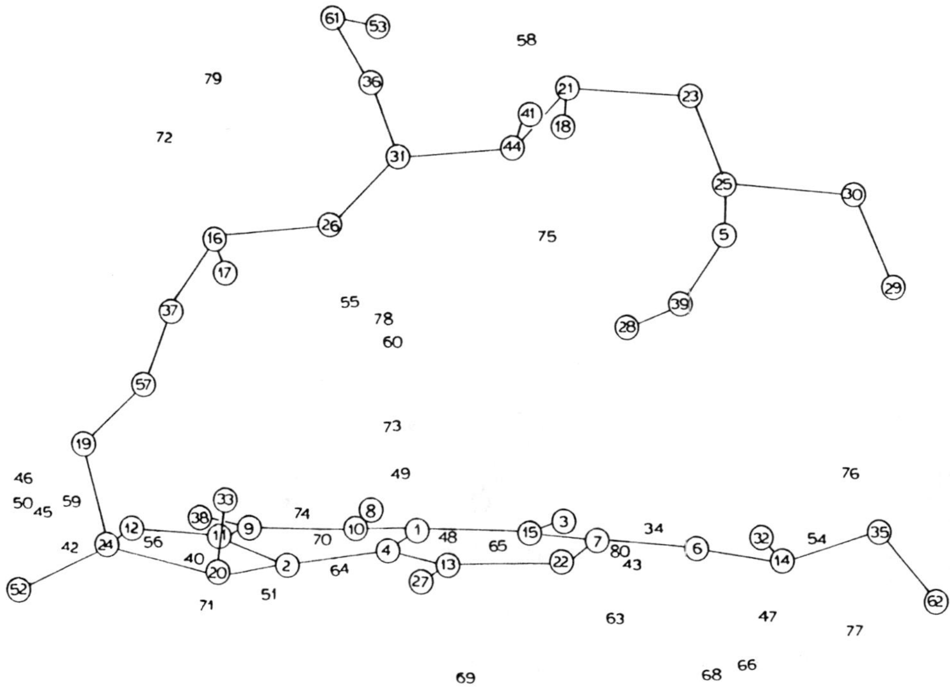

Figure   1

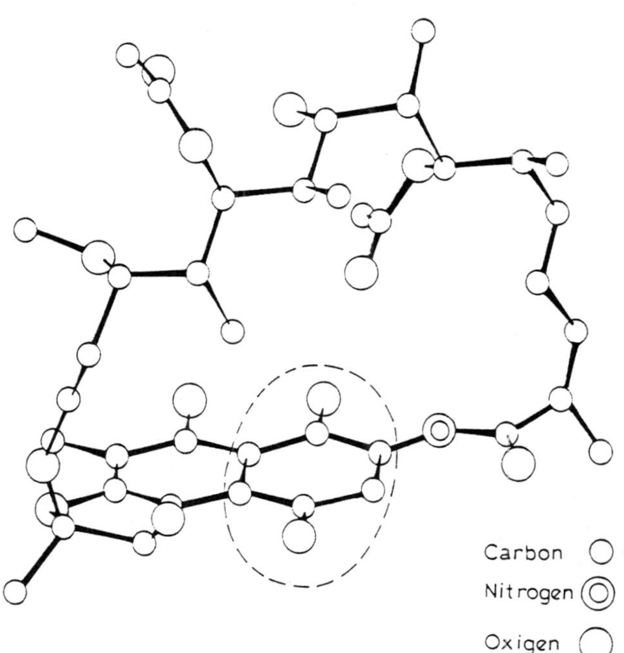

Figure 2 - RIFOL

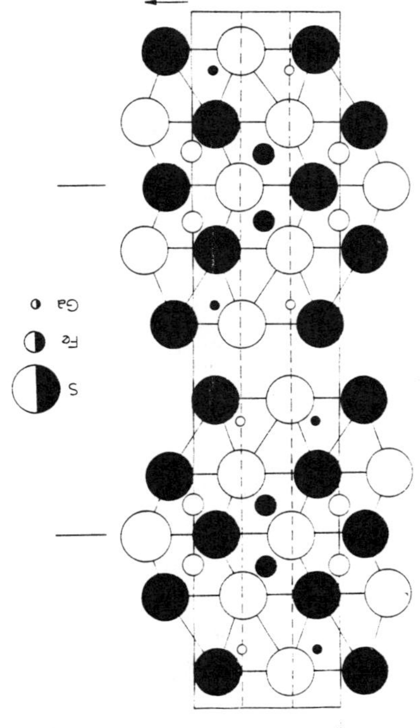

Figure 3 - FEGAS

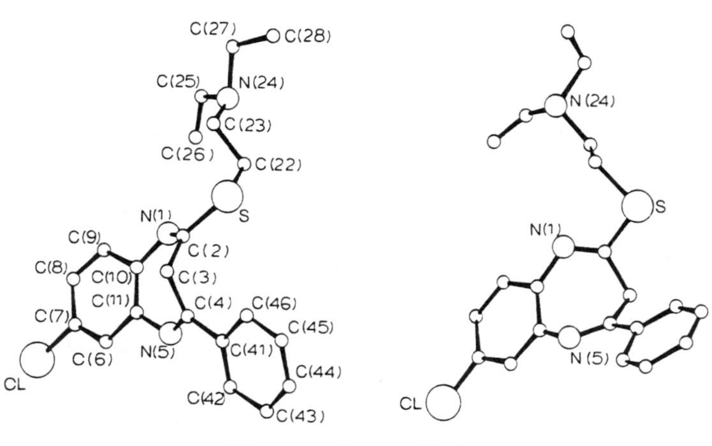

Figure 4 - BZDIA

The RIFOL structure was also used to test the partial structure procedure; starting from the 8 atom fragment (11% of the total number of electrons) surrounded by a dotted line in Figure 2, the complete structure was recovered.

As a second example we shall consider, in less detail, the solution of the structure of the *polytype 2H of the* $Fe_2Ga_2S_5$ *family (FEGAS)* (Cascarano, Dogguy-Smiri & Nguyen-Huy-Dung, 1987) crystallizing in the space group $P6_3/mmc$ with Z=2. This type of very regular structure is known to be troublesome for direct methods, and this particular one, shown in Figure 3, had resisted several attempts of solution also by Patterson methods. The normalization routine of SIR identifies the presence of a pseudotranslational symmetry with **u**=c/3. When this information is used in phase determination, the E-map computed with the highest CFOM reveals the coordinates of the Fe and S atoms; the structure is then completed by standard Fourier methods.

The use of the P10 formula to improve the estimates of triplets is one of the key feature of SIR. As an example of its importance we report here the case of the structure of *7-chloro-2-(diethylamino-ethylthio)-3H,1,5-benzodiazepine*, $C_{21}H_{24}$ N_3SCl (Pennini et Al., 1988) (*BZDIA* ,space group $Pca2_1$, Z=8). Previous attempts at solving it by other programs or by SIR using only Cochran's estimates failed to give a solution; only the use of the second representation estimates of triplets yielded an interpretable map.

R e f e r e n c e s

- Beurskens P.T., Bosman W.P., Doesburg H.M., Gould R.O., Van den Hark Th.E.M., Prick P.A.J., Noordik J.H., Beurskens G., Parthasarathi V., Bruins Slott H.J. & Haltiwanger R.C. (1985), "Program System DIRDIF", Tech. Report, Crystallography Laboratory, University of Nijmegen (The Netherlands).

- Burla M.C., Camalli M., Cascarano G., Giacovazzo C., Polidori G., Spagna R. & Viterbo D. (1989a), J. Appl. Crystallogr., **22**, 389.

- Burla M.C., Cascarano G., Fares E., Giacovazzo C., Polidori G. & Spagna R. (1989b), Acta Crystallogr., **A45**, 781.

- Burla M.C., Cascarano G., Giacovazzo C., Nunzi A. & Polidori G. (1987), Acta Crystallogr., **A43**, 370.

- Burla M.C., Giacovazzo C., & Polidori G. (1980), Acta Crystallogr., **A36**, 573.

- Burla M.C., Giacovazzo C. & Polidori G. (1987), Acta Crystallogr., **A43**, 797.

- Burla M.C., Giacovazzo C. & Polidori G. (1989), Acta Crystallogr., **A45**, 99.

- Burla M.C., Nunzi A., Polidori G., Busetta B. & Giacovazzo C. (1980), Acta Crystallogr., **A36**, 573.

- Burzlaff H. & Hountas A. (1982), J. Appl. Crystallogr., **15**, 464.

- Busetta B., Giacovazzo C., Burla M.C., Nunzi A., Polidori G. & Viterbo D. (1980), Acta Crystallogr., **A36**, 68.

- Camalli M., Giacovazzo C. & Spagna R. (1985), Acta Crystallogr., **A41**, 605.

- Cascarano G. (1990), Doctorate Thesis and references therein, University of Bari (Italy).

- Cascarano G. & Giacovazzo C. (1983), Z. Kristallogr., **165**, 169 (1983).

- Cascarano G., Dogguy-Smiri & Nguyen-Huy-Dung (1987), Acta Crystallogr., **C43**, 2050.

- Cascarano G., Giacovazzo C., Camalli M., Spagna R., Burla M.C., Nunzi A. & Polidori G. (1984) , Acta Crystallogr., **A40**, 278.

- Cascarano G., Giacovazzo C., Calabrese G., Burla M.C., Nunzi A., Polidori G. & Viterbo D. (1984), Z. Kristallogr., **167**, 34.

- Cascarano G., Giacovazzo C. & Luic M. (1985), Acta Crystallogr., **A41**, 544.

- Cascarano G., Giacovazzo C. & Luic M. (1987), Acta Crystallogr., **A43**, 14.

- Cascarano G., Giacovazzo C. & Luic M. (1988a), Acta Crystallogr., **A44**, 176.

- Cascarano G., Giacovazzo C. & Luic M. (1988b), Acta Crystallogr., **A44**, 183 (1988).

- Cascarano G., Giacovazzo C., Polidori G., Spagna R. & Viterbo D. (1982), Acta Crystallogr., **A38**, 663.

- Cascarano G., Giacovazzo C. & Viterbo D. (1987), Acta Crystallogr., **A43**, 22.

- Cerrini S., Lamba D., Burla M.C., Polidori G. & Nunzi A. (1988), Acta Crytallogr., **C44**, 489.

- Germain G., Main P. & Woolfson M.M. (1970), Acta Crystallogr., **B26**, 274.

- Giacovazzo C. (1978), Acta Crystallogr., **A34**, 562.

- Giacovazzo C., Spagna R., Vickovic I. & Viterbo D. (1979), Acta Crystallogr., **A35**, 401.

- Karle J. (1968), Acta Crystallogr., **B24**, 182.

- Pennini R., Cerri A., Tajana A., Nardi D. & Giordano F. (1988), J. Heterocyclic Chem., **25**, 305.

SIMPEL83, A SYMBOLIC ADDITION PROGRAM SYSTEM FOR DIRECT METHODS BASED ON HIGHER INVARIANTS

Henk Schenk

Laboratory for Crystallography, University of Amsterdam
Nieuwe Achtergracht 166, 1018 WV Amsterdam
The Netherlands

Introduction

The program system SIMPEL83 is a complete direct methods system, which starts from $|F|$-values, produces an E-map and finally refines and correct the structure. The unique part of SIMPEL83 is formed by a series of routines, successively gathering the relationships, determining a starting set, carrying out a symbolic addition, finding the correct numerical values for the symbols and executing a numerical phase extension and refinement. The other parts of the program system consist of a normalization program and a Fourier program, peak search and map interpretation program. Usually the latter parts are the same as those used in MULTAN, however, they may also belong to other systems, such as the XRAY or XTAL system or a local system. Our system ends with a very fast diagonal least squares structure factor program, which enables us to check interactively solutions from E-maps and refine structures at minimal computing-costs from R=45% down to R=15%. At present this program AMALS also contains error-recovering routines to delete erroneous atom positions and to introduce missing positions in the model (Shi and Schenk, 1988).

In present direct method procedures and in particular in SIMPEL83 and SIMPEL88 (see chapter of Peschar in this book) higher invariants play a more and more significant role. They serve in nearly all stages of the phase determination from the starting set procedures to the determination of the correct solution by means of figures merit and also in extension and refinement processes. Higher invariants are known from the beginning of direct methods. One of the Σ-relations of Hauptman and Karle (1953) is at present known as the quartet relation

$$\varphi_4 = \varphi_H + \varphi_K + \varphi_L + \varphi_{-H-K-L} \tag{1}$$

Because the indices of the four main reflections H,K,L and −H−K−L sum to zero it is a structure invariant, and because it involves more than three reflections it is called a higher invariant. To estimate φ_4 Hauptman and Karle considered the $|E|$'s of the reflections H, K, L and −H−K−L only which make this relation inferior compared with the triplet relation. The first useful approach to higher invariants was invented when the $|E|$ magnitudes of the seven reflections H, K, L, −H−K−L, H+K, H+L and K+L were used to estimate the quartet phase sum φ_4 (Schenk, 1973). It was shown that when the cross terms H+K, H+L and K+L are used in addition to the main terms, more reliable estimates of φ_4 are obtained. Later, these empirical observations were theoretically funded and also other higher invariants (such as quintets) and seminvariants were investigated. The latter are generally estimated via higher invariants.

Direct Methods of Solving Crystal Stuctures
Edited by H. Schenk, Plenum Press, New York, 1991

It is the purpose of this paper to present applications of higher invariants as they are implemented in SIMPEL83 and some more. Higher invariants are in particular useful when the normal direct methods fail as a result of the symmetry of the space group. The first paragraphs will be devoted to a brief description of SIMPEL83.

Collecting phase relationship in SIMPEL83

The different phase relationships are all collected by means of similar algorithms based on the use of two arrays, one (array A) in which all strong reflections the phases of which are aimed at are taken up and another (array B) in which the information of all reflections is stored three dimensionally on the basis of the crystallographic indices h, k and l. All relations are searched for by taking the relevant reflections from array A, calculating indices for the remaining reflections and looking those up in array B. For instance, quartets are gathered by taking H, K and L from array A, calculating $-H-K-L$ and looking this reflection up in array B. If it is potentially a useful quartet with a quartet product $N^{-1}|E_H E_K E_L E_{-H-K-L}|$ above a specified limit, also the information about the cross terms H+K, H+L and K+L is looked up in array B and the quartet with all its information is stored on a disk file.

SIMPEL83 searches for triplets, quartets (negative, positive and enantiomorph sensitive ones), two-dimensional quartets also called 'Harker-Kasper' relations (H,H,H+K and H-K), and Σ_1-relationships. It is very straightforward to add routines for calculating other relationships such as quintets and in our research version of SIMPEL quintet and seminvariant routines are present. The algorithms have been described in more detail elsewhere (Overbeek, 1985)

The starting set procedure, convergence and divergence

The determination of the starting set begins with a convergence procedure similar to that of Germain, Main and Woolfson (1970). The procedure searches for that reflection which is weakest linked to the other reflections by means of the phase relations and then this reflection is removed from the set of reflections. At the same time all its phase relations are removed from the collection of phase relationships. This process is repeated until no reflections are left in the set. Any time a reflection is removed without having phase relationships linking it to phases still within the set, this reflection is taken up in the starting set. Our procedure uses triplets and seven magnitude quartets for a relative small number of strong reflections; usually the number is in between 30 and 60. This procedure goes back to the first paper on seven magnitude quartets (Schenk, 1973) when it was found that many of the strong quartets are found within the group of very strongest reflections. Since in a starting set reflections with large $|E|$-values are to be preferred, the use of triplets and quartets generally leads to very good starting sets. After completing the convergence procedure the origin is fixed and to all other starting set reflections symbolic phases are attached.

The second step in our procedure is a few cycles of symbolic addition applied to the largest E-values only, using triplets and quartets, employing very strict acceptance criteria and not allowing relations among the symbolic phases for reflections to be taken up in the list of known phases. This leads to a much larger starting set, hereafter referred to as large starting set.

Then finally in the third step the large starting set is subjected to an accessibility test generally referred to as the divergence procedure. Now it is checked, without using symbols, whether the phases of all reflections can be evaluated from the starting set using the adopted strict accepting scheme. If this is not the case, a warning is given because weak links between phases are found and an additional symbolic starting reflection may be neccessary. As a rule, however, the starting sets produced by SIMPEL are quite satisfactory, and no additional starting phases are necessary.

When SIMPEL does not provide the solution of the phase problem this could be caused by a bad starting set. In that case one can very easily develop an own starting set by using a triplet list produced by SIMPEL and feed this starting set to the symbolic addition procedure. Facilities for doing this are present in the system.

Two alternative procedures to create starting sets have been developed in our laboratory, both more reliable and more timeconsuming compared with the convergence–divergence procedure as described above. The one uses mathematical techniques other than that of the convergence procedure to reduce a group of relationships in order to find the minimal number of unknown phases, necessary to calculate all others (J.D. Schagen, 1986). This is a reasonable fast alternative for the convergence procedure and leads to reliable starting sets. When used with a large number of relationships the procedure often ends with one solution. The other procedure (Peschar, 1986) is based on the mathematical theory of dynamic programming and builds up reliable phasing sequences from the strongest and most reliable relationships. This process, which is called optimal symbolic addition, is slower, because when handling e.g. a group of 60 reflections, it might happen that around 250 different optmally chosen symbolic addition trials are performed simultaneously in order to trace in the end those few ones for which all phases are determined in the most reliable way. Starting sets produced in this way are excellent and contain much smaller phase errors than the normal SIMPEL starting sets, thus forming an attractive alternative in case of a SIMPEL83 failure. Moreover since it is a stronger and more reliable procedure it is also a better starting point for datasets which have uncertainties such as powderdiffraction intensities (Jansen, 1991). The Optimal Symbolic Addition procedure forms the basis of the program SIMPEL88 (Peschar, these proceedings).

The symbolic addition procedure of SIMPEL83

The next step in the process is the extension of the large starting set. This is done in SIMPEL83 using triplet relations only. It is essential that no errors are included in the set of reflections with known phase in the course of this process and therefore the criteria used to accept a symbolic phase for a reflection have to be very strict, in particular in the beginning of the process. In general, no single indication will be accepted unless it belongs to the ten to fifteen strongest triplet relationships. For multiple indications giving rise to the same symbolic phase, high acceptance limits are applied and when a reflection gets more than one different symbolic phase via its triplets, this reflection will not be accepted in the set of reflections with known phase and as a result not being used as phase generating reflection in the further symbolic addition. This rule prevents relations occuring from different phase indications for the same reflection to multiply themselves trivially and to dominate the FOM calculations in the final steps.

In general the above precautions make sure that the resulting set of symbolic phases contains the correct solution, that is to say that when the correct values for the symbols are substituted, the phases of the reflections are good enough to image the structure. In the centrosymmetric case about 20 cycles of extension are carried out and in the noncentrosymmetric case 10. In the end a summary is given of all used and unused information. From those reflections having more than one different phase indication and neglected in the symbolic addition itself, now relations between the symbols are collected and stored to use in the successive step.

Numerical values for the symbols in SIMPEL83: Figures of Merit

With the extended group of phased reflections as input numerical values for the symbols are calculated using a number of Figures of Merit (FOM's). The general tactic to calculate FOM's is first to substitute the symbolic phases into the relationships of the FOM, then to combine all identical symbolic phase sums by summing their weights and finally calculating the FOM's by replacing the symbols by numerical trial values.

In SIMPEL83 the following FOM's are calculated:
1) The Σ_1–consistency FOM Q (Schenk, 1971), which can be calculated very fast;

2) The Positive Quartet Criterion (PQC, Schenk and Kiers, 1984);
3) The Negative Quartet Criterion (NQC, Schenk, 1974, weights from Schenk, 1975);
4) The Σ_1–criterion (Overbeek and Schenk, 1976);
5) The Harker–Kasper Criterion (HKC, Schenk and De Jong, 1973, Schenk, 1973b).
The FOM's 2, 3 and 5 are based on higher invariants. In noncentrosymmetric spacegroups with centric projections a next FOM is being used, the so-called Projection Criterion based on the reflections with restricted phases and used to determine symbol values modulo π only.

Apart from the separate FOM's also a Combined FOM is calculated, which mostly discriminates the correct solution without difficulties. In general few correlation exists between the five or six individual Figures of Merit, thus resulting in a strong Combined FOM.

For different space groups it is advisable to rely on groups of different FOM's as is indicated in the author's chapter on symbolic addition. For instance, whereas in space groups like $P2_1/c$ the Q and PQC are superior, in space groups like $P\underline{1}$ (for typographical reasons the triclinic centrosymmetric space group P1bar will be referred to as $P\underline{1}$) they are unreliable as a result of the lack of translational symmetry. In this space group NQC and HKC are the ones which generally guide the phase determination to a successful end. The projection criterion is particularly successful in space groups like $P2_12_12_1$, in which many reflections have restricted phases.

<u>Numerical phases for all strong reflections</u>

After determining the symbols all symbolically phased reflections can get a numerical phase. These phases are then used as starting point for a numerical phase extension, because in general the number of phased reflections is yet unsufficient to calculate a good E-map. In the centrosymmetric case this is achieved by means of a fast Σ_2 refinement and extension, in the noncentrosymmetric case by means of a usual 'tangent' or exponetial refinement (see also the author's chapter on symbolic addition, expression 6).

<u>Implementation and performance of SIMPEL83</u>

SIMPEL now exists in three modifications. The first one is a CDC Cyber version, in which by means of wordpacking all 60 bits of the memory words are used to store as much information as possible in a small part of core. The second is the PDP11/VAX version, which forms part of the ENRAF-NONIUS Structure Determination Package (Note: in MOLEN SIMPEL88 is included). Both systems are very similar if not identical and user-friendly. The PDP-11 version runs also on the Digital P350/380 microcomputers. The third SIMPEL83 version is the one taken up in the XTAL system. It is nicely interfaced to the GENSIN and GENTAN direct method programs as written for XTAL by Hall (1985). It uses as many of the already present routines such as the generation of relationships, however, in particular the starting set procedure and the symbolic addition are very similar to the two other versions (Schenk and Hall, 1986).

Because many of the procedures used in SIMPEL83 differ completely from all other existing direct methods systems like MULTAN, GENTAN, Mithril and Shelx, it is well possible that a failure of another system is solved in a default SIMPEL run. Of course, the reverse may also be the case. However, in any case SIMPEL will present its results in a relative short time compared to the other systems, because SIMPEL carries out only two phase extensions, one symbolical and one numerical extension, whereas all multisolution programs do 'multi' extensions, in which 'multi' is large and to be defined by the user. Therefore people familiar with SIMPEL will try that system first, just because of its speed.

The rate of success of SIMPEL in the space groups $P\underline{1}$ and $P2_1/c$ is larger than 90%, provided that the appropriate FOM's are used. The same rate of success can also be claimed for $P2_12_12_1$, although the number of structures tried in this space group is smaller. In $P2_1$ the rate is less, however, this applies to other programs as well. The other spacegroups behave in a similar way, regarding their symmorphic and/or polar nature. A number of results is reviewed in Schenk (1983), and others in Overbeek (1985).

<u>Difficult structures, symmetry and induced problems.</u>

In principle a Direct Method application should lead directly to the image of the structure. In practice however, quite a few structures remain unsolved. This may have several reasons, such as space group symmetry, size of the structure and special features in the atom positions. The effect of symmetry is the most serious reason and will be dealt with here in some more detail.

It can easily be seen that the trivial set of phases with all $\varphi_H=0$ fulfills all triplet relations

$$\varphi_H + \varphi_K + \varphi_{-H-K} = 0 \tag{2}$$

exactly; this implies that in unfavourable cases no solution of the phase problem will be found. Fortunately in many cases the space group symmetry adds constraints to the phases. E.g. in space group $P2_1/c$ the phase constraints $\varphi_{hkl}=0$ or π, $\varphi_{hkl}=\varphi_{-h-k-l}=\varphi_{-hk-l}$ for $k+l=2n$ and $\varphi_{hkl}=\varphi_{-h-k-l}=\pi+\varphi_{-hk-l}$ for $k+l=2n+1$ follow from the symmetry and notably the latter condition implies that the set of all phases $\varphi_H=0$ is no longer a solution of the relations (2). In practice, in this space group the most consistent solution under (2) gives nearly always a correct image of the structure. However, also in the triclinic centrosymmetric space group $P\underline{1}$, where no symmetry constraints are present, the triplet relations (2) are reliable enough to carry out a phasing procedure and thus, by using a good starting set of origin defining and symbolic phases, a phase set can be built up containing among other solutions both the trivial and the correct one. Since the triplet relations (2) tend to indicate the trivial solution as the correct one, a FOM based on these relations will not find the correct solution, and this implies that other FOM's must be used for this purpose.

In this chapter it is not possible to deal with the symmetry aspects of the use of triplets (2) in all space groups in detail, for that the reader is referred to the original paper (Schenk, 1972). Nevertheless, one other serious problem must be mentioned, namely that of the refinement of phases in polar non-centrosymmetric space groups. As an example space group $P2_1$ is chosen, in which the origin may ly on any point of the twofold screw axis. This implies that there is an infinite number of correct phase sets in stead of just one and they are related by the expression:

$$\varphi_{hkl} = a \times k + \varphi_{hkl} \tag{3}$$

inwhich a is a constant related to the shift of the origin. Now, a φ_H derived from a triplet (2) carrying an error, will in general be in error as well. However, when used in further phase extension cycles this wrong phase will be interpreted as being correct with respect to another origin or even as being correct to the enantiomorph of the structure. As a result solutions of the phase problem using the set of triplets (2) often consist of several partial structures with respect to different origins in both enantiomorphs. Because of this one finds that in space groups such as $P2_1$ application of (2) often leads to centric phases, whereas similar refinements in $P2_12_12_1$, where the origin is fixed and eight permissible origins are present only, show stable non-centric phases (e.g. Schenk, 1982). It needs no argument that this behaviour of the triplets (2) in space group $P2_1$ induces extra problems in phase-determination processes.

In summary there exist two categories of centrosymmetric space groups: a) the symmorphic space groups (no translational symmetry, e.g. $P\underline{1}$) for which the most consistent solution under (2) is the trivial one with all $\varphi_H=0$ and b) the other space groups (e.g. $P2_1/c$) for which the most consistent solution usually is the correct one. For the noncentrosymmetric space groups there are three categories: c) the symmorphic space groups (e.g. P1), for which the relations (2) produce also the $\varphi_H=0$ solution, but also less consistent solutions of (2) mostly having centric phases; d) the polar space groups with an infinite number of possible origins (e.g. P2$_1$), for which the phasing process usually ends with centric phases and e) the other space groups (e.g. P2$_1$2$_1$2$_1$) with an origin restricted to a finite number of positions for which the relations produce the correct phases. To overcome the above difficulties, higher invariants play an important role.

<u>Probability distributions for quartet phase relationships</u>

Once the reflections controlling the value of the phase sum of a (sem)invariant are identified, a conditional joint probability distribution can be derived, describing the probability that the phase sum has a certain value, given the $|E|$'s of all reflections involved. Throughout the years many scientists have been contributing to this part of direct methods, e.g. Hauptman, Karle, Cochran, Kitaigorodsky, Klug, Bertaut, Naya, Nitta, Oda, Giacovazzo and Heinerman, and more recently, Bricogne and Peschar, some of whom have been contibuting to this proceedings. Starting point of the derivation of a joint probability distribution is a process to mimic structures. Then given such a structure the formalism of probability theory is used to derive a joint probability distribution. Finally the conditional joint probability distribution is found by precizing some parameters of the joint probability distribution.

For the process of mimicing structures in general randomly chosen coordinates are given to all atoms. Thus for all positions in the unit cell there is an equal chance to find an atom. Of course, for real structures this is an incorrect assumption, because all atoms are situated at sites which satisfy chemical and physical rules like chemical bonding and Van der Waals forces. However, inclusion of these rules in the process of mimicing structures complicates the procedure to arrive at a conditional joint probability distribution substantially. In the second step probability theory is used. In general also approximations are applied in order to make the derivation feasible. Different authors use different approximations and may thus arrive at different joint probability distributions for the same phase relationship.

This paper is not intending to give a full account of theoretical direct methods, other chapters of this proceedings will deal with that part of Direct Methods. In stead, here some results in the field of quartet relations are compared in view of their usefulness in practical applications. After the first empirical observations concerning 7 magnitude quartets (Schenk, 1973a) theories were developed by Hauptman (1975) and Giacovazzo (1976) along the lines described above. The conditional probability distribution of Giacovazzo (1976) is given by

$$P(\varphi_4,R_1,R_2,R_3,R_4,R_5,R_6,R_7) =$$

$$C^{-1}\exp[-2N^{-1}R_1R_2R_3R_4(R_5^2-1)(R_6^2-1)(R_7^2-1)] \qquad (4)$$

and the one of Hauptman (1975) by

$$P(\varphi_4,R_1, \ldots,R_7) = C^{-1}\exp[-2N^{-1}R_1R_2R_3R_4\cos\varphi_4] \times$$

$$\times I_0(2N^{-1/2}R_5Z_5)I_0(2N^{-1/2}R_6Z_6)I_0(2N^{-1/2}R_7Z_7) \qquad (5)$$

with $Z_5 = [R_1{}^2R_2{}^2 + R_3{}^2R_4{}^2 + 2R_1R_2R_3R_4\cos\varphi_4]^{1/2}$

and similar expressions for Z_6 and Z_7. In all expressions R_1 is the random variable associated to $|E_H|$, R_2 to $|E_K|$, R_3 to $|E_L|$, R_4 to $|E_{-H-K-L}|$, R_5 to $|E_{H+K}|$, R_6 to $|E_{H+L}|$, and R_7 to $|E_{K+L}|$.

The form of expression (4) shows that the maximum of P is either found at $\varphi_4=0$ or at $\varphi_4=\pi$, depending on the values of R_5, R_6 and R_7. For large values of the cross terms $\varphi_4=0$ and for small values the sign of the exponential part changes and consequently $\varphi_4=\pi$. The formula of Hauptman is capable to predict all values in the range $0 \leqslant \varphi_4 \leqslant$

π, and has therefore a wider range of possible applications. Nevertheless, also this expression has shortcomings. In particular the fact, that in real structures the predicted φ_4-values show systematic errors for both the modes and the means of the conditional probability distributions, is a limiting factor for successful applications.

The Giacovazzo- and Hauptman-expressions have been derived neglecting higher order terms. In order to judge whether this leads to the systematic errors of the φ_4 estimates, in my laboratory Peschar (1987) followed another route to the joint probability distribution. This computerized route enables the use of higher order terms to any specified order of N. Therefore the characteristic function C of the joint probability distribution (jpd) is brought into a special form and then expanded in a Taylor series. The result can be Fourier-transformed term by term. In the final form of the jpd products of the Hermite and Laguerre polynomials play an important role. In order to make this process feasible a computer program has been developed, which does all the hard work from both the mathematical and administrative point of view. The program derives the expression for the conditional jpd to any specified order of N and delivers the result as a series of terms on a disk file. Typically for quartet phase relations the conditional jpd correct to the order of N^{-3} contains 1010 terms and that correct to N^{-4} 9133 terms. These numbers show the impossibility to derive these expressions by hand.

The fact that the pd is available as a computer disk file makes it easy to test it, which has been done extensively. Convergence is normally reached using the N^{-3} expression, only when the reflections involved are very strong the N^{-4} expression has to be used (Peschar and Schenk, 1987). The estimates of φ_4 do not contain any systematic error with respect to the true phase sum whereas the systematic errors in the case of expression (5) are as big as $15°$ on average. Moreover, there is no systematic error variation over the whole range of $0 < \varphi_4 < \pi$ whereas the differences in the case of (5) vary from 20 to $10°$ for different φ_4 averages. A last advantage of the new formula is its lower error level; the absolute errors of the estimated φ_4 with respect to the true φ_4 are reduced by approximately 15%. In conclusion it has been shown by Peschar that the incorporation of higher order terms give rise to superior jpd's for the quartet phase relationship. Along the same route, taking advantage of similar computer programs, also other jpd's are derived with similar results.

Application of higher invariants in practical Direct Methods

In practical direct methods the applications of estimates of phase relations based on higher invariants vary from well-established widely used ones to applications which have not been developed yet beyond the research stage.

Among the well-established applications are the use of quartets in the starting set determination and the use of figures of merit based on quartets, quintets and seminvariants. These applications are implemented in several Direct Method programs, including our own SIMPEL programs. A number of other applications make use of the fact that for many invariants and seminvariants absolute values of sums of phases can be estimated which differ from 0 and π. These estimates are therefore enantiomorph sensitive, and are used for enantiomorph specific procedures. However, over all these procedures are in a research state and as such only useful at the places where they were invented. Some applications will be dealt with hereafter in more detail.

Figures of Merit

As was shown before, in space group P$\underline{1}$ the most consistent solution under (2) corresponds to the set of trivial phases. Nevertheless, among the other solutions the set of correct phases may wel be present, but, of course, this set will only be found by using appropriate FOM's. FOM's, which in recent years proved to be very valuable, are those based on phase relations equal to π, e.g. the negative quartets

$$\varphi_H + \varphi_K + \varphi_L + \varphi_{-H-K-L} \approx \pi \qquad (6)$$

and the special negative quartets

$$2\varphi_H + \varphi_{-H+K} + \varphi_{-H-K} \approx \pi \tag{7}$$

When applied in space group P$\underline{1}$, these FOM's make the number of successful phase determinations as large as that reached by using the triplet consistency in spacegroup P2_1/c. The first FOM of this kind was HKC (Schenk and De Jong, 1973), based on the special quartets, soon followed by the so-called Negative Quartet Criterion (NQC (Schenk, 1974). Later other FOM's were defined on the basis of the same relations, which give rise to similar and not better results. To get an idea of the strength of these FOM's, results of 12 structures are collected in table 1 (taken from Schenk, 1983). It can easily be seen that NQC is the best FOM of all, the second best are φ_0 and HKC. The triplet consistency Q is not very reliable at all in this particular space group. In may be remarked, however, that in nonsymmorphic space groups such as P2_1/c the Q FOM is one of the most reliable FOM's of all.

Unpublished results from our laboratory show that the Negative Quartet Criterion is correlated strongly with the ψ_0 check (Cochran and Douglas, 1957), as employed in other Direct Methods systems. However, as can be seen from table 1 NQC has much more discriminating power than ψ_0, and therefore has to be used preferably. This difference in discriminating power can easily be explained. The ψ_0 check is the application of Sayre's equation to the weak reflections, thus when all triplets which a particular weak reflection makes with all strong reflections are collected and from these triplets the associated quartets are constructed, half of them will be positive and half of them negative, although all share the weak reflection as one of the cross terms. The appearence of negative quartets explain why NQC and ψ_0 are correlated. However, in the associated quartets of ψ_0 the other two cross terms have to discriminate between the positivity and negativity of the particular quartet but in the ψ_0 check these cross terms are not playing any explicit role. In the Negative Quartet Criterion only reliable negative quartets are explicitly included and therefore its superiority above the ψ_0 test is obvious. Moreover, in the φ_0-test there are some adjustable parameters, which makes its application in routine phasing more troublesome, in particular in symbolic addition programs.

Table $\underline{1}$. Figures of Merit for 12 structures in space group P$\underline{1}$.
In the FOM columns the rank of the correct solution after the FOM in question have been given; in this respect it can be noted that in general only a few highest ranked solutions are considered for a calculation of a Fourier–summation. This implies that a rank of 5 or 6 means that a structure will not be solved using that particular FOM.
In total 5 different FOM's are given: 1) the triplet consistency Q, 2) the special negative quartet criterion HKC, 3) the negative quartet criterion NQC, 4) the φ_0–check and 5) the sigma–1 criterion.

Structural code	Number of atoms in unit cell	Number of solutions of (1)	Rank of the correct solution after FOM				
			Q	HKC	NQC	ψ_0	SI
ADEN	54	32	>10	2	1	2	1
LLOYD3	108	2048	1	13	1	4	>20
FUKS	76	128	2	1	1	5	8
BAR2	40	16	13	7	1	1	8
FABRE4	104	512	1	>20	1	17	5
LAB26	46	16	8	3	1	1	10
LAB25	58	64	3	1	2	5	2
MARIN	30	64	3	1	1	2	16
SEVIL	28	16	6	11	1	1	16
VITA	44	8	7	1	1	4	7
QUINOL	80	128	3	3	1	1	>25
BORAAN	58	64	13	7	1	1	1

We also developed a FOM (Kiers and Schenk, 1984) based on the positive quartet relation:

$$\varphi_H + \varphi_K + \varphi_L + \varphi_{-H-K-L} = 0 \tag{8}$$

Of course in symmorphic space groups the trivial solution is indicated by this FOM to be most probably correct like the Σ_2-consistency. However, the positive quartet criterion appears to be more reliable than the Σ_2-consistency FOM.

Enantiomorph specific procedures

In polar space groups (e.g. $P2_1$) the triplet consistency tends to produce centric phases and in these space groups it is therefore not a very reliable FOM either. Nevertheless, the symbolic phases may be correct but, in order to find the structure, it is necessary to use an enantiomorph specific FOM. It was shown that reliable estimates of $\pi/2$ can be obtained for quintet phase sums (Van der Putten and Schenk, 1977): and the same applies for quartet phase sums (Hauptman, 1975). On the basis of these estimates two enantiomorph specific FOM's were developed (Van der Putten and Schenk, 1979):

$$ENQUAC = \sum_i W_i \left| \varphi_{H_i} + \varphi_{K_i} + \varphi_{L_i} + \varphi_{-H_i -K_i -L_i} + S_i |p_i| \right| \tag{9}$$

and

$$ENQUIC = \sum_i W_i \left| \varphi_{H_i} + \varphi_{K_i} + \varphi_{L_i} + \varphi_{M_i} + \varphi_{-H_i -K_i -L_i -M_i} + S_i |q_i| \right| \tag{10}$$

in which p_i and q_i are the estimated values of quartet and quintet phase sums respectively.
To calculate an ENQUAC value all phases are substituted on the right-hand side and the signs S_i are chosen such that each individual term in expression (10) is closest to zero. ENQUAC is zero for the correct phases and the correct $|p|$-values. In practice as long as the errors in $|p|$ have a random character, the FOM will be enantiomorph specific and it can be expected that the smallest value of ENQUAC corresponds to the set of approximately correct phases. Any set of centric phases will have much larger ENQUAC values. Similar reasonings and results are valid for the FOM ENQUIC.

Also enantiomorph conserving refinement procedures have been developed employing estimates $|p|$ and $|q|$ for quartet and quintet phase sums. E.g. for quartets a modified tangent formula (Van der Putten and Schenk, 1979) is given by:

$$\tan \varphi_H = \frac{\sum_K \sum_L E_4 \sin (\varphi_K + \varphi_L + \varphi_{H-K-L} + s|p|)}{\sum_K \sum_L E_4 \cos (\varphi_K + \varphi_L + \varphi_{H-K-L} + s|p|)} \tag{11}$$

in which s is chosen such that $-\varphi_H + \varphi_K + \varphi_L + \varphi_{H-K-L} + s|p|$ is closest to 0. This refinement will maintain the enantiomorph and introduces random errors in the phases φ_H only. The strongest practical procedure combines the use of triplets, quartets and quintets in a way similar to (11). (Van der Putten et al., 1979). This procedure has also been adopted in Mithril by Gilmore (see these proceedings).

Seminvariants and higher invariants

Other applications of higher order phase relations are their use in the estimation of lower order phase relations. This is a very broad field in theoretical direct methods with up to the present only a few applications for routine phase determination. In particular in the SIR system one- and two-phase seminvariants are used extensively. (See these proceedings). Also in SIMPEL83 a FOM is based on one phase seminvariants: the Σ_1 relation.

This most simple phase relation correlates in space group P$\underline{1}$ the phase of 2H with $|E_H|$ and $|E_{2H}|$: if both $|E|$'s are large the phase φ_{2H} is most probably zero; this is one of the so called sigma 1 relations. Sigma 1 relations in space groups of higher symmetry (Hauptman & Karle, 1953) can be very reliable although they are always few in number. Some authors advise the use of sigma 1 relations in the starting set procedure, however, we have better experiences by using them in a FOM (Overbeek and Schenk, 1976). As an example, in table 1 the rank of the correct solution with the sigma 1 criterion has also been given. It is the weakest FOM in that table; in monoclinic and orthorhombic space groups, however, its reliability approximates the strength of the negative quartet criterion.

Estimating $|E|$'s of non-observed reflections

A completely different application of quartets is the estimation of structure factor amplitudes of non-observed reflections. This approach is based on the observation that if $|E_H|$, $|E_K|$, $|E_L|$, $|E_{-H-K-L}|$, $|E_{H+K}|$ and $|E_{H+L}|$ are large, as a rule $|E_{H+K}|$ is also large, and that if the first four are large and $|E_{H+K}|$ and $|E_{H+L}|$ are small, then often $|E_{K+L}|$ is also small. In case a non-observed reflection acts as K+L in many quartets with the other six reflections in each quartet observed, in general a rather firm statement can be made about the $|E|$ of that reflection. Using these observations as starting point a procedure had been developed, which enables rather good estimates of a number of $|E|$-values of non-observed reflections (Van der Putten, Schenk, and Tsoucaris, 1982). Of course, this successful procedure does not imply that the resolution of a structure determination is improved; this can only be realized by measuring more data. The importance of the procedure lies in the fact that a re-ordering of the information is realized which sometimes may facilitate the phase determination appreciably.

The above reasoning has also been used to develop an intensity-deconvolution procedure in powder diffraction for completely overlapping reflections (Jansen et. al., 1990). From a powder diffraction pattern first in the normal way the indexes of the reflections and the cell constants are extracted and then by means of least squares profile fitting as many uniquely determined intensities as possible are collected. As a rule many reflections are overlapping heavily in clusters and cannot be determined in the above way. For those clusters a deconvolution technique has been developed which enables to estimate the intensities of the individual reflections on the basis of among others the quartet concept as presented in the preceding paragraph. The resulting intensities are of that quality that a successive run of SIMPEL88 in which provisions are included to keep track of the original reliabilities of the intensities generally leads to a succesful structure determination (Jansen, 1991)

Conclusion

In conclusion the field of Direct Methods has been enriched by the possibilities of higher invariants substantially. Moreover, since not all potential applications are yet available in the form of user programs and secondly since research in this field is still proceeding, the future Direct Methods procedure will gain more from higher invariants.

References

Germain, G., Main, P. and Woolfson, M.M. (1970). Acta Cryst. $\underline{B26}$, 274.
Giacovazzo, C, (1976). Acta Cryst. $\underline{A32}$, 91.
Giacovazzo, C. (1977). Acta Cryst. $\underline{A33}$, 933.
Hall, S.R. (1985). J. Appl. Cryst. $\underline{18}$, 263-264.
Harker, D and Kasper, J.S. (1948). Acta Cryst. $\underline{1}$, 70.

Hauptman, H. (1975). Acta Cryst. A31, 680.

Hauptman, H. and Karle, J. (1953). A.C.A. Monograph No. 3. Pittsburgh, Polycrystal.

Jansen, J., Peschar, R. and Schenk, H. (1990). Acta Cryst. A46, C-57, submitted to J. Appl. Cryst.

Jansen, J., forthcoming thesis (1991)

Karle, J. and Hauptman, H. (1950). Acta Cryst. 3, 181.

Kiers, C.T. and Schenk, H. (1984), unpublished results.

Kitaigorodsky, A.I. (1954). Dokl. Acad. Nauk. SSSR 94, 225.

Overbeek, A.R. (1985). Description of SIMPEL, Amsterdam.

Overbeek, A.R. and Schenk, H. (1976). Proc. K. Ned. Akad. Wet. B79, 341.

Peschar, R. (1987). Thesis, Amsterdam.

Peschar, R. and Schenk, H. (1987). Acta Cryst. A43, 751-763.

Sayre, D. (1952). Acta Cryst. 5, 60.

Schagen, J.D. (1986). Thesis, Amsterdam.

Schenk, H. (1971). Acta Cryst. B27, 2039.

Schenk, H. (1972), Acta Cryst. A28, 412.

Schenk, H. (1973a). Acta Cryst. A29, 77.

Schenk, H. and de Jong, J.G.H. (1973). Acta Cryst. A29, 31.

Schenk, H. (1973b). Acta Cryst. A29, 480.

Schenk, H. (1974). Acta Cryst. A30, 477.

Schenk, H. (1975). Acta Cryst. A31, 257.

Schenk, H. (1982) in Computational Crystallography, ed. D. Sayre, Oxford, p. 231.

Schenk, H. (1983). Recl. Trav. Chim. Pays-Bas 102, 1.

Schenk, H. and de Jong, J.G.H. (1973). Acta Cryst. A29, 480.

Schenk, H. and Hall, S.R. (1986). in Users Manual XTAL 2.0, edited by J.M. Stewart et al., Maryland.

Schenk, H. and Kiers, C.T. (1984) in Methods and Applications in Crystallographic Computing, edited by S.R. Hall and T. Ashida, Clarendon Press, Oxford, p 96-105.

Schenk, H. and Overbeek, A.R. (1984) in Methods and Applications in Crystallographic Computing, edited by S.R. Hall and T. Ashida, Clarendon Press, Oxford, p 91-95.

Van der Putten, N. and Schenk, H. (1977), Acta Cryst. A33, 378.

Van der Putten, N. and Schenk, H. (1979), Acta Cryst. A35, 381.

Van der Putten, N., Schenk, H. and Tsoucaris, G. (1982), Acta Cryst. A38, 98.

DIRECT METHODS FOR MODULATED STRUCTURES

Fan Hai-fu

Institute of Physics
Chinese Academy of Sciences
Beijing 100080, P.R. China

INTRODUCTION

Modulated structures belong to that kind of crystal structure, in which the atoms suffer from certain occupational and/or positional fluctuation. If the period of fluctuation is commensurate with that of the three-dimensional unit cell then a superstructure results, otherwise an incommensurate modulated structure is obtained. Modulated phases have been found in many important inorganic and organic solids. In many cases, the transition to the modulated structure corresponds to a change of certain physical properties. Hence it is important to know the structure of modulated phases in order to understand the mechanism of the transition and properties in the modulated state. Up to the present most procedures used for solving modulated structures, especially those for solving the incommensurate ones, depend on some preliminary assumption to the form of modulation. It is worthwhile to find a straightforward way to solve this important kind of structures. Both commensurate modulated structure (superstructure) and incommensurate modulated structure possess pseudo-translational symmetry. They have a corresponding subcell-structure called average structure or main structure, which is often relatively simple. In the reciprocal space, there will be two subsets of reflections: one includes the 'main reflections' contributed mainly from the average structure; the other includes the 'systematically weak reflections' which are due to the modulation. The phases of main reflections can easily be solved by traditional direct methods, however phases of the systematically weak reflections are rather difficult to determine. Special direct method procedures have been developed to extend phases from the main (strong) subset to the weak one. Unlike the others, direct methods do not rely on any assumption about the form of modulation. They solve the phase problem for the systematically weak reflections directly and reveal the modulation objectively. In this paper, theoretical and practical aspects on solving superstructures and incommensurate modulated structures are described.

WHAT IS A MODULATED STRUCTURE

A modulated structure can be regarded as the result of applying a periodic modulation to a regular structure. Fig. 1 shows two simplified

examples. The modulation function in the figure represents the fluctuation
of atomic occupancy. When it is applied to the regular structure, the
'height' of atoms are modified. A commensurate modulated structure
(superstructure) will result (Fig. 1a), if the period $\underline{T}$ of the modulation
function is commensurate with the period $\underline{t}$ of the structure, i.e. $\underline{T}/\underline{t}=n$,
where n is an integer. The resulting superstructure now has a true
period $\underline{T}$ and a pseudo period $\underline{t}$, respectively corresponding to a true
unit cell and a pseudo unit cell. On the other hand, if $\underline{T}$ is
incommensurate with $\underline{t}$ (Fig. 1b), i.e. $\underline{T}/\underline{t}=r$, where r is not an integer, we
obtain an incommensurate modulated structure, in which no exact
periodicity can be found, however $\underline{t}$ remains a pseudo period. A
modulation function can also represent the fluctuation of atomic position.
In this case the structure will be modulated in atomic positions. The
positional modulation can also be either commensurate or incommensurate.
In practice a modulated structure can include simultaneously different
kinds of occupational and/or positional modulations.

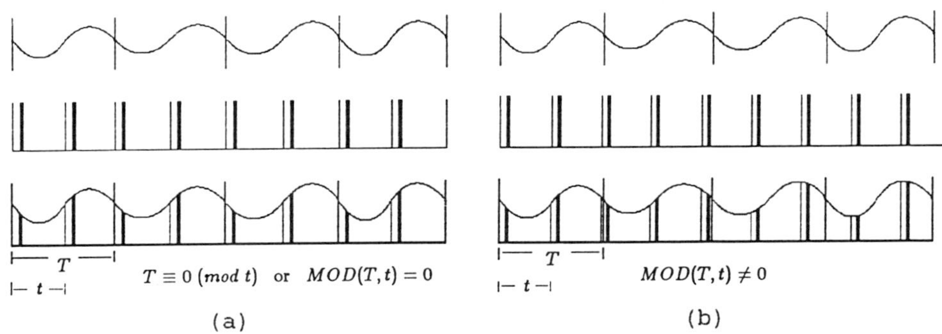

$T \equiv 0 \ (mod\ t)$ or $MOD(T,t) = 0$ $MOD(T,t) \neq 0$

(a)   (b)

Fig. 1. Occupational modulation of one-dimensional structure.
(a) Commensurate modulation; (b) incommensurate modulation.
Top: modulation function with a period equal to $\underline{T}$; middle:
one-dimensional structure with atoms shown as thick vertical
lines and with a period equal to $\underline{t}$. bottom: the resulting
modulated structures.

DIRECT METHODS FOR SUPERSTRUCTURES

Pseudo systematic extinction due to pseudo periodicity

In a superstructure there exists at least one pseudo-translation
vector $\underline{t}=\underline{T}/n$, where n is an integer, $\underline{T}$ specifies the direction and period
of the modulation wave and is the shotest lattice vector parallel to $\underline{t}$.
The structure factor can then be written approximately as

$$\mathbf{F}(\mathbf{H}) \sim \sum_{j=1}^{N/n} f_j e^{i2\pi \mathbf{H} \cdot \mathbf{r}_j} \left[1 + e^{i2\pi \mathbf{H} \cdot \mathbf{t}} + e^{i2\pi \mathbf{H} \cdot 2\mathbf{t}} + \dots + e^{i2\pi \mathbf{H} \cdot (n-1)\mathbf{t}} \right] \ . \quad (1)$$

The sum of the series inside the brackets of (1) is given by

$$S = \left[exp(i2\pi \mathbf{H} \cdot n\mathbf{t}) - 1 \right] \Big/ \left[exp(i2\pi \mathbf{H} \cdot \mathbf{t}) - 1 \right] = \begin{cases} n, & if \ \mathbf{H} \cdot \mathbf{t} = integer \\ \\ 0, & if \ \mathbf{H} \cdot \mathbf{t} \neq integer \end{cases} \quad (2)$$

Notice that $\underline{H} \cdot n\underline{t} = \underline{H} \cdot \underline{T}$=integer. Hence all reflections with $\underline{H} \cdot \underline{t}$ not equal to an
integer will be systematically very weak, leading to an effect of pseudo-

systematic extinction. The difficult problem of solving a superstructure is to derive phases for the systematically weak reflections. Attempts have long been made to solve this problem since the 1970's. For more details the reader is referred to the original papers (Fan Hai-fu, 1975; Fan Hai-fu, He Luo, Qian Jin-zi & Liu Shi-xiang, 1978; Gramlich, 1975, 1978; Boehme, 1982; Prick, Beurskens & Gould, 1983 and papers in the proceedings of the Pre-meeting Workshop on Direct methods and their Application to Structures Showing Superstructure Effects, 9th European Crystallographic Meeting, Torino, Italy, 1985).

<u>Modified Sayre equation for superstructures</u>

A superstructure can be described by superimposing a 'difference structure' on to an average structure, which fulfils exactly the pseudo-translational symmetry. The 'difference structure' here serves as a modulation function. We thus have

$$\rho(\mathbf{r}) = \rho(\mathbf{r})_{av} + \Delta\rho(\mathbf{r}) \quad . \tag{3}$$

By squaring both sides of (3) and neglecting the term $\Delta\rho^2(\mathbf{r})$, it follows that

$$\rho^2(\mathbf{r}) = \rho^2(\mathbf{r})_{av} + 2\rho(\mathbf{r})_{av}\Delta\rho(\mathbf{r}) \quad . \tag{4}$$

From (4), by following the derivation of Sayre (1952), we finally obtain

$$\mathbf{F(H)}_{wk} = 2\frac{\Theta}{V} \sum_{\mathbf{H'}} \mathbf{F(H')}_{st}\mathbf{F(H - H')}_{wk} \quad . \tag{5}$$

Where $F(H)_{wk}$ denotes the structure factor of the systematically weak reflections, while $F(H)_{st}$ denotes the structure factor of the systematically strong reflections, which are contributed mainly from the average structure; Θ is an atomic form factor; V is the volume of the true unit cell. (5) can be used to extend phases from the systematically strong reflections to the systematically weak ones. It forms the base of direct-method determination of superstructures. Since the phase problem for the average structure can easily be solve by traditional direct methods. A procedure has been developed for *ab-initio* determination of super-structures by direct methods (Fan Hai-fu, Yao Jia-xing, Main & Woolfson, 1983). The procedure has been further improved and optimized in the latest version of the program system SAPI (Fan Hai-fu, Yao Jia-xing & Qian Jin-zi, 1988; Fan Hai-fu, Qian Jin-zi, Zheng Chao-de, Gu Yuan-xin, Ke Heng-ming & Huang Sheng-hua, 1990). Examples of using SAPI to solve superstructures are given in the following paragraph.

<u>Examples</u>

<u>Freielebenite</u>. (Ito & Nowacki, 1974). Formula: $PbAgSbS_3$; space group: $P2_1/a$; unit cell: a=7.518, b=12.809, c=5.940 A, beta=92.25° and Z=4. This is a typical superstructure and had been a test structure of the 'key shift method' (Ito, 1973). There are two pseudo-translation vectors $\underline{t}_1=\underline{a}/2$ and $\underline{t}_2=\underline{b}/3$ in the unit cell. The asymmetric unit contains only six atoms. However, Patterson or conventional direct methods led only to the avrage structure. The original results on solving the complete structure by Ito (1973) are listed in Table 1(a). In a default run, SAPI indicated that the structure possesses pseudo-translational symmetry. The reflections were classified automatically by the program into six index groups and normalized separately. Phase derivation resulted in an E map

with peaks shown in Table 1(b). The top six peaks correspond to the six independent atoms with positional shifts consistent with those in Table 1(a).

Table 1. Atomic positions of Freielebenite. (a) Results by Ito (1973); (b) results by a default run of SAPI.

(a)

Atom	Average position			True position			Shift		
	X	Y	Z	X	Y	Z	dX	dY	dZ
Pb	0.375	0.417	0.250	0.350	0.415	0.253	−0.025	−0.002	+0.003
Ag	0.375	0.750	0.250	0.378	0.760	0.212	+0.003	+0.010	−0.038
Sb	0.375	0.083	0.250	0.365	0.087	0.272	−0.010	+0.004	+0.022
S_1	0.125	0.250	0.250	0.138	0.219	0.346	+0.013	−0.031	+0.096
S_2	0.125	0.583	0.250	0.135	0.622	0.131	+0.010	+0.039	−0.119
S_3	0.125	0.917	0.250	0.148	0.942	0.266	+0.023	+0.025	+0.016

(b)

Peak No	Atom Height	Atom	Position			Shift		
			X	Y	Z	dX	dY	dZ
1	4134	Pb	0.358	0.416	0.256	−0.017	−0.001	+0.006
2	3502	Ag	0.390	0.752	0.216	+0.015	+0.002	−0.034
3	2957	Sb	0.366	0.084	0.284	−0.009	+0.001	+0.034
6	1161	S_1	0.132	0.213	0.362	+0.007	−0.037	+0.112
5	1196	S_2	0.131	0.621	0.139	+0.006	+0.038	−0.111
4	1247	S_3	0.145	0.944	0.257	+0.020	+0.027	+0.007

Hwangheite. (Qian Jin-zi, Fu Ping-qui, Kong You-hua & Gong Guo-hong, 1982). Formula: $BaCeF(CO_3)_2$; space group: R3; unit cell: a=5.070, c=38.408 A and Z=6. This is a superstructure with 14 independent atoms in the asymmetric unit. The two independent Ba atoms have their sites close to (0,0,0) and (0,0,1/2), while the two Ce atoms are close to (0,0,1/4) and (0,0,3/4). Hence there is a pseudo-translation vector $\underline{t}=\underline{c}/4$ in the unit cell. Patterson or conventional direct methods failed to give the true structure, since they could not find the positional shifts of Ba and Ce atoms correctly. A Fourier map phased with the average positions of Ba and Ce atoms yields eight rather than two indepemdent F atoms and six instead of three O atoms for each CO_3 group. This makes the structure analysis much more difficult. However, a default run of SAPI automatically solved all these problems. The reflections were classified automatically into four index groups and normalized separately. The phase derivation procedure resulted in an E map, of which the largest four peaks correspond to the four heavy atoms with correct positional shifts. Among the 17 next largest peaks, ten correspond to the light atoms and others correspond to pseudo translation images of some light atoms or diffraction ripples of the heavy atoms. Among the above peaks, those corresponding to true atoms are obviously larger than peaks of the corresponding pseudo-images of atoms. Hence there will be no difficulty in assigning an atom to a proper peak and rejecting the others which correspond to the pseudo-images of atoms. The result so obtained is given in Table 2(a). in comparison with the result obtained by Qian Jin-zi et al. (1982) after least-squares refinement (Table 2b).

Table 2. Atomic positions of hwangheite. (a) Results by a default run of SAPI; (b) results from least-squares refinement by Qian *et al.* (1982).

(a)

Peak No	Height	Atom	Position X	Y	Z	Shift dX	dY	dZ
1	4994	Ce1	0.000	0.000	0.258	0.000	0.000	+0.008
2	4946	Ce2	0.000	0.000	0.749	0.000	0.000	-0.001
3	4330	Ba1	0.000	0.000	0.000	0.000	0.000	0.000
4	4101	Ba2	0.000	0.000	0.506	0.000	0.000	+0.006
11	726	F1	0.000	0.000	0.318			
9	775	F2	0.000	0.000	0.688			
10	762	O1	0.275	0.146	0.132			
12	708	O2	0.275	0.147	0.388			
21	577	O3	0.150	0.292	0.617			
8	789	O4	0.148	0.291	0.873			
6	820	C1	0.000	0.000	0.130			
17	615	C2	0.000	0.000	0.388			
14	658	C3	0.000	0.000	0.618			
7	818	C4	0.000	0.000	0.886			

(b)

Atom	Position X	Y	Z	Shift dX	dY	dZ
Ce1	0.000	0.000	0.255	0.000	0.000	+0.005
Ce2	0.000	0.000	0.745	0.000	0.000	-0.005
Ba1	0.000	0.000	0.000	0.000	0.000	0.000
Ba2	0.000	0.000	0.501	0.000	0.000	+0.001
F1	0.000	0.000	0.317			
F2	0.000	0.000	0.685			
O1	0.294	0.150	0.127			
O2	0.298	0.149	0.395			
O3	0.139	0.286	0.608			
O4	0.148	0.290	0.872			
C1	0.000	0.000	0.127			
C2	0.000	0.000	0.394			
C3	0.000	0.000	0.609			
C4	0.000	0.000	0.874			

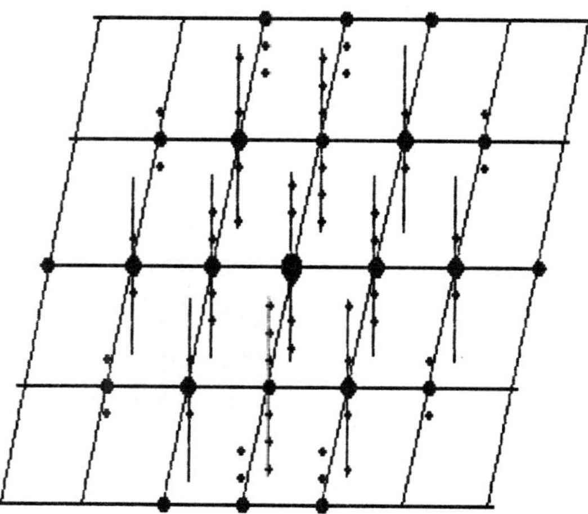

Fig. 2. Schematic diffraction photograph of an incommensurate modulated structure. The vertical line segments indicate the projection of lattice lines parallel to the fourth dimension.

DIRECT METHODS FOR INCOMMENSURATE MODULATED SRTUCTURES

The fourth dimension in structure analysis

During the determination of the incommensurate modulated structure γ-Na$_2$CO$_3$ (Aalst, Hollander, Peterse & Wolff, 1976; Wolff, 1977), de Wolff introduced an extra-dimension into reciprocal space for indexing the satellite reflections. This extra dimension in fact specifies the direction and period of the modulation wave. An incommensurate modulated structure produces a three-dimensional diffraction pattern, which contains satellites round the main reflections. A section of the three-dimensional diffraction pattern is shown schematically in Fig. 2. The main reflections are consistent with a regular three-dimensional reciprocal lattice. However the satellites can not fit the same lattice, or some of the translation vectors will be very short, leading to an unreasonably large unit cell in real space. On the other hand as is seen in Fig. 2. although the satellites are not commensurate with the main reflections, they have their own periodicity. Hence, in this example, it can be imagined that the three-dimensional diffraction pattern is the projection of a four-dimensional reciprocal lattice, in which the main and the satellite reflections are all regularly situated at the lattice nodes. According to the convolution theorem, the incommensurate modulated structure here considered can be regarded as a three-dimensional 'section' of a four-dimensional periodic structure. This representation greatly simplified the structure analysis of γ-Na$_2$CO$_3$. The above example belongs to a one-dimensional modulation. For an n-dimensional (n=1,2,...) modulation, it needs a (3+n)-dimensional description.

(3+n)-dimensional description and the modified Sayre equations

A (3+n)-dimensional reciprocal vector is expressed as

$$\mathbf{H} = h_1\mathbf{b}_1 + h_2\mathbf{b}_2 + h_3\mathbf{b}_3 + ... + h_{3+n}\mathbf{b}_{3+n} \quad (n = 1, 2, ...)$$

where $\underline{b}_i$ is the ith translation vector defining the reciprocal unit cell. The structure factor formula is written as

$$\mathbf{F}(\mathbf{H}) = \sum_{j=1}^{N} f_j(\dot{\mathbf{H}}) \exp\left[i2\pi(h_1\bar{x}_{j1} + h_2\bar{x}_{j2} + h_3\bar{x}_{j3})\right] \ , \qquad (6)$$

where

$$f_j(\hat{\mathbf{H}}) = f_j(\mathbf{H}) \int_0^1 d\bar{x}_4 \ ... \ \int_0^1 d\bar{x}_{3+n} \ P_j(\bar{x}_4, \ ... \ , \bar{x}_{3+n})$$

$$exp\left\{i2\pi\left[(h_1 U_{j1} + h_2 U_{j2} + h_3 U_{j3}) + (h_4 x_{j4} + ... + h_{3+n}x_{j(3+n)})\right]\right\} \ . \qquad (7)$$

$f_j(\underline{H})$ on the right hand side of (7) is the ordinary atomic scattering factor; P_j is the occupational modulation function; U_j describes the deviation of the jth atom from its average position. For more details on (6) and (7) the reader is referred to the papers by de wolff (1977), Yamamoto (1982) and Hao Quan, Liu Yi-wei & Fan Hai-fu (1987). What should be emphasized here is that, according to (6) a modulated structure can be regarded as a set of 'modulated atoms' situated at their average positions in three-dimensional space. The 'modulated atom' is in turn defined by a 'modulted atomic scattering factor' expressed as (7). With

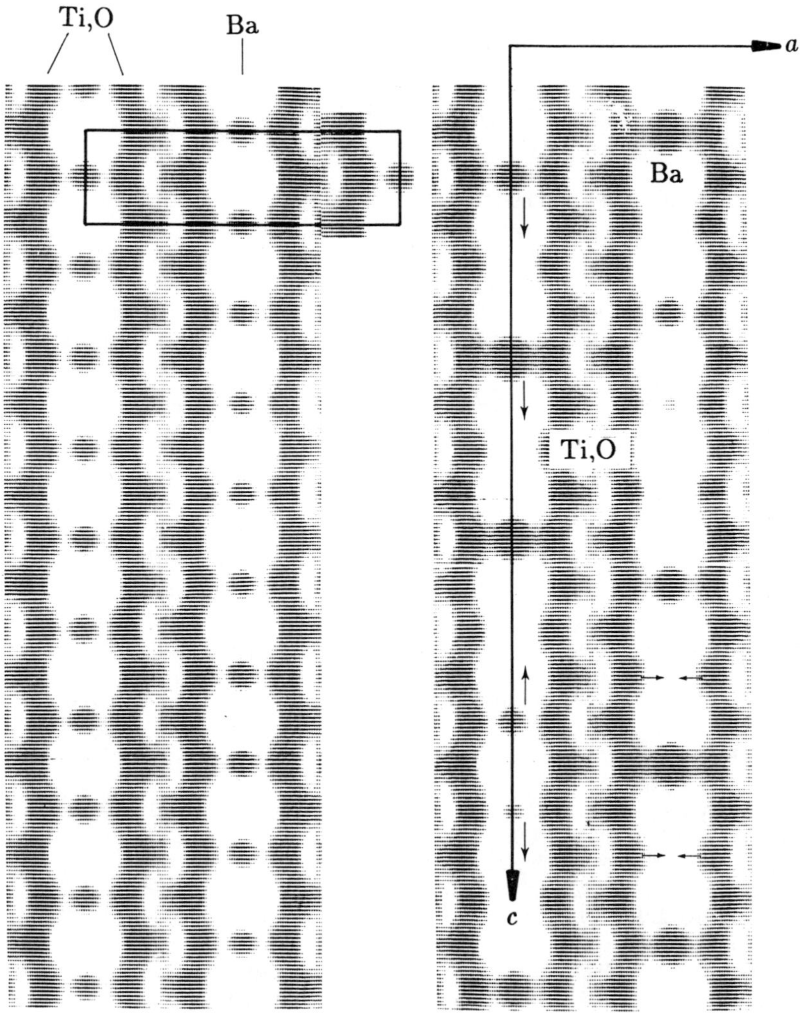

Fig. 3. Electron potential distibution of ankangite projected along the $\underline{b}$ axis. (a) The average structue calculated using 24 main reflections of the type $(h_1,0,h_3,0)$, a unit cell of which is indicated by the rectangle; (b) the incommensurate modulated structure calculated using 24 main reflections of the type $(h_1,0,h_3,0)$ and 36 satellite reflections of the type $(h_1,0,h_3,1)$ and $(h_1,0,h_3,2)$. The occupational modulation of Ba atoms is clearly seen, it appears as the ordered vacancies along the $\underline{c}$ direction and disordered vacancies perpendicular to the $\underline{c}$ direction. In addition positional modulations of Ba as well as Ti,O atoms are also observed as indicated by the short arrows.

this description, two kinds of modified Sayre equations were obtained (Hao Quan, Liu Yi-wei & Fan Hai-fu, 1987):

$$\mathbf{F}_m(\hat{\mathbf{H}}) = \frac{\Theta}{V} \sum_{\hat{\mathbf{H}}'} \mathbf{F}_m(\hat{\mathbf{H}}') \ \mathbf{F}_m(\hat{\mathbf{H}} - \hat{\mathbf{H}}') \quad , \tag{8}$$

$$\mathbf{F}_s(\hat{\mathbf{H}}) = 2\frac{\Theta}{V} \sum_{\hat{\mathbf{H}}'} \mathbf{F}_m(\hat{\mathbf{H}}') \ \mathbf{F}_s(\hat{\mathbf{H}} - \hat{\mathbf{H}}') \quad . \tag{9}$$

Where $F_m(\underline{H})$ denotes structure factors of the main reflections; $F_s(\underline{H})$ denotes those of the satellites. (8) indicates that the phases of main reflections can be derived by a conventional direct method neglecting the satellites. While (9) can be used for phase extension from the main reflections to the satellites. The latter provides a way to determine directly the modulation function, without preliminary assumptions.

<u>Example on solving an unknown incommensurate modulated structure</u>

Ankangite, $Ba_{0.8}(Ti,V,Cr)_8O_{16}$, is a newly found mineral. The average structure belongs to space group I4/m with unit cell a=b=10.12 A, and c=2.96 A. The modulation wave is parallel to the <u>c</u> axis with a period 2.27 times longer than c. An electron diffraction pattern containing reflections of the type $(h_1,0,h_3,h_4)$ was used to determine the modulation of Ba atoms (Xiang Shi-bin, Fan Hai-fu, Wu Xiao-jing, Li Fang-hua & Pan Qing, 1990). Starting from the known phases of the main reflections $(h_1,0,h_3,0)$, phases of the satellites were derived by making used of (9). The resulted Fourier map is shown in Fig. 3b revealing clearly the occupational modulation of Ba atoms. Positional modulations are also observed.

REFERENCES

Aalst, W. van, Hollander, J. den, Peterse, W.J.A.M. & Wolff, P.M. de (1976). *Acta Cryst.* B<u>32</u>, 47-58.
Boehme, R. (1982). *Acta Cryst.* A<u>38</u>, 318-326.
Fan Hai-fu (1975). *Acta Phys. Sin.* <u>24</u>, 57-60.
Fan Hai-fu, He Luo, Qian Jin-zi & Lui Shi-xiang (1978). *Acta Phys. Sin.* <u>27</u>, 554-558.
Fan Hai-fu, Qian Jin-zi, Zheng Chao-de, Gu Yuan-xin Ke Heng-ming & Huang Sheng-hua (1990). *Acta Cryst.* A46 (in the press).
Fan Hai-fu, Yao Jia-xing, Main, P. & Woolfson, M.M. (1983). *Acta Cryst.* A<u>39</u>, 566-569.
Fan Hai-fu, Yao Jia-xing & Qian Jin-zi (1988). *Acta Cryst.* A<u>44</u>, 688-691.
Gramlich, V. (1975). *Acta Cryst.* A<u>31</u>, S90.
Gramlich, V. (1978). *Acta Cryst.* A<u>34</u>, S43.
Hao Quan, Liu Yi-wei & Fan Hai-fu (1987). *Acta Cryst.* A<u>43</u>, 820-824.
Ito, T. (1973). *Z. Krist.* <u>137</u>, 399-411.
Ito, T. & Nowacki, W. (1974). *Z. Krist.* <u>139</u>, 85-102.
Prick, P.A.J., Beurskens, P.T. & Gould, R.O. (1983). *Acta Cryst.* A<u>39</u>, 570-576.
Qian Jin-zi, Fu Ping-qui, Kong Yuo-hua & Gong Guo-hong (1982). *Acta Phys. Sin.* <u>31</u>, 577-584.
Sayre, D. (1952). *Acta Cryst.* <u>5</u>, 60-65.
Wolff, P.M. de (1977). *Acta Cryst.* A<u>30</u>, 777-785.
Xiang Shi-bin, Fan Hai-fu, Wu Xiao-jing Li Fang-hua & Pan Qing (1990). *Acta Cryst.* (submitted)
Yamamotb, A. (1982). *Acta Cryst.* A<u>38</u>, 87-92.

RESULTS ON THE APPLICATION OF DIRECT METHODS

TO PROTEIN CRYSTALLGRAPHY AND ELECTRON MICROSCOPY

Fan Hai-fu

Institute of Physics
Chinese Academy of Sciences
Beijing, China.

Direct Phasing of OAS Data from A Small Protein

Multiple isomorphousreplacement is now dominating the structure analysis of proteins with no structural precedent. It may occur that the derivatives are not isomorphous with the native protein. In this case multi-wavelength anomalous scattering (MAS) can in principle be used, if there are some suitable heavy atoms in the native protein or its non-isomorphous derivative. However MAS technique suffers from the difficulty of collecting and scaling data at different wavelengths accurately. OAS technique does not have this difficulty but it leads to the problem of phase ambiguity. A direct method has been proposed to solve this problem [Fan et al. (1984). Acta Cryst. A40, 489-495; 495-498.]. The method has been tested with the Hg-derivative of a known protein, avian pancreatic polypeptide [Glover et al. (1985). Adv. Biophys. 20, 1-12.]. It resulted in an interpretable Fourier map. A part of which together with the true structure model is shown in Fig. 1.

Resolution Enhancement of An Electron Micrograph

The high resolution electron micrograph of chlorinated copper phthalocyanine (Fig. 2a) [Uyeda et al. (1978-1979). Chem. Scripta 14, 47.] though provides useful structural information, it is not able to resolve individual atoms. On the other hand the corresponding electron diffraction pattern offers much higher resolution but the phase problem has been proved difficult to solve by traditional direct methods. A phase extension technique with starting phases obtained from the electron micrograph at 2A resolution has been used to derive the phases for reflections between $2A^{-1}$ and $1A^{-1}$ on the diffraction pattern. This led to an image (Fig. 2b) much closer to the true structure model (Fig. 2c).

Fig. 1

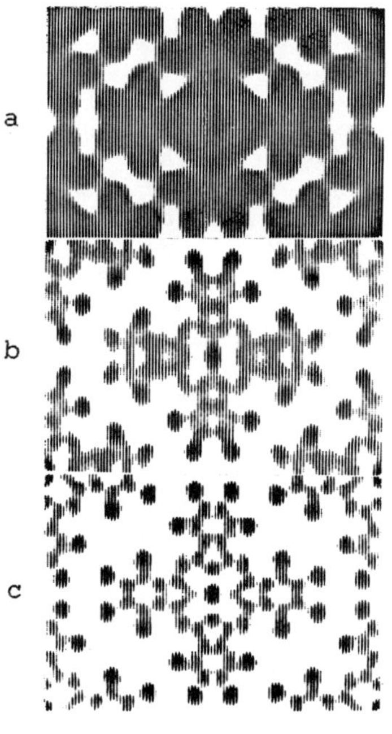

Fig. 2

DIRECT PHASE DETERMINATION IN THE ELECTRON CRYSTALLOGRAPHY OF ORGANIC

COMPOUNDS

Douglas L. Dorset

Medical Foundation of Buffalo, Inc.
73 High Street
Buffalo, New York   14203   U.S.A.

ABSTRACT.   Although other, more traditional, methods for determination
of crystallographic phases have been often used with electron
diffraction intensity data from organic crystals, direct phasing based
on the probabilistic estimate of structure invariants has been found
recently to be generally applicable for such structure analyses. New
structure solutions include the lamellar packings of several
phospholipids, a previously undetermined polymorph of an odd-chain
paraffin, the solid solution of two paraffins, and the three-dimensional
crystal structure of polyethylene. In some cases, low-dose high-
resolution electron microscope images provide phase information useful
for the elucidation of the total crystal structure, especially when the
diffraction pattern is dominated by the scattering from a sublattice
structure.

INTRODUCTION

The use of electron diffraction intensity data for quantitative
determination of organic crystal structures has been an interesting
alternative to x-ray crystallographic methods, especially for materials
which are most easily grown as thin microcrystals. This technique has
been used at least since the initial attempt to solve an n-paraffin
crystal structure in the 1930's by Rigamonti[1]. With the derivation of
reasonably accurate atomic scattering factors for electrons[2], the
technique was adapted to other small molecule problems[3-5], in addition
to the n-paraffin structure determination[6]. Most recently it has been
often used for the elucidation of linear polymer crystal structures[7],
since the lamellar crystals are most easily obtained by self-seeding
using a dilute solution in a poor solvent.

Various techniques have been used to obtain crystallographic phase
information to carry out the structure analysis. Often a model is
suggested by a related x-ray crystal structure. For linear polymers, an
oligomer structure can be concatenated to form a model of the infinite
polymer so that a conformational search around specified linkage bonds
can be tested both against the crystallographic residual and also a
minimized non-bonded potential energy[8]. For simpler structures, the
interpretation of Patterson functions[9] has enabled the construction of
phasing models. In some optimal cases, details of the atomic

Direct Methods of Solving Crystal Stuctures
Edited by H. Schenk, Plenum Press, New York, 1991

arrangement are visible in electron microscope "lattice images" of the crystalline array[10].

The major objection to using model structures for phasing electron data is that the number of refineable parameters has the same order of magnitude as the number of observed data. As is well known, the crystallographic residual in this case is a very imprecise figure of merit so that models with slightly different atomic arrays may not be distinguished at the confidence level often assumed in x-ray crystal structure determinations[11]. High-resolution electron microscope images would overcome this problem but the resolution limits imposed by the transfer function of the objective lens, and also by radiation damage, can be a serious problem.

DIRECT PHASING WITH STRUCTURE INVARIANTS

Polymethylene Chains in Projection Along the Long Axis

The first actual use of traditional direct phasing techniques based on probabilistic estimates of three- and four-phase structure invariants was reported by Dorset and Hauptman[12] in 1976. In their analysis of an n-paraffin crystal structure in a projection down the long chain axis, the phases of 24 of 42 measured reflections were correctly determined from Σ_2-triples and quartets. The resulting electrostatic potential map clearly showed the positions of the carbon atoms (Fig. 1) and these were found to be close to the positions found earlier from the interpretation of the Patterson function.

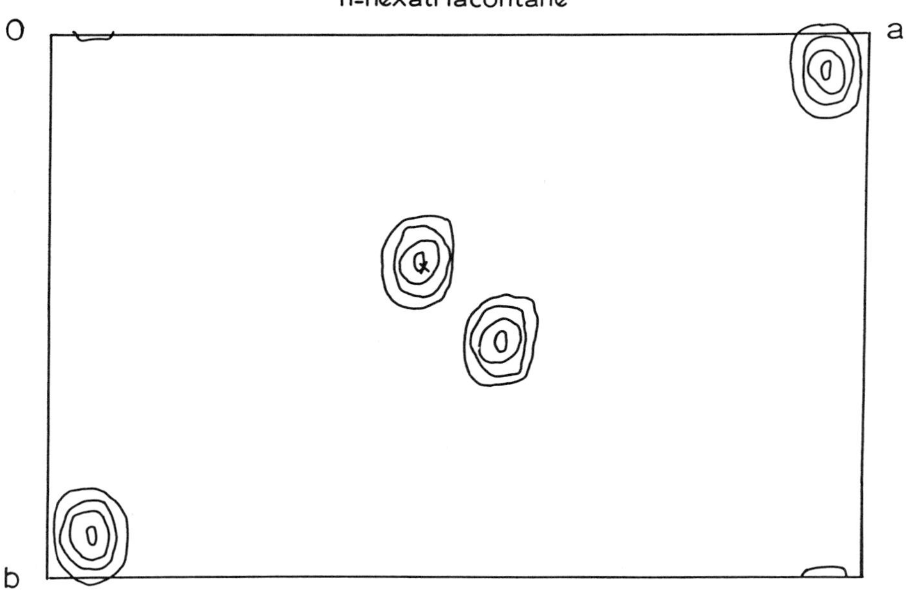

Figure 1. Crystal structure of a n-hexatriacontane monolamellar layer determined by direct phasing.

A second methylene chain packing was analyzed using electron diffraction data from a phospholipid . In this case 29 of 42 phases were determined unambiguously from the structure invariants and three additional phases were inter-related by an unresolved ambiguity. Calculation of two maps allowed the correct structure to be discerned.

Model Studies on Aromatics

In order to test the effect of dynamical scattering on the success of
structure analysis by direct phasing, multislice n-beam calculations of
structure factors were carried out for two aromatic structures[13] at
various simulated crystal thicknesses and at two electron accelerating
voltages. For the smaller crystal structure of cytosine, a 76Å thick
crystal scattering 100kV electrons was found to be correctly analyzed by
direct phasing to produce an interpretable structure map. At 1000kV,
data from a 300Å crystal could be used. A similar thickness limit was
found at 100kV for a larger molecule, disodium 4-oxypyrimidine-2-
sulfinate hexahydrate, but, in this instance, no improvement was found
at 1000kV, in order to be equivalent to the cytosine example. The
maximum number of phase errors allowed for the calculation of a
structurally reasonable electrostatic potential map was 4 in 69
reflections where $|E|>1.0$.

A similar analysis was carried out for the cytosine crystal
structure simulating the perturbation of the diffraction data from bend-
deformed crystals[14]. For a projected unit cell axis dimension c=3.82Å,
corresponding, perhaps, to an epitaxially crystallized aromatic, a
crystal bend deformation of 7.5° could be tolerated and still allow a
correct structure to be determined. The situation would be less
favorable for structures with larger projected unit cell axial
dimensions.

More recent simulations have been carried out with calculated image
and electron diffraction data from epitaxially-oriented
phthalocyanines[15]. Hence, the likelihood of phase extension beyond the
resolution limit defined by the electron microscope objective lens
transfer function was evaluated by using the crystallographic phases
determined from the image as a basis set to phase the higher resolution
electron diffraction data. The phase refinement used constraints on
real and reciprocal space defined by Gassmann and Zechmeister[16].

Lamellar Structures

The effectiveness of phase extension realized by the use of structure
invariants was demonstrated with measured lamellar data from epitaxially
oriented phospholipids[17]. In one case, 1,2-dihexadecyl-*sn*-
glycerophosphoethanolamine (DHPE), direct low dose high resolution
lattice images photographed using an electron microscope with a helium-
cooled superconducting objective lens were found to diffract to 6Å
resolution on an optical bench[18]. Image analysis demonstrated that the
phases of the first seven lamellar orders correspond exactly to the
values found in an earlier crystal structure analysis based on a
molecular packing model. Calculation of three- and four-phase structure
invariants permitted the phases of the remaining 9 reflections to be
determined, resulting in a one-dimensional electrostatic potential map
at 3.4Å resolution (Fig. 2) with features very similar to those seen in
electron density maps determined from the phasing of x-ray data by
various means (e.g. swelling of lamellae in water or model building).

If the electron microscope image is assumed to be unavailable and
if one specifies the value of one phase to define the unit cell origin,
the analysis of structure invariants correctly phases 13 of 16
reflections for DHPE. In this case the electrostatic potential map
again closely resembles the complete Fourier transform. With this
encouragement, electron diffraction data from four other phospholipids
were directly phased to produce maps from which the structure can be
interpreted. These analyses include two structures for which no x-ray
determination has been reported. It is also possible to analyze similar
lamellar x-ray data from phospholipids with equally positive results.

Epitaxially-Oriented n-Paraffins

When n-paraffins are epitaxially-oriented on substrates such as benzoic acid, it is possible to collect three-dimensional electron diffraction data suitable for *ab initio* crystal structure analysis[19]. The three-dimensional data from n-hexatriacontane was analyzed by combining phase information from high resolution electron microscope images[18] and the use of structure invariants. The electron microscope images obtained on the cryomicroscope mentioned above diffract on an optical bench to 5Å resolution, enabling the phases of lamellar reflections to

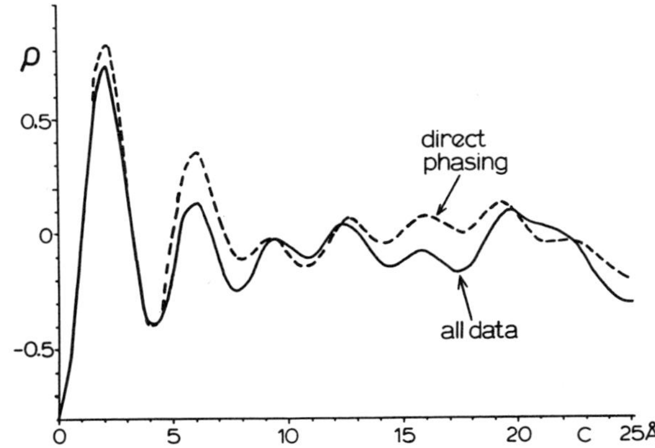

Figure 2. Lamellar structure of epitaxially-crystallized 1,2-dihexadecyl-*sn*-glycerophosphoethanolamine. ("All data" combines both electron microscopy and direct phasing.)

be determined by image analysis. Structure invariant relationships phase the so-called "polyethylene" reflections, permitting the major reflections of the unit cell transform to be determined and thus the electrostatic potential map can be compared favorably to the x-ray crystal structure determined by Teare[20]. A similar analysis carried out for the B-form of the odd-chain paraffin, n-tritriacontane, determines the phase values of seventeen key $0k\ell$ reflections to produce the structure map in Figure 3. The positions of all 33 carbon atoms in a chain can be discerned. Along with the earlier use of a packing model[21] this represents the first quantitative structure analysis of this polymorphic crystal form.

More interesting perhaps is the determination of the crystal structure of a 1:1 solid solution of $nC_{32}H_{66}/nC_{36}H_{74}$ with structure invariant relationships. An earlier use of a superlattice model[22] was found to be somewhat incorrect, since the crystal structure analysis distinctly requires an average lamellar packing like that of $nC_{34}H_{70}$, except that the outer chain atoms have non-unitary occupancies. Refinement of this occupancy factor lowers the R-value to 0.23.

Polyethylene

Recently the crystal structure analysis[23] of polyethylene based on three-dimensional data collected from epitaxially-oriented and solution-crystallized samples was reported. Of the 50 measured reflections with intensity significantly above background, use of structure invariants enables one to determine 40 unique phase values. Calculation of the electrostatic potential map supports the notion that the chain setting angle is nearly the same as the value found for even-chain n-paraffins.

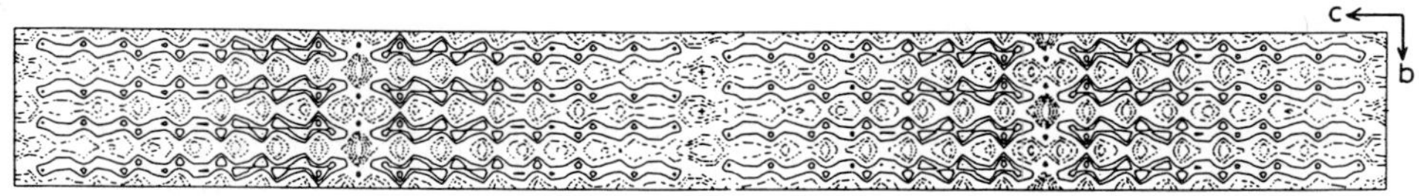

Figure 3. Crystal structure of epitaxially-oriented n-tritriacontane in a projection down the unit cell a=7.5Å axis.

CONCLUSIONS

Although electron diffraction data have been demonstrated to be prone to perturbations resulting from multiple scattering and crystal bend deformation, conditions can be manipulated to permit *ab initio* analyses yielding new crystal structures which are chemically reasonable. Among the arsenal of phasing techniques, use of direct methods is now found to be as useful potentially as in their application to x-ray diffraction data. As demonstrated here, a variety of different structure types can be analyzed including polymethylene compounds, the structures of which are often difficult to determine due to the dominance of the diffraction pattern by the strong sublattice scattering. In this case, electron microscope images provide the complementary phase information needed to permit the elucidation of the total crystal structure.

REFERENCES

1. R. Rigamonti, 1936, La struttura della catena parafinnica studiata mediante i raggi di elettroni, Gazz. Chim. Ital., 66:174-182.
2. B.K. Vainshtein, 1964, "Structure Analysis by Electron Diffraction", Pergamon Press, Oxford.
3. B.K. Vainshtein, 1955, Elektronograficheskoe issledovanie diketopiperazina, Zh. Fis. Khim., 29:327-344.
4. A.N. Lobachev and B.K. Vainshtein, 1961, An electron diffraction study of urea, Sov. Phys.-Crystallogr., 6:313-317.
5. I.A. D'yakon, L.N. Kairyzk, A.V. Ablov, and L.F. Chapurina, 1977, Atomic structure of the copper salt of two different amino acids. Cu(DL-Ala) (DL-Ser), Dokl. Akad. Nauk. SSSR, 236:103-105 (translation journal p. 481-483).
6. B.K. Vainshtein, A.N. Lobachev, and M.M. Stasova, 1958, Electron diffraction determination of the C-H distance in some paraffins, Sov. Phys. Crystallogr., 3:452-459.
7. D.L. Dorset, 1989, Electron diffraction from crystalline polymers, in "Comprehensive Polymer Science, Vol. 1," Sir. G. Allen, ed., Pergamon Press, Oxford, Chapter 29, pp. 651-668.
8. S. Perez, M. Roux, J.F. Revol, and R.H. Marchessault, 1979, Dehydration of nigeran crystals: Crystal structure and morphological aspects, J. Mol. Biol., 129:113-133.
9. D.L. Dorset, 1976, Aliphatic chain packing in three crystalline polymorphs of a saturated racemic phosphatidylethanolamine. A quantitative electron diffraction study, Biochim. Biophys. Acta, 424:396-403.
10. N. Uyeda, T. Kobayashi, K. Ishizuka, and Y. Fujiyoshi, 1979, High voltage electron microscopy for image discrimination constituent atoms in crystals and molecules, Chem. Scripta, 44:47-61.
11. W.C. Hamilton, 1964, "Statistics in Physical Science," Ronald Publ. Co., New York, p. 157-162.
12. D.L. Dorset and H.A. Hauptman, 1976, Direct phase determination for quasi-kinematical electron diffraction intensity data from organic microcrystals, Ultramicroscopy, 1:195-201.

13. D.L. Dorset, B.K. Jap, M.-H., Ho, and R.M. Glaeser, 1979, Phasing of electron diffraction data from organic crystals. The effect of n-beam dynamical scattering, Acta Cryst., A35:1001-1009.

14. B. Moss and D.L. Dorset, 1982, Effect of crystal bending on direct phasing of electron diffraction data from cytosine, Acta Cryst., A38:207-211.

15. K. Ishizuka, M. Miyazaki, and Uyeda, N., 1982, Improvement of electron microscope images by the direct phasing method, Acta Cryst., A38:408-413.

16. J. Gassmann and K. Zechmeister, 1972, Limits of phase expansion in direct methods, Acta Cryst., A28:270-280.

17. D.L. Dorset, 1987, Electron diffraction structure analysis of phospholipids, J. Electron Microsc. Techn.,7:35-46.

18. F. Zemlin, E. Beckmann, and D.L. Dorset, in preparation.

19. B. Moss, D.L. Dorset, J.C. Wittmann, and B. Lotz, 1984, Electron crystallography of epitaxially-grown paraffin, J. Polym. Sci.-Polym. Phys. Ed., 22:1919-1929.

20. P.W. Teare, 1959, The crystal structure of orthorhombic hexatriacontane $C_{36}H_{74}$, Acta Cryst., 12:294-300.

21. D.L. Dorset, 1986, Electron diffraction structure analysis of epitaxially crystallized n-paraffins, J. Polym. Sci.-Polym. Phys. Ed., 24:79-87.

22. D.L. Dorset, 1985, Crystal structure of n-paraffin solid solutions: an electron diffraction study, Macromolecules, 18:2158.

23. H. Hu and D.L. Dorset, 1989, Three-dimensional electron diffraction structure analysis of polyethylene, Acta Cryst., B45:283-290.

FROM PARTIAL STRUCTURE TO COMPLETENESS -

DIRECT METHODS APPLIED TO DIFFERENCE STRUCTURE FACTORS

Paul T. Beurskens and Cornelia Smykalla

Crystallography Laboratory, Toernooiveld
6525 ED Nijmegen, The Netherlands

Abstract

When part of a structure is known, direct methods can be used to solve the unknown part of that structure. Often the known part of the structure consists of one or more heavy atoms, either on general or on special or pseudo-special positions. The known part of the structure may also consist of a molecular fragment, found by ab-initio direct methods or by Patterson vector search techniques.

The difference structure factors, phased by the partial structure, are used as input to a weighted tangent-refinement procedure for phase extension and for the refinement of input phases and amplitudes. If the known atoms do not uniquely determine the structure, symbolic addition techniques are introduced to solve the (pseudo-symmetry) ambiguities.

The method is referred to as 'DIRDIF'. It is most useful when the known part is only marginally sufficient to solve the structure, when the known atoms lie on special or pseudo-special positions (origin ambiguity), or when, for noncentrosymmetric structures, the known atoms form a centrosymmetric arrangement (enantiomorph ambiguity).

The computer program system is also referred to as 'DIRDIF' (Beurskens, Bosman, Doesburg, Gould, van den Hark, Prick, Noordik, Beurskens, Parthasarathi, Bruins-Slot, Haltiwanger, Strumpel, Smits, García-Granda, Smykalla, Behm, Schäfer, and Admiraal, 1990). DIRDIF consists of a collection of individual programs, written in FORTRAN77, which are designed to run fully automatically. The actual calculations are controlled by a control-data file which is generated by an interactive start-up procedure. A simple computer-dependent job-control procedure is used to activate various individual programs whenever their execution is requested by the control-data file.

Vector search rotation functions, and translation functions in reciprocal space, as well as heavy-atom Patterson interpretation procedures are incorporated in the program system.

1. Introduction

The excitement in the early days of direct methods was caused not only by the discovery of many new crystallographic methods, but also by the availability of computers that allowed the new methods to be used by non-expert crystallographers. But, *vica versa*, the computer allowed of the development of more complicated methods. Thus the intelligent but laborious examples given by I. L. Karle (e.g. Karle et al., 1957) paved the way to the development of the first widely used automated direct method programs (sign correlation, Beurskens, 1963, 1964).

Four decades have gone by, and we are no longer so heavily dependent on direct methods: Patterson methods and reciprocal space correlation methods have caught up. Fully automatic computer programs (e.g. ORIENT and TRACOR, see section 11) can use structural information which is available through crystallographic data bases (more than 100 000 known crystal structures) or which can be generated by very sophisticated molecular modelling programs.

It is the purpose of this chapter to show how known chemical information can be incorporated in various 'direct methods' procedures such as the tangent formula, symbolic addition, multisolution, quartets, etc. The DIRDIF procedure, described in this chapter, allows the expansion of a small known fragment, and the solution of various types of phase ambiguities and symmetry problems.

2. Notation and definitions

$\mathbf{h}$	reciprocal space vector (h, k, l)

$\mathbf{h}$ reciprocal space vector (h, k, l)
In most formulae the indices $\mathbf{h}$ will be omitted to simplify the notation.

φ phase of the reflection $\mathbf{h}$. In principle this phase is not known; for all practical purposes φ represents the best available estimate for the phase.

$|F_o|$ observed structure factor for reflection $\mathbf{h}$ on absolute scale!

F a complex number representing the true structure factor, or the 'best' phased value for the observed structure factor:

$$F = |F_o| \cdot \exp(i\,\varphi) \qquad (1)$$

F_p partial structure factor; sum over N_p known atoms

φ_p phase of F_p

F_r structure factor of the remainder; sum over N_r remaining unknown atoms.

F_r in principle is not known; for all practical purposes F_r represents the best available estimate for the true value.

φ_r phase of F_r

Σ_p, Σ_r summation over all N_p known atoms, and all N_r unknown atoms

Z atomic number of an atom

p^2 scattering fraction of the partial structure:

$$p^2 = \Sigma_p Z^2 / (\Sigma_p Z^2 + \Sigma_r Z^2) \qquad (2)$$

r^2 scattering fraction of the rest structure,

$$p^2 + r^2 = 1 \qquad (3)$$

The relation from these definitions:

$$F = F_p + F_r \qquad (4)$$

gives the fundamental DIRDIF equation:

$$|F| = |F_p + F_r| \qquad (5)$$

where F_r is to be determined by 'direct methods'.

E, E_p, E_r normalized values for F, F_p and F_r.

$$E_p = (\varepsilon \Sigma_p f_p^2)^{-\frac{1}{2}} \cdot F_p \cdot \exp(B_p s^2) \qquad (6)$$

$$E_r = (\varepsilon \Sigma_r f_r^2)^{-\frac{1}{2}} \cdot F_r \cdot \exp(B_r s^2) \qquad (7)$$

s $\sin \vartheta / \lambda$

ε conventional symmetry enhancement factor

B_p, B_r overall temperature factor parameter of the known atoms, and of the remaining unknown atoms

From the definition of normalisation, it follows then:

$$E = pE_p + rE_r \qquad (8)$$

The normalisation of the data is achieved by solving the conditional equations:

$$Sc^2 \langle I \rangle = |F_p(B_p{=}0)|^2 \cdot \exp(-2B_p s^2) + \Sigma_r f_r^2 \cdot \exp(-2B_r s^2) \qquad (9)$$

$\langle I \rangle$ averaged intensities over all reflections in ranges of $|E_p|$ and s

Sc scale factor that brings the observed structure factor $|F_o|$ on absolute scale (Gould, van den Hark and Beurskens, 1975).

ΔF the conventional difference Fourier coefficient:

$$\Delta F = (|F_o| - |F_p|) \cdot \exp(i\,\varphi_p) \qquad (10)$$

Note that the phase of ΔF is either φ_p or $\varphi_p + \pi$

E_1 normalised value of ΔF:

$$E_1 = (\varepsilon \Sigma_r f_1^2)^{-\frac{1}{2}} \cdot \Delta F \cdot \exp(B_r s^2) \qquad (11)$$

φ_1 is the phase of E_1, i.e. $\varphi_1 = \varphi_p$ or $\varphi_1 = \varphi_p + \pi$

W_1 weight associated with the reliability of φ_1 (Van den Hark, Prick and Beurskens, 1976)

φ_t, W_t subscript t refers to tangent formula results.

Pictograms: structure factors are represented by vectors in the complex plane. The known F_p is given by a heavy line. The observed $|F_o|$ is represented by a *circle* with radius $|F_o|$. Fig. 1 represents eq. (4) and (5) for a given vector F_r.

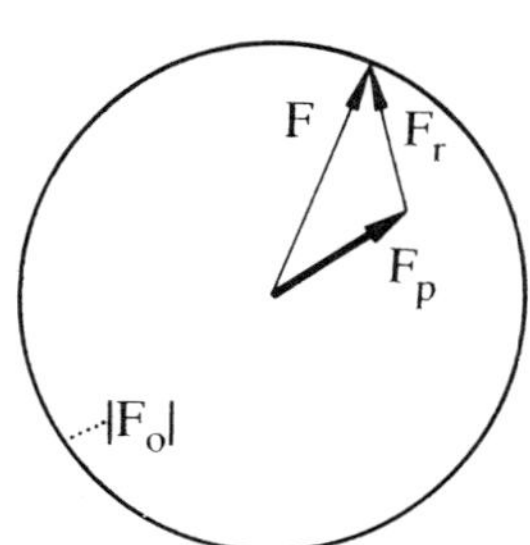

Fig. 1. Vector diagram for F, F_p and F_r.

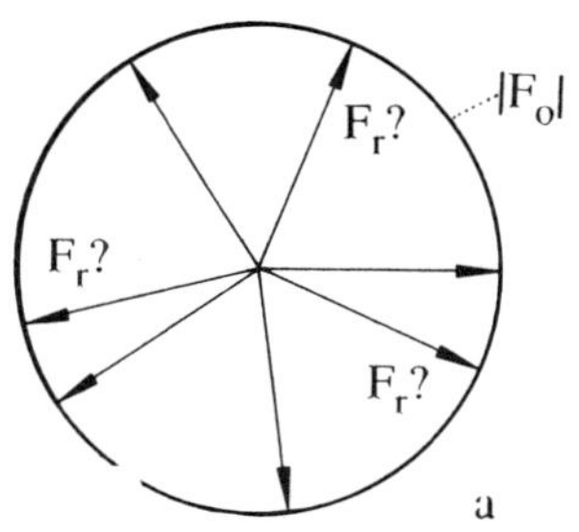
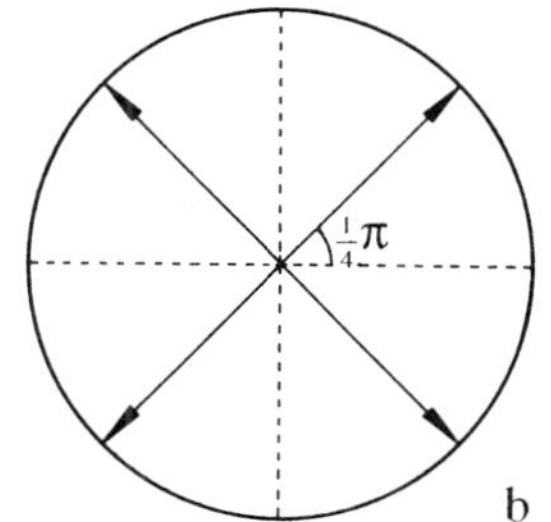
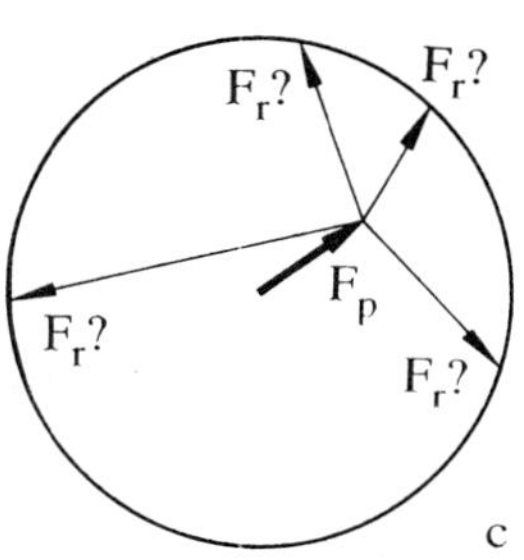

Fig. 2. a: Ab initio direct methods: find direction of F
b: Fourfold ambiguity phases $\pm\frac{1}{4}\pi$, $\pm\frac{3}{4}\pi$.
c: Structure factor F_p, known fragment: find remainder of structure; i.e. F_r

3. Preliminary considerations. The phase problem

Ab-initio direct methods can be employed to determine the direction of the vector F in the above diagram, Fig. 2a. Multisolution techniques, for instance, may be initiated by assigning a four-fold ambiguity to some selected reflections, Fig. 2b.

The situation changes if a part of the structure is known. Then, the calculated structure factor of the partial structure may be drawn in the diagram (Fig. 2c), but the phase problem is not yet solved. From the end-point of the known vector F_p an unknown vector F_r, representing the structure factor of the remaining part of the structure, is drawn with its end-point lying on the $|F_o|$-circle.

Comparing Fig. 2a and 2c we notice two important differences:

Fig. 2a: *a priori*: all phases are equally probable, all possible vectors F have length equal to $|F_o|$.

Fig. 2c: known fragment: the length of the unknown vector F_r is not fixed; it depends on φ_r (i.e. the phase of F_r) according to equation (4) and, consequently, possible phases φ_r do not have equal probability.

In fact, when the known fragment is sufficiently large, one can calculate a conventional difference Fourier synthesis using the assumption that the most probable phase of F_r is equal to φ_p , ie. the calculated phase of F_p (equation (9)): see Fig. 3a.

Distribution functions for φ_r have been given by Sim (1960); we will use these later, but for the time being it is sufficient to associate a weight W_1 with the ΔF value. The Fourier transform of $W_1 \Delta F$ is the weighted difference electron density map. But this is certainly not the same as the electron density map of the difference structure!

When the known fragment is relatively large, the (difference) Fourier synthesis can be used to find the rest of the structure, however if the known fragment is small (or exhibits higher symmetry than the space group symmetry!) most of the ΔF values are a poor estimate of the true F_r values.

Nevertheless, the ΔF's contain valuable phase information, and we shall see that direct methods (DIRDIF) can be used to improve these phases (Fig. 3b). Note that a change of phase (φ_r) implies an adjustment of the magnitude ($|F_r|$, see Fig. 3c).

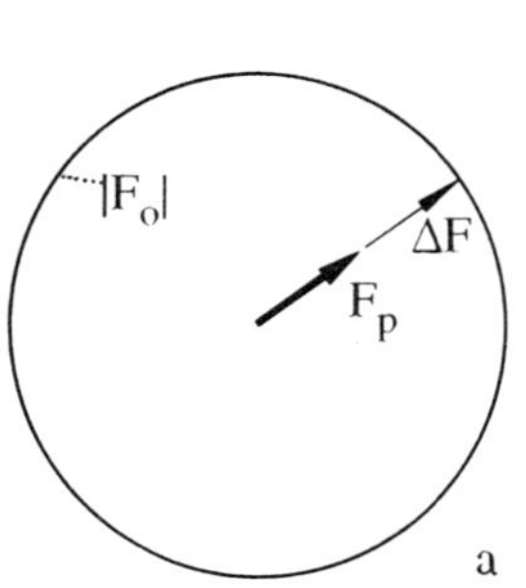
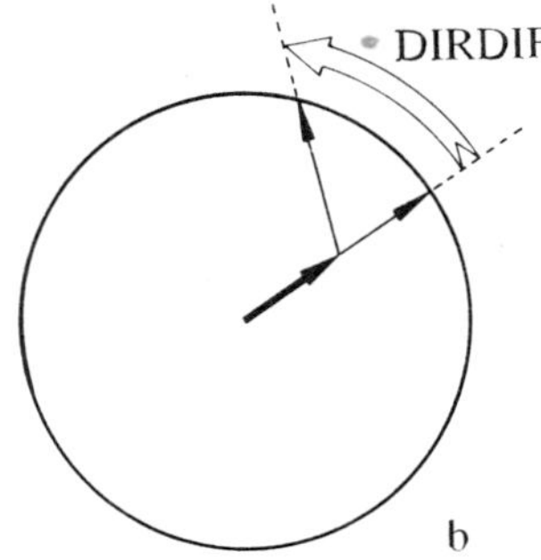
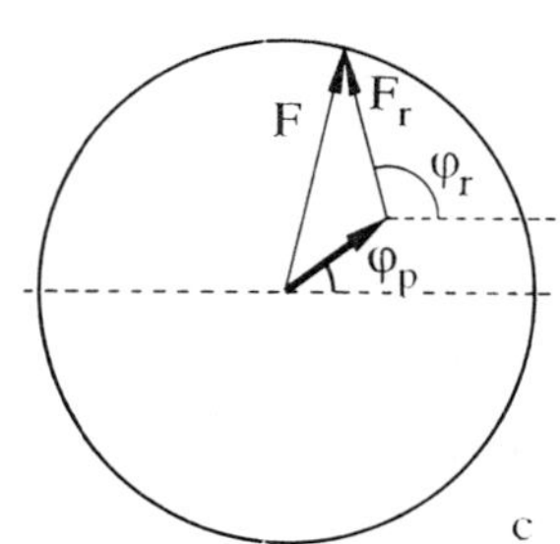

Fig. 3. a: Difference Fourier synthesis: use ΔF
b: DIRDIF: input ΔF, output F_r
c: The phase φ_r obtained by direct methods is used to construct F_r and F.

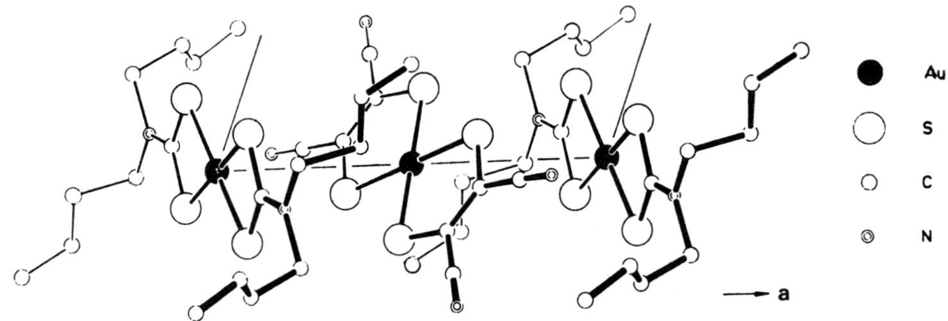

Fig. 4. Projection of part of the structure of NORA.

4. A special example: NORA

The compound NORA (Beurskens and Noordik, 1971), $Au(S_2CN(C_4H_9)_2)_2$ $Au(S_2C_2(CN)_2)_2$, centrosymmetric space group $P2_1/c$, with $Z = 2$, consists of two ionic units, see Fig. 4, both having an approximately square planar AuS_4 moiety. The positions of the gold atoms were easily found by inspection of the Patterson synthesis, but due to heavy overlap in the Patterson, not enough S positions could be assigned to allow the solution of the structures by conventional difference Fourier syntheses.

The problem at hand was a superstructure problem. The gold positions are:

Au(1): 000 (and by symmetry: $0\frac{1}{2}\frac{1}{2}$)

Au(2): $\frac{1}{2}00$ (and by symmetry: $\frac{1}{2}\frac{1}{2}\frac{1}{2}$)

Thus the gold atoms form a subcell due to the pseudo-translation symmetry vectors $\frac{1}{2}00$ and $0\frac{1}{2}\frac{1}{2}$. As a consequence we have to distinguish between two sets of data:

'strong' reflections $h = 2n$, $k+l = 2n$, the gold atoms scatter in phase, $F_p = 4f_{Au}\cdot exp(-B_p\, s^2)$.

$F_r = |F_o| - F_p$ (note: $F = |F_o|$ is always positive, because of the dominant scattering power of the gold atoms; F_r, however, can be positive or negative).

Therefore all reflections in this category had measurable intensities, and the structure factors of the rest structure F_r (both magnitude and phase) are known.

'weak' reflections: $h = 2n+1$ and/or $k+l = 2n+1$, $F_p = 0$.

The contributions of the gold atom cancel for the weak reflections, and no phase information is obtained: $|F_r| = |F_o|$

At this point we wish to point out that conventional direct methods cannot be applied to the F values because all sign triplets are biased. (In centrosymmetric structures phases are represented by signs: $S = +1$ if $\varphi = 0$ and $S = -1$ if $\varphi = \pi$.) For instance: the sign relation

$$S_{h+h'} \approx S_h \cdot S_{h'} \tag{12}$$

where h and h' are 'weak' reflections, cannot be used to find the sign of reflection $h+h'$ because this is a 'strong' reflection and its sign is always positive.

However, we are able to apply eq. (11) to the data of the rest structure.

After scaling (done separately for 'strong' and 'weak' reflections) and normalisation (using the temperature factor B_r for the rest structure, which could be obtained from a Wilson plot, using the weak reflections only!) the following data resulted:

365 'strong' reflections: F_r (and E_r) with known sign;

270 'weak' reflections: $|F_r|$ (and $|E_r|$) with unknown sign.

The normalised data is not hampered by the super-symmetry since the substructure consisting of all the gold atoms has been subtracted, and there is no reason why direct methods cannot be applied to the data.

So far we have seen for this special case the scaling procedure which enables us to find B_r, and the subtraction of F_p, which leads to the determination of a large number of signs. But that is not enough to solve the structure.

All reflections with known signs are in parity groups eee and eoo (e = even, o = odd). None of the weak reflections had a known sign, and like in ab-initio direct methods, we must now make two arbitrary assignments to fix the origin.

Having done so, we are in a position to generate signs for all reflections. In order to make sure that aberrant sign relationships do not lead us to a false solution, we use the sign-correlation procedure (Sax et al., 1965), which is described below. In this way we determined the signs of all reflections (with $|E_r| > 1.20$) without one single error.

5. The tangent formula applied to difference structure factors

The application of direct methods to the solution of a heavy-atom superstructure problem (see previous section) led to the concept of the DIRDIF method, which has been extensively developed over 20 years of practical experience. It has been intuitively assumed, that direct methods are applicable to a hypothetical structure consisting of the complete structure minus the known part of the structure. The tangent formula for the difference structure is

$$E_r(\mathbf{h}) \approx c \, \Sigma_{\mathbf{k}} E_r(\mathbf{k}) \cdot E_r(\mathbf{h}-\mathbf{k}) \tag{13}$$

where c is a positive scaling factor defined to satisfy (4) (for infinite resolution: $c = N_r^{-\frac{1}{2}}$). Except for very-heavy-atom structures (where a difference map is preferred) we usually wish to see the electron density function for the entire structure. Substitution of (13) in (8) gives:

$$E(\mathbf{h}) = pE_p(\mathbf{h}) + c \, r \, \Sigma_{\mathbf{k}} E_r(\mathbf{k}) \cdot E_r(\mathbf{h}-\mathbf{k}) \tag{14}$$

The Fourier transformation of E($\mathbf{h}$) (or of its de-normalised equivalent F($\mathbf{h}$)) gives all the known atoms (Fourier transform of E_p) as well as all the unknown atoms (Fourier transform of E_r).

This procedure, described by Beurskens and Noordik in 1971, was put on firm statistical grounds by Camalli et al., 1985 (see also Beurskens, 1987).

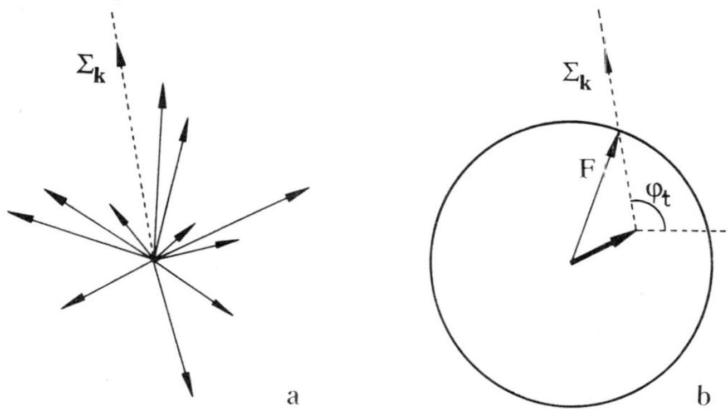

Fig. 5. a: shows a graph of the tangent formula for E_r
arrows: all terms $E_r(\mathbf{k}) \cdot E_r(\mathbf{h}-\mathbf{k})$; dotted line: sum of all arrows (eq. 13)
b: shows a construction of F; dotted line: direction of F_r (see Fig. 5a)

The application of (13) can be initiated by using phases which, in principle, may be obtained from either the known fragment, or by arbitrarily specifying multi-solution trial phases, or by assigning ambiguity symbols (such as letter symbols in the original symbolic addition method, Karle and Karle, 1963). The initial phases are denoted φ_1. In this section we assume that the known fragment does not exhibit special or additional pseudo-symmetry, and that it is sufficiently strong to guarantee a problem-free phase generation and phase refinement procedure. This implies that we can use φ_1 = phase of ΔF.

In DIRDIF the normalised structure factors $E_r(\mathbf{k})$ and $E_r(\mathbf{h}-\mathbf{k})$ in equation (13) are multiplied with their weights W_1

$$W_1 = (2P_1 - 1)^2 \tag{15}$$

where P_1 gives the probability that the input phase φ_1 is correct relative to the other extreme value

$\varphi_1 + \pi$ (Van den Hark, Prick and Beurskens, 1976). This is a measure to select E_1, and to use it as a first estimate for E_r. The weight assigned to the phases that result from the tangent formula (13) is W_t:

$$W_t = (2P_t - 1)^2 \qquad (16)$$

(Van den Hark et al., 1976) where P_t is given by:

$$P_t = \tfrac{1}{2} + \tfrac{1}{2}\tanh\left[\frac{\sigma_3}{\sigma_2^{3/2}} |E_h| \Sigma_k W_k W_{h-k} E_k E_{h-k}\right] \qquad (17)$$

with $\quad \sigma_n = \Sigma_r Z^n, \quad$ Z being the atomic number of the atoms of the rest structure.

In the first cycle of the tangent refinement, only E_1 values with phases φ_1 are used as first estimates for E_r for the reflections $\mathbf{k}$ and $(\mathbf{h-k})$. Whether or not the output phases φ_t are accepted depends on the corresponding weights. If $W_t > W_1$ then the new phase $\varphi_r = \varphi_t$ is accepted with weight W_t. If $W_t < W_1$ then the calculated φ_t value is only partially accepted when $|\varphi_t - \varphi_1|$ is less than $\pi/2$: $\varphi_r(\text{new}) = \varphi_1 + (W_t/W_1)[\varphi_t - \varphi_1]$. The new φ_r value is used to calculate a new value for E_r, which may be used as input for the next tangent refinement cycle.

An example of the quality of the tangent refinement is described for the structure MONOS (Noordik et al., 1978), $C_{15}H_{16}N_2O_2S$, space group $P2_12_12_1$, with $Z = 4$, Fig. 6. The sulphur position was input as: $x = 0.02$, $y = 0.09$, $z = 0.14$. From a conventional weighted Fourier synthesis only 11 out of the 20 non-hydrogen atoms can be found among the 23 highest peaks. If the phases are refined in three cycles with a tangent refinement, the resulting Fourier synthesis shows all 20 non-hydrogen atoms among the 23 highest peaks.

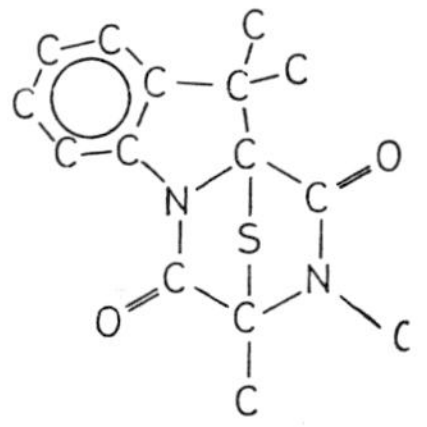

Fig. 6. The molecule MONOS.

6. Symbolic addition and phase correlation

In the original symbolic addition method (Karle and Karle, 1963) as few as possible 'letter symbols' were used to represent unknown phases. Alternatively, in the sign-correlation method (Beurskens, 1963, 1965) many (up to 26) letter symbols were used, which enabled the statistical analysis of a large number of more or less consistent phase relationships. For the present purpose (symbolic addition applied to difference structure factors) we found that about ten letter symbols are sufficient.

Let us assume that this technique is used for the solution of the example NORA (section 4). The two reflections 221 and $34\bar{8}$ that were arbitrarily given a + sign could have been given letter symbols: $\varphi_r(221) = A$ and $\varphi_r(34\bar{8}) = B$. The signs of all 'weak' reflections would have been expressed in A and B, but the signs of the 'strong' reflections would have been obtained from these by sign products AA (= ++ or --) = +.

At the end of the procedure the origin choice $A = B = +$ would have given exactly the same solution of the phase problem for NORA.

However, if the known part of the structure does not dominate as much as in NORA, one should expect errors in the initial phases φ_1 as well as inconsistencies in the phase relationships employed, which may lead to a wrong set of phases and failure to solve the structure. This problem is easily overcome by introducing more letter symbols for the application of the phase correlation procedure.

Let us rewrite (13) for one term in the same form as (12):

$$\varphi(\mathbf{h+h'}) \approx \varphi(\mathbf{h}) + \varphi(\mathbf{h'}) \qquad (18)$$

The primary set H is obtained by selecting the best phased reflections (i.e. with the highest $W_1 |E_1|$ values) and assigning phase symbols X (= A, B, C...) to ten selected unphased reflections.

Substitution of these phases in (18) gives the set H+H' where the phases $\varphi(\mathbf{h+h'})$ are expressed by numerical values and symbols. The most reliable results are selected to constitute the secondary set K. Application of (18) now gives phases for sets H+K and K+K'. From the many multiple phase indications symbolic relations are obtained. (For instance if $\varphi(\mathbf{h})$ is found as A, and via another route as B+π, then A $\approx$ B+π is a symbolic relation). These relations are analysed as shown below.

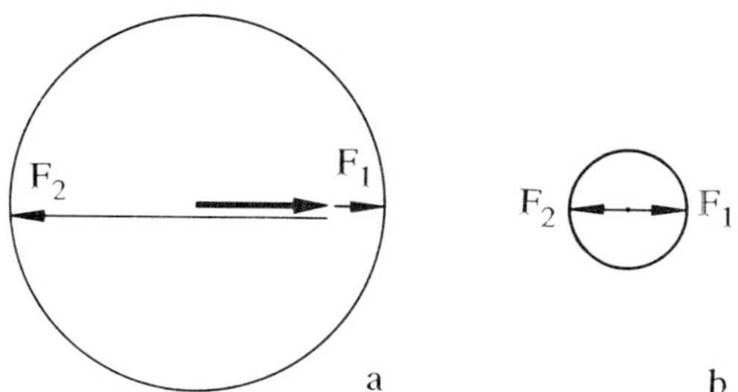

Fig. 7. a: The 'strong' reflections have large $|F_p|$: $F_r \approx F_1 = \Delta F$ is more probable than $F_r \approx F_2$.
b: The 'weak' reflections have $F_p = 0$: F_1 and F_2 have equal probability.

7. Origin fixation (centrosymmetric)

As a first and simple example we consider the centrosymmetric superstructure problem NORA.

F_r is either F_1 or F_2. The 'weak' reflections (i.e. the superstructure reflections) have no (or negligible) contribution from the partial structure (see Fig. 7b). The ten strongest reflections (i.e. having the largest $|E_1|$ values) will be given letter symbols X = A, B, C ... and many relations between the symbols are obtained.

The symbolic relations are easily solved by the fast symbol analysis procedure (Beurskens and Prick, 1981) using the symmetric mode:

$$X = 0 \text{ or } \pi, \ X + X = 0, \ -X = +X.$$

The most probable solution (ie. the assignment of numerical values to the letters A, B, C ... which gives the highest consistency with respect to (13) or (18)) is assumed to be correct. Substitution of the value in all symbolic phases fixes the origin and solves the phase problem. Several cycles of tangent refinement (13) are used before calling the Fourier and peak search routine.

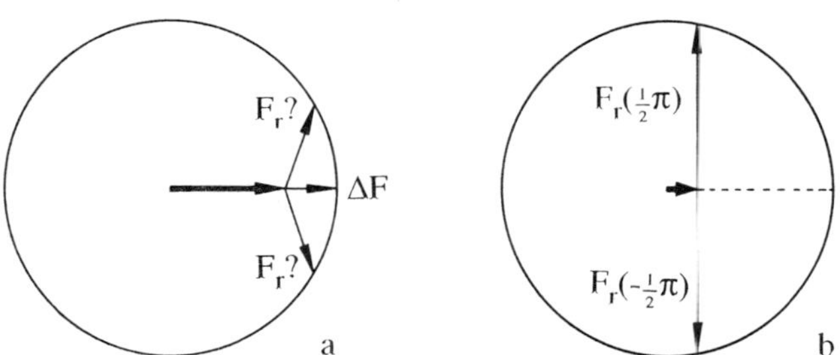

Fig. 8. a: Large F_p: F_r is probably close to ΔF.
b: Small F_p: φ_r may be $+\frac{1}{2}\pi$ or $-\frac{1}{2}\pi$? Enantiomorph sensitive?: input to selection procedure.

8. Enantiomorph fixation

The structure of MONOS (see section 5) may serve as an example of an enantiomorph problem. The S position was found from a Patterson synthesis as 0.0, 0.09, 0.14 . The corresponding S atoms form a centrosymmetric arrangment around $\frac{1}{4}\frac{1}{4}0$. By arbitrarily shifting the S position along x over a short distance, the enantiomorph problem may be solved, as was shown in the previous example (section 5). We now assume that this has not been done. The origin is shifted over a vector $\frac{1}{4}\frac{1}{4}0$, and the symmetry relations are changed accordingly. All calculated phases then are either 0 or π. This constitutes the enantiomorph problem. We need to find reflections of which the phases deviate substantially from the value 0 or π.

Enantiomorph sensitive reflections are selected by inspection of the tangent formula summation (13). Reflections which are found most inconsistently as 0 and π, are expected to be either $\frac{1}{2}\pi$ or $-\frac{1}{2}\pi$. Ten of these reflections are selected and given letter symbols X = A, B, C ... . The symbolic relations are solved by the fast symbol analysis procedure, using the anti-symmetric mode:

$$X = \tfrac{1}{2}\pi \text{ or } -\tfrac{1}{2}\pi, \ X + X = \pi, \ -X = X + \pi.$$

The most probable solution is accepted. The second solution, with the same probability (!), refers to the enantiomeric structure.

9. Super structures (general)

If neither origin nor enantiomorph is fixed, the same phase correlation procedure is used. The letter symbols X = A, B, C ... are assumed to represent phases $\pm\frac{1}{4}\pi, \pm\frac{1}{4}\pi$: It is easily seen (if we represent phases X by vectors $\exp(i\,X)$) that the symbolic relations have to be satisfied for the real part (projection of the vectors on the real axis) as well as for the imaginary part (projection on the imaginary axis). The best solution is thus found by taking the vector sum of the symmetric mode and the anti-symmetric mode, (eg. if A = 0 and A = $\frac{1}{2}\pi$ for the two modes, then A = $\frac{1}{4}\pi$).

10. The use of the weak reflections

In DIRDIF weak reflections are generally used if $|F_p| > |F_o|$ and $|E_1| > 0.9$, because these reflections have very reliable phases ($\varphi_1 = \varphi_p + \pi$). In addition, the weak reflections are used in figures of merit (FOMs) in case the known part of the structure is rather small (Smykalla and Beurskens, 1989).

If the relative scattering power of the input fragment (2) is sufficiently large, DIRDIF expands the partial structure automatically with the tangent refinement procedure, but for small fragments the tangent refinement may fail, because the input phases are less reliable (Parthasarathi, Beurskens and Bruins Slot, 1983). Therefore the symbolic addition method is used to generate a multiple of phase sets for cases where the scattering fraction is less than, say, 0.15.

The symbolic phasing procedure is initiated as before, but the solution with the highest consistency from the fast symbol analysis (Beurskens and Prick, 1981) is not necessarily the best one as described in section 7.

We take at most 25 different solutions with highest consistencies, and select the 100 weakest reflections ($|E_o| \approx 0$ and $|E_1| \approx 0$). To select the most probable phase set two types of FOMs are used which use triplets and negative quartets: a Ψ_0-FOM (Cochran and Douglas, 1955) and two negative-quartet FOMs (De Titta et al., 1975; Schenk, 1974, 1984). A combined figure of merit is calculated based on the consistency of the original symbol relations, the consistency after substition of numerical values of symbols, Ψ_0-FOM , and the two negative-quartet FOMs. This gives the best phase set, which is then input to the tangent formula for phase refinement and phase expansion.

11. How to get started

Apart from heavy-atom Patterson interpretations, or getting a chemically reasonable fragment from a direct methods failure, we can get started by applying Patterson techniques. The program ORIENT (Beurskens, Beurskens et al., 1987) uses vector search methods for finding the orientation of a known molecular fragment, and the program TRACOR (Beurskens et al., 1987) uses reciprocal space correlation functions for finding the position of the fragment. These programs work well if about 10% or more of the independent part of the scattering matter is used as input. The resulting atomic positions constitute the partial structure, to be expanded by the direct method procedure described in this chapter.

12. Conclusions

A summary of various applications is given in Fig. 9.

Comparison with *a priori* direct methods:

- Heavy atoms or planar molecules are 'subtracted' from the data and do not hamper the use of direct methods.
- Special supersymmetry problems have been taken care of before entering the tangent refinement procedure.
- In contrast to multisolution methods, the phase refinement is performed on one phase set only.
- If DIRDIF is used for recycling of direct methods fragments, different reflections are used and there is no danger that the original direct methods failure is reproduced.
- Many reflections (several hundreds) are used to initiate the refinement: 'weak links' do not exist.
- The input phases need not to be accurate: the large numbers of terms in eq.(8) compensates for the individual inaccuracy, and an overall improvement of phases results.
- The computing algorithm differs from that used in *a priori* methods; in each cycle all phase changes lead to changes in |E| values.

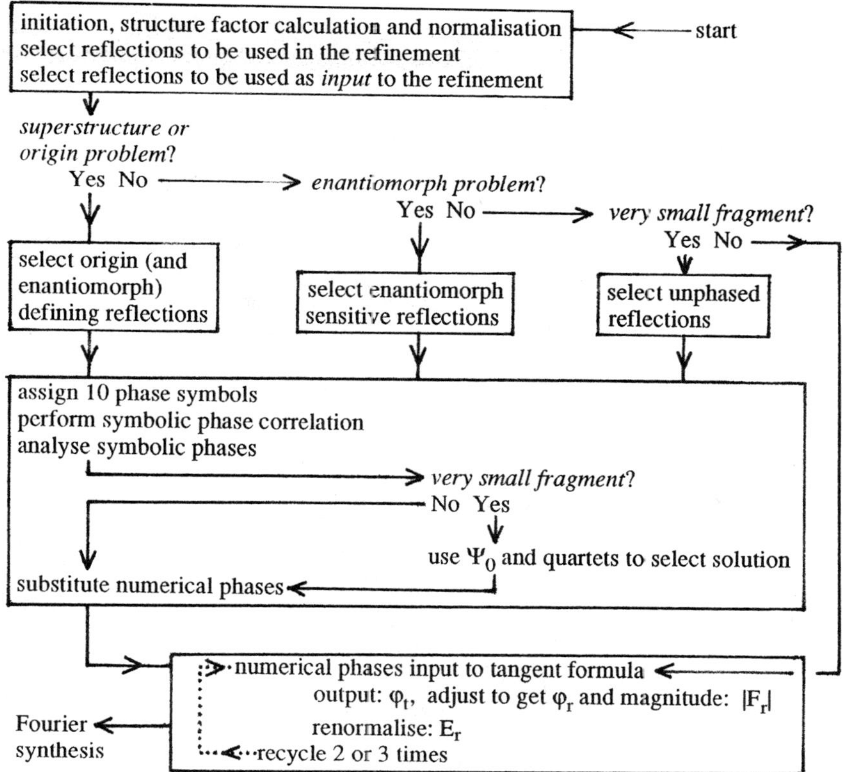

Fig. 9. Flow diagram of the application of various direct methods techniques to difference structure factors.

References

Beurskens, P.T., 1963, Technical report on *Sign Correlation by the Sayre Equation*. The Crystallography Laboratory, Univ. of Pittsburgh, Pennsylvania.

Beurskens, P.T., 1964, A computer procedure for the systematic application of Sayre's equation for solving the phase problem for the analysis of centrosymmetric structures, Acta Cryst., 17:462.

Beurskens, P.T., 1965, Thesis, University of Utrecht, The Netherlands.

Beurskens, P.T., 1987, Comments on *From a partial to the complete crystal structure. II. The procedure and its applications*, by M. Camalli, C. Giacovazzo & R. Spagna, 1985, Acta Cryst., A43:283.

Beurskens, P.T., Beurskens, G., Strumpel, M., and Nordman, C.E., 1987, Automation of rotation functions in vector space, in *Patterson and Pattersons*, edited by Glusker, J.P., Patterson, B.K., and Rossi, M., Oxford University Press, New York.

Beurskens, P.T., Bosman, W.P., Doesburg, H.M., Gould, R.O., van den Hark, Th.E.M., Prick, P.A.J., Noordik, J.H., Beurskens, G. Parthasarathi, V., Bruins-Slot, H.J., Haltiwanger, R.C., Strumpel, M.K., Smits, J.M.M., García-Granda, S., Smykalla, C., Behm, H.J.J., Schäfer, G., and Admiraal, G., 1990, The DIRDIF90 program system, University of Nijmegen, in preparation.

Beurskens, P.T., Gould, R.O., Bruins Slot, H.J., and Bosman, W.P., 1987, Translation functions for the positioning of a well oriented molecular fragment, Z. Kristallogr., 179:127.

Beurskens, P.T., and Noordik, J.H., 1971, The application of direct methods to centrosymmetric structures containing heavy atoms, Acta Cryst., A27:187.

Beurskens, P.T., and Prick, P.A.J., 1981, A fast algorithm for determining phase values for symbols from weighted symbol relations in direct methods, Acta Cryst., A37:180.

Camalli, M., Giacovazzo, C., and Spagna, R., 1985, From a partial to the complete crystal structure. II. The procedure and its applications, Acta Cryst., A41:605.

Cochran, W., and Douglas, A.S., 1955, Proc. Roy. Soc., A227:486.

De Titta, G.T., Edmonds, J.W., Langs, D.A. and Hauptman, H., 1975, Use of negative quartet cosine invariants as a phasing figure of merit : NQEST, Acta Cryst., A31:472

Gould, R.O., van den Hark, Th.E.M., and Beurskens, P.T., 1975, The application of direct methods to centrosymmetric structures containing heavy atoms II, Acta Cryst., A31:813.

Noordik, J.H., Beurskens, P.T., Ottenheijm, H.C.J., Herscheid, J.D.M., and Tijhuis, M.W., 1978, 9,9a-dihydro-1,2,9,9-tetramethyl-2,9a-epitho-3,10-diketopiperazino [1,2-a]indole, $C_{15}H_{16}N_2O_2S$, Absolute Configuration, Cryst. Struct. Comm., 7:669.

Karle, I.L., Hauptman, H., Karle, J., and Wing, A.B., 1957, Crystal and molecular structure of p,p'-dimethoxybenzophenone by the direct probability method, Acta Cryst., 10:481.

Karle, I.L., and Karle, J., 1963. An application of a new phase determination procedure to the structure of cyclo(hexaglycyl) hemihydrate, Acta Cryst., 16:969.

Parthasarathi, V., Beurskens, P.T., and Bruins Slot H.J., 1983, A note on practical aspects of the application of *DIRDIF*, a procedure of structure elucidation when a part of the structure is known, Acta Cryst., A39:860.

Sax, M., Beurskens, P.T., and Chu, S., 1965, The crystal structure of o-nitroperoxybenzoic acid, Acta. Cryst., 18:252.

Schenk, H., 1974, On the use of negative quartets, Acta Cryst., A30:477.

Schenk, H., 1984, Methods and applications in crystallographic computing, by S.R. Hall and T. Ashida, Summer school on crystallographic computing 1983, Clarendon Press, Oxford

Sim, G.A., 1960, A note on the heavy-atom method, Acta. Cryst., 13:511.

Smykalla, C. and Beurskens, P.T., 1989, The use of weak reflections for the expansion of small known fragments by the symbolic addition method applied to difference structure factors (DIRDIF), Z. Kristallogr., 186:276.

Van den Hark, Th.E.M., Prick, P. and Beurskens, P.T., 1976, The application of direct methods to non-centrosymmetric structures containing heavy atoms, Acta Cryst., A32:816.

Mariken van Nieumeghen

PHASE ANNEALING

George M. Sheldrick

Institut für Anorganische Chemie der Universität
Tammannstraße 4, D-3400 Göttingen, Federal Republic of Germany

1. INTRODUCTION

Simulated annealing (Kirkpatrick, Gelatt & Vecchi, 1983) is a very generally applicable method of combinatorial optimization which has been successful in such diverse applications as the design of computer chips, protein structure refinement, spin glasses and the (inevitable) travelling salesman problem. By means of a thermodynamic analogy it provides an efficient search for a global minimum in the presence of many local minima for a system with many degrees of freedom; it is frequently used in combination with multiple random starting positions.

Before discussing the application of simulated annealing to direct methods, which we shall call *phase annealing*, a brief historical digression is necessary.

2. THE MODIFIED TANGENT FORMULA

The direct methods program *SHELXS-86* employs multiple random starting phase sets and a modified tangent formula incorporating negative quartets (NQRs) as well as triplet phase relations (TPRs):

$$\text{new } \phi_h = \text{phase of } [\, \underset{\sim}{\alpha} - \underset{\sim}{\eta} \,]$$

Direct Methods of Solving Crystal Stuctures
Edited by H. Schenk, Plenum Press, New York, 1991

where $\alpha = 2 |E_h| E_k E_{h-k} / N^{\frac{1}{2}}$ and $\eta = g |E_h| E_k E_l E_{h-k-l} / N$

where N is the number of atoms (assumed equal here for simplicity) per (primitive) unit cell. Note that α and η are complex numbers in this equation and in the definition of NQUAL (below). g is a positive constant which takes the E-values for the cross-terms into account; it is often set to a value larger than the theoretical value to compensate for the smaller absolute value of η compared with α. It is important to use only those quartets for which *all three* cross-terms have been measured and found to be weak. The programs *MITHRIL* and *SIR* provide similar options; the *SAYTAN* approach can be reformulated in the above form, but in general only *one* crossterm is known to be weak for the quartet contributions. *Phase annealing* could equally well be incorporated into these methods.

The introduction of NQRs is a major advance, but it is still possible for iterative phase refinement to produce an over-consistent solution, with loss of enantiomorph resolution or a single dominant peak in a subsequent Fourier synthesis. In *SHELXS-86*, this degradation of the tangent refinement is handled by special action for reflections which have α greater than its expected value $\langle\alpha\rangle$ for a correct solution. A correction of $\delta\phi$ is applied to the pure tangent formula phase ϕ_h (calculated without NQRs), by projecting α onto its expected value, so that:

$$\cos(\delta\phi) = (\langle\alpha^2\rangle)^{\frac{1}{2}} / |\alpha|$$

and the sign of $\delta\phi$ is determined by choosing the value so that the resulting ϕ_h is most consistent with the NQR sum (if no NQRs contribute to a particular reflection, the sign is set randomly). Thus even if there are not many NQRs and they are not very reliable, they can make a useful contribution. This technique is very effective at driving the solution to a small R_α and a negative NQUAL (see below), but can only be applied to non-centrosymmetric structures.

It seems that optimum results are obtained when a limiting resolution sphere of about 1 Å is employed. This increases the

number of TPRs for a fixed number of reflections, and even more
markedly increases the number and reliability of the NQRs, even
though it means that some reflections with lower E-values are
refined.

3. IDENTIFYING THE CORRECT SOLUTION

The consistency of the TPR contributions can be judged by the
figure of merit R_α (Roberts, Pettersen, Sheldrick, Isaacs & Kennard,
1973):

$$R_\alpha = \sum_h w(\alpha - \langle\alpha\rangle)^2 \; / \; \sum_h w\alpha^2$$

where w is a suitable weight, e.g. $1/(\alpha+5)$. [in *SHELXS-86* and
previous versions the denominator was $\sum w\langle\alpha\rangle^2$; the new version is
more selective provided that it is used in conjunction with the
α-projection technique (see above) for non-centrosymmetric
structures].

Rather than the usual criterion NQEST which represents the
weighted mean cosine of the strongest NQRs, we prefer to use the
correlation between $\underset{\sim}{\alpha}$ and $\underset{\sim}{\eta}$:

$$\text{NQUAL} = \sum_h |\underset{\sim}{\alpha} \cdot \underset{\sim}{\eta}| \; / \; \sum_h |\alpha| \cdot |\eta|$$

We find that NQUAL is both more negative and also more sensitive
than NQEST for large structures, for which the larger number of NQRs
involved in the summations for a given h can to a large extent
compensate for the $1/N$ term in the probability expression. The two
criteria can be combined to give:

$$\text{CFOM} = R_\alpha \quad \text{if } Q > \text{NQUAL, or}$$

$$\text{CFOM} = R_\alpha + (\text{NQUAL} - Q)^2 \quad \text{otherwise.}$$

where Q is usually set to a value about 0.1 more negative than the
expected NQUAL. Finally the phase set with the best (i.e. *smallest*)
CFOM is used to calculate a Fourier synthesis with coefficients E
and the final direct methods phases (E-map) which is searched to
find potential atoms. An iterative procedure (Sheldrick, 1982), in
which potential atoms are eliminated to reduce an R-index
(calculated for E-values assuming point atoms) and the remaining

atoms are used to phase the next E-map, provides a further figure of merit (R_E), and facilitates the chemical interpretation of the solution.

4. PROBLEM STRUCTURES

The approach outlined above has been extensively tested by users of the program *SHELXS-86*, with several successes - often in symmorphic space groups - which had previously proved resistant to direct methods. There seemed to be three general areas where problems remained:

1. Pseudosymmetry problems, usually characterized by a number of non-equivalent solutions with approximately equally good figures of merit. Such problems are particularly prevalent in the space group $P\bar{1}$, for fused aromatic systems and for structures with heavy atoms on (almost) special positions. Lowering the E-threshold and thereby increasing the number of reflections refined often improves the discrimination of the figures of merit, but the simplest approach is to calculate E-maps for all the good solutions, not just the best.

2. Resolution problems. Experience with a large number of structures has led us to formulate the empirical rule that if less than half the number of theoretically measurable reflections in the range 1.1 to 1.2 Å are "observed" [i.e. have $F > 4\ \sigma(F)$], it is very unlikely that the structure can be solved by direct methods. This critical ratio may be reduced somewhat for centrosymmetric structures, and for structures containing heavier atoms. This rule simply reflects the assumption of resolved atoms, which is often invoked in direct methods. It may still be possible to solve such structures by molecular replacement methods if a sufficiently large and accurate search fragment is available. Alternatively the data collection can be repeated more carefully with a larger crystal at lower temperature.

3. Large structures. The problem may be likened to that of finding a needle in a haystack. Even when using random starting phases rather than generating explicit phase permutations, the computer time per correct solution rises steeply with the number of atoms in the unit-cell. The rest of this talk is concerned with this class of structure.

A technique used in *SHELXS-86* to reduce the computer time required for large structures involved dividing the phase refinement into two stages. The first stage involved about half the number of reflections used in the second, but much less than half the number of phase relations (and hence time) since the number of TPRs and NQRs rises rapidly with the number of reflections used. On the basis of the figures of merit at the end of the first stage, only the best (say) 10% phase sets were refined further. The phase relations for the first stage could be stored in the main computer memory and did not need to be read in from the disk. With a smaller number of reflections, the reduced multimodality of phase space should increase the chances of finding the solution, but at the cost of increasing the mean phase error. For straightforward test structures this *filter* procedure did indeed reduce the average computer time per correct solution appreciably. For large structures (including some of those subsequently solved by phase annealing) *it often resulted in repetition of the same wrong solution.* One explanation is that the correct solution may have figures of merit at the end of the first stage that do not place it amongst the best 10%. A further explanation (if we assume that phase refinement corresponds to the minimization of some target function) is that false minima are usually broader but shallower than true minima, because the phase relations are unlikely to have their individual minima so closely in phase for an accidental solution. Assuming that the minimization procedure cannot escape from a minimum, then the chance of random starting phases falling in the catchment area of a broad but shallow minimum will be greater than of a narrow but deep minimum. *Simulated annealing* should be effective for such problems.

5. PHASE ANNEALING

The *simulated annealing* procedure (Kirkpatrick, Gelatt & Vecchi, 1983), may be described by analogy with statistical thermodynamics. It is necessary to define an algorithm by which the system can be brought into a state of *thermodynamic equilibrium* described by a temperature T. The annealing is then achieved by lowering the temperature slowly. The energy of the system is given by the sum of the potential energy (the function to be minimized) and the kinetic energy. The form taken by the kinetic energy depends on the system being considered, but involves some random behaviour of the individual contributors. The presence of the kinetic energy makes it more likely that the system will escape from broad shallow minima than from narrow deep minima, so slow cooling increases the chance of finishing in a global rather than a local minimum, consistent with the annealing analogy.

Simulated annealing has recently (Kuriyan, Brünger, Karplus & Hendrickson, 1989) established itself as an effective method for the intermediate stages of refinement of protein structures using a potential energy function which includes the X-ray data. The kinetic energy is provided naturally by a molecular dynamics simulation of the motion of the individual atoms, and enables side-chain conformations to escape from false local minima. For systems which cannot be described in terms of continuous motion, the Metropolis algorithm (Metropolis, Rosenbluth, Rosenbluth, Teller & Teller, 1953) enables the system to approach thermodynamic equilibrium by imposing a Boltzmann distribution. Each time a parameter is changed, the energy E of the system is calculated. If the energy decreases, the jump is always made. If ΔE is positive, the probability P of the new state relative to the old is calculated from the Boltzmann formula $\exp(-\Delta E/kT)$, where k is a constant. If P is greater than a random number in the range 0 to 1, the jump is made, but if P is less the original state is retained.

To see how simulated annealing can be applied to direct methods (we shall name this *phase annealing*) it is necessary to reformulate Cochran & Woolfson's (1955) formula for centrosymmetric structures:

$$P_+ = \tfrac{1}{2} + \tfrac{1}{2} \tanh(\alpha/2)$$

where P_+ is the probability that E_h takes the sign of α. For positive α, P_+ measures the probability of the lower energy state, and $P_- = 1-P_+$ is the probability of the higher state. Thus the ratio of the two probabilities is given by:

$$P_-/P_+ = \exp(-\alpha)$$

so all that is needed is to divide α by kT, and we hae the required Boltzmann distribution. In a slight variation of the Metropolis algorithm we take the higher energy P_- state only if $(P_-/2P_+)$ (which must be less than ½) is greater than a random number R in the range 0-1, otherwise we choose the lower energy state (i.e. the phase given by the modified tangent formula). By incorporating this simple modification into a phase refinement program, we can set up a thermodynamic equilibrium in which α is associated with the "potential energy" of the system.

In the classical applications of the Metropolis algorithm, a *random* change is made; for example a spin is "flipped" (in the spin glass problem) or two paths are interchanged (in the travelling salesman problem and in computer chip optimization). The probability of this change is then estimated via the Boltzmann formula and tested against a random number to see whether the change should be made or not. Often in such applications a direct refinement (i.e. $T = 0$ in the Boltzmann formula) is known to be ineffective because of the large number of local minima; practical tests have shown that for such applications the simulated annealing algorithm is computationally much more efficent than this $T = 0$ "steepest descent" method combined with many random starting configurations. The direct methods problem (for centrosymmetric structures) is conceptually very similar, but with a comparatively small number of local minima - otherwise multisolution tangent refinement would not be so successful - so the *phase annealing* approach appears promising. On the other hand if we are designing a computer chip, a local minimum which is almost as good as a global minimum is an entirely acceptable solution; for direct methods this may not be good enough, and indeed the "correct" solution does not necessarily correspond to the global minimum of any function which can be calculated quickly! Thus multiple phase starting sets are still a *sine qua non*.

The above centrosymmetric formula can also be applied to restricted phases in non-centrosymmetric space groups. After testing a variety of formulae we have chosen to calculate general phases in non-centrosymmetric structures by adding $\Delta\phi$ to the phase obtained by the modified tangent formula, where:

$$\cos(\Delta\phi) = \{ 4\alpha/kT + \ln(R) \} / \{ 4\alpha/kT - \ln(R) \}$$

where the sign of $\Delta\phi$ may be assigned at random and R is a random number between 0 and 1. Thus $\cos(\Delta\phi)$ is in the range -1 to +1 and approaches +1 in the limit of large α/kT, as required. It should be emphasised that this formula is purely empirical, and that the numerical factor of 4 has been chosen to give similar mean absolute phase shifts for general and restricted phases at the start of the phase determination (random phases). In the most recent version of this algorithm we have assigned the sign of $\Delta\phi$ which leads to the best agreement with the NQRs involving the reflection in question, instead of using the NQRs directly in the modified tangent formula, and the $\underset{\sim}{\alpha}$-projection method (see section 2) is used to determine the magnitude of $\Delta\phi$ if it results in a larger absolute value.

If this phase annealing is performed with zero kT, it will correspond to a phase refinement using the modified tangent formula. If kT is infinite, the phases will remain random (*unconstrained maximum entropy*). If the temperature is slowly lowered, the well defined phases (high α) will be subject to smaller fluctuations than those with low α. Thus phases linked by strong phase relations will tend to become established earlier in the phase determination procedure, whilst the remaining phases are still free to explore phase space. If a "good" (but not necessarily correct) solution is found it will remain more stable than a phase set with low mean α.

Since the appropriate value of kT depends on the structure, it is more convenient to discuss the performance of the phase annealing in terms of B, the estimated initial Boltzmann ratio for random starting phases. $1/kT$ is estimated as $-\ln(B)$ divided by the mean (over all reflections employed) of $(\langle\alpha^2\rangle)^{\frac{1}{2}}$. In a first test of *phase annealing*, CFOM values were compared for various B values after 50 and 250 phase annealing cycles at constant kT for the Loganin test structure. For $B=0$ (which corresponds to $T=0$) the correct solutions are clearly separated from the rest, but the number of correct solutions only increases a little between 50 and 250 cycles. That it increases at all suggests that it is an oversimplification to regard tangent refinement as a pure minimization procedure. As B increases all the CFOM values increase, and the percentage of correct solutions grows with the number of cycles performed. Finally for $B = 0.3$ or higher it is no longer possible to distinguish the correct solutions from the rest because all CFOM values are high (high kinetic energy). In addition, the fluctuations of the distribution from one cycle to the next increase appreciably with B. This is indeed the behaviour we would expect from the thermodynamic analogy.

6. COMPUTATIONAL ASPECTS

The reliability of individual phase relations decreases with increasing N, so we need to use more phase relations per reflection in the final phase refinement to obtain the same phase accuracy and quality of figures of merit for larger structures. Since the phase annealing stage will be rate determining and we are deliberately introducing noise into it, it makes sense to perform it using only the 50 - 60% of reflections with the largest $<\alpha>$ values. In contrast to the filter procedure used in *SHELXS-86*, *all* phase sets are then finally refined for 3 cycles (centrosymmetric) or 4 (non-centrosymmetric) with the full number of phases and relations. The selectivity of R_α can be improved further by including summations for reflections additional to those used for the phase refinement.

The modified tangent formula and phase annealing procedure are particularly suitable for vector processing on modern computers with parallel architectures. All that is necessary is to process a suitable number M (we usually use 128) of phase sets in parallel. All rate determining stages can be formulated as vectorizable loops over the M phase sets. Even on scalar machines, this brings an appreciable bonus, because various overheads such as reading phase relations from the disk occur only once per M phase sets.

7. PRACTICAL TESTS

Table 1 illustrates some practical tests using seven structures selected from a test bank of direct methods structures that is available from the author. The first three were chosen to illustrate the effect of varying the size of the structure for a given (well behaved) space group, the remaining four to illustrate

symmorphic and/or polar space groups. In all seven cases it was
known from extensive tests that the correct solution could be
recognised on the basis of the figures of merit alone. Chance can
play a considerable role in direct methods, so to obtain a
statistically significant number of solutions about 10000 (or more)
phase sets were refined for each of the entries. For each entry in
Table 1, 25 cycles of phase annealing were performed starting with
the given Boltzmann ratio B (for random phases) and multiplying the
temperature by 0.95 after each cycle. For these tests the NQRs were
included in the summations throughout, so that $B = 0$ corresponds
exactly to the modified tangent formula.

For the test structures in Table 1 a "correct" solution is one
which (a) had a CFOM value which clearly separated it from the
"incorrect" solutions, and (b) after the usual Fourier recycling
gave a peaklist in which all atoms were higher that any spurious
peaks (criterion (b) was relaxed a little for PEP1). Almost all one-
phase seminvariants had identical signs for different "correct"
solutions of a given structure.

Table 1

Structure code	LOG	SUOA	PEP1	NEWQB	BHAT	MBH2	HOPS
space group	$P2_12_12_1$	$P2_12_12_1$	$P2_12_12_1$	$P\bar{1}$	Pc	$P1$	$R3(hex)$
N/cell	108	188	340	124	84	54	243
$n(E)$ for p.a.	139	190	257	254	184	263	156
$n(E)$ full	219	320	455	448	304	467	253
final R_α	0.080	0.095	0.098	0.078	0.070	0.036	0.052
final NQUAL	-0.70	-0.36	-0.62	-0.84	-0.74	-0.89	-0.80
Correct solutions per 10000 at B=0	206	6	1	1	26	469	56
Correct solutions per 10000 at B=3	644	50	25	20	97	825	184

For the P1 test structure (MBH2) the increase in the number of solutions obtained with phase annealing was only sufficient to compensate for the extra computer time required (compared with the $B=0$ run). For the other six structures the phase annealing produced significant dividends, and for the largest structure studied (PEP1) increased the chance of finding a correct solution by more than an order of magnitude.

8. CONCLUDING REMARKS

In addition to performing at least as well as our previous programs on each of the structures in the full test data bank, the phase annealing method has already been successful in solving eight large structures with up to 180 atoms in the asymmetric unit which had resisted all previous efforts using a variety of programs including *SHELXS-86*.

Phase annealing is also appropriate for the expansion of a partial structure which is too small for successful elucidation by tangent expansion followed by *E*-Map recycling. Preliminary tests indicate that, instead of using a small number of fixed partial structure phases plus random starting values for the rest, it is more effective for all attempts to start from the *same* partial structure phases for all reflections. The inherent randomness of the phase annealing approach then ensures that each phase set follows a different refinement path, leading to effective exploration of phase space in the vicinity of the partial structure solution. The cooling schedule should be just "warm" enough to ensure that not too many solutions are the same.

A more detailed account of the *SHELXS-86* philosophy and its extension (to *SHELXS-90*) by *phase annealing* should appear in Acta Crystallographica at about the time of this School.

REFERENCES

Cochran, W. & Woolfson, M.M. (1955). Acta Cryst. **8**, 1-12.

Kirkpatrick, S., Gelatt C.D. & Vecchi, M.P. (1983). Science 220, 671-680.

Kuriyan, J., Brünger, A.T., Karplus, M. & Hendrickson, W.A. (1989). Acta Cryst. A45, 396-409.

Metropolis, N., Rosenbluth, A.W., Rosenbluth, M.N., Teller, A.H. & Teller, E. (1953). J. Chem. Phys. 21, 1087-1092.

Roberts, P.J., Pettersen, R.C., Sheldrick, G.M., Isaacs, N.W. & Kennard, O. (1973). J. Chem. Soc. Perkin II, 1978-1984.

Sheldrick, G.M. (1982). In *Computational Crystallography*, edited by D. Sayre, pp. 506-514. Oxford: Clarendon Press.

SOLVING DIFFICULT STRUCTURES INCLUDING THOSE FROM POWDER AND
ELECTRON DIFFRACTION DATA

C.J. Gilmore

Department of Chemistry, University of Glasgow
Glasgow G12 8QQ, Scotland

1.0 INTRODUCTION

This chapter outlines some of the options available to you if routine
direct methods have failed to solve a structure. I am going to include data
from powder and electron diffraction, since the *ab-initio* solution of crystal
structures from these experiments is becoming a topic of increasing
importance. The possibilities outlined in the following sections are not quoted
in any order of preference, except perhaps for normalisation where the most
scope exists for easy ways to alter the phasing path taken by the direct
methods program you are using.

The first necessity is to have available *several* direct methods
packages and not just one. All programs have their strengths and weaknesses,
and one structure may yield readily to one program whilst being quite difficult
for another. It is also important to *read the output* generated by these
programs. This may sound facetious, but such is the level of automation in
direct methods, there is a temptation to try as many options as quickly as
possible without scanning the output carefully.

Why are some structures difficult? There are some obvious reasons
and these are usually known *before* data collection begins:

(i) The structure is large and/or there are several molecules in the asymmetric
unit. The latter are often particularly difficult; the presence of multiple
molecules in the asymmetric unit gives rise to phase relationships between
certain phase angles which can distort the phasing process.
(ii) The crystal diffracts weakly and there is a paucity of high angle
reflections – in particular those which violate Sheldrick's Rule:

"If less than 50% of the theoretically observable reflections in the resolution
range 1.1-1.2 A are observed $(F > 4.0\sigma(F))$ then the structure will be difficult to
solve by conventional direct methods."

(iii) The data are sparse – this applies to powder and electron diffraction data.
In the former case peak overlaps effectively reduce resolution, and in the
latter the
use of tilted slices and the methods by which the sample is prepared limits
the accessibility of reciprocal space.

Direct Methods of Solving Crystal Stuctures
Edited by H. Schenk, Plenum Press, New York, 1991

(iv) Pseudosymmetry may be present giving rise to subsets of reflections which are systematically weak.
(v) Structures with a high degree of internal regularity.
(v) The sheer peversity of nature.

Many structures give problems in most of these categories – macromolecules for example. In general, however, these problems are known *before* data collection, and it is important to use this information from the outset; in other words data collection becomes an important aspect of solving the structure.

2.0 DATA COLLECTION

The rules are simple to write, but often difficult in practice:

(i) Spend as much time as possible to get good intensity measurements of the *weak* reflections. They intervene in phasing in three ways: as data for normalisation, in the ψ_0 and NQEST figures of merit, and in the estimation of negative quartets for active use of four-phase invariants. If you have pseudosymmetry present collect the systematically weak reflections separately from the strong and scale the two data sets. Never use options that pre-scan peaks, and to use this information as a filter by which the weak reflections are not measured. The Bayesian approach pioneered by French and Wilson (French & Wilson, 1978) provides an optimal approach to data processing.
(ii) Use low temperature if possible. It lowers the background and hence improves signal to noise ratios and so can extend the data resoution.
(iii) Collect equivalent reflections
(iv) In powder diffraction the highest possible resolution is of paramount importance; obviously synchrotron data is optimal.

3.0 NORMALISATION

(i) Check the data. If there are many duplicates or systematic absences list them all. Then investigate this list with suspicion-are you sure about the space group? Have you collected at least the unique data? How good is the resolution of the data? It is important no to put duplicate reflections into a direct methods program which does not test for them.

(ii) Read the normalisation output very carefully. Are the statistics sensible? Do the large E-magnitudes form a readily identifiable subset where some parity groups are missing? If some of the E-magnitudes are very large (say >3.5-4.0), it is often useful to use a to reduce or remove them. The variation of $E^2 - 1.0$ as a function of $\sin\vartheta/\lambda$ is a good guide to the applicability of the calculated temperature factor. If there is a fall-off in this average as $\sin\vartheta/\lambda$ increases, use the relevant command to input a larger value of B than that calculated by the normalisation module. If some or all the stereochemistry is known, then input this as a randomly oriented, randomly positioned fragment if your program permits. This is especially important with molecules containing planar or nearly planar fused rings.

(iii) If you have generated the best set of E-magnitudes that you can, and the structure still will not solve, then a systematic distortion of the E's can be successful. Changes in the E- magnitudes cause changes in relative weights of the invariants, and these in turn give rise to drastic modifications of the convergence map, and the subsequent phasing path. There are several ways of doing this:

(a) Modify the unit cell contents-doubling the contents is the best starting point.
(b) Use artificially raised or lowered temperature factors . Often only a small change in B gives rise to drastic changes in the E's. This is especially recommended for situations where the resolution is less than the Cu sphere.
(c) Insert a molecular fragment that does not correspond to any group expected in the molecule.

(iv) Downweight or remove the high angle E-magnitudes. These are often the E's with the largest associated standard deviations. Triplets involving three such E's are especially prone to error.

(v) Try using another normalising program that has different facilities. For example the NORMAL module in the X-ray system permits the use of an overall anisotropic temperature factor. Nixon (1978), has described an alternative approach to normalisation via the Patterson function.

The problems of normalising sparse or restricted data are very acute since there is an implicit assumption in the process that all non-input data is zero. Under these circumstances it is usually better to define a temperature factor and let the program scale although this still introduces errors. A Bayesian approach to normalisation can be much more powerful here especially when prior knowledge is used. This is an option in the new version of MITHRIL (Gilmore and Brown, 1988)

4.0 INVARIANTS AND SEMINVARIANTS

There are two problems which can arise here. One concerns a paucity of suitable relationships which can be common in situations of low symmetry. The other concerns the accuracy of the invariants themselves.

(i) If there is a paucity of triplets:

(a) Use quartets as well actively in the phasing process. Try just the negative ones first, then add the positives if this is unsuccessful but remember the correlations which exist between triplets and positive quartets for which allowance must be made.
(b) Increase the number of reflections for which triplets are generated.
(c) Reduce the minimum $\varkappa$-value ($\varkappa = |E_{\underline{h}} E_{\underline{k}} E_{\underline{l}}|/\sqrt{N}$) for which triplets are used. This will, however, introduce a number of very unreliable relationships.

(ii) If there is a paucity of quartets:

(a) Increase the number of reflections for which quartets are generated. This is usually the best way.
(b) Invoke the third neighbourhood although this increases the required computer time.
(c) Ask for positive quartets as well. This can also be useful when triplets are scarce.
(d) Allow more missing second (and third) neighbours; in this case the missing E-magnitudes are given values of unity.

(iii) If there is still a paucity of good negative invariants, then it can be worth considering quintets but it is very time consuming and of dubious value.

(iv) Even if the space group is not symmorphic, quartets often have a very beneficial effect on a direct methods analysis, and can be recommended as an option to try early in the list of weapons in the armoury. Only the negative quartets should be tried first, since they are independent of the triplets. Even a few four-phase invariants can drastically alter the phasing path.

(v) The MDKS or similar formulae which attempt to estimate the triplet cosine, coupled with triplet weighting also has a drastic effect. However, the MDKS estimates themselves are very time consuming to calculate, and are very inaccurate. The P_{10} formula (Cascanaro, Giacovazzo, Camalli, Burla, Nunzi & Polidori 1984) is more reliable.

In the case of powder or electron diffraction, quartets are essential, and it may prove necessary to attempt the phasing of those reflections having $|E_{\underline{h}}| < 1.0$.

5.0 THE STARTING SET AND CONVERGENCE MAPPING

Convergence mapping lies at the very heart of the multisolution approach to direct methods. With the rise of random phasing procedures, it is less significant that hitherto, but it is still a very informative procedure, since it shows how individual reflections are linked *via* the invariants. It is important to examine the convergence map carefully in cases of difficult, even if random phasing is being used.

(i) Make sure that the starting set is a good one with all the starting set reflections used early in the phase determination. This can be checked by examining the bottom of the convergence map. If a starting set reflection is not used at all early on, then a better starting point can often be obtained by including at least one other reflection whose phase depends on that of the late starter. Introducing quartets or quintets may also have a similar effect.

(ii) If there are gaps near the bottom of the convergence map (i.e. reflections with a zero estimated α and no invariants contributing), or the map is very 'thin' with many phases determined by only one or two relationships, then the phasing often fails. This can be remedied by increasing the size of the starting set or introducing higher invariants, particularly quartets.

(iii) Be wary of the Σ_1 determined phases. If they play a major role in the early stages of phasing, it is often worthwhile excluding them. They can be given permuted phases if really necessary) If the MDKS or a similar checking equation is available, perform a Σ_1 triplet analysis. The special triplets should have estimated cosines close to unity.

(iv) If it still proves impossible to obtain a suitable convergence map without a massive amount of computer time, then several options are possible:

(a) Run MAGEX or symbolic addition. Apart from origin and enantiomorph definition, they can largely ignore the convergence map.
(b) Run YZARC. Try both least-squares and steepest descents-they give different results.
(c) Run a random tangent procedure instead of regular tangent refinement.

(v) Check to see if all the reflections at the bottom of the map have something in common e.g. they all have h even or k+l divisible by 3. If so, then make sure that the average value of $E^2 - 1.0$ is unity for such reflections. Renormalisation may be necessary. It may be possible to use editing facilities at normalisation time to juggle these magnitudes. Try introducing new reflections into the starting set which do not belong to these groups.

(vi) Altering the origin and enatiomorph is often unsuccessful, particularly if only small changes are made. The same relationships are still used in the early stages of phasing, but in a different form. For example, the triplet $\varphi_1 - \varphi_2 + \varphi_3$ may appear in one map generating φ_1 from φ_2 and φ_3. If the origin is partially re-defined by the user, this triplet may well appear again in a critical place but this time generating φ_2 from φ_1 and φ_3 . If this triplet is erroneous it will be erroneous however it is used. This said, juggling with the starting set can be successful on some occasions, and is worth a try.

(vii) Symbolic addition, even in a limited form, can give rise to possible relationships between phases, and these can be introduced into the convergence map if your program permits . The relationships linking two phases (the pair relationships) are the most valuable. The inclusion of only one or two with high associated $\varkappa$-value will drastically alter a convergence map. There is the added bonus that symbolic addition can give valuable insights into the causes of phasing difficulties. (Karle and Karle 1966). Do not use the convergence map for symbolic addition; get a list of triplets and work with this. The convergence procedure has too many weak relationships early in the phasing path. Symbolic addition can be very useful for heavy atom structures where it can prevent the overconsistency problem of the tangent formula.

6.0 PHASE EXPANSION AND REFINEMENT

The same formula is used for both phase extension and refinement. This is usually the tangent formula, but Woolfson and co-workers have pioneered the use of Sayre's equation in the SAYTAN package, and Sheldrick has developed a method based on simulated annealing which is a powerful optimisation technique now finding increasing use in a wide variety of such problems. There is also the maximum entropy method which is discussed in section 8.

The only way to monitor phase expansion and refinement in traditional packages is by inspecting the final figures of merit, so it is important to examine these closely in difficult cases. In particular do not just inspect the final combined figure of merit (CFOM), but look also at its individual contributors:

(a) ABSFOM is the least reliable. If the ABSFOM values all tend to be large then the refined phases are over-consistent. Using Hull-Irwin weights will often give better results.
(b) NQEST and ψ_0 are the most reliable figures of merit provided that the weak reflections have been accurately measured. Do not expect NQEST to be very negative, particularly if quartets are to be used actively in phase refinement. In these circumstances values around -0.1 are often satisfactory.
(c) Heavy atom cases often give extreme figures of merit. The correct solution may well be present even if the figures of merit seem unrealistic.
(d) If all the phase sets have similar figures of merit too few invariants may be present. See section 2 for a remedy.

In the case of pseudo-symmetry, the presence of heavy atoms or substantial planar moieties in the structure, use the Hull-Irwin scheme. Do not forget weighting schemes. If your system has more than one, try them all - if one does not work, the other may, even using the same starting set and convergence map.

Be careful of early stop options in which the program stops tangent refinement finds a solution with an obviously optimal set of figures of merit. It is often found that the set selected by the tangent refinement modules during refinement as being an obvious solution is not the correct one. Similarly, if you are using early figures of merit, and most sets are getting rejected, it is possible that the correct set is also being discarded. Turn off early figures of merit for all difficult structures.

Do not confine your attentions to the one or two phase sets with the highest CFOM's. It may be necessary to examine maps with quite low associated CFOM's. The CFOM's themselves are dependent on the relative weights of the individual figures of merit. Adjusting the weights to reflect your own intuition concerning the contributing figures of merit will often result in a drastic re-ranking of the solutions.

Sparse data sets will give figures of merit that are often worthless; it is necessary to examine all the solutions produced. An alternative in all these cases is to use a maximum entropy-likelihood method.

7.0 MAPS

Look at the resulting E-maps carefully. Remember that the interpretation assumes that you have well-resolved peaks, and this may not be the case. The routines which perform the chemical interpretation may be quite sophisticated, but they are never as good as a trained crystallographer. Do not, therefore, accept the given interpretations as the only possibilities. Some other points to note are:

(a) If the map contains one or two large peaks and no heavy atoms are expected, the phases are probably incorrect - but not always. Sometimes something can be salvaged. If heavy atoms are present direct methods will probably only produce these atoms.
(b) If the E-maps show pseudosymmetry switch to Hull-Irwin weights in tangent refinement or use simulated annealing or the Sayre approach.
(c) One or two missing peaks coupled with one or more spurious ones, can quickly make a map uninterpretable. Increasing the number of peaks can make parts of the map more readily interpretable at the expense of producing more noise peaks.

In the case of powder and electron diffraction data the contoured maps themselves must be examined. It is worthwhile taking a leaf from the protein crystallographer's book and transferring maps to a graphics device capable of running FRODO/TOM to examine them and to carry out model fitting exercises. This is of obvious applicability to small single crystal structures as well.

In very difficult cases be tenacious. If just a small portion of the expected structure is found, stick with it through all the possible recycling schemes. It may be correctly oriented but misplaced, so the use of the relevant group type (3 in MULTAN) useful, but this procedure is not infallible. It sometimes does not work even with correct information. In addition, the use of a rotation-translation function package will prove indispensible. The use of vector methods in SHELX has been discussed by Eggert (Eggert 1983). It is an approach that could be very relevant when dealing with sparse data sets. C.C. Wilson has discussed the limits of data quality in vector methods (Wilson 1989)

As an object lessson in tenacity and persistence see Karle, Karle, Mastropaolo,
Camerman and Camerman, 1983.

8.0 MAXIMUM ENTROPY

Bricogne and Gilmore have shown in two recent papers (Bricogne & Gilmore 1990; Gilmore, Bricogne & Bannister, 1990) how Bayesian methods can be used to solve crystal structures by using a maximum entropy prior combined with likelihood which serves, in part, as a figure of merit in a multisolution environment of great power. One feature of the method is stability at all ranges of data resolution and data sampling. The method has been used to solve $KAlP_2O_7$ from its powder diffraction pattern using *ca.* 60 non-overlapped reflections in routine calculations. This stability extends to other data sets, indeed the method is stable with just meridional reflections from fibre diffraction. It seems likely that the method offers the best approach to sparse data sets.

REFERENCES

Bricogne, G. and Gilmore, C.J. (1990) Acta Cryst. **A46**, 284-297.
Burla, M.C., Cascanaro, G., Giacovazzo, C., Nunzi, A. and Polidori, G. (1987) **A43,** 370-374.
Cascanaro, G., Giacovazzo, C., Camalli, M., Burla, M.C., Nunzi, A. and Polidori, G. (1984) Acta Cryst. **A40,** 278-283.
Eggert, E. (1983) Acta Cryst. **A39,** 936-940.
French, S. and Wilson, K. (1978) Acta Cryst. **A34**, 517-525.
Gilmore, C.J., Bricogne, G and Bannister, C. Acta Cryst. (1990) **A46** 297-308.
Karle, J. and Karle, I.L. (1966) Acta Cryst. **21,** 849-859. Karle, I.L., Karle, J., Mastropaolo, D., Camerman, A. and Camerman, N. (1983) Acta Cryst. **B39,** 625-637.
Nixon, P.E., (1978) Acta Cryst. **A34,** 450- 453.
Wilson, C.C. (1989) Acta Cryst. **A45,** 833-839.

DIRECT METHODS FOR MACROMOLECULAR CRYSTALLOGRAPHY

Suzanne Fortier

Department of Chemistry
Queen's University, Kingston
Canada K7L 3N6

INTRODUCTION

The term "direct methods", as it is used in a small molecule
context, refers to crystal structure determination methods that take
advantage of relationships among the structure factors to extract missing
phase information from the observed intensities. An initial model of the
structure can thus be calculated with little or no *a priori* chemical
information. While in general the chemical content of the unit cell is
known, even this information is not essential for the successful
application of direct methods. Indeed there are numerous examples of
crystal structures solved with unknown or presumably known, but
erroneous, chemical formulae.

Attempts to use direct methods phasing techniques, in their
traditional format, were made early on in protein crystallography. For
example, in 1952 Kendrew applied sign relationships to projection data
from myoglobin, an experiment repeated in 1954 by Perutz and Scatturin on
horse methaemoglobin (Perutz and Scatturin, 1954). From these early
tests, as well as some subsequents ones, it was concluded, however, that
traditional direct methods had only limited applicability to *ab initio*
determination of macromolecular structures.

Several phasing methods have been developed for macromolecular
crystallography, the most commonly used being isomorphous replacement,
anomalous scattering, molecular replacement and density modification. In
the course of a protein structure determination project, it is very
common to use a combination of phasing techniques. Whereas small

Direct Methods of Solving Crystal Stuctures
Edited by H. Schenk, Plenum Press, New York, 1991

molecule direct methods phasing procedures are very often a single-pass solution process, macromolecular phasing procedures are usually iterative and rely on the exploitation of several sources of phase information. In the direct methods, it is the common feature of crystal structures, the random atomic distribution in their asymmetric unit, that is exploited. This allows the solution of the phase problem to be formulated in a general and universal way. In contrast, protein phasing tools exploit the characteristic features of the problem at hand - the heavy atom/anomalous scatterer substructure, the presence of non-crystallographic symmetry, the definition of the molecular envelope, structural information from homologous structures, etc. With such fundamental differences, it is questionable whether the term "direct methods" could ever be used in a strict sense in a macromolecular context. However, while the standard machinery of direct methods, as it is applied so successfully to small molecules, cannot be applied directly to macromolecular problems, some of the infrastructure or shell of the direct methods, with its broad and well developed theoretical basis, can be usefully exported to such problems. In this paper, we will discuss precisely those phasing tools that have evolved from the techniques of direct methods and have been adapted to address structure determination problems specific to the macromolecular context.

There have been considerable efforts to make use of direct methods in four areas of macromolecular phasing: 1) the determination of the heavy atom/anomalous scatterer substructure, 2) the primary phasing of single isomorphous replacement (SIR) and single wavelength anomalous scattering (SAS) data, 3) the refinement and extension of phases, and 4) the phasing of multiple - wavelength anomalous dispersion (MAD) data. The goal of this paper is not to offer a comprehensive review of these various topics but rather to examine some of the problems commonly encountered in protein crystal structure determination exercises and to look at examples of solutions offered by direct methods. The topics discussed here have been reviewed and discussed by several other authors (e.g. Karle, 1984a, 1989; Giacovazzo, 1980; Schenk, 1982). These publications are highly recommended, as they offer complementary information and a different perspective. The topic of maximum-entropy based phasing techniques will not be discussed here, as it is the subject of a separate article (Bricogne, 1991). Phasing tools based on solvent modification, which are sometimes described as direct space direct methods (Schevitz et al, 1981), are not discussed in this paper, which focuses on tools derived from the traditional direct methods approach.

1. THE DETERMINATION OF THE HEAVY-ATOM /ANOMALOUS SCATTERER SUBSTRUCTURE

Most of the protein structures determined to date have relied on the successful application of the multiple isomorphous replacement method (MIR). For new structures, that is structures for which a previously determined homologous structure does not exist, MIR remains the method of choice. In this approach, the determination of the heavy-atom substructure is one of the critical first steps. Such a determination is usually accomplished through Patterson techniques and, in particular, through the computation of a difference Patterson map.

Assuming a perfectly isomorphous pair of structures, P and PH - such that the atomic content of the derivative, PH, equals the atomic content of the native, P, plus the heavy atom content, H - we have,

$$\vec{F}_{PH} = \vec{F}_P + \vec{F}_H \tag{1.1}$$

To compute a Patterson function that contains information on the heavy-atom substructure, one needs the structure factor magnitudes F_H's. These are not available from the experimental data, which consist of the native and derivative structure factor magnitudes, F_P's and F_{PH}'s respectively. Their difference, ΔF_{PH-P}, can be expressed as :

$$\Delta F_{PH-P} = \left| F_{PH} - F_P \right| = \left| F_H \cos (\alpha_{PH} - \alpha_H) - 2 F_P \sin^2 \left\{ \frac{\alpha_P - \alpha_{PH}}{2} \right\} \right| \tag{1.2}$$

When the degree of heavy-atom substitution is small, the difference $\alpha_P - \alpha_{PH}$ is usually small as well and equation (1.2) reduces to :

$$\Delta F_{PH-P} \approx \left| F_H \cos (\alpha_{PH} - \alpha_H) \right| \tag{1.3}$$

Therefore, ΔF_{PH-P} is a good approximation to F_H when all three vectors $\vec{F}_{PH}$, $\vec{F}_P$ and $\vec{F}_H$ are nearly colinear; otherwise it systematically underestimates the value of F_H. The difference Patterson map is usually computed with the coefficients ΔF^2_{PH-P}.

Since $\Delta F^2_{PH-P} \approx \frac{F_H^2}{2} + \frac{F_H^2}{2} \cos 2 (\alpha_{PH} - \alpha_H) \tag{1.4}$

such maps contain a high level of noise, often referred to as intrinsic noise, due to the second term on the right hand side of (1.4). Equation (1.4) also shows that the peak heights are roughly half of their values from a map calculated with F_H^2. Thus, the normal problems associated with Patterson map interpretation are exacerbated by a decrease of peak heights against higher background noise. In addition, errors in the measurements and errors arising from the scaling of native and derivative data, as well as errors due to imperfect isomorphism, contribute to an already high noise level. Finally, it is often not known *a priori* how many heavy-atom sites there are or what their occupancies are. This greatly adds to the complexity of the problem of map interpretation, especially since one can rarely confirm or discard solutions on the basis of their chemical plausibility. For all these reasons, it is very useful to have alternative or complementary methods to aid the solution of the heavy-atom substructure.

Steitz (1968) was the first to use direct methods as a tool for locating heavy atoms in macromolecular crystals. He applied Sayre's equation to reflections in centrosymmetric zones and was able to locate heavy atom positions in the resulting ΔE_{PH-P} maps. The advantage of restricting the phasing exercise to centrosymmetric data is that, for this case, the relationship $\Delta F_{PH-P} = F_H$ holds (except in the few cases where there is "cross over"), since the three vectors $\vec{F}_{PH}$, $\vec{F}_P$ and $\vec{F}_H$ are new exactly colinear. Neidle (1973) extended the work of Steitz to three-dimensional data and showed that useful ΔE_{PH-P} maps could be calculated from general phases extended and refined by the tangent formula. Many successful applications of traditional direct methods to heavy-atom substructure determination problems have been reported (e.g. Schevitz et al, 1972; Steitz, Fletterick and Hwang, 1973; Anderson, Fletterick and Steitz, 1974; Navia and Sigler, 1974; Wilson, 1978; Artymiuk, Blake, Rice and Wilson, 1982; Westbrook, Piro and Sigler, 1983). Particularly noteworthy is the systematic investigation by Wilson (1978) on the application of MULTAN to the analysis of isomorphous derivatives.

From these studies, it is clear that direct methods are a very useful tool for determining heavy-atom substructure in macromolecular crystals. There are now several examples where direct methods were able to reveal correctly heavy-atom positions in multiple site derivatives, a problem not easily solved by Patterson techniques. Yet direct methods have not proven sufficiently powerful to supplant the Patterson approach. Rather, they are viewed as a useful complementary technique. The main difficulty in the application of direct methods to heavy-atom substructure determinations appears to lie in the selection of the best or most promising phase set. Although in many instances the best sets, as rated by the figures of merit, point to correct solutions, it is often the case that the figures of merit cluster around similar values and thus are not discriminatory. The figures of merit also appear to be less effective as the number of heavy-atom sites increases.

Some of the difficulties encountered in the direct methods applications are well understood. As in the Patterson case, the data needed, the E_H's, are not available. Again it is an approximation, ΔE_{PH-P}, derived from ΔF_{PH-P} that is used, and again the approximation is valid only when the native, derivative and heavy-atom substructure vectors are nearly colinear. ΔE_{PH-P} systematically underestimate the value of E_H. As pointed out by Wilson (1978), this leads to a statistical distribution of the normalized structure factors that is biased towards the centrosymmetric distribution. The errors in the estimation of the structure factors are then carried, in a direct methods procedure, to the normalized structure factors, and in turn to the estimate of the reliability of the phase invariants, to the determination of the best phasing path and finally to the figures of merit. These intrinsic errors, when added to experimental errors, severely undermine the power of direct methods.

Clearly, a possible way to improve on the present methods is to improve on the starting point, that is, on the normalized structure factor estimates. We are presently assessing the performance of integrated direct methods - SIR phasing tools (Hauptman, 1982a), which contain terms similar to Sim's weights (Sim, 1960) that take into account the expected phase difference between native and derivative. Such formulae thus take into account, at least in a statistical way, the non-colinearity of the native, derivative and heavy-atom vectors. We are also investigating the usefulness of a Bayesian approach to structure factor normalization which, in the present case, would include prior information on the heavy-atom substructure intensity distribution. Our studies on intensity statistics of heavy-atom substructures in large unit cells, as is the case in protein derivatives, show intensity distributions that are nearly ideal Wilson (1949) distributions.

When Friedel pair data are available, a difference Patterson map using $(F_{hkl} - F_{\overline{hkl}})^2$ as coefficients can be calculated to locate the position of the anomalous scatterers. Again, though, the difference used, $\Delta F_{FR} = |F_{hkl} - F_{\overline{hkl}}|$ is only an approximation to F_H. Indeed, it can be shown that,

$$\Delta F_{FR} \approx \frac{2}{k} \left| F_H \sin (\alpha_{PH} - \alpha_H) \right| \tag{1.5}$$

where $k = F_H / F_H''$ $\tag{1.6}$

with F_H' and F_H'' being the real and imaginary part respectively of the anomalous scatterers structure factor.

There have been far fewer attempts to make use of Friedel pair structure factor magnitude differences in a direct methods approach, presumably because the ΔF_{FR}'s are usually small and, until recently, very difficult to measure reliably. Recently, Mukherjee, Helliwell and Main (1989) described an application of MULTAN on Friedel pair differences in a variety of metalloproteins, cytochrome c_4 (MW $\approx$ 19,000), Pea lectin (MW $\approx$ 49,000) and a Hg derivative of α-amylase (MW $\approx$ 45,000) containing respectively two, two and one anomalous scatterer sites. In all three cases, they were able to locate correctly the position of the metal atoms in the map calculated from the phase set with the best combined figure of merit.

Fused direct methods - SAS three phase invariant distributions (Hauptman, 1982b) have also been used by Furey et al. (1986) to locate a 4-Cd cluster in the crystal structure determination of Cd, Zn Metallothionein (MW $\approx$ 10,000).

2. THE PRIMARY PHASING OF SIR AND SAS DATA

Because of the difficulties involved in crystallizing several derivatives with different heavy-atom substitution sites, there has always been much interest in devising techniques that could be used with data from a native and a single derivative (SIR).

In the SIR case, from the knowledge of the heavy-atom substructure and therefore of its structure factors, amplitudes and phases, and the measured amplitudes of the native and derivative structure factors, two possible values are calculated for the native or derivative phases. These two solutions are enantiomorphic with respect to the heavy-atom substructure vector. In the absence of any other information, the two solutions are equally probable.

Blow and Rossman (1961) proposed to compute a Fourier map using the average of the two possible phases and multiplying the structure factor amplitudes by suitable weights. Although they showed that the method could be used successfully, it has severe limitations, largely due to the high level of background noise. Most of the efforts to make use of SIR data have thus been directed at breaking down the two-fold ambiguity. Among the methods proposed are the use of anomalous scattering (Bijvoet,

1954), non-crystallographic symmetry (Bricogne, 1976), noise filtering techniques (Wang, 1985) and direct methods.

The first direct methods attempts to resolve the SIR phase ambiguity made use of the tangent formula (Coulter, 1965; Weinzierl, Eisenberg and Dickerson, 1969). These initial calculations were encouraging but not conclusive. Tests on model structures yielded a small average error for the phases associated with the largest E's but the error was seen to increase rapidly as the magnitude of the normalized structure factors decreased.

In recent years a great deal of effort has been directed, instead, at combining the techniques of direct methods and isomorphous replacement at the theoretical level, with formulae specifically derived for the SIR case. Of the methods proposed, the probabilistic approach of Hauptman (1982a) and algebraic approach of Karle (1983) go from intensities to phases by the most "direct" route with a minimum requirement of *a priori* chemical information. Other approaches (Fan Hai-fu et al, 1984, Fan Hai-fu and Gu Yuan-xin, 1985; Fortier, Moore and Fraser, 1985; Langs, 1986; Karle, 1986; Klop, Krabbendam and Kroon, 1987; Hao Quan and Fan Hai-fu, 1988) make use of heavy-atom substructure information which is integrated into the formulae. To our knowledge none of the methods has yet been used successfully to determine an unknown macromolecular structure.

The initial testings on error-free data have, however, been very promising. For example, the first calculations published by Hauptman, Potter and Weeks (1982), on the protein cytochrome C_{550}, (MW $\approx$ 14,500), and its $PtCl_4^{2-}$ derivative, showed that several tens of thousands of three-phase invariants could be reliably estimated as having values close to $0°$ or $180°$. In a set of 25,000, generated from 1000 native and 1000 derivative phases, an error of $30°$ is calculated. It was further shown (Fortier, Moore and Fraser, 1985) that the error could be decreased by a factor of two after incorporating heavy-atom substructure information into the formulae, which allowed cosine estimates in the full range of -1 to $+1$. Karle (1986) also tested his algebraic formulae on the protein cytochrome C_{550}. After introducing heavy-atom substructure information, he showed that some 19,000 or so invariants, generated from the 400 acentric reflections having the largest $\mid F_{PH} - F_P \mid$ values in the 2.5 Å resolution set could be estimated with an average error of $35°$. He also did calculations in which random errors were introduced in the F_{PH},

F_P and F_H magnitudes and demonstrated that the method could tolerate small levels of random errors. Test calculations were also carried on error-free insulin (MW $\approx$ 12,000) data. Langs (1986) showed that the phases associated with the 1000 largest E's, in the 1.9 Å resulution set, could be determined and refined to an average error of 6°, provided that perfect modular estimates of +1 or -1 were available for the 65,000 or so triples formed by these phases. Hao Quan and Fan Hai-fu (1988) also tested on insulin data their procedure for resolving the SIR ambiguity from Hauptman distributions in which heavy-atom information had been incorporated. They were able to resolve the two-fold ambiguity correctly in 75% of the 1000 phases associated with the largest E's in the 1.9 Å resolution set. The average error was 18°.

In order to put these results in perspective and assess the scope and nature of direct methods in the solution of the phase problem in the SIR case, we will examine the formulae obtained by Hauptman (1982a). They are particularly interesting because they were derived with a minimum of *a priori* information. In fact the only assumption made is that diffraction data are available for a pair of isomorphous structures for which the unit cell content is known. In addition these formulae are at the basis of the work of Fortier, Moore and Fraser (1985), Klop, Krabbendam and Kroon (1987) and Hao Quan and Fan Hai-fu (1988).

For a pair of isomorphous structures, there exist two normalized structure factors E_H and G_H for each reciprocal-lattice vector H. For a triplet of reciprocal lattice vectors H, K, L satisfying H + K + L = 0, there exist eight structure invariants

$$
\begin{aligned}
\omega_1 &= \phi_H + \phi_K + \phi_L \quad , \\
\omega_2 &= \phi_H + \phi_K + \psi_L \quad , \\
\omega_3 &= \phi_H + \psi_K + \phi_L \quad , \\
\omega_4 &= \psi_H + \phi_K + \phi_L \quad , \\
\omega_5 &= \phi_H + \psi_K + \psi_L \quad , \\
\omega_6 &= \psi_H + \phi_K + \psi_L \quad , \\
\omega_7 &= \psi_H + \psi_K + \phi_L \quad , \\
\omega_8 &= \psi_H + \psi_K + \psi_L \quad ,
\end{aligned}
\tag{2.1}
$$

where the ϕ's and the ψ's are the phases associated with the isomorphous pair of structures.

318

Let

$$|E_H| = R_1, \quad |E_K| = R_2, \quad |E_L| = R_3$$
$$|G_H| = S_1, \quad |G_K| = S_2, \quad |G_L| = S_3 \qquad (2.2)$$

The conditional probability distributions of the three-phase structure invariants ω_i given the six magnitudes $|E_H|$, $|E_K|$, $|E_L|$, $|G_H|$, $|G_K|$, $|G_L|$ in their first neighborhood are given by

$$P_i(\Omega_i | R_1, \ R_2, \ R_3, \ S_1, \ S_2, \ S_3) \approx (1/K_i) \, \exp(A_i \, \cos \Omega_i) \ , \qquad (2.3)$$

$$i = 1, 2, .., 8,$$

where

$$K_i = 2\pi I_0(A_i) \qquad (2.4) .$$

and I_0 is the modified Bessel function (Hauptman, 1982a).

The A_i values are given by

$$
\begin{aligned}
A_i \ = \ & 2\{\beta_1 \tau_1 R_1 R_2 R_3 \\[4pt]
& + \ \beta_2 [\tau_{21} R_1 R_2 S_3 + \tau_{22} R_1 S_2 R_3 + \tau_{23} S_1 R_2 R_3] \\[4pt]
& + \ \beta_3 [\tau_{31} R_1 S_2 S_3 + \tau_{32} S_1 R_2 S_3 + \tau_{33} S_1 S_2 R_3] \\[4pt]
& + \ \beta_4 \tau_4 \, S_1 S_2 S_3 \} \ , \qquad (2.5)
\end{aligned}
$$

where the β's are functions of the atomic scattering factors, and $\tau = C_1 C_2 C_3$ is obtained by comparing the ith structure factor associated with the coefficient of τ with the ith structure factor associated with the invariant. If they are of the same type, i.e. of both R or both S, then $C_i = 1.0$, $i = 1,2,3$. If one is of type R and the other of type S, then

$$C_i = I_1(2 \Upsilon R_i S_i)/I_0(2 \Upsilon R_i S_i), \qquad i = 1,2,3, \qquad (2.6)$$

where I_1 and I_0 are the modified Bessel functions and Υ is a function of the atomic scattering factors. If for the special case of a native protein and a heavy-atom isomorphous derivative, we assume that the atomic content of the derivative equals the atomic content of the native protein (P) plus the heavy-atom content (H) then

$$\Upsilon = \alpha_{20}^{1/2} \alpha_{02}^{1/2} / (\alpha_{02} - \alpha_{20}), \qquad (2.7)$$

$$\beta_1 = \frac{-(\alpha_{03} - \alpha_{30})\alpha_{20}^{3/2}}{(\alpha_{02} - \alpha_{20})^3} \ , \qquad (2.8)$$

$$\beta_2 = \frac{(\alpha_{03} - \alpha_{30})\alpha_{20}\alpha_{02}^{1/2}}{(\alpha_{02} - \alpha_{20})^3} , \qquad (2.9)$$

$$\beta_3 = \frac{-(\alpha_{03} - \alpha_{30})\alpha_{20}^{1/2}\alpha_{02}}{(\alpha_{02} - \alpha_{20})^3} , \qquad (2.10)$$

$$\beta_4 = \frac{(\alpha_{03} - \alpha_{30})\alpha_{02}^{3/2}}{(\alpha_{02} - \alpha_{20})^3} \qquad (2.11)$$

$$\alpha_{mn} = \sum_{j=1}^{N} f_j^m g_j^n \qquad (2.12)$$

where f_j and g_j denote atomic structure factors for a corresponding pair of isomorphous structures.

From equation (2.7) to (2.11) it is seen that the distribution does not depend, as in the case of the traditional three-phase invariant, on the total number of atoms per unit cell but rather on the scattering difference between the native protein and the derivative - that is, on the scattering of the heavy atoms in the derivative. For the special case in which the heavy atoms in the derivative are of equal weight and equal occupancy, the distribution depends on the number of heavy atoms in the derivative. Since this number is usually small, it becomes clear that the distribution is capable of yielding extremely reliable estimates, as demonstrated by the calculations of Hauptman, Potter and Weeks (1982). When the τ functions approach 1.0, i.e. when the $2\pi R_i S_i$'s are large, it can be shown (Fortier, Weeks and Hauptman, 1984) that the predominant term in the distribution involves the product $(S_1 - R_1) (S_2 - R_2) (S_3 - R_3)$. Thus a result comparable to the simple rule of Karle (1983) is obtained: if that product is positive, than the invariant is estimated as having a cosine value close to +1, while if it is negative the cosine value is estimated as being close to -1.

The τ functions used in estimating the A values are either equal to 1.0 or the Bessel-function ratio as defined in (2.6). This Bessel-function ratio is the expected value of the cosine of the phase difference $(\psi_i - \phi_i)$ associated with the two magnitudes R_i and S_i (Sim, 1960). A τ function is thus equal to 1.0 when the phase difference between the native and derivative is expected to be 0°. Thus, the case for which the τ functions all approach 1.0 corresponds to the special case in which the phases of all three reflections participating in the invariants have approximately the same value for the native and derivative.

In this special case,

$$\phi_P = \phi_{PH} = \phi_H \qquad\qquad \text{if } (S-R) > 0$$

$$\phi_P = \phi_{PH} = \phi_H + 180° \qquad \text{if } (S-R) < 0 \qquad\qquad (2.13)$$

Since the heavy-atom substructure normally consists of a small number of atoms, it has associated phases that form three-phase invariants with values very close to $0°$. The corresponding native and derivative invariants will thus have estimates of $0°$ or $180°$ depending on whether the product $(S_1-R_1)(S_2-R_2)(S_3-R_3)$ is larger or smaller than zero. It is possible to obtain very accurate estimates of this special class of invariants. It should be noted, however, that the phases that participate in these invariants are precisely those for which the two-fold ambiguity vanishes.

To look at the cases for which the τ functions differ from 1.0, we can substitute in the distribution the equivalent cosine functions. Let

$$\psi_i - \phi_i = \pm\alpha_i \qquad\qquad (2.14)$$

Then, for example, the conditional probability distribution of the ω_1 invariant can be written as

$$P_1(\Omega_1) = \frac{1}{K_1} \exp(A_1 \cos \Omega_1) \qquad\qquad (2.15)$$

where

$$\begin{aligned}
A_1 \cos \Omega_1 = 2 \cos \Omega_1 \{ &\beta_1 R_1 R_2 R_3 + \beta_2 [R_1 R_2 S_3 \cos \alpha_3 \\
&+ R_1 S_2 R_3 \cos \alpha_2 + S_1 R_2 R_3 \cos \alpha_1] + \beta_3 [R_1 S_2 S_3 \cos \alpha_2 \cos \alpha_3 \\
&+ S_1 R_2 S_3 \cos \alpha_1 \cos \alpha_3 + S_1 S_2 R_3 \cos \alpha_1 \cos \alpha_2] \\
&+ \beta_4 S_1 S_2 S_3 \cos \alpha_1 \cos \alpha_2 \cos \alpha_3 \}.
\end{aligned} \qquad (2.16)$$

For simplicity, let us assume that $\alpha_2 = \alpha_3 = 0$ and $|\alpha_1| \neq 0$. Equation (2.16) then becomes

$$A_1 \cos \Omega_1 = 2 \cos \Omega_1 \{\beta_1 R_1 R_2 R_3 + \beta_2 [R_1 R_2 S_3 + R_1 S_2 R_3]$$

$$+ \beta_3 R_1 S_2 S_3\} + \{\cos(\Omega_1 + \alpha_1) + \cos(\Omega_1 - \alpha_1)\} \times$$

$$\{\beta_2 S_1 R_2 R_3 + \beta_3 [S_1 R_2 S_3 + S_1 S_2 R_3] + \beta_4 S_1 S_2 S_3\}. \qquad (2.17)$$

While the expected value of the phase difference $\psi_1 - \phi_1$ can be estimated, its sign is not known. In the form of the distribution shown in equation (2.17), both signs are considered equally probable and their contributions are averaged, as is done in the standard SIR technique (Blow and Rossmann, 1961). Once the heavy-atom substructure has been determined, the $\cos \alpha_1$ functions appearing in quation (2.16) can be calculated rather than estimated. Several approach s (Fortier, Moore and Fraser, 1985; Klop, Krabbendam and Kroon, 1987; Hao Quan and Fan Hai-fu, 1988) have made use of this information to improve the phase estimates.

Several important points emerge from a careful analysis of the formulae obtained through the joint probability distribution approach. Firstly, it is seen that even without any heavy-atom substructure information, the method is able to identify very accurately those reflections for which the native and derivative have the same phase angle and furthermore to determine whether this value is the same as, or 180° away from the heavy-atom substructure phase. An estimate of the absolute value of the derivative-native phase difference - and, consequently, of the heavy-atom substructure vector - falls out of the derivation. These results are obtained simply from the knowledge of the unit cell content of the isomorphous pair of structures. Secondly, it is seen that the "inherent" SIR two-fold ambiguity does not disappear magically. Rather, it is integrated in the formulae and, through a weighting scheme similar to that used by Blow and Rossman (1961), unique invariant estimates are obtained. It is encouraging to see that many of the standard SIR practices, which have emerged from careful experimental observations, fall out naturally in the mathematical derivation. On the other hand, this may indicate that the reliability of unique invariant estimates will decrease as the protein and derivative phases diverge from one another and thus as they diverge from their standard SIR values. This trend has in fact been observed by Fraser (1987). In order to demonstrate whether or not a given approach has indeed resolved the SIR two-fold ambiguity, it is extremely important to compare the direct methods results against standard SIR results.

As in the SIR case, the SAS experiment yields estimates of phases bearing a two-fold ambiguity. Unlike the SIR case, however, the two estimates are not equally probable. This is because the two phase estimates are enantiomorphic with respect to the _imaginary_ part of the anomalous scatterer substructure vector. The phase ambiguity can thus be resolved by selecting the phase closest to the anomalous scatterer substructure phase. However, as the ratio of heavy-atom to light-atom scattering decreases, the probability that this choice will be the correct one decreases. This method has been used successfully in the solution of the lysine hydrochloride structure (Raman, 1959) and, more recently, in a probabilistic fashion, in the solution of the protein crambin (Hendrickson and Teeter, 1981).

The first results on the use of the integrated direct methods-anomalous dispersion technique for the estimation of the three-phase structure invariant were presented by Kroon, Spek and Krabbendam (1977). By extending the method proposed by Peederman and Bijvoet (1956), Ramachandran and Raman (1956) and Okaya and Pepinsky (1956) to the three-phase structure invariant, they showed that three-phase sine invariants could be estimated from the observed intensities. This work was then extended by Heinerman, Krabbeddam, Kroon and Spek (1978), who used a probabilistic approach to improve the three-phase sine invariant estimates.

Using the method of joint probability distribution, Hauptman (1982b) and, subsequently, Giacovazzo (1983) obtained formulae which give unique estimates of the two-phase and three-phase structure invariants and thus unique estimates of the phases themselves. Unique estimates have also been obtained by Karle (1984b and 1985) by use of formulae derived through an algebraic approach. As in the probabilistic approach of Hauptman (1982b) and Giacovazzo (1983), this is done without any information about the anomalous scatterer substructure. Several direct methods approaches that make use of anomalous scatterer substructure information to resolve the SAS phase ambiguity have also been proposed (Fan Hai-fu et al, 1984; Fan Hai-fu and Gu Yuan-xiu, 1985; Qian Jin-zi, Fan Hai-fu and Gu Yuan-xiu, 1985; Langs, 1986).

In the initial applications made by Hauptman (1982b), Giacovazzo (1983) and Karle (1984b) on macromolecular data, it was seen, however, that although unique estimates were obtained substantial errors persisted, even when the calculations were done using error-free data. In the probabilistic formulae, it was also observed that the error magnitudes were was not well predicted, as the variances of the distributions were systematically underestimated. It was later shown that certain approximations appearing in the formulae, which are responsible for the ability to obtain the unique estimates, are also largely responsible for the errors observed (Fortier, Fraser and Moore, 1986). More specifically, a probabilistic estimate of the two-phase invariant $\phi_{hkl} + \phi_{\overline{hkl}}$ that appears in the distribution assigns a positive value to the sum of $\phi_{hkl} + \phi_{\overline{hkl}}$. This corresponds to the solution for which the phase angles are closest to the anomalous scatterer substructure phase angles. Thus the SAS ambiguity is resolved essentially in the usual manner. It is important to note, though, that this is done without any anomalous scatterer substructure information. Large errors are observed in invariants that contain a phase angle whose value is not closest to the anomalous scatterer substructure phase angle.

Fan Hai-fu and Gu Yuan-xiu (1985) incorporated anomalous scatterer substructure information through the product of the Cochran (1955) and Sim (1959) distributions. They tested their formula (Qian Jin-zi, Fan Hai-fu and Gu Yuan-xin, 1985) on the Hg derivative of avian pancreatic polypeptide (MW $\approx$ 5,000) using both calculated and observed data. Their calculations on error-free data resolved the SAS two-fold ambiguity in 87% of the phases associated with the 1000 largest E's out of the 2100 reflections in the 2.1 Å resolution data set. The average error was $11°$. With experimental data the average error increased to $37°$ but the percentage of resolved phases remained the same. The method was also tested on error-free data from rice ferricytochrome C (MW $\approx$ 12,000). The contribution of the anomalous scatterers is far smaller in this case. The SAS ambiguity was resolved for 68% of the phases associated with the 1000 largest E's out of the 6000 reflections in the 2.0 Å resolution sphere. The average error was $25°$.

Hao Quan and Woolfson (1989) also tested the application of the P_s function (Okaya, Saito and Pepinsky, 1955) on observed data from the Hg derivative of avian pancreatic polypeptide. They were able to phase the 2109 reflections of the 2.1 Å resolution set with an average unweighted error of $40°$ and a $|F_oF_c|$ weighted error of $29°$. They also showed that the resultant electron density map contained the characteristic features of the correct model and could lead to a complete structure determination.

3. THE REFINEMENT AND EXTENSION OF PHASES

In principle, the method of multiple isomorphous replacement (MIR) leads to a unique solution for the phases and should thus be sufficient for the complete determination of the native protein structure. There are, however, several problems in the applications and the method is rarely applied in such a straightforward manner.

Some of the difficulties arise from errors in the observed structure factor magnitudes. These errors can be random, when they result from the counting statistics, but also systematic, when they result, for example, from absorption or radiation damage problems or from the relative scaling of multiple crystal data. A further problem, and a very serious one, comes from imperfect isomorphism. It is therefore common for MIR phases to have average errors in the order of $50°$ or so. Furthermore, because problems associated with lack of isomorphism increase with increasing resolution, the quality of the phases, and of the resultant electron density maps, decreases with increasing resolution. It is thus often necessary to improve and extend the MIR phases. The most successful methods for doing so have been real space density modification type techniques, including the use of non-crystallographic symmetry. There have been also many attempts to use direct methods for refining and extending macromolecular phases and, in particular, through the use of the tangent formula, Sayre's equation and higher-order determinants. A few examples are given below, although the list is by no means exhaustive.

The tangent formula, in its standard form, was used early on for the extension and refinement of macromolecular phases (e.g. Weinzierl, Eisenberg and Dickerson, 1969; Reeke and Lipscomb, 1969; Coulter, 1971; Hendrickson and Karle, 1973). Many of these calculations were done on observed data, e.g. from cytochrome c (MW $\approx$ 12,400), carboxy peptidase A (MW $\approx$ 34,600), sperm myoglobin (MW $\approx$ 18,000) and carp muscle calcium-binding protein (MW $\approx$ 11,500). Some encouraging results were obtained. For example, the tangent formula was able to improve the low resolution phases of carboxypeptidase A and perhaps, more impressively, appeared capable of improving the electron density map of carp muscle calcium-binding protein sufficiently so as to reveal a second calcium-binding site. The method, however, had many shortfalls. For example, Reeke and Lipscomb (1969) found that their results deteriorated rapidly with increasing resolution. It also appeared difficult to assess the success of the technique by the use of figures of merit. Finally, the results seemed very much dependent on the particular procedure used, the number of phases extended per cycle, the number of cycles of refinement, etc. More recently, Olthof and Schenk (1982) proposed the use of a modified tangent formula which preserves enantionorph discrimination. They applied their formula to observed data from metmyoglobin (MW $\approx$ 20,000) and showed significant improvements over the results obtained through the use of the unmodified tangent formula. They used their formula to extend and refine phases from 2.8 Å to 2.0 Å resolution and obtained phase sets with average errors ranging from 37° to 47°, depending on the specific procedure used. They also found that the results deteriorated significantly when similar calculations were initiated with 3.0 Å rather than 2.8 Å data and recommended that the process be applied to limited ranges only.

Sayre (1972) proposed to use a least-square refinement procedure based on a formula he had previously derived (Sayre, 1952). The method was tested on rubredoxin (MW $\approx$ 6,000) (Sayre, 1974) and insulin (Cutfield et al., 1975) to extend phases from 2.5 Å to 1.5 Å and 1.9 Å to 1.5 Å respectively. The method was found to be smooth and reliable but computer intensive.

Woolfson and Yao Jia-xing (1988) have used the Sayre-equation tangent formula to extend and refine phases. They have applied their technique to observed data from the small protein avian pancreatic polypeptide. They selected the 1500 largest E value reflections in the 1 Å resolution sphere for phase determination. Of these, the 129 reflections within the 3 Å resolution sphere constituted the initial set of known phases. The remaining 1,371 phases were assigned random values. The resultant phase set with the best combined figure of merit had an average phase error of 33°. These results are very encouraging, particularly since they were obtained with observed data and with a relatively small initial set of known phases.

Much use has been made also of determinental methods (Tsoucaris, 1980) for the expansion and refinement of macromolecular phases. At one end of the spectrum, the method was used successfully to expand relatively high resolution phases to further resolution by de Rango et al. (1985). Using observed data from 2 Zn Insulin, they used determinental methods to refine and extend the phases from 1.9 Å to 1.5 Å resolution. Improved 1.9 Å resolution phases were obtained, as well as a large number of new phases with good accuracy. These yielded significantly better electron density maps. At the other end of the spectrum, determinental methods were used by Podjarny, Schevitz and Sigler (1981) on yeast tRNA$_f^{Met}$ (MW $\approx$ 25,000) to extend MIR phases to a lower resolution. Very low resolution phases are often poorly estimated, either because of measurement problems at low angle or because of the solvent contribution to the scattering. Podjarny, Schevitz and Sigler used matricial direct methods to phase 28 strong low resolution (100 - 19 Å) reflections from a starting set consisting of 107 reflections in the resolution range 32 - 14 Å. The resultant phased data allowed them to define the molecular envelope which they felt provided the basis of a successful structure determination.

4. THE PHASING OF MULTIPLE WAVELENGTH ANOMALOUS DISPERSION (MAD) DATA

It was realized early on that multiple wavelength anomalous dispersion data could solve the phase problem (Okaya and Pepinsky, 1956). Indeed, the technique mimics that of multiple isomorphous replacement (MIR) with the added advantage that, in the case of MAD, isomorphism is perfect. However, the difficulties of measuring with precision the small differences between the intensities of Friedel pair members, coupled with difficulties of measuring data at several wavelengths using a conventional X-ray tube, have delayed the application of the technique.

With the advent of readily tunable X-rays from synchrotron sources and the availability of array detectors, the MAD technique has come of age. Following the design of appropriate experimental procedures and analytical tools, MAD has indeed been used successfully for the structure determination of several macromolecules, for example streptavedin (MW $\approx$ 14,000) ((Hendrickson et al., 1989) and a "blue" copper-protein isolated from cucumber seedlings (MW $\approx$ 10,000) (Guss et al., 1988). According to Hendrickson (1990) the technique is now reaching maturity. Karle (1980) and Hendrickson (1985, 1988) use an algebraic approach to retrieve phase information from multiple wavelength anomalous dispersion data. Their analysis is well documented in the previously mentioned publications and thus will not be repeated here. They have shown that, with a single kind of anomalous scatterer, provided that the scattering factors are known, two-wavelength data will yield a system consisting of five equations and three unknowns, $|{}^{\circ}F_T|$, the normal scattering contributions from all atoms, $|{}^{\circ}F_A|$, the normal scattering contribution from the anomalous scatterers and $\Delta\phi = {}^{\circ}\phi_T - {}^{\circ}\phi_A$, their phase difference. In order to compute an electron density for the structure, both the magnitude $|{}^{\circ}F_T|$ and the phase ${}^{\circ}\phi_T$ are needed. The analysis is thus broken down into two main steps. In the first one, the $|{}^{\circ}F_A|$ are used, either via Patterson or direct methods techniques, to determine the anomalous scatterer substructure. Once this information has been gained, it becomes possible to compute ${}^{\circ}\phi_A$ values and, therefore, the ${}^{\circ}\phi_T$ phase angles which leads directly to the desired electron density map. Hendrickson (1988) has developed a full system of computer programs, MADSYS, to implement the MAD phasing procedure. Although the procedure differs greatly from the small molecule direct methods, it probably represents the most direct route, from measured intensities to the three-dimensional image of the structure, currently available in protein crystallography.

An alternative approach to that of Hendrickson is to use the information concerning the normal scattering structure factor magnitudes and phase difference as input to the estimation of three-phase structure invariants. This can be done either through an algebraic approach (Karle, 1984c) or probabilistic approach (Klop, Krabbendam and Kroon, 1989). By the use of three-phase invariants, the determination of the anomalous substructure is bypassed. The method was tested by Klop (1989), who used error-free data from ferredoxin (MW $\approx$ 6000) to estimate three-phase invariants and then individual phase values through the use of a convergence procedure and generalized tangent formula. He was able to show convincingly that the distributions could lead to fairly accurate estimates of the invariants and that this information could be used in a standard MULTAN procedure to produce reliable phases. The 2225 reflections at 2 Å resolution were phased with an average error of 12° while the 304 reflections at 4 Å resolution were phased with an average error of 10°.

CONCLUDING REMARKS

Several applications of direct methods to problems of macromolecular crystallography have been described. Although some of the results presented here are clearly encouraging, it is fair to say that direct methods, except for the case of multi-wavelength anomalous scattering (MAD), have not evolved yet into significant methods for the phasing of macromolecular data. The best results obtained so far have been largely in areas where alternative techniques meet with equal success. It is worth mentioning, however, the recent application of direct methods to the moduli of the envelope transform, which allowed the computation of a map from which the molecular envelope of the tryptophanyl-t-RNA-synthetase dimer could be defined (Carter et al., 1990).

Many of the direct methods formulae, as shown above, mimic the standard macromolecular phasing procedures. The fact that such procedures fall out of the mathematical derivations quite automatically is actually very encouraging. It is an indication that the extensive theoretical base of the direct methods may constitute and appropriate infrastructure within which the macromolecular phasing problem can be phrased and solved.

Although there have been many attempts to use direct methods in a macromolecular context, these have far from exhausted the full power of the technique. Many of the procedures used so far have been restricted to a fairly narrow domain and applied in a rigid fashion. For example, many of the applications are committed to a particular kind of invariant used in very specific experimental conditions, e.g. three-phase invariants for the SIR case, etc. As has been realized already by macromolecular crystallographers, the best route and, ultimately the most direct route to the solution of macromolecular structures may be one that allows the utilization of every source of phasing information. The direct methods can accommodate such an approach. Indeed, we already have in place the theoretical basis needed to integrate various kinds of phasing information in a general joint probability distribution framework e.g. diffraction data such as SIR and SAS, partial structure information, etc. (Bricogne, 1988; Fortier and Nigam, 1989). Bricogne (1988) has recently presented a Bayesian statistical theory of the phase problem in which such a unified approach to the solution of the phase problem is proposed. Our own efforts have focused on the design of a knowledge-based system in which an artificial intelligence infrastructure is used to combine direct methods phasing strategies with crystallographic data base information (Fortier, Glasgow and Allen, 1990). While both of these approaches are well rooted in the direct methods, they are a significant departure from the traditional ways in which direct methods are used. The simplicity of a general algorithmic solution is replaced by the flexibility of a context driven solution process which attempts to make use of all available information. Thus, in the knowledge based approach, solving a crystal structure becomes a more fluid process in which the search space is scanned by mathemetical tools afforded by direct methods, while it is guided by pattern recognition techniques derived from chemical and crystallographic reasoning.

ACKNOWLEDGEMENTS

Financial assistance from the Natural Sciences and Engineering Research Council of Canada is gratefully acknowledged.

REFERENCES

Anderson, W.F. Flitterick, R.J. and Steitz, T.A. (1974) J. Mol. Biol., 86, 261.

Artymiuk, P.J., Blake, C.C.F., Rice, D.W. and Wilson, K.S. (1982) Acta Cryst. B38, 778.

Bijvoet, J.M. (1954) Nature (London), 173, 888.

Blow, D.M. and Rossman, M.G. (1961) Acta Cryst., 14, 195.

Bricogne, G. (1976) Acta Cryst., A32, 832.

Bricogne, G. (1988) Acta Cryst., A44, 517.

Bricogne, G. (1991). In *Direct Methods of Solving Crystal Structures*, H. Schenk, Ed., Plenum Press, London.

Carter, W. Jr., Crumley, K.V., Coleman, D.E., Hage, F. and Bricogne, G. (1990) Acta Cryst., A46, 57.

Cochran, W. (1955) Acta Cryst., 8, 473.

Coulter, C.L. (1965) J. Mol. Biol., 12, 292.

Coulter, C.L. (1971) Acta Cryst., B27, 1730.

Cutfield, J.F., Dodson, E.J., Dodson, G.G., Hodgkin, D.C., Isaacs, N.W., Sakabe, K. and Sakabe, N. (1975) Acta Cryst., A31, S21.

Fan Hai-fu and Gu Yuan-xin (1985) Acta Cryst., A41, 280.

Fan Hai-fu, Han Fu-son, Qian Jin-zi and Yao Jia-xing (1984) Acta Cryst., A40, 489.

Fortier, S., Fraser, M.E. and Moore, N.J. (1986) Acta Cryst., A42, 149.

Fortier, S., Glasgow, J. and Allen, F.H. (1991), In *Direct Methods of Solving Crystal Structures* , H. Schenk, ed., Plenum Press, London.

Fortier, S., Moore, N.J. and Fraser, M.E. (1985) Acta Cryst., A41, 571.

Fortier, S. and Nigam, G.D. (1989) Acta Cryst., A45, 247.

Fortier, S., Weeks, C.M. and Hauptman, H. (1984) Acta Cryst., A40, 544.

Fraser, M.E. (1987) Ph.D. Thesis, Queen's Univ., Kingston, Ontario, Canada.

Furey, W.F., Robbins, A.H., Clancy, L.L., Winge, D.R., Wang, B.C. and Stout, C.D. (1986) Science, 231, 704.

Giacovazzo, C. (1980) *Direct Methods in Crystallography*, Academic Press, London.

Giacovazzo, C. (1983) Acta Cryst., A39, 585.

Guss, J.M., Merritt, E.A., Phizackerley, R.P., Hedman, B., Murata, M., Hodgson, K.O. and Freeman, H.C. (1988) Science, 241, 806.

Hao Quan and Fan Hai-fu (1988) Acta Cryst., A44, 379.

Hao Quan and Woolfson, M.M. (1989) Acta Cryst., A45, 794.

Hauptman, H. (1982a) Acta Cryst., A38, 289.

Hauptman, H. (1982b) Acta Cryst., A38, 632.

Hauptman, H., Potter, S. and Weeks, C.M. (1982) Acta Cryst., A38, 294.

Heinerman, J.J.L., Krabbendam, H., Kroon, J. and Spek, A.L. (1978) Acta Cryst., A34, 447.

Hendrickson, W.A. (1985) In *Transactions of the ACA*, Vol. 21.

Hendrickson, W.A. (1988) In *Crystallographic Computing 4*, N.W. Issacs and M.R. Taylor, eds., pp. 97-108, Oxford University Press, New York.

Hendrickson, W.A. (1990) Proceedings of the American Crystallographic Association Meeting, 8-13 April 1990, New Orleans, Louisiana, U.S.A. Abstract M01.

Hendrickson, W.A. and Karle, J. (1973) J. Biol. Chem., 248, 3327.

Hendrickson, W.A., Pähler, A., Smith, J.L., Satow, Y., Merrit, E.A. and Phizackerley, R.P. (1989) Proc. Natl. Acad. Sci. USA, 86, 2190.

Hendrickson, W.A. and Teeter, N.M. (1981) Nature, 290, 107.

Karle, J. (1980) Int. J. Quantum Chem. Symp., 7, 357.

Karle, J. (1983) Acta Cryst., A39, 800.

Karle, J. (1984a) In *Methods and Applications in Crystallographic Computing*, S.R. Hall and T. Asheda eds., pp. 120-140, Clacendon Press, Oxford.

Karle, J. (1984b) Acta Cryst., A40, 4.

Karle, J. (1984c) Acta Cryst., A40, 526.

Karle, J. (1985) Acta Cryst., A41, 387.

Karle, J. (1986) Acta Cryst., A42, 246.

Karle, J. (1989) Acta Cryst., A45, 765.

Klop, E.A. (1989) Ph.D. Thesis, U. of Utrecht, Utrecht, The Netherlands.

Klop, E.A., Krabbendam, H. and Kroon, J. (1987) Acta Cryst., A43, 810.

Klop, E.A., Krabbendam, H. and Kroon, J. (1989) Acta Cryst., A45, 203.

Kroon, J., Spek, A.L. and Krabbendam, H. (1977) Acta Cryst., A33, 382.

Langs, D.A. (1986) Acta Cryst., A42, 362.

Murkherjee, A.K., Helliwell, J.R. and Main, P. (1989) Acta Cryst., A45, 715.

Navia, M.A. and Sigler, P.B. (1974) Acta Cryst., A30, 706.

Neidle, S. (1973) Acta Cryst., B29, 2645.

Okaya, Y. and Pepinsky, R. (1956) Phys. Rev., 103, 1645.

Okaya, Y., Saito, Y. and Pepinsky, R. (1955) Phys. Rev., 98, 1857.

Olthof, G.J. and Schenk, H. (1982) Acta Cryst., A38, 117.

Peederman, A.F. and Bijvoet, J.M. (1956). Proc. Koninkl Ned. Akad. Wetenschap., B59, 312.

Perutz, M.F. and Scatturin, V. (1954) Acta Cryst., 7, 799.

Podjarny, A.D., Schevitz, R.W. and Sigler, P.B. (1981) Acta Cryst., A37, 662.

Qian Jin-zi, Fan Hai-fu and Gu Yuan-xin (1985) Acta Cryst., A41, 476.

Ramachandran, G.N. and Raman, S. (1956) Curr. Sci. (India), 25, 348.

Raman, S. (1959) Zeitschrift für Kristallographie, 111, 301.

Rango, C. de, Mauguen, Y., Tsoucaris, G., Dodson, E.J., Dodson, G.G. and Taylor, D.J. (1985) Acta Cryst., A41, 3.

Reeke, G.N. and Lipscomb, W.N. (1969) Acta Cryst., B25, 2614.

Sayre, D. (1952) Acta Cryst., 5, 60.

Sayre, D. (1972) Acta Cryst., A28, 210.

Sayre, D. (1974) Acta Cryst., A30, 180.

Schenk, H. (1982) In *Computational Crystallography*, D. Sayre ed., pp. 231-241, Clarendon Press, Oxford.

Schevitz, R.W., Navia, M.A., Bartz, D.A., Cornick, G., Rosa, J.J., Rosa, M.D.H. and Sigler, P.B. (1972) Science, 177, 429.

Schevitz, R.W., Podjarny, A., Zwick, M., Hughes, J.J. and Sigler, P.B. (1981) Acta Cryst., A37, 669.

Sim, G.A. (1959) Acta Cryst., 12, 813.

Sim, G.A. (1960) Acta Cryst., 13, 511.

Steitz, T. (1968) Acta Cryst., B24, 504.

Steitz, T.A., Flitterick, R.J. and Hwang, K.J. (1973) J. Mol. Biol., 78, 551.

Tsoucaris, G. (1980) In *Theory and Practice of Direct Methods in Crystallography*, M.F.C. Ladd and R.A. Palmer eds., pp. 287-359, Plenum Press, New York.

Wang, B.C. (1985) In *Methods in Enzymology Vol. 115*, H.W. Wyckoff, C.H.W. Hirs and S.N. Timasheff eds., Academic Press, London.

Weinzierl, J.E., Eisenberg, D. and Dickerson, R.E. (1969) Acta Cryst., B25, 380.

Westbrook, E.M., Piro, O.E. and Sigler, P.B. (1984). J. of Biol. Chem., 259, 9096.

Wilson, A.J.C. (1949) Acta Cryst., 2, 318.

Wilson, K.S. (1978) Acta Cryst., B34, 1599.

Woolfson, M.M. and Yao Jia-xing (1988) Acta Cryst., A44, 410.

SIR, SAD, MAD, DM: INTEGRATED METHODS

H. Krabbendam

Laboratorium voor Kristal- en Structuurchemie
Bijvoetcentrum voor Biomoleculair Onderzoek
Rijksuniversiteit Utrecht
The Netherlands

1. THE METHOD OF SINGLE ISOMORPHOUS REPLACEMENT

For a long time the method of isomorphous replacement has
been the only method used to solve the phase problem in protein
crystallography. Starting with a native protein crystal, of
which a reflection data set has been measured (leading to $|F_i{}^P|$,
an abbreviation for $|F_{h_i}{}^P|$), this crystal is soaked in a solution
of a heavy-atom salt for a day, a week, or perhaps for even
longer. What one hopes for is that a heavy atom will attach to a
protein molecule in a specific way (i.e. identically in a great
number of unit cells) without changing (i.e. less than 0.5%) the
geometry of the unit cell. Subsequently a reflection dataset of
the alleged heavy-atom derivative is collected (leading to
values for $|F_i{}^{HP}|$).

The diffraction pattern of the native protein and of the
derivative will show small differences in intensities, provided
the attached heavy atom is heavy enough and its occupation
number is sufficiently large. An average intensity difference of
15 to 20% is usually regarded as suitable for a subsequent phase
determination. The two datasets are brought on the same scale by
Wilson statistics (determining the relative scale factor and
ΔB).

We can write:

Direct Methods of Solving Crystal Stuctures
Edited by H. Schenk, Plenum Press, New York, 1991

$$F_i{}^{HP} = F_i{}^{P} + F_i{}^{H} \qquad\qquad (1.1)$$

in which

$\qquad HP \equiv$ derivative (heavy atom + protein)

$\qquad P \ \equiv$ native protein

$\qquad H \ \equiv$ heavy atom configuration

and $F_i{}^{HP} = |F_i{}^{HP}|\exp i\varphi_i{}^{HP}$, $F_i{}^{P} \equiv |F_i{}^{P}|\exp i\varphi_i{}^{P}$, $F_i{}^{H} = |F_i{}^{H}|\exp i\varphi_i{}^{H}$ are the derivative, protein and heavy-atom structure factors respectively (the subscript i denotes $\mathbf{h}_i$).

Relation (1.1) can be represented by a triangle in the complex plane (Fig. 1.1). The cosine rule applied to the triangle gives

$$|F_i{}^{HP}|^2 = |F_i{}^{P}|^2 + |F_i{}^{H}|^2 + 2|F_i{}^{P}||F_i{}^{H}|\cos(\varphi_i{}^{P}-\varphi_i{}^{H}) \qquad (1.2)$$

In this equation $|F_i{}^{HP}|$ and $|F_i{}^{P}|$ are known from measurement, so the unknowns are $|F_i{}^{H}|$, $\varphi_i{}^{H} - \varphi_i{}^{P}$. The value of $\varphi_i{}^{P}$ is what we are aiming at, so there must be found a way to determine $|F_i{}^{H}|$ and $\varphi_i{}^{H}$, i.e. to determine the heavy-atom configuration. What one does is to make a difference-Patterson synthesis with $||F_i{}^{HP}| - |F_i{}^{P}||^2$ (a rough estimate of $|F_i{}^{P}|^2$) as coefficients. This synthesis should show the vectors between the heavy atoms only (with a lot of noise). From this difference-Patterson function the heavy-atom positions are deduced; then these positions are refined, $|F_i{}^{H}|$ and $\varphi_i{}^{H}$ are calculated, and $\varphi_i{}^{P}$ is solved from (1.2). However, as $\varphi_i{}^{P}$ is contained in the cosine function, two solutions for $\varphi_i{}^{P}$ are possible. This can be seen by writing (1.2) explicitly as:

$$\varphi_i{}^{P} = \varphi_i{}^{H} + \cos^{-1}\left(\frac{|F_i{}^{PH}|^2-|F_i{}^{P}|^2-|F_i{}^{H}|^2}{2|F_i{}^{P}||F_i{}^{H}|}\right) = \varphi_i{}^{H} \pm |\varepsilon_i| \qquad (1.3)$$

where ε_i is defined in Fig. 1.1.

This so-called phase ambiguity is also apparent from geometrical considerations: In the triangle of Fig. 1.1 two sides are known (in length) and of the third side ($F_i{}^{H}$) both length and direction are known; this will fix the orientation of the triangle, but nor entirely as, in fact, two orientations are possible (Fig. 1.2). The ambiguity can, in principle, be resolved by a "double

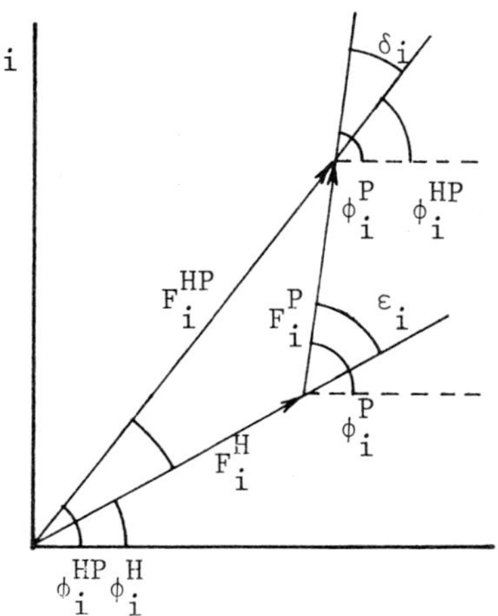

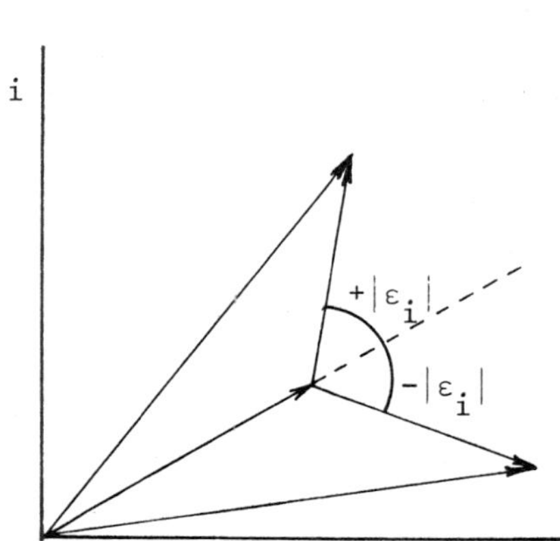

Fig. 1.1. The relation
$F_i^{HP} = F_i^P + F_i^H$
as a triangle in
the complex plane

Fig. 1.2. The phase ambiguity in the
method of Single Isomor-
phous Replacement (SIR)

phased synthesis", this is by making a Fourier with both possible complex vectors F_i^P as coefficients. Then for each structure factor the correct one will contribute to the structural image, while the false one will only contribute to the noise. The result, however, is a noisy map, which is difficult to interpret. Attempts to apply the tangent formula at this stage, to refine the phases, were rather unsuccessful. In practice, protein crystallographers use a number of heavy-atom derivatives, and their measured data sets, to resolve the phase ambiguity (often in combination with measurements of the anomalous effect). This has the additional advantage that the phase determination is then based on more data and is less likely to be affected by chance errors in the data. This method of Multiple Isomorphous Replacement (MIR) has, however, one drawback, i.e. the possibility of accumulation of systematic errors caused by non-isomorphy of the derivatives. Since a few years there is an alternative to MIR, that is: use of SIR in combination with a density modification method (e.g. the solvent flattering method); this is employed for those cases where only one derivative can be obtained or where the native protein contains a heavy atom (e.g. ferredoxine). The application of direct methods is in theory another possibility to solve the phase ambiguity, or even to find the phases without determining the heavy-atom configuration first. This will be dealt with in the next chapter.

2. SINGLE ISOMORPHOUS REPLACEMENT AND DIRECT METHODS

SIR and direct methods are both phase determining methods, the former based on the availability of additional physical (i.e. experimental) information, the latter based on a very general kind of additional information concerning the shape of the electron-density function (the very existence of atoms). If we could find a way to integrate the two methods then we could end up with a more powerful phasing procedure.

The theoretical framework for the integration of the two methods was done by Hauptman (1982a). He derived the joint probability distribution (j.p.d.)

$$P(|E_i^{PH}|, |E_i^{P}|, \varphi_i^{PH}, \varphi_i^{P}, i=1,2,3)$$

for reflections $\mathbf{h}_i$ (i=1,2,3) for which $\mathbf{h}_1+\mathbf{h}_2+\mathbf{h}_3\equiv0$.
Without writing down the expressions for the j.p.d. we know that it can contain the variables only in the form of structure-invariant phase sums, i.e. phase sums of the type

$$\varphi_i^{HP}-\varphi_i^{P} \quad (i=1,2,3,)$$
$$\varphi_1^{HP}+\varphi_2^{HP}+\varphi_3^{HP}$$
$$\varphi_1^{P}+\varphi_2^{P}+\varphi_3^{P}$$
$$\varphi_1^{HP}+\varphi_2^{HP}+\varphi_3^{P} \qquad (2.1)$$
$$\varphi_1^{HP}+\varphi_2^{P}+\varphi_3^{P}$$

Hauptman derived from the j.p.d., by integration over phases such that the sum over three phases is fixed, the conditional probability distribution of each of the possible triplet invariants, e.g. of $\Phi=\varphi_1^{P}+\varphi_2^{P}+\varphi_3^{P}$.
The result is a Von Mises distribution (circular distribution):

$$P(\Phi||F_i^{HP}|, |F_i^{P}|, i=1,2,3) = C \exp\{A \cos\Phi\} \qquad (2.2)$$

where A is a function of the six known amplitudes and of the scattering factors. (A is a measure for the sharpness of the distribution.)

Test calculations by Hauptman, Potter & Weeks (1982b) and Fortier (1983) using artificial (errorless) data on a protein indicated that a large number of Φ-estimates with large A-values can be obtained. However, as is apparent from the shape of the expression, (2.2) can only give estmates $\Phi=0$ (if A>0) or $\Phi=\pi$ (if A<0). It must be noted that values for all kinds of triplet relations, as given in (2.1), can be estimated in this way. Only those estimates are retained to be used in a subsequent determination of structure-factor phases that are sufficient accurate (i.e. have large $|A|$'s). However, the last step, determination of structure-factor phases, has not been reported yet.

In the j.p.d. $P(E_i^{HP}, E_i^P, i=1,2,3)$, which is an abbreviation for $P(|E_i^{HP}|, |E_i^P|, \varphi_i^{HP}, \varphi_i^P, i=1,2,3)$, the quantities E_i^{HP}, E_i^P are complex vectors which can take on any direciton or angle. This means that E_i^H is not fixed either, i.e. the heavy-atom positions are not assumed to be known. However, if we do know the positions of the heavy atoms, so if we do know E_i^H, then severe restrictions are imposed on E_i^{HP} and E_i^P (by relation (1.1)), and sharper distributions should result. Such distributions have been derived by Fortier, Moore & Fraser (1985) and Klop, Krabbendam & Kroon (1987).

In order to show how information about the positions of the replacement atoms can be introduced into the j.p.d. we will introduce new variables in two ways (Klop et al. 1987):

(i) Transform the variables φ_i^{HP}, φ_i^P $(i=1,2,3)$ into

$$\Phi = \varphi_1^P + \varphi_2^P + \varphi_3^P$$
$$\delta_i = \varphi_i^P - \varphi_i^{HP} \quad (i=1,2,3)$$

All invariants of the type (2.1), that occur in the expression for $P(|E_i^{HP}|, |E_i^P|, \varphi_i^{HP}, \varphi_i^P, (i=1,2,3))$, can now be expressed in the new variables, e.g.:

$$\varphi_1^{HP} + \varphi_2^{HP} + \varphi_3^P = (\varphi_1^P + \varphi_2^P + \varphi_3^P) - (\varphi_1^P - \varphi_1^{HP}) - (\varphi_2^P - \varphi_2^{HP})$$

$$= \Phi - \delta_1 - \delta_2$$

By introducing the new variables in the j.p.d. a new expression, now for

$$P(\Phi, \delta_1, \delta_2, \delta_3 \mid |E_i^{HP}|, |E_i^P|, i=1,2,3),$$

is easily obtained.

Note that δ_i is an angle of the triangle in the complex plane (see Fig.1.1). As we have seen in paragraph 1 the triangle can have two orientations (if the positions of the replacement atoms are known), so δ_i can have two possible values $\pm|\delta_i|$, where $|\delta_i|$ is known. By summing the expression for $P(\Phi\,|\,\delta_i,\,|E_i^{HP}|,\,|E_i^P|,\,i=1,2,3)$ over all 8 sign combinations of $\delta_1,\delta_2,\delta_3$ and introducing the known values of $|\delta_1|,|\delta_2|,|\delta_3|$, an expression for the conditional distribution $P(\Phi\,|\,|\delta_i|,\,E_i^{HP}|,\,|E_i^P|,\,i=1,2,3)$ is obtained. Again this expression appears to be a circular distribution, of the type (2.2), with a maximum on $\Phi=0$ or $\Phi=\pi$.

(ii) Transform the variables $\varphi_i^{HP},\varphi_i^P$ $(i=1,2,3,)$ into

$$\theta = \varphi_1^H + \varphi_2^H + \varphi_3^H$$
$$\varepsilon_i = \varphi_i^P - \varphi_i^H \quad (i=1,2,3,)$$

If the heavy-atom configuration is known, then θ and $|\varepsilon_i|$ $(i=1,2,3)$ are known. Then the following scheme can be applied:

$$P(|E_i^{HP}|,|E_i^P|,\varphi_i^{HP},\varphi_i^P,\ i=1,2,3)$$
$$\downarrow$$
$$P(\theta,\varepsilon_1,\varepsilon_2,\varepsilon_3\,|\,|E_i^{HP}|,\,|E_i^P|,\ i=1,2,3)$$
$$\downarrow$$
$$P(\varepsilon_1,\varepsilon_2,\varepsilon_3\,|\,\theta,\,|E_i^{HP}|,\,|E_i^P|,\ i=1,2,3)$$
$$\downarrow$$
$$P(S_1,S_2,S_3\,|\,|\theta_i|,\,|\varepsilon_i|,\,|E_i^P|,\,|E_i^{PH}|,\ i=1,2,3)$$

leading to an expression for the j.p.d. of the signs S_1,S_2,S_3 $(S_i=\pm1)$ of $\varepsilon_1,\varepsilon_2,\varepsilon_3$.

3. ANOMALOUS DISPERSION

If the frequency of the incident X-ray beam comes close to the absorption edge of an atom, then a resonance phenomenon will occur. As a result an extra phase shift is found between the scattered and incident beam, accompanied by a change in the magnitude of the scattering factor. This phenomenon, called anomalous dispersion, can be formalized by adding to the normal scattering factor $f_i(h)$ (subscript i labels the atom) a wavelength-dependent real part f'_{ji} and an imaginary part if''_{ji} (subscript j indicates use of wavelength λ_j). The resulting complex scattering factor $f_{ji}(h)$ then is

$$f_{ji}(h) = f_i(h) + f_{ji}' + if_{ji}''$$

The anomalous effect finds its origin in the scattering by the electrons of the inner shell; these are tightly bound to the nucleus. The inner shell occupies such a small volume that hardly any path differences between the scattered waves exist. This is the reason that usually the dependence of f' and f" on h ($=2\sin\theta/\lambda$) can be neglected. If the anomalous contributions to the structure factor are taken into account in the structure-factor formalism, then Friedel's law is violated (to what extent depends on the value of f' and f"). This break down of Friedel's law is demonstrated in the following example.

We consider a simple structure that consists of an atom at the origin of the unit cell with a normal scattering factor $f_1(h)$ and an atom at position $\mathbf{r}_2$ with an anomalous scattering factor $f_2(h)+f'_2(h)+if''_2(h)$ (omitting the index j). Then:

$$F_\mathbf{h} = f_1(h) + [f_2(h)+f_2'+if_2'']\exp2\pi i\mathbf{h}.\mathbf{r}_2$$
$$F_{-\mathbf{h}} = f_1(h) + [f_2(h)+f_2'+if_2'']\exp-2\pi i\mathbf{h}.\mathbf{r}_2$$

Fig.3.1 gives a representation of the two equations in the complex plane. It is then apparent that $|F_\mathbf{h}| \neq |F_{-\mathbf{h}}|$ and $\varphi_\mathbf{h} \neq -\varphi_{-\mathbf{h}}$ which is a violation of Friedel's law.

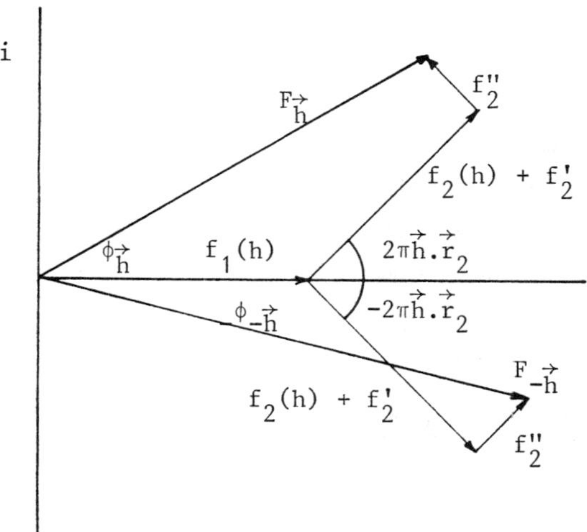

Fig. 3.1. A violation of Friedel'a law.

The fixed wavelengths of the conventional X-ray sources are usually far from the absorption edge of any atom, resulting in small anomalous contributions to the structure factor, and consequently in small deviations from Friedel's law. In fact, only for the heavier atoms a substantial effect can be expected. Nowadays, using radiation from a synchrotron source, it is possible to tune the wavelength close to the absorption edge of a large range of elements, resulting in appreciable values for f' and f". Even the collection of datasets (of the same crystal) at several wavelengths is now feasible and has, in fact, been done.

Anomalous dispersion effects play an important role in X-ray crystallography. Its application allowed the determination of absolute configuration (Bijvoet (1949)). It is also an important tool in phase-determining methods, especially for proteins. With the advent of tunable sychrotron radiation the anomalous effects can be optimised and new phasing methods have emerged. Some of them will be presented in the next paragraph.

4. ANOMALOUS DISPERSION AND STRUCTURE DETERMINATION

We consider the case of one kind of anomalous-scattering atoms in the cell (anomalous contribution to the scattering factor f'+if").
Then:

$$F(\mathbf{h}_i) = \sum_{k=1}^{N} f_k(h_i) \, \exp 2\pi \, i\mathbf{h}_i.\mathbf{r}_k + \sum_{k=1}^{p} (f'+if") \exp 2\pi i\mathbf{h}_i.\mathbf{r}_k \quad (4.1)$$

In (4.1) the first term on the right-hand side is a summation over all N atoms in the cell, only taking their normal scattering factors $f_k(h_i)$ into account; the second term is a summation over the p anomalously scattering atoms; this second term only contains the anomalous contributions to the scattering factor.
(4.1) can be written as:

$$F(\mathbf{h}_i) = F^N(h_i) + \frac{f'+if"}{f(h_i)} \sum_{k=1}^{p} f(h_i) \exp 2\pi \, i\mathbf{h}_i\mathbf{r}_k$$

$$= F^N(h_i) + \frac{f'+if}{f(h_i)} F^H(\mathbf{h}_i) \quad (4.2)$$

Here $F^N(\mathbf{h_i})$ is the "normal" structure factor (contributions from *all* atoms) and $F^H(\mathbf{h_i})$ is the normal contribution from the anomalous scatterers to the structure factor ($f(h_i)$ is the normal scattering factor of the anomalously scattering atoms). Looking at (4.2) it is apparent that for reflection $-\mathbf{h_i}$ we have

$$(F(-\mathbf{h}_1))^* = F^N(\mathbf{h}_i) + \frac{f'-if''}{f(h_i)} F^H(\mathbf{h}_i) \qquad (4.3)$$

The wavelength-dependent quantities in these equations are $F(\mathbf{h_i})$, f' and f''; we give these quantities a subscript j to indicate the wavelength λ_j ($j=1,\ldots,U$; U is the number of wavelengths for which $|F(\mathbf{h_i})|$ and $|F(-\mathbf{h_i})|$ have been measured).

Notation:

    $F(\mathbf{h_i})$ at wavelength λ_j : F_{ij}^+

    $F(-\mathbf{h_i})^*$ at wavelength λ_j: F_{ij}^-

    f' at wavelength λ_j : f_j'

    f'' at wavelength λ_j : f_j''

    $f(h_i)$ at wavelength λ_j : f

Using the above notation, (4.2) and (4.3) can be combined into one equation:

$$F_{ij}^\pm = F_i^N + \frac{f_j' \pm i f_j''}{f} F_i^H \qquad (4.4)$$

Put

$$f_j' + i f_j'' = \left(f_j'^2 + f_j''^2 \right)^{1/2} \exp i\delta_j$$

with

$$\delta_j = \arctan \frac{f_j''}{f_j'}$$

then (4.4) is

$$F_{ij}^\pm = F_i^N + \frac{(f_j'^2 + f_j''^2)^{1/2}}{f} (\exp \pm i\delta_j) F_i^H \qquad (4.5)$$

From this, after some calculation:

$$\left|F_{ij}^{\pm}\right|^2 = \left|F_i^N\right|^2 + \frac{f_j'^2 + f_j''^2}{f^2}\left|F_i^H\right|^2 + \frac{2(f_j'^2 + f_j''^2)^{1/2}}{f}$$

$$\times \left|F_i^N\right|\left|F_i^H\right|\cos(\phi_i^N - \phi_i^H \mp \delta_j) \tag{4.6}$$

Here ϕ_i^N and ϕ_i^H are the phase of F_i^N and F_i^H respectively. (4.6) is the basis for the following observations.

(i) $U = 1$

In the one-wavelength case there are two equations, one for $\left|F_{i1}^{+}\right|^2$ and one for $\left|F_{i1}^{-}\right|^2$. The unknowns in the equations are: $\left|F_i^N\right|$, $\left|F_i^H\right|$ and $\phi_i^N - \phi_i^H$. In the classical phase-determining method using one-wavelength anomalous diffraction (Peerdeman & Bijvoet (1956)), the heavy-atom configuration is determined with aid of the Patterson function. From the coördinates of the heavy atoms (anomalous scatterers) the values of $\left|F_i^H\right|$ and ϕ_i^H are calculated. Then the quantities $\left|F_i^N\right|$ and $\phi_i^N - \phi_i^H$ are solved from the quadratic equations, from which ϕ_i^N follows. There is a problem, however: the quadratic equations admit two solutions for $\phi_i^N - \phi_i^H$. One can solve this problem by just taking the smallest value for $\phi_i^N - \phi_i^H$; it is then required that the heavy-atom structure factor F_i^H has a large contribution to the total structure factor and determines its direction. For this reason the one-wavelength method cannot, on its own, be applied to determine protein phases.

(ii) $U = 2$

In this two-wavelength case there are four measured amplitudes, i.e. $|F_{i1}^{+}|$, $|F_{i1}^{-}|$, $|F_{i2}^{+}|$, $|F_{i2}^{-}|$. From the four corresponding equations the 3 unknowns, $|F_i^N|$, $|F_i^H|$, $\phi_i^N - \phi_i^H$ plus a interset scale factor can be solved. Of course, only the structure invariant $\phi_i^N - \phi_i^H$ can be determined, not the separate values of ϕ_i^N and ϕ_i^H, and there will be two solutions (as we are dealing with quadratic equations). The ambiguity can be solved by taking the solution with the smallest value of $|F_i^H|$ (the HLE estimate, i.e. the heavy atom lower estimate, which is used in protein crystallography in another context). This choice is, from a statistical point of view, the best we can do.

From test calculations it appears that the HLE estimate is largely correct and that the twofold ambiguity does not play a role in this two-wavelength case. After solving the equations, what we can do is to calculate the Patterson function with the $|F_i{}^H|^2$ as the Fourier coefficients; this Patterson function contains only the vectors between the anomalous scattering atoms and can, in general, be solved. The next step is to calculate, from the coordinates of the anomalous scatterers, the $F_i{}^H$ ($= |F_i{}^H|\exp i\varphi_i{}^H$). As the value of $\varphi_i{}^N - \varphi_i{}^H$ is known, $\varphi_i{}^N$ follows.

> Two methods are known, for U=2, to solve the set of quadratic equations: the "Bijvoet-difference" method and the "Bijvoet-ratio" method. In the first method the set of equations is solved by subtracting and adding the equations (4.6) in pairs; in the second method the ratio of pairs of equations are used. De difference method was introduced by Singh & Ramaseshan (1968). For application in neutron diffraciton, the ratio method was proposed by Unangst, Müller, Müller & Keinert (1967), a modern version is by Cascarano, Giacovazzo, Peerdeman & Kroon (1982). Klop (1989) showed that both methods are identical, provided certain requirements about scaling of the data sets are fulfilled.

(iii) U = 3

In the three-wavelength case there are 6 equations and 5 unknowns ($|F_i{}^N|, |F_i{}^H|, \varphi_i{}^N - \varphi_i{}^H$, two interset scale factors), so the twofold ambiguity can now be solved. If the scale factors are known from experiment, then the set of equations is overdetermined (see IV).

(iv) U > 3

Now the problem is overdetermined; the wavelength-independent structure invariants ($|F_i{}^N|, |F_i{}^H|, \varphi_i{}^N - \varphi_i{}^H$) can be determined by a least-squares procedure (Black (1965) for γ-diffraction, Karle (1980)). Hendrickson (1988) put the method into operation via the program system MADSYS; in his group an increasing number of protein structures are determined by this method.

5. ANOMALOUS DISPERSION AND DIRECT METHODS

An algebraic method, based on a combination of direct methods and anomalous dispersion, was introduced by Kroon, Spek & Krabbendam (1977), followed by a statistical version (Heinerman, Krabbendam, Kroon & Spek (1978)). The method, based on the conditional probability distribution $P(\Phi||E_hE_kE_{-h-k}|, |E_{-h}E_{-k}E_{h+k}|)$, with $\Phi=\phi_h+\phi_k+\phi_{-h-k}$, worked well for small molecules but failed for macromolecules. Possibly correlations were missed by using the triplet amplitudes as conditional information instead of the separate structure factor amplitudes. An algebraic method that does make use of the separate structure-factor amplitudes was given by Karle (1984); this method can also be applied to the isomorphous replacement method. Heinerman et al.(1978) suggested that the best way to solve the problem was to derive:

$$P(|E_h|, |E_k|, |E_{-h-k}|, |E_{-h}|, |E_{-k}|, |E_{h+k}|, \phi_h, \phi_k, \phi_{-h-k}, \phi_{-h}, \phi_{-k}, \phi_{h+k}).$$

This distribution has been derived by Hauptman (1982) and Giacovazzo (1983). From this j.p.d. a number of conditional distributions was derived of the general shape:

$$P(\Phi||E_h|, |E_k|, |E_{-h-k}|, |E_{-h}|, |E_{-k}|, |E_{h+k}|) = C\,\exp\{A\,\cos(\Phi-\xi)\}$$

$$(5.1)$$

with $\Phi \equiv \phi_h+\phi_k+\phi_{-h-k}$ (or any of the many other (mixed) triplet invariants, e.g. $\phi_h-\phi_{-k}+\phi_{-h-k}$). De expressions for A and ξ are complicated and extensive; they contain the structure-factor amplitudes and the scattering factors. Numerical tests show that expressions of this type can give a large number of reliable estimates for triplet phases for the case of a macromolecular structure. The method has been used with some success as an aid in solving the crystal structure of Cd, Zn metallothionein (Furey, Robbins, Clancy, Winge, Wang & Stout (1986)).

Corresponding joint probability distributions of structure factors for two or more wavelengths will be difficult to derive with the traditional methods. Peschar (1989) adapted his computer program, originally devised to derive the j.p.d. of any number of structure factors, to include Friedel-related reflections. Klop, Krabbendam & Kroon (1986, 1987) followed a completely different approach. They showed that the problem can be simplified to a great extent by splitting it up into two steps: an algebraic step (i.e. the calculation of wavelength independent quantities) and a statistical step (use of Hauptman's (1982) j.p.d. of wavelength, independent structure factors for the case of two isomorphous structures). As it turns out, the resulting distribution is valid for structure-factor amplitudes measured at any number of wavelengths; only the calculation of wavelength independent quantities (in the algebraic step) will be more accurate according as the number of wavelengths (i.e. the number of independent data sets) increases.

Test calculations, based on artificial protein data contaminated with random errors, show that: (i) a large number of triplet phases can be obtained with high accuracy (ii) an ordinary multisolution program, used to solve small-molecule structures, can handle the large amount of triplet phases and can give accurate structure-factor phases, whithout determining the heavy-atom configuration first (iii) the figures of merit, that are in common use , are able to indicate the correct set of structure factor phases.

The derivation of $P(\Phi|12$ amplitudes$)$ $(U=2)$ is divided into two steps:

(i) An algebraic step

From two-wavelenght data values for the wavelength-independent structure invariants $|F_i^N|$, $|F_i^L|$, $|F_i^H|$ and $\phi_i^H-\phi_i^N$ can be obtained via the (modified) algebraic method of Singh & Ramaseshan (1968) using the equations (4.6) (F_i^L is the contribution of the light atoms to the structure factor).

(ii) A statistical step

The treatment is based on the formal similarity between

$$P(|E_i^N|, |E_i^L|, \phi_i^N, \phi_i^L, i=1,2,3) \tag{5.2}$$

connected to the triangle relation

$$F_i^N = F_i^L + F_i^H \tag{5.3}$$

and

$$P(|E_i^{HP}|, |E_i^P|, \phi_i^{HP}, \phi_i^P, \quad i=1,2,3) \tag{5.4}$$

based on

$$F_i^{PH}=F_i^P+F_i^H \tag{5.5}$$

which is the triangle relation as it occurs in the theory of single isomorphous replacement. As a consequence Hauptman's (1982) j.p.d. (5.4) can be applied to the wavelength-independent quantities $|E_i^N|, |E_i^L|, \varphi_i^N, \varphi_i^L$, resulting in an expression for (5.2).

After the transformation

$$\Phi^N=\varphi_1^N+\varphi_2^N+\varphi_3^N \qquad (\mathbf{h}_1+\mathbf{h}_2+\mathbf{h}_3\equiv0)$$

$$\nu_i=\varphi_i^N-\varphi_i^L \qquad (i=1,2,3)$$

(5.2) transforms to

$$P(\Phi^N|, |E_i^N|, |E_i^L|, \nu_i, i=1,2,3)$$

which leads to

$$P(\Phi^N|E_i^N|, |E_i^L|, \nu_i, i=1,2,3)= c \exp \kappa \cos (\Phi^N-\zeta)$$

This is a circular distribution round ζ. The quantities κ and ζ are relatively simple functions of $|E_i^N|, |E_i^L|, \nu_i$ $(i=1,2,3)$ and the scattering factors. Note that here, contrary to the case of SIR, the two-phase invariants ν_i are not only known in absolute value but also in sign. The remarkable result of this derivation in two steps is, that the number of wavelength-dependent datasets (i.e. the number of wavelengths) used in the first step is immaterial to the distribution in the second step as long as U>1. However, the accuracy of the calculated values of $|E_i^N|, |E_i^L|, \nu_i$ will increase with the number of wavelengths used.

In a test calculation with artificial two-wavelength data (2Å resolution) of the protein ferredoxine (MW≈6000) a total of 1707567 triplet phases were estimated with an average phase error of 35.1° (Table 5.1). From the top 80000 invariants 96 structure-factor phase

Table 5.1 Average phase-error in the estimated values of
triplet invariants arranged according to
decreasing values of $\langle\kappa\rangle$ for ferredoxine at 2Å
resolution.

Number of invariants	κ_{max}	κ_{min}	$\langle\kappa\rangle$	$\langle\Phi^{init}(error)\rangle$
1000	27.14	7.12	9.15	11.7
1000	7.12	5.88	6.41	15.5
1000	5.88	5.26	5.54	17.6
1000	5.26	4.85	5.04	19.2
1000	4.84	4.53	4.68	20.4
20000	27.14	2.87	4.14	23.8
80000	27.14	1.70	2.63	35.1

Table 5.2 Tangent formula phasing for ferredoxine at 2Å
resolution for different magnitudes of error
(s.d.)

s.d.(%)	0	4	8
number of reflections	2225	2210	2283
number of triplets generated	1707567	1750232	1973111
number of triplets used	80000	80000	80000
κ_{min}	1.70	1.78	2.23
$\langle\kappa\rangle$	2.63	2.76	3.30
$\langle\Phi^{init}(error)\rangle°$	35.1	41.4	57.2
number of phase sets	96	96	96
number of reflections	1800	1843	1934
$\langle\varphi^{tan}(error)\rangle°$	11.7	13.0	24.2
$\langle\Phi^{tan}(error)\rangle°*$	12.6	15.5	32.0
ABSFOM	1.0110	0.9099	0.6475
PSIZERO	2.648	2.211	2.53
RESID	9.32	11.38	32.34
COMFOM	2.9796	2.9813	2.9269
Rank of correct set	1	2	1

* refined structure-factor phases re-introduced into triplets.

sets were generated by a multisolution procedure (the program SULTAN, an adapted version of MULTAN). The set with the best "combined" figure of merit had an average phase-error of 9.32˚. These results, referring to errorless data, are compiled in Table 5.2. This table also shows that the method is robust in regard to random errors in the data.

Multi-wavelength methods have become of practical importance since the advent of (tunable) synchrotron radiation. Up till now only the algebraic equations (4.6) have been used to determine protein structures (Hendrickson, 1988). Conclusions about the practical value of the application of direct methods to the multi-wavelength case must wait for tests with real data.

REFERENCES

Bijvoet, J.M.(1949) Proc.Acad.Sci.Amst.52, 313

Black, P.J.(1965). Nature. 206, 1223-1226

Blundell, T.L. & Johnson, L.N. (1976). "Protein Crystallography", Academic Press, p. 154

Cascarano, C, Giacovazzo, C., Peerdeman, A.F. & Kroon, J. (1982). Acta Cryst. A38, 710-717

Giacovazzo, C. (1983). Acta Crystl A39, 585-592

Fortier, S., Moore, N.J. & Fraser, M.E. (1985). Acta Cryst. A41, 571-577

Furey, W.F., Robbins, A.H., Clancey, L.L., Winge, D.R., Wang, B.C. & Stout, C.D. (1986). Science 231, 704-710

Hauptman, H.A. (1982[a]). Acta Cryst. A38, 289-294

Hauptman, H.A. (1982[b]). Acta Cryst. A38, 632-641

Hauptman, H.A., Potter, S. & Weeks, C.M. (1982). Acta Cryst. A38, 294-300

Heinerman, J.J.L., Krabbendam, H., Kroon, J. & Spek, A.L. (1978). Acta Cryst. 447-450

Hendrickson (1988). In "Crystallographic Computing 4" eds. Isaac & Taylor, pp. 97-108

Karle, J. (1980). Intern. J. of Quantum Chemistry 7, 357-367

Karle, J. (1984). Acta Cryst. A40, 374-379

Klop, E.A., Krabbendam, H. & Kroon, J. (1986). Collected Abstracts, ECM10, Wroclaw, Poland, 3D-15

Klop, E.A., Krabbendam, H. & Kroon, J. (1987). Acta Cryst. A43, 810-820

Klop, E.A., Krabbendam, H. & Kroon, J. (1989)

Klop, E.A. (1989). Thesis University of Utrecht

Kroon, J., Spek, A.L. & Krabbendam, H. (1977). Acta Cryst.
     A33, 382-385
Peerdeman, A.F., Bijvoet, J.M. (1956). Proc. Koninkl. Ned.
     Akad. Wetenschap. B59, 312
Singh, A.K. & Ramaseshan, S. (1968). Acta Cryst. B24, 35-39
Templeton, L.K., Templeton, D.H., Phizackerly, R.P. & Hodgson,
     K.O. (1982). Acta Cryst. A38, 74-78
Unangst, D., Müller, E., Müller, J. & Keinert, B. (1967). Acta
     Cryst. 23, 898-901

NOTE ON "SUPERLARGE" STRUCTURES AND THEIR PHASE PROBLEM

David Sayre

IBM Research Center
Yorktown Heights, NY 10598, USA

INTRODUCTION

Work carried out over the last few years[1] suggests that it may be
possible to extend x-ray diffraction techniques to the determination of
structures of non-crystalline specimens of micron size, such as single
small biological cells. Work to date has concentrated on showing that the
diffraction from microscopic non-crystalline objects, while extremely
faint, can be observed by employing high-intensity synchrotron x-ray
sources, shifting to longer wavelengths, and increasing the coherence of
the radiation. Assuming undulator radiation with high spatial coherence
it is mainly the temporal coherence of the radiation which is at issue;
assuming radiation of monochromaticity M the maximum resolution to which
diffraction from a specimen of diameter a will normally extend is of the
order of a/M. Using radiation with M=150 and individual cells of
Minutocellus polymorphus with a=3µm we have observed pattern in 18A radiation to approximately 200A resolution, in good agreement with the above.
Fig. 1 shows a portion of such a pattern. In certain special cases we have
observed diffraction to 70A resolution.

Efforts are now proceeding to allow work with M=1000 and with smaller
cells having a=1µm, at which time we hope (subject to radiation damage
considerations) that pattern will be observable to 10A, or essentially the
diffraction limit of 9A achievable with 18A radiation.

Should this hope be realized, this form of non-crystalline diffraction
could supply diffraction intensity data on general micron-size biological
and non-biological specimens which would permit in principle the imaging
of these 10^{12} dalton structures at resolutions approximating 10A. For this
to occur, however, it will be necessary to phase the intensity data.

THE PHASE PROBLEM

At first sight the phase problem in this case seems a very difficult
one, in view of the extreme size of the structures involved. Curiously,
on second sight, it appears that there may be almost no phase problem at
all, due to the property of the non-periodic specimen that its Fourier
transform can be observed continuously in reciprocal space and not simply
at the discrete Bragg locations (see again Fig. 1). It was shown several

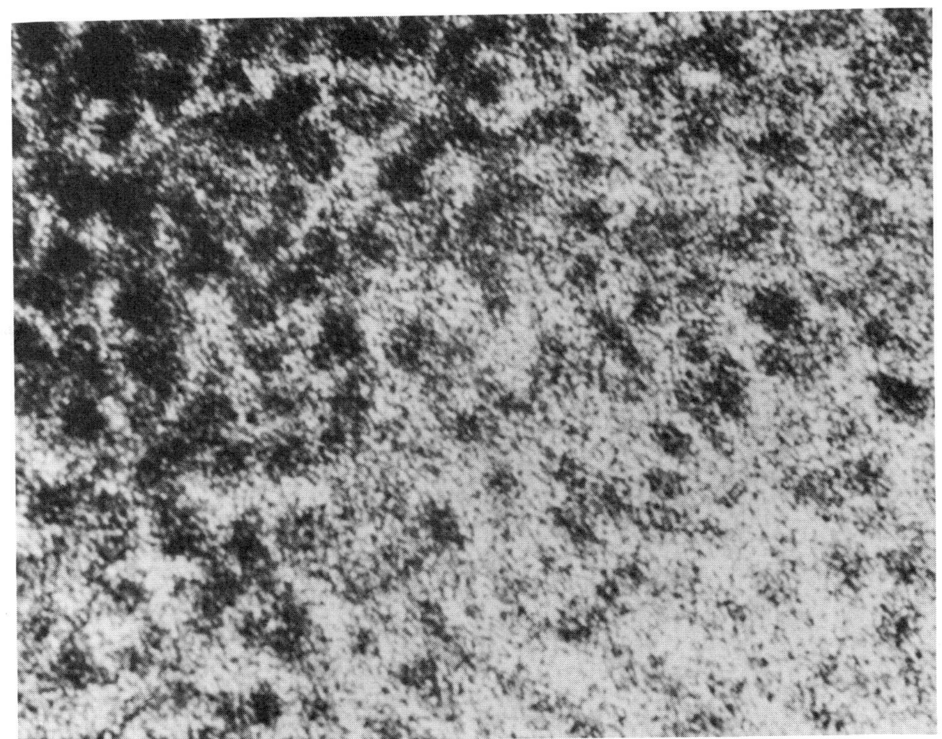

Fig. 1. Portion of a diffraction photograph of a single cell of
Minutocellus polymorphus, taken with 18A x-rays from the soft
x-ray undulator at the National Synchrotron Light Source at
Brookhaven. Exposure time 20 minutes. The specimen was
stationary during the exposure, and the pattern seen is the trace
of the stationary reflecting sphere through the continuous-valued
Fourier transform of the specimen. The pattern was recorded on
silver halide film and is shown here magnified 250 times; the
film graininess can be seen. The amount of structural information
contained even in this small portion (containing approximately
1000 Bragg points) of the pattern is striking.

decades ago, for the centrosymmetric case, that if magnitudes could be
measured at the half-integral as well as the integral values of the Miller
indices (8-fold oversampling in the 3-dimensional case) the phase problem
would disappear[2], and more recently it has been shown that in the non-
centrosymmetric case as well, possession of the magnitude data at the integral and half-integral values uniquely determines any 2- or 3-dimensional
structure having a finite envelope[3,4]. Following on this result there has
grown up a sizeable signal recovery literature concerned with numerical-
processing techniques for actually reaching the uniquely determined
structure[5]. This latter area has had its counterpart in crystallography
also, principally in cases where a solvent-expanded unit cell makes possible some degree of oversampling of the Fourier magnitudes; a unified
treatment of this and other crystallographic techniques has been given[6].
Thus there seems to be a very strong theoretical basis, and a fairly strong
practical one, for the phasing of diffraction data from superlarge struc-

tures, when and if such data become available.* An additional point is that
diffraction with soft x-rays offers interesting anomalous dispersion effects with low-Z atoms. An example is the virtual disappearance of carbon
atoms as coherent scatterers at a wavelength of 44.6A.

Finally, on third sight, it is wise to retain a significant degree
of caution on the subject of superlarge structure phasing. The actual
signal recovery experiments to date (ref. 5) are not, in the author's view,
especially impressive in terms of image quality; they are computationally
expensive; and the experiments have been on an extremely small scale compared with the 10^9 picture elements which will exist in the case of
3-dimensional 10A imaging of a 1μm object. Thus, while the possibilities
exist, a large amount of work will be needed as well.

RELATION TO OTHER PHASING METHODS

It is of interest to note that the direct phasing methods which are
the main subject of this school consist in combining a minimal amount of
intensity data (data at Bragg positions only) with a strong structural
constraint (atomicity). At the other end of the spectrum is the case presented here: 8 times more intensity data, but with a lesser structural
assumption (finite envelope). Between the two are the techniques of
macromolecular crystallography involving enlarged intensity data sets based
on variations in site occupancy, crystal form, wavelength, solvation, etc.,
in each case with an appropriate special structural assumption. It is appropriate to note here that there is a growing recognition that these
methods can be thought of from a common point of view.

ACKNOWLEDGEMENTS

Colleagues in the work described in the introductory part of the paper
have been Prof. Janos Kirz, Dept. of Physics, State University of New York
at Stony Brook; Dr. Wen Bing Yun, Argonne National Laboratory; and Prof.
Mark Sharnoff, Dept. of Physics, University of Delaware. Dr. Gerard
Bricogne has played an important part in the discussions to date on the
problem of phasing, and it is hoped will continue to play an active part
in future work.

---

*The continuity of the diffraction pattern may also be exploited through
the method of holography, a technique which normally fails in
crystallography because of the practical impossibility of maintaining accurate phase relationships between signal and reference beams while carrying out the crystal rotation which is necessary to bring new Bragg
reflections into reflecting position. The problem does not arise in the
present case where an extensive amount of diffraction pattern can be recorded with the specimen stationary. Experiments in mixing reference
signals with patterns like that of Fig. 1, with subsequent image reconstruction, have been successfully carried out[7-9]. The diffraction methods
will be much slower than the holographic methods, but are expected to yield
higher resolution.

REFERENCES

1. D. Sayre, W.B. Yun, and J. Kirz, Experimental observation of diffraction patterns from micro-specimens, in: "X-Ray Microscopy II", Springer Series in Optical Sciences Vol 56 (1988).

2. D. Sayre, Some implications of a theorem due to Shannon, Acta Cryst. 5:843 (1952).

3. M.H. Hayes, The reconstruction of a multidimensional sequence from the phase or magnitude of its Fourier transform, IEEE ASSP-30:140 (1982).

4. R.H.T. Bates, Fourier phase problems are uniquely solvable in more than one dimension. I: Underlying theory, Optik 61:247 (1982).

5. J. Fienup and C. Rushforth, eds., J.Opt.Soc.Am. A4:102-304 (1987).

6. G. Bricogne, A Bayesian statistical theory of the phase problem. I. A multichannel maximum-entropy formalism for constructing generalized joint probability distributions of structure factors, Acta Cryst A44:517 (1988).

7. S. Aoki and S. Kikuta, X-ray holographic microscopy, Jpn.J.App.Phys. 13:1385 (1974).

8. C. Jacobsen, J. Kirz, M. Howells, K. McQuaid, S. Rothman, R. Feder, and D. Sayre, Progress in high-resolution x-ray holographic microscopy, in: "X-Ray Microscopy II", Springer Series in Optical Sciences Vol 56 (1988).

9. D. Joyeux, S. Lowenthal, F. Polack, and A. Bernstein, "X-ray microscopy by holography at LURE", ibid., p. 246 (1988).

SOME EXERCISES IN DIRECT METHODS

Henk Schenk and Yuan-fang Wang

Laboratory for Crystallography, University of Amsterdam
Nieuwe Achtergracht 166, 1018 WV Amsterdam
The Netherlands

Introduction

This chapter consists primarily from a table of triplet relations which should be used in practical exercises. The triplets belong to a triclinic centrosymmetric structure (space group P1bar) of code name WILLES. It is a steroid with 20 'heavy' atoms and formula $C_{17}H_{23}NO_2$. The cell constants are a = 8.505, b = 8.688, c = 13.513 A, α =125.00, β = 96.51, and γ = 106.46°.

The triplets are collected by the program SIMPEL83 and reproduced here in a reduced mode such that only the triplets of the strongest 60 reflections have been printed. Per reflection the following numbers are given: the reference number of the particular reflection (marked with REFL1), a star, the three indices h, k and l and the $|E|$ value and then for each triplet of that reflection the reference numbers of the second and third reflection (REFL2 and REFL3) of the triplet and the triple product E_3 (marked with E3). WILLES is a centrosymmetric structure so that in the exercises the product of the signs will be used rather that the sum of the phases. For ease of use the tables are organised such that
$$sign(REFL1) = sign(REFL2) * sign(REFL3).$$
So the first line of the table implies that reflections 1, 16 and 21 form a triplet such that the sign s(1) is given by s(1)=s(16)*s(21). Check the existence of this triplet through the indices of the three reflections. Note that the same triplet is also present in the partial lists of the reflections 16 and 21 at pages 3 and 4 of the list. Check some more triplets in the list.

The first part of this paper describes the experiments which can be carried out with the help of the tables. The second part handles about a Computer Assisted Symbolic Addition program which enables the teaching of Direct Methods through a PC or other microcomputer.

The exercises

For ease of working four worksheets are added for four different exercises refering to the first chapter (Introduction) and the chapter on Symbolic Addition. The following exercises can be carried out:

1. Sign progagation, to be worked out after reading the introductory chapter. In the first worksheet seven reflections have been starred, those reflections form a starting set. The three origin defining reflections have been given the + sign (check whether have

Direct Methods of Solving Crystal Stuctures
Edited by H. Schenk, Plenum Press, New York, 1991

have been chosen according the rules), the other four are ambiguities and can be either
− or +. Choose for each sign either + or − (donot choose all reflections to be +, why
not?) and write their signs in the worksheet. Try to find new signs by systematically
checking the triplet list following the not known phases in the worksheet. So reflection 6
is the first one to be checked because 1, 2, 3, 4 and 5 are already in the starting set.
For 6 there is a triplet with 5 and 8 which are both known and as a result the phase
of 6 can be calculated and taken up in the worksheet. Continue with reflection 7, 9, 10,
11, 13 and so on and take each determined sign up in the worksheet as known and use
it for determining new signs. When reaching 60 start again with the first not known
reflection and go on until all signs have been found. Is it possible to determine all signs?
If not, are the remaining reflections correlated through triplets and what can be done to
phase them?

In case more than one person works simultaneously on this exercise different sets of
starting signs should be choosen by each of them. In case 15 personss are working in
parallel all 15 non−trivial solutions can be worked out simultaneously, thus mimicing the
phase−determining process of a multisolution procedure.

2. Origin choice and starting set determination followed by sign propagation. For the
sign propagation the rules of exercise 1 can be followed and origin and starting set
choice have to be done following the receipes of the chapters on Origin and
Enantiomorph definition (Krabbendam) and Starting Set (Beurskens) respectively. Are
there essential differences between the outcomes of exercises 1 and 2?

3. With the same starting set as used in exercise 1 now symbolic addition must be
carried out following the rules of the chapter on Symbolic Addition by the author. In
particular notice expression 5 of that paper and the way this expression should be used
as described in the paragraphs on The Handling of Symbols in Phase Propagation and on
Symbolic Phase Extension. Very strict acceptance criteria should be used indeed based on
the actual E_3 values, so e.g. 6.5 for a single indication, 8.5 for the sum of two identcal
indications, 11 for a triple one and so on. The ambiguities in the starting set get now
the symbolic phases a, b, c, and d. It will be clear that by using symbols all possible
solutions of the phase problem are now worked out in one run (compare with exercise 1).

4. Similar to exercise 3 with the difference that now an own starting set has to be
determined following the receipes of the relevant chapters (see also exercise 2) or, as an
alternative, following the last lines of the paragraph on Symboloc Phase Extension of the
chapter on Symbolic Addition.

The first worksheet follows on the next page and the other three follow the seven
and a half page table of triplets.

A Symbolic Addition Teaching Program for PC's

Since Direct Methods are still of growing importance as a tool of solving crystal
structures from single crystal data and most program systems for Direct Methods are not
very transparent, in our teaching practice to chemistry students we developed a small
program system for the TRS80 model I microcomputer, in which the students are guided
to work with the symbolic addition method (A. van der Voort, B.O. Loopstra and H.
Schenk, course material in Dutch). The main task of the computer is to teach the
student direct methods and to do the administration, while the students learn to take the
essential decisions in the course of a phase extension process.

With these experiences as background we developed a new program system which
allows more flexibility to the user. It has been written for a CBM64 with floppy disk
unit and recently ported to PCDOS and MSDOS computers. The program system is
menu driven and the manual has been integrated. Input data for projections of a few
structures are provided on disk as well, however, it is easy to create a file of a suitable
structure. Structures in which the overlap in the projection is minimal are to be

preferred as Direct Methods function better at atomic resolution. The procedure can handle projections with triclinic centrosymmetric symmetry only, and as a result structures with one short axis can be used successfully.

The menu gives the following options:
- Manual:
A short explanation of the method and how to handle it. All input and output parameters in detail.
- Input data:
The student can choose between an input by keyboard or by disk. Keyboard data can be written on disk as well, for future use.
- Generation of triplets:
This part can only be executed after data input. It is fully automatic and generates triplets with an E3 value higher than a limit value. This limit is given by the student.
- Symbolic addition:
In this part the computer does only the administration; the student makes the decisions, i.e. she/he chooses the sign/symbol of the the reflections in thestarting set and decides whether the calculated phase of a reflection is being accepted.
- E-map:
The student defines the values of the symbols and the system, then the program calculates on the basis of the phases an E-map. The user of the program finally decides from the diffrent E-maps which one resembles most the chemical reality.
- For the teacher there are possibilities to by-pass the decision-making process in a semi-automated and a full-automated mode. These options are valuable for error correction and demonstration.

The program is accompanied by a programmed text, which teaches the triplet relation along the lines of chapter 1 of the proceedings of this conference and uses a similar approach to teach the priciples of the symbolic addition method. The work on the teaching programs has been sponsored in part by the Erasmus Scheme of the European Community.

WORKSHEET 1. Worksheet for the generation of new signsfrom a given starting set (exercise 1). Starred reflections are the reflections of the starting set, signs imply the reflection defining the origin.

*1		11	21	31	41	51
*2	+	*12	22	32	42	52
*3	+	13	23	33	43	53
*4		14	24	34	44	54
*5		15	25	35	45	55
6		16	26	36	46	56
7		17	27	37	47	57
*8	+	18	28	38	48	58
9		19	29	39	49	59
10		20	30	40	50	60

TRIPLET LISTING FOR: WILLES

REFL1	H	K	L	E	REFL2	REFL3	E3
1 *	6	-5	9	4.15	16	21	7.247
					17	19	7.343
					23	26	6.618
					24	49	5.528
					31	57	4.996
2 *	7	-5	7	3.98	3	29	7.522
					10	38	6.175
					11	17	7.341
					13	28	6.903
					31	47	5.075
					46	55	4.318
3 *	6	-6	9	3.87	2	29	7.522
					6	55	5.885
					8	13	7.537
					10	24	6.868
					11	21	6.877
					26	48	5.179
					31	53	4.770
4 *	-1	5	1	3.82	7	29	6.962
					8	11	7.442
					10	48	5.693
					13	21	6.790
					14	31	6.230
					15	19	6.898
					22	41	5.278
					23	49	5.087
					24	26	6.091
					37	45	4.556
					39	41	4.584
					43	60	4.096
5 *	-4	-3	10	3.82	6	8	8.035
					12	43	5.699
					13	55	5.382
					14	22	6.627
					28	46	5.104
					29	37	5.210
					30	32	5.472
					31	41	4.961
6 *	2	-2	9	3.76	3	55	5.885
					5	8	8.035
					9	12	7.306
					10	31	6.312
					14	48	5.439
					15	33	5.875
					23	34	5.368
					24	53	4.898
					29	46	4.924
					38	47	4.430
					40	41	4.431

7	*	0	-6	1	3.73		
						4	29	6.962

13	*	0	-7	10	3.48	2	28	6.903
						3	8	7.537
						4	21	6.790
						5	55	5.382
						7	17	6.881
						12	30	5.867
						19	36	5.131
						22	26	5.586
						24	41	4.779
						40	53	3.896
14	*	-3	5	3	3.43	4	31	6.230
						5	22	6.627
						6	48	5.439
						11	40	5.153
						15	43	5.025
						18	29	5.592
						20	45	4.806
						26	55	4.399
15	*	5	1	1	3.42	4	19	6.898
						6	33	5.875
						12	22	6.025
						14	43	5.025
						21	36	4.874
						26	30	5.250
						31	60	4.070
16	*	-7	3	2	3.42	1	21	7.247
						19	29	5.579
						23	41	4.695
						30	38	4.601
						34	40	4.195
						35	54	3.903
						39	49	3.982
17	*	0	-1	9	3.35	1	19	7.343
						2	11	7.341
						7	13	6.881
						21	29	5.285
						30	49	4.296
						38	39	4.093
18	*	-2	6	1	3.34	7	31	5.924
						14	29	5.592
						22	37	4.716
						46	48	3.778
						49	58	3.510
19	*	-6	4	0	3.34	1	17	7.343
						4	15	6.898
						9	40	5.089
						12	41	5.018
						13	36	5.131
						16	29	5.579
						24	30	5.138
						27	35	4.721
						31	43	4.303
20	*	-5	-2	10	3.28	14	45	4.806
						25	27	5.232
						31	37	4.357

21	*	-1	-2	11	3.23	1	16	7.247
						3	11	6.877
						4	13	6.790
						15	36	4.874
						17	29	5.285
						24	39	4.511
						26	41	4.420
						38	42	3.845
22	*	-1	-8	7	3.20	4	41	5.278
						5	14	6.627
						8	48	4.779
						9	33	5.162
						12	15	6.025
						13	26	5.586
						18	37	4.716
						58	60	3.177
23	*	5	-6	6	3.18	1	26	6.618
						4	49	5.087
						6	34	5.368
						9	46	4.759
						12	29	5.408
						16	41	4.695
						37	43	3.803
						45	60	3.398
24	*	2	-4	2	3.18	1	49	5.528
						3	10	6.868
						4	26	6.091
						6	53	4.898
						13	41	4.779
						19	30	5.138
						21	39	4.511
						29	38	4.317
						31	55	3.870
						46	47	3.615
25	*	-8	-1	2	3.18	8	51	4.661
						20	27	5.232
						37	52	3.677
26	*	1	1	3	3.17	1	23	6.618
						3	48	5.179
						4	24	6.091
						13	22	5.586
						14	55	4.399
						15	30	5.250
						21	41	4.420
27	*	3	-1	8	3.17	19	35	4.721
						20	25	5.232
						29	54	3.965
						31	52	3.952
						40	44	3.699
28	*	-7	-2	3	3.15	2	13	6.903
						5	46	5.104
						7	11	6.471
						8	29	5.447

29	*	-1	-1	2	3.09	2	3	7.522
						4	7	6.962
						5	37	5.210
						6	46	4.924
						8	28	5.447
						9	34	4.899
						12	23	5.408
						14	18	5.592
						16	19	5.579
						17	21	5.285
						24	38	4.317
						27	54	3.965
						39	42	3.680
						47	53	3.395
30	*	-4	0	2	3.06	5	32	5.472
						12	13	5.867
						15	26	5.250
						16	38	4.601
						17	49	4.296
						19	24	5.138
						36	41	3.689
						43	55	3.357
31	*	-2	0	2	3.01	1	57	4.996
						2	47	5.075
						3	53	4.770
						4	14	6.230
						5	41	4.961
						6	10	6.312
						7	18	5.924
						8	40	4.596
						12	58	4.156
						15	60	4.070
						19	43	4.303
						20	37	4.357
						24	55	3.870
						27	52	3.952
32	*	0	-3	8	2.96	5	30	5.472
						8	50	4.372
						12	55	4.168
33	*	-3	-3	8	2.89	6	15	5.875
						9	22	5.162
						10	60	4.036
						43	48	3.305
34	*	-3	4	3	2.84	6	23	5.368
						9	29	4.899
						16	40	4.195
35	*	-3	3	8	2.82	9	44	4.254
						16	54	3.903
						19	27	4.721
						43	52	3.167
36	*	-6	-3	10	2.79	13	19	5.131
						15	21	4.874
						30	41	3.689

37	*	-3	-2	8	2.79	4	45	4.556
						5	29	5.210
						8	46	4.189
						18	22	4.716
						20	31	4.357
						23	43	3.803
						25	52	3.677
						44	51	3.125
						49	60	2.925
						56	59	2.804
38	*	-3	3	0	2.78	2	10	6.175
						6	47	4.430
						16	30	4.601
						17	39	4.093
						21	42	3.845
						24	29	4.317
39	*	-3	2	9	2.78	4	41	4.584
						10	11	5.402
						16	49	3.982
						17	38	4.093
						21	24	4.511
						29	42	3.680
40	*	4	1	1	2.73	6	41	4.431
						8	31	4.596
						9	19	5.089
						11	14	5.153
						13	53	3.896
						16	34	4.195
						27	44	3.699
41	*	-2	-3	8	2.73	4	22	5.278
						4	39	4.584
						5	31	4.961
						6	40	4.431
						8	10	5.394
						11	48	4.012
						12	19	5.018
						13	24	4.779
						16	23	4.695
						21	26	4.420
						30	36	3.689
						43	58	2.937
42	*	-4	1	11	2.71	21	38	3.845
						29	39	3.680
43	*	-8	4	2	2.71	4	60	4.096
						5	12	5.699
						8	9	5.353
						14	15	5.025
						19	31	4.303
						23	37	3.803
						30	55	3.357
						33	48	3.305
						35	52	3.167
						41	58	2.937

44	*	-1	-2	7	2.70	8	52	3.961
						9	35	4.254
						27	40	3.699
						37	51	3.125
45	*	-2	-7	7	2.70	4	37	4.556
						14	20	4.806
						23	60	3.398
46	*	3	-1	7	2.68	2	55	4.318
						5	28	5.104
						6	29	4.924
						8	37	4.189
						9	23	4.759
						18	48	3.778
						24	47	3.615
						51	52	2.911
47	*	5	-5	9	2.68	2	31	5.075
						6	38	4.430
						24	46	3.615
						29	53	3.395
48	*	5	-7	6	2.67	3	26	5.179
						4	10	5.693
						6	14	5.439
						8	22	4.779
						11	41	4.012
						18	46	3.778
						33	43	3.305
49	*	4	-1	7	2.65	1	24	5.528
						4	23	5.087
						7	12	5.440
						16	39	3.982
						17	30	4.296
						18	58	3.510
						37	60	2.925
50	*	-6	-4	9	2.64	8	32	4.372
51	*	-2	0	1	2.62	8	25	4.661
						37	44	3.125
						46	52	2.911
52	*	5	-1	6	2.62	8	44	3.961
						25	37	3.677
						27	31	3.952
						35	43	3.167
						46	51	2.911
53	*	4	-6	11	2.59	3	31	4.770
						6	24	4.898
						13	40	3.896
						29	47	3.395
54	*	4	0	6	2.56	16	35	3.903
						27	29	3.965

55	*	-4	4	0	2.56		2	46	4.318
							3	6	5.885
							5	13	5.382
							12	32	4.168
							14	26	4.399
							24	31	3.870
							30	43	3.357
56	*	1	-2	12	2.53		37	59	2.804
57	*	4	-5	11	2.53		1	31	4.996
58	*	6	-7	6	2.51		9	10	4.947
							12	31	4.156
							18	49	3.510
							22	60	3.177
							41	43	2.937
59	*	4	0	4	2.51		37	56	2.804
60	*	-7	-1	1	2.50		4	43	4.096
							10	33	4.036
							15	31	4.070
							22	58	3.177
							23	45	3.398
							37	49	2.925

WORKSHEET 2. Worksheet for the generation of new signs from a free starting set (exercise 2).

1	11	21	31	41	51
2	12	22	32	42	52
3	13	23	33	43	53
4	14	24	34	44	54
5	15	25	35	45	55
6	16	26	36	46	56
7	17	27	37	47	57
8	18	28	38	48	58
9	19	29	39	49	59
10	20	30	40	50	60

WORKSHEET 3. Worksheet for the generation of new symbolic signs from a given starting set (exercise 3). Starred reflections are the reflections of the starting set, signs with starred reflections indicate reflections defining the origin.

*1	11	21	31	41	51
*2	*12	22	32	42	52
*3 +	13	23	33	43	53
*4 +	14	24	34	44	54
*5	15	25	35	45	55
6	16	26	36	46	56
7	17	27	37	47	57
*8 +	18	28	38	48	58
9	19	29	39	49	59
10	20	30	40	50	60

WORKSHEET 4. Worksheet for the generation of new symbolic signs from a free starting set (exercise 4).

1	11	21	31	41	51
2	12	22	32	42	52
3	13	23	33	43	53
4	14	24	34	44	54
5	15	25	35	45	55
6	16	26	36	46	56
7	17	27	37	47	57
8	18	28	38	48	58
9	19	29	39	49	59
10	20	30	40	50	60

THE DIRDIF PROGRAM SYSTEM -

PRACTICAL SESSION AND EXERCISES

Paul T. Beurskens, Gezina Beurskens, W.P. Bosman,
S. García-Granda, R.O. Gould, J.M.M. Smits and C. Smykalla

Crystallography Laboratory, Toernooiveld
6525 ED Nijmegen, The Netherlands

1. Job control in FORTRAN

The new version of the DIRDIF system consists of several programs which are interrelated.
The program flow is controlled by

1. user-defined instructions,
2. control data (new, or from preceeding runs),
3. system control commands and, of course,
4. the outcome of various calculations.

All decisions regarding the calculations are prepared by (or based on) programs written in
FORTRAN77, which are truly computer independent and run equally well on a PC as on a main
frame computer. The one and only computer dependent procedure (which is to link the various
FORTRAN programs in the requested order) can be defined in a few statements.
DD is the name of this job procedure file: it is activated by entering 'DD' via the keyboard. The
interactive program DDSTART prepares the DDSYST file and the CONDA file. The procedure then
continues as follows: read LABEL from this DDSYST file, and go to execute the program identified
by LABEL; see Fig. 1.

Programs:

PROGi (i = 1, 2, ...) are various individual and independent programs.

DDSTART is the first program to be executed in any run (computer job). The decision over what is
to be done is made here. The program is an interactive program for the preparation of all
control data files. It has access to all crystal data, reflection data, atomic parameters and
control data files. If necessary, the program will prompt the user to supply such data.

Files:

CONDA is a file which contains control data and user supplied executional parameters for the current
run. It serves as communication between the various stages of the calculation.
DDSYST is a file which contains a list of labels (program names) which control the program flow in
the current run.

Direct Methods of Solving Crystal Stuctures
Edited by H. Schenk, Plenum Press, New York, 1991

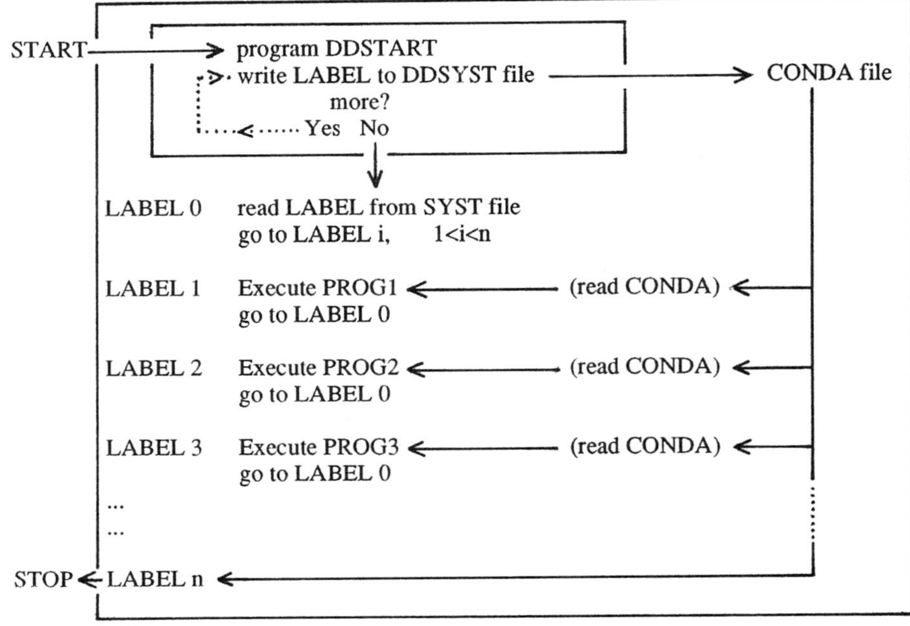

Fig. 1. Flow scheme of the job control procedure.

2. Flow diagram of the DIRDIF program system

DIRDIF is a composition of several interrelated computer programs, which are finely tuned into a program system for solving crystal structures when part of the structure is known. DIRDIF is also the name of its main component, a program which applies DIRect methods to DIFference structure factors for phase refinement and phase extension.

The most important programs of the DIRDIF system are:

ORIENT = orientation functions in vector space (minimum average function, Nordman)
TRACOR = translation functions in reciprocal space (correlation functions)
DIRDIF = direct methods including phase refinement (subprogram DIFTAN)

The interrelationship is shown in Fig. 2.
When you
- have molecules with known geometry .... ===> use ORIENT - TRACOR - DIRDIF
- ended up with a misplaced fragment .... ===> use TRACOR - DIRDIF
- know a small fragment or a heavy atom .... ===> use DIRDIF

The main program flow for these entries, as given in Fig. 2, is controlled by the DDSYST and CONDA files. During the course of the analysis, several short formatted files are generated which store permanent data (e.g. CRYSDA), variable data (e.g. ATOMS) or control data. Other files are defined as needed for the individual programs PROGi. All are defined or generated by a FORTAN77 program, and only a few files may be modified manually by the user.

At the start of a new DIRDIF run, the program DDSTART is executed interactively for the preparation of the control data files. If the CRYSDA file is absent, the program system will link automatically to the program CRYSDA to create this file (where some user input will be required) before continuing.

If the BINFO file (permanent reflection data file) is absent, the program MERBIN will be called to read, merge and sort the formatted reflection data file FREFA to create the BINFO file before continuing. Thereafter the user may choose from the available options.

Note: DDMAIN is a collection of subprograms which may be needed for the execution of other programs (to be denoted PROGi).

370

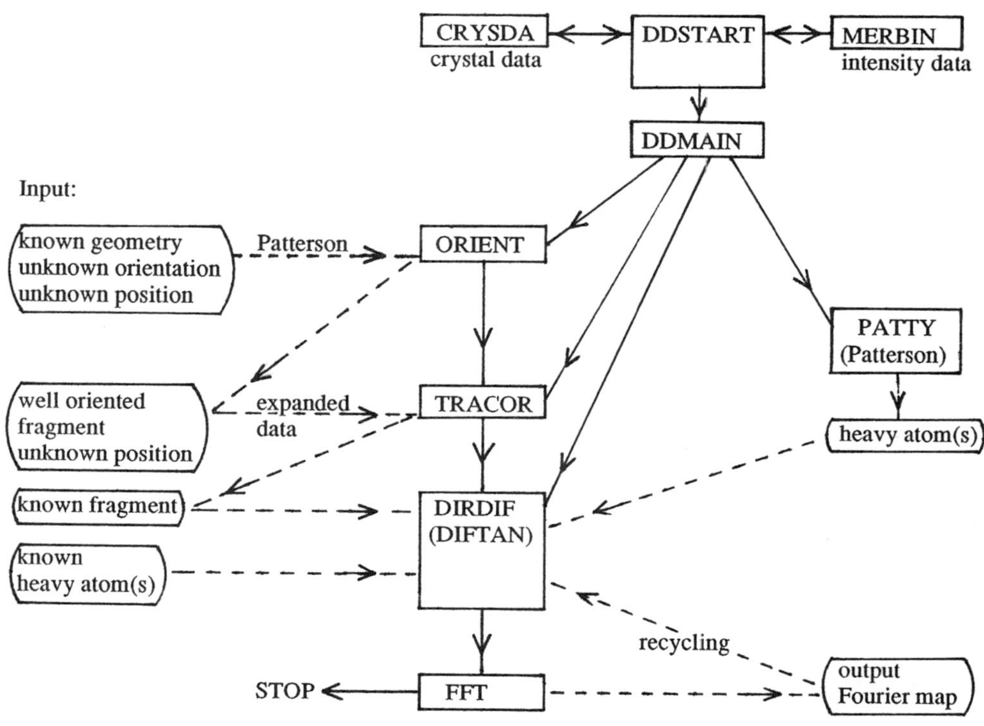

Fig. 2. Simplified flow diagram of the DIRDIF program system.

3. How to get started (terminal session)

The following programs and/or system options are available:

PATTY   for a sharpened heavy atom Patterson
ORIENT   for the consecutive execution of ORIENT, TRACOR and DIRDIF
         (when the molecular geometry is known)
TRACOR   for the consecutive execution of TRACOR and DIRDIF
         (when a misplaced fragment is known)
DIRDIF   for the solution of (pseudo-)symmetry problems and/or
         for the expansion of a small fragment
DIRP1   for DIRDIF with expanded data in space group P1
         (for heavy atom problem structures)
FOUR   for a weighted Fourier synthesis (when nearly all atoms are known).

File definitions and file requirements
VAX file-definitions consist of

$$\text{node::device:[directory]filename.extension;version}$$

where default values may be left out.

The command procedure file assumes that the local node is used, that appropriate alterations have been made by the system manager when data are stored on any other device than DUA0 or in any other directory than ['compound-code'].

File definitions for all applications always have the extension DAT. This leaves the filename as the one and only discriminator and indicator for the contents or purpose of a file. The compound to which files refer is not expressed in their filenames. (This means that equivalent files for different compounds have identical file definitions, except for the directory in which they are stored.)

371

File names for the most important files:

CONDA.DAT	Control data file, created and maintained by the system.
CRYSDA.DAT	Crystal data of the present compound may be created by system.
FREFA.DAT	Formatted reflection data file in ASCII; must be present.
BINFO.DAT	Permanent binary data file with F_o, created by the system.
IPR2.DAT	Listing file for most important results.
IPR3.DAT	Auxiliary listing file.
ATOMS.DAT	Input atomic parameters.
ATOUT.DAT	Output atomic parameters.
ATMOD.DAT	Atomic parameters of a search model, for input to ORIENT.
ATOLD.DAT	Concatenation of all previous ATOMS files.

How to run DIRDIF

Logon and enter 'DD' or 'DD CCODE'. By CCODE is meant a compound-code of at most six characters. ===> This CCODE must correspond to the directory in which the data are stored! <===

You should only use capitals when choosing options or entering alphanumeric data.

The program DDSTART is executed interactively for the preparation of the control data files. If the CRYSDA.DAT file is absent, the program system will link automatically to the program CRYSDA to create this file (where some user input will be required) before continuing. Also, if other necessary files are absent (e.g. input atomic parameters), the user will be asked to supply the data at the keyboard.

Extensive help-files are available to facilitate the use of the system. At any point in DDSTART where input is required, the user may enter 'H' to get relevant help information.

The first run of any option or program should always be the default run: option ORIENT, or TRACOR, or DIRDIF without special control data. Have a look at the print (IPR2: scaling, statistics, $<E^2>$, etc... all okay?) and see if the structure is solved. If a rerun is necessary, one should study the print carefully and consider the use of special control data.

Selected control data for ORIENT:

MCEL	= unit cell of the atomic parameters (not needed if the parameters are Cartesian, or if the model cell is the cell of the present compound 'CCODE')
VMAX	= maximum vector length (only for an ill-defined model)

Selected control data for DIRDIF:

BBB	= temperature factors
SCALE	= scale factor (if there are scaling problems) .
STLMAX	= maximum value of $\sin\vartheta/\lambda$ to be used

Selected control data for FFT (Fourier and peak search)

PEAKS	= number of peaks
DMAX	= maximum bonding distance

After the execution of DDSTART all files are ready. The user may now choose to 'QUIT', in which case the files generated by DDSTART will be saved for use later on.

The output atoms file ATOUT.DAT must be edited to prepare a new version of file ATOMS.DAT for a rerun of DIRDIF (when the structure did not come out completely), or for the local structure refinement program.

4. Example of a DIRDIF structure solution with enantiomorph fixation

MONOS, space group $P2_12_12_1$, a = 8.166(4), b = 11.405(3), c = 15.936(4) Å, $C_{15}H_{16}N_2O_2S$, Z = 4, 1684 reflections, is taken as an example of an organic molecule with one medium-heavy atom.

From a sharpened Patterson synthesis the sulphur position is found at 0.0, 0.09, 0.14. This position is input to DIRDIF which starts with programs DDSTART and DDMAIN. The scattering power of the sulphur atom is 0.253. First, the structure factors F_p for sulphur are calculated with an overall temperature factor $B_{ov} = 4.937$ Å^2 determined by a conventional Wilson plot.

A Wilson-Parthasarathy plot is used to improve the temperature factors and the scale factor Sc:

$$Sc^2 <I>_s = <\Sigma_r f_r^2 \cdot \exp(-s^2 (B_r + \Delta B))>_s + <|F_p(B_p)|^2 \cdot \exp(-s^2 \Delta B)>_s$$

where the average is taken over reflections $\mathbf{h}$ within a given interval of s. Initially $B_p = B_r = B_{ov}$; the new temperature factors are found by $B_p(new) = B_p + \Delta B$, and $B_r(new) = B_r + \Delta B$. This results in $B_p = B_r = B_{ov} = 4.918$ Å^2 and Sc = 0.3979.

A better estimate of the temperature factors is determined by refining simultaneously the temperature factors for the partial and the rest structure seperately. This two-dimensional refinement depends on the reflections being distributed over a two-dimensional array in ranges of s and $|E_p|$. Now define:

$$G_o = <I / \Sigma f^2>$$
$$G_p = K <|F_p(B_p)|^2 \cdot \exp(-\Delta B_p\, s^2) / \Sigma f^2>$$
$$G_r = K <\Sigma_r f_r^2 \cdot \exp(-B_r \cdot s^2) \cdot \exp(-\Delta B_r\, s^2) / \Sigma f^2>$$

where the average is taken over the appropriate range of s and $|E_p|$. The quantity to be minimized is

$$\Sigma_{ranges}(G_o - G_p - G_r)^2$$

and the parameters to be refined are $K = Sc^{-2}$, ΔB_p and ΔB_r (Gould et al., 1975).

After four cycles the refinement converged. The new scale factor is Sc = 0.4677 and the temperature factors are $B_p = 5.389$ Å^2 and $B_r = 3.350$ Å^2. The maximum difference between the overall and the individual temperture factors is restricted to 1 Å^2. In this case, the difference is greater for B_r and resetting is required: Sc = 0.4353, $B_p = 5.123$ Å^2 and $B_r = 4.123$ Å^2.

Now the subprogram FCALC of program DDMAIN is finished, and the preparation of the data for the expansion and refinement of phases and amplitudes begins.
Two extreme possible values of $|E_r|$ are calculated:

$$|E_1| = |[|F_o| \cdot Sc - F_p \cdot \exp(-B_p\, s^2)] / [(\varepsilon \Sigma f_r^2)^{\frac{1}{2}} \cdot \exp(-B_r\, s^2)]|$$
$$\text{with phases } \varphi_1 = \varphi_p \text{ or if } |F_p| > |F_o|,\ \varphi_1 = \varphi_p - \pi$$

$$|E_2| = |[|F_o| \cdot Sc + F_p \cdot \exp(-B_p\, s^2)] / [(\varepsilon \Sigma f_r^2)^{\frac{1}{2}} \cdot \exp(-B_r\, s^2)]|$$
$$\text{with phases } \varphi_2 = \varphi_p - \pi.$$

The condition $|E_1| \leq |E_2|$ is implied, and E_1 is the most probable value (except in non-centrosymmetric space groups if $|E_1|$ is very small), and its weight W_1 is calculated. For reflections with $|F_p| > |F_o|$ the weight $W_1 = 1.0$, because the phases are known and can be used to determine other phases.

The scattering power is recalculated as:

$$p^2 = \Sigma |F_p(B_p)|^2 / \Sigma(|F_o| \cdot Sc)^2 \qquad\qquad \text{(here } p^2 = 0.283\text{)}$$

along with a rescale factor to make the expectation value of $|E_r|^2 = 1$:

$$RESC = <|E_r|^2_{exp}>^{-2} \qquad\qquad \text{(here RESC = 0.958)}$$

where $|E_r|^2_{exp}$ = expectation value for $|E_r|^2$ determined by $|E_1|$ and $|E_2|$, and the average is taken over all reflections. (If this rescale factor is less than 0.95 or greater than 1.10 a rescaling procedure starts, which calculates a new scale factor and repeats the procedure to calculate normalised structure factors.)

The following program is DIFTAN, which expands and refines the reflections with $|E_1| > 0.9$, here 482 reflections. First the program finds the existence of a center of symmetry at $\frac{1}{4}\frac{1}{4}0$, which leads to an enantiomorph problem: a conventional difference Fourier synthesis will lead to a double image, i.e. a superposition of the structure and its enantiomorph (Prick, Beurskens and Gould, 1983). Therefore the enantiomorph must be fixed. First the origin is shifted over a vector $\frac{1}{4}\frac{1}{4}0$ to situate the centre of symmetry at the origin. All calculated phases are $\varphi_p = 0$ or $\varphi_p = \pi$. These phases cannot solve the enantiomorph problem; for this, phases which substantially deviate from the values of 0 or π are required. Therefore enantiomorph-sensitive reflections are determined by inspecting sigma-2 interactions:

$$\varphi_h \approx \varphi(\Sigma_k E_k \cdot E_{h-k})$$

and by selecting the reflections which are most inconstistent with respect to the sigma-2 relationship (which means that φ_r is often found as 0 and also as π). The true value is expected to be in between, which can be about $\frac{1}{2}\pi$ or $-\frac{1}{2}\pi$. There are 334 enantiomorph sensitive reflections, from which 10 reflections are chosen. These reflections get symbol phases X (= A, B, ..., J), representing the phase values $\frac{1}{2}\pi$ or $-\frac{1}{2}\pi$, so that the possible error is less than $\frac{1}{4}\pi$. A starting set of numeric reflections is chosen with $|E_1| > 0.9$ and $W_1 > 0.953$, i.e. reflections for which a sufficiently accurate value of the numerical phase φ_r is known ($\varphi_r = \varphi_1 = 0$ or π).

The enantiomorph discrimination reflections and the numerical reflections of the starting set are input to a symbolic-phase generation procedure to find symbolic phases for as many reflections as possible. The 60 most probable results of 1098 possible phases with symbols are used as a secondary set and from these 5796 symbolic phases are generated.
The fast symbol analysis procedure (Beurskens and Prick, 1981) using the antisymmetric mode determines consistencies for numerical values of the symbols. The enantiomorph is fixed by assigning one numerical value (either $\pi/2$ or $-\pi/2$) to one of the symbols. The best solution (= highest consistency) of the procedure for the symbolic phases is:

A = $-\pi/2$, B = $-\pi/2$, C = $-\pi/2$, D = $\pi/2$, E = $\pi/2$, F = $\pi/2$, G = $\pi/2$, H = $\pi/2$, I = $-\pi/2$, J = $\pi/2$.

These numerical values are substituted in all symbolic phases. One reflection is found without a symbolic phase, 13 reflections are expressed by only one symbol and 468 reflections are expressed by more than one symbols (the average number of symbols is 8.2).

The new numeric starting set consists of 221 reflections, which are input to the tangent formula. A measure for both the quality of the symbolic addition results and the tangent refinement are the expected $<|E_r|^2>$ values for all reflections:

	general refl. (1223)	special refl. (461)
a priori expectation values	1.120	1.008
calculated at start	0.998	0.950
calculated after symbol analysis	1.074	0.912
calculated before cycle 2	1.119	1.026
calculated after cycle 2	1.119	1.031

Program DDMAIN is called again to prepare Fourier coefficients: 1290 reflections are accepted for the Fourier synthesis. The Fourier program FFT gives 23 peaks of which 17 (of total 20) are atomic positions.

References

Beurskens, P.T., and Prick, P.A.J., 1981, A fast algorithm for determing phase values for symbols from weighted symbol relations in direct methods, Acta Cryst., A37:180.

Gould, R.O., van den Hark, Th.E.M., and Beurskens, P.T., 1975, The application of direct methods to centrosymmetric structures containing heavy atoms II, Acta Cryst., A31:813.

Prick, P.A.J., Beurskens, P.T., and Gould, R.O., 1983, The application of direct methods to difference structures. Non-centrosymmetric structures containing heavy atoms III, Acta Cryst., A39:570.

5. Example of a superstructure problem

We now present an example which shows several interesting aspects of program performance and user-intervention. The example is TRICO, $CoCl_3(P(C_2H_5)_2)$ (Van Enckevort, Hendriks and Beurskens, 1977, Cryst. Struct. Comm., 6:531). Triclinic, spacegroup $P\bar{1}$, Z = 8, a = 7.775(3) Å, b = 28.26(2) Å, c = 18.029(1) Å, α = 90.97(2)°, β = 97.88(3)° γ = 90.19(3)°. A superficial study of the Patterson synthesis revealed the four independent cobalt atoms near pseudo special positions, (Table 1, set 1) forming a sub-cell of $\frac{1}{8}$ the unit cell (the 'strong' reflections are: k=2n, l=2n, h+$\frac{1}{2}$k+$\frac{1}{2}$l=2n); see Fig. 3.

Table 1. Cobalt positions for TRICO.

	set 1			set 2			set 3		
Co(1)	0.0	0.0	0.25	-0.02	0.01	0.25	0.028	0.014	0.231
Co(2)	0.0	0.5	0.25	0.02	0.49	0.25	0.015	0.497	0.252
Co(3)	0.5	0.25	0.0	0.44	0.26	0.0	0.471	0.253	0.993
Co(4)	0.5	0.25	0.5	0.56	0.24	0.50	0.567	0.239	0.502

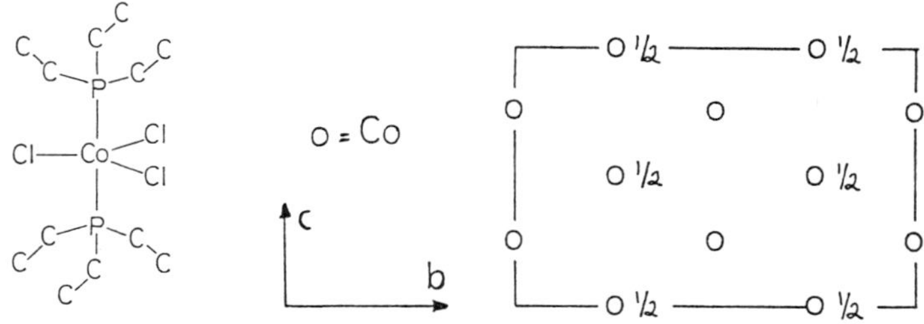

Fig. 3. The TRICO molecule and the cobalt structure.

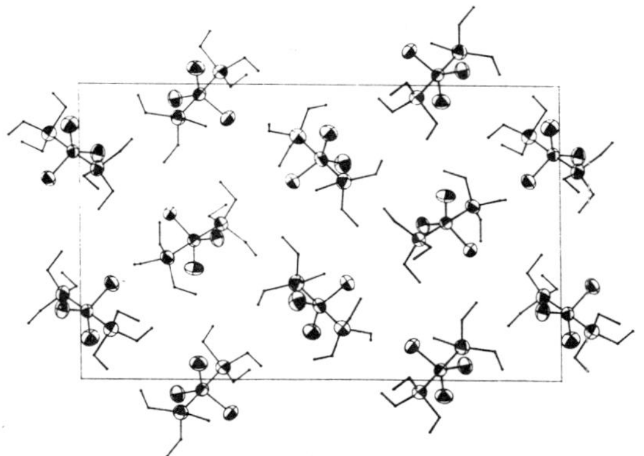

Fig. 4. Projection of TRICO along the a-axis.

As the Patterson Co-Co peaks appeared around the special positions, and other peaks (Co-P, Co-Cl) were highly anisotropic, we expected that these Co positions would not be very accurate. (In fact the deviations were found to be 0.58, 0.15, 0.26 and 0.61 Å). For that reason we did not use these Co parameters as input for DIRDIF.

More accurate cobalt positions were determined from the Patterson. (The shifts of the Co positions were based upon the anisotropy of the Co-P and Co-Cl peaks.) The arrangement of the cobalt atoms (Table 1, set 2), however, still had a center of symmetry at $0, \frac{1}{4}, \frac{1}{4}$, and formed an A-centered lattice (sub-cell of $\frac{1}{2}$ the unit cell). Because of the Patterson overlap, a more detailed solution was not possible, and we decided to put this parameter set into the DIRDIF procedure. An automatic run of the computer program (normalization, origin fixation, 3 cycles of phase refinement, and Fourier synthesis) showed all of 20 independent P and Cl atoms and the misplacement of one of the Co atoms. (The deviations of the input positions were found to be 0.07, 0.05, 0.22 and 0.06 Å.) The remaining C atoms were easily found by the DIRDIF phase refinement procedure, using all Co, Cl, and P atoms as the partial structure. The final Co parameters are given in Table 1, set 3; the structure is given in Fig. 4.

Recently we have done some experiments with the original 8-fold supercell cobalt parameters. It was found that (a) using all data ($\sin\vartheta/\lambda < 0.5$ Å^{-1}), the structure could not be solved: after the origin fixation, the phase refinement did not converge; (b) using only data up to ($\sin\vartheta/\lambda < 0.3$ Å^{-1}), the first 19 peaks did include 15 correct peaks (Co, P or Cl). It is easily understood, that for low order data the atomic displacements do not cause such gross phase errors.

6. Exercises

Answers are given at the end of the section.

Exercise 1

Prove: $E = pE_p + rE_r$ (eq. 8) using $F = F_p + F_r$, and verify (8) using $\langle |E|^2 \rangle = 1$.

Exercise 2

	Partial Structure	Rest Structure
Space group P1		
Consider a structure of 100 equal atoms,		
of which 10 atoms are at known position:	$N_p = 10$	$N_r = 90$
Temperature parameters are:	$B_p = 3$	$B_r = 5$ Å^2
Consider two reflections, both with $s = \sin\vartheta/\lambda = 0.32$ ($s^2 = 0.1$)		
Temperature factors: $\exp(-0.1\,B) = T = $	0.74	0.61
Atomic form factor: $f = 3.2$ ($f^2 = 10$)		
Scattering power: $\Sigma f^2 = $	100	900
Scattering fraction of known fragment:		$p^2 = 0.1$
Normalization factor: $g = |F| / |E| = \sqrt{\Sigma f^2} \cdot T$:		$g = 18.3$

$|E_1|$ is the smallest possible $|E_r|$ value: $|E_1| = ||F| - |F_p|| / g$

$|E_2|$ is the largest possible $|E_r|$ value: $|E_2| = (|F| + |F_p|) / g$

The weight W_1 is expressed in terms of $|E_1|$ and $|E_2|$ by:

$$W_1 = \left[\frac{2P'_1}{P'_1 + P'_2} - 1 \right]^2 , \quad \text{with } P'_i = |E_i| \exp(-|E_i|^2)$$

Mathematical table:	$|E|$	= 1.0	1.2	2.0
P' for some selected $|E|$ values.	P'	= 0.368	0.284	0.037

Data for the two reflections are given in the following table:
(F_p is calculated using B_p)
($|F| = |F_o|$)

Refl.	$|F|$	$|F_p|$	φ_p	‖	φ_r
(i)	20.15	1.85	50°	‖	230°
(ii)	29.3	7.3	30°	‖	120°

Calculate $|E_1|$, $|E_2|$, and W_1 for input to the tangent refinement.

The output of the tangent refinement for these two reflections, φ_t, is considered reliable ($W_t > W_1$), so $\varphi_r = \varphi_t$, which is also given in the data table.

Calculate F_r using $|F|^2 = |F_r|^2 + |F_p|^2 + 2\,|F_r F_p| \cos(\varphi_r - \varphi_p)$. Make construction in $|F_o|$ circle!

Exercise 3

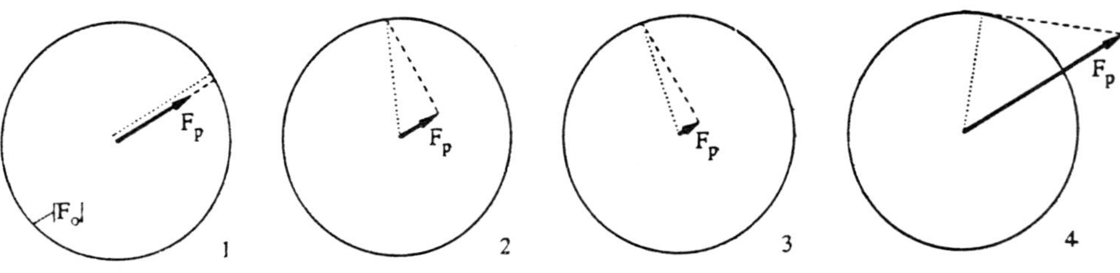

Fig. 5. Exercise 3.

Consider the four circles given Fig. 5. These circles represent $|F| = |F_o| = 100$ (absolute scale).

Consider four reflections, each with $|F_o| = 100$, and calculated phase $\varphi_p = 30°$. The normalization factor for $g = |F| / |E|$ for the rest structure is $g = 50$ (i.e.: a vector F_r of length 50 represents $|E_r| = 1.0$). For these reflections, the calculated $|F_p|$ values are: 75, 40, 20 and 160. For each of the reflections:
(a) Estimate $|E_1|$ and give its phase.
(b) Should E_1 be input to the tangent formula refinement? If so, with small, medium, or large weight?
(c) After tangent formula refinement, a phase $\varphi_r = 120°$ is calculated in each case in which the reflection was refined. Construct F_r, if possible, and estimate its value.
(d) What values should be used for the DIRDIF Fourier synthesis?

Exercise 4

A palladium complex crystallizes is space group P2$_1$/c, with four palladium atoms per unit cell. There are three outstanding peaks in the asymmetric unit of the Patterson:

u	v	w	rel. height
0	0	0.5	100
0.42	0.5	0.64	50
0.42	0.5	0.14	50

Interpret this Patterson. Is the origin of the unit cell fixed by the palladium positions? What should be input to DIRDIF, and what will happen?

Exercise 5

Consider a structure in space group Cc, unit cell: a = 16.3, b = 9.8, c = 26.5 Å, $\alpha = 90°$, $\beta = 109°$, $\gamma = 90°$, containing C, H, N and O atoms, and two independent Cl atoms. Good data was measured up to $\sin\vartheta/\lambda = 0.56$. Cl positions, obtained from the Patterson synthesis, are
        Cl(1) = 0, 0, 0 and Cl(2) = 0.22, 0.17, 0.26.
These two chlorine atoms were fed into DIRDIF, but the resulting peak positions of the Fourier map could not be interpreted in terms of a chemically reasonable structure. The six highest peaks are listed below. Question: what to do next?

peak	pk height	x	y	z
1	2901	1.001	0.018	0.499
2	2796	0.219	0.167	0.261
3	327	0.139	0.040	0.613
4	301	1.001	0.208	0.763
5	294	0.361	0.006	0.160
6	275	1.227	0.526	0.848

Exercise 6

The common space group P2$_1$2$_1$2$_1$ has equivalent postions
x, y, z; $\frac{1}{2}+x, \frac{1}{2}-y, -z$; $-x, \frac{1}{2}+y, \frac{1}{2}-z$; $\frac{1}{2}-x, -y, \frac{1}{2}+z$.
There are no special positions in this space group. Will the origin and/or the enantiomorph need to be fixed if a single heavy atom is located in one of these positions?
a) $0, y, z$;   b) $\frac{1}{4}, y, z$;   c) $\frac{1}{4}, 0, z$;   d) $\frac{1}{4}, \frac{1}{4}, z$.

Answer 1

Use $F = F_p + F_r$, and normalise: $E \sqrt{\Sigma f^2} = E_p \sqrt{\Sigma_p f^2} + E_r \sqrt{\Sigma_r f^2}$

Divide by $\sqrt{\Sigma f^2}$ and substitute $p^2 = \Sigma_p f^2 / \Sigma f^2$ and $r^2 = \Sigma_r f^2 / \Sigma f^2$. Result: $E = pE_p + rE_r$.

Verification: $\langle |E|^2 \rangle = \langle (pE_p + rE_r)(pE_p^* + rE_r^*) \rangle = \langle p^2|E_p|^2 + r^2|E_r|^2 + \text{cross-terms} \rangle$

Substitution of $\langle |E_p|^2 \rangle = \langle |E_r|^2 \rangle = 1$ and $\langle \text{cross-terms} \rangle = 0$ gives: $\langle |E|^2 \rangle = p^2 + r^2 = 1$.

Answer 2

	$|\Delta F_1|$	$|\Delta F_2|$	$|E_1|$	$|E_2|$	W_1
Note: $|\Delta F_1| = \||F| - |F_p|\|$	18.3	22.0	1.0	1.2	0.017
$|\Delta F_2| = \||F| + |F_p|\|$	22.0	36.6	1.2	2.0	0.592

Note: reflection 1 (small weight) is not input to the tangent formula.
Output: Refl. (i): $(\varphi_r - \varphi_p) = 180°$: $|F_r| = |\Delta F_2| = 22.0$, $\varphi_r = 230°$ (from data table)

Refl. (ii): $(\varphi_r - \varphi_p) = 90°$: $|F_r| = \sqrt{(29.3^2 - 7.3^2)} = 28.4$, $\varphi_r = 120°$ (from data table)

Note: the final Fourier synthesis will usually be made with $F = F_p + F_r$ as coefficients ($|F| = |F_0|$).

Answer 3

		(a)		(b)	(c)		(d)	
	$|F_p|$	$|E_1|$	φ_1	W_1	$|F_r|$	φ_r	F	φ
Refl. 1	75	0.5	30°	no	25?	30°	<100	30°
Refl. 2	40	1.2	30°	medium	92	120°	100	97°
Refl. 3	20	1.6	30°	small	98	120°	100	108°
Refl. 4	160	1.2	210°	large	125	171°	100	81°

Note: in the pictogram: F_r is given by a dashed line, F is given by a dotted line.
Answer (c): Refl. 1: unrefined!, $\Delta F = 25$, which is an unreliable estimate for $|F_r|$.

Refl. 4: no match possible. The tangent result $\varphi_t = 120°$ must be reset: $\varphi_r > \varphi_t$ (see pictogram).

Calculation gives: $|F_r| = \sqrt{|F_p|^2 - |F|^2} = 125$, $\varphi_r(min) = \varphi_1 - \arcsin(|F_o| / |F_p|) = 171°$.

Answer (d): for all reflections, $|F| = 100$, but for refl. 1 the Fourier amplitude will be less because $|F|$ should be multiplied by Sim's weight (< 1).
For reflections 2 - 4: all have refined phases. See pictogram.

Answer 4

There are two different solutions! x, y, z = 0.21, 0.00, 0.32 or 0.21, 0.00, 0.07. Both results are a super structure (A-cell); the origin is not fixed. Use each solution as input for DIRDIF; the origin will be fixed by the program, and only one of the DIRDIF runs will show the structure.

Answer 5

DIRDIF shows that Cl(1) is to be moved ($\Delta y \approx 0.2$ Å). Rerun DIRDIF with the new parameters before attempting to solve the light atom structure.

Answer 6

The enantiomorph need to be fix in all cases.
a) $\bar{1}$ at: $\frac{1}{4}\ \frac{1}{4}\ 0$ The origin is fixed by the heavy atom.
b) $\bar{1}$ at: $0\ \frac{1}{4}\ 0$ The origin is fixed by the heavy atom.
c) $\bar{1}$ at: $0\ \frac{1}{4}\ 0$ and $0\ \frac{1}{4}\ \frac{1}{4}$ The origin need to be fixed ($0\ 0\ \frac{1}{2}$ supertranslation).
d) There is a 4 fold origin ambiguity!

THE USE OF KARLE–HAUPTMAN MATRICES IN STRUCTURE DETERMINATION

Rudolf A. G. de Graaff

Gorlaeus Laboratories
University of Leyden
The Netherlands

INTRODUCTION

In the majority of structures determined using Direct Methods the phase problem is solved by the application of the well known Σ_2-relation. Modern programme systems produce satisfactory results in about 95% of all problems. Various attempts to include higher order phase relations in the actual calculation of the phases have been made, with some measure of success. In modern Direct Methods packages higher order relations are commonly used for the calculation of figures of merit.

In 1950 Karle and Hauptman described thr properties of hermitian matrices having structure factors as elements. In particular, the so called Karle-Hauptman matrices, which are uniquely determined by the composition of the first row, are useful for the determination of phases.

KARLE–HAUPTMAN MATRICES

Definitions

$\mathbf{A}$	The KH-matrix of elements $E(\mathbf{H})$
$\mathbf{H}$	a reciprocal lattice vector
m	The order of $\mathbf{A}$
$\mathbf{B}$	The inverse of $\mathbf{A}$
a_{ij}	an element of $\mathbf{A}$ $(i,j=1,2,............,m)$
α_{ij}	the phase of a $(i,j=1,2,............,m)$
b_{ij}	an element of $\mathbf{B}$ $(i,j=1,2,............,m)$
β_{ij}	the phase of β $(i,j=1,2,............,m)$
n	the number of independent reflections in $\mathbf{A}$
a_i	the ith independent reflection $(i=1,2,............,n)$
b_i	the element of $\mathbf{B}$ having the same row and column number as a_i
α_i	the phase of a_i
β_i	the phase of b_i
$\mathbf{a}$	a vector of elements $a_1,a_2,............,a_n$
$\mathbf{b}$	a vector of elements $b_1,b_2,............,b_n$
$\boldsymbol{\alpha}$	a vector of elements $\alpha_1,\alpha_2,............,\alpha_n$

Direct Methods of Solving Crystal Stuctures
Edited by H. Schenk, Plenum Press, New York, 1991

β a vector of elements $\beta_1, \beta_2, \ldots\ldots\ldots, \beta_n$

π a vector of elements π

N the number of atoms in the unit cell

detA   the value of the determinant of **A**

Properties

The element a of **A** is equal to $E(\mathbf{H}_j - \mathbf{H}_i)$, where i and j denote the i*th* and j*th* elements of the first row of the matrix. From this property the fact that **A** is hermitian follows immediately. In 1952 Goedkoop showed **A** to be positive semi-definite, *i.e.* all eigenvalues of **A** are greater than or equal to zero. Goedkoop and Kitaigorodsky proved that detA > 0 if $m \leq N$ and detA $= 0$ if $m > N$ (Goedkoop, 1952 and Kitaigorodsky, 1950).

In 1970 Tsoucaris formulated and derived the Maximum Determinant Rule. When a given KH-matrix **A** of order m, containing structurefactors of known magnitude and phase, is bordered – to obtain a new KH-matrix **A'** of order $m + 1$ – by a row and a column containing elements of unknown phase and known magnitude, the most probable set of phases maximizes detA'/detA. This probabilistic rule was generalized by Tsoucaris to read : For a given KH-matrix **A** the most probable set of phases will maximize detA. Heinerman, Kroon & Krabbendam (1979) were able to prove this Generalized Maximum Determinant Rule neglecting terms of the fifth and higher orders and certain rather strict conditions on the reflections contained in **A**.

USING KARLE–HAUPTMAN MATRICES FOR PHASE DETERMINATION

General

The use of KH matrices in the refinement of phases of proteines will not be discussed here. For a KH-matrix to be useful in *ab initio* phase determination three criteria have to be met :

1. The matrix **A** must have many large off-diagonal elements (to minimize detA) in order to increase the selectivity of the local maxima.
2. Independent (groups of) rows and columns in **A** must be avoided.
3. There is an effective upper as well as a lower limit for m : m must not be larger than approximately $N/2$, whereas smaller matrices of order fifteen or less are not selective enough.

History

Several attempts to use the Maximum Determinant Rule for the determination of *ab initio* phases are reported in the literature, *e.g.* Main 1975 and Taylor, Woolfson & Main, 1978. Main proposed the maximization of a few of the largest eigenvalues of the KH-matrix as a method to refine approximate phases obtained using so called 'magic determinants', KH-matrices where the phases are expressed in a few symbols *via* magic integers. Taylor *et al.* reported on the results obtained by this method. By that time the work on matrix-methods had started in Leyden as well and promising results were obtained (Vermin & de Graaff, 1978 and de Graaff & Vermin, 1982).

In the following paragraphs the different aspects of the use of KH matrices in phase-determination will be discussed.

The construction

A general algorithm for the construction of KH matrices was first proposed by Main in 1975. Roughly the idea is to construct a very large matrix and then to chip a small block of the desired properties from that. In practice this method does not yield satisfactory results in an automatic way, a fact pointed out by Taylor *et al.*

De Graaff and Vermin proposed a different approach in 1982. Instead of building a large matrix and contructing from it a KH-matrix of suitable dimensions, a matrix of the desired order is constructed and optimized iteratively. The algorithm proposed was aimed at constructing starting sets for Multan. The reflections identified by Converge as being critical to the phase extension process, were included in the matrix with a high priority. Control of the optimization process proved possible using exponential functions to limit the number of unobserved reflections and to avoid matrices containing a subset of all parity groups only. The authors used five Multan-failures as teststructures. Reasonably large starting sets of around 25 reflections were obtained for all five tests.

Recently an alternative procedure was developed by de Gelder (de Gelder, de Graaff & Schenk, 1990).

A KH-matrix of order m is defined by a top row of $m-1$ non identical reflections (element a_{ji} is by definition $E(000)$). This poses the question: which set of top row reflections defines a KH-matrix of optimum properties.

In terms of mathematical graph theory a KH-matrix may be regarded as a graph formed by the top row elements. The branches of the graph linking all possible pairs of top row elements are identical to KH-matrix elements lower in the matrix. Reformulating the above question in terms of graph theory: Does a graph exist (formed by top row elements) where all branches are reflections of the desired type (*e.g.* all branches are strong reflections).

Formulating the problem in terms of graph theory leads directly to a method which combines elements of both methods discussed earlier. In a topological matrix representing the graph formed by a given set of top row elements – reflections – each element corresponds to an element of a KH-matrix. Replacing the value of the elements in the topological matrix, normally either zero or one, by a value related to the $|E|$-value of the reflections leads to a modified topological matrix. The top row maximizing the sum of the elements of this matrix, henceforth Quality matrix, is considered to be the top row yielding the best KH-matrix. This idea is the basis for the new algorithm in which one tries to find the best top row out of a large Quality matrix.

The algorithm consists of the following steps:

1) A large Quality matrix (order 500-1000) is constructed from a top row chosen from all large $|E(H)|$'s including symmetry-related reflections. Only strong reflections are used in the top row with a reciprocal vector of a length smaller than a given maximum, avoiding the generation of reflections outside the sphere of measurement in the rest of the KH-matrix and stimulating the number of symmetry dependent reflections in the resulting matrix.

2) Rows (and columns) are sorted in descending order based on criteria involving the $|E|$-values of the reflections appearing in the matrix. Criteria based on the maximum values of $\langle |E|^2 \rangle$, $\langle ||E|^2 -1| \rangle$, $\langle |E|^3 \rangle$ and $\langle ||E|^3 -1| \rangle$ give satisfactory results.

3) A block inside the Quality matrix constructed from the first $m-1$ top row elements (situated in the upper lefthand corner) is optimized iteratively by swapping rows (and columns) for ones further down in the Quality matrix. During this optimization, elements inside the chosen block of order $m-1$ are compared to only those elements outside the block entering the block by swapping. The order of the KH-matrix desired (m) determines directly the course of the construction process.

4) Step 3 may be repeated for blocks constructed from top row elements $m+1$, $m+2$,..., $2m$; $2m+1$, $2m+2$,...,$3m$ etc. Top row elements from matrices previously optimized are not used again.

5) From the collection of optimized blocks of order m-1 the block with the largest sum of the elements is accepted as the most suitable one. In the selection of this block the same criterium is used as in steps 2) and 3). This block is moved to the upper left hand corner of the Quality matrix and is optimized again, now using top row elements from other optimized blocks as well. This last step may again improve the best block because top row elements previously unused may now be included.

The advantage of this method compared to earlier ones is the availability of all reflections during the entire construction process. Moreover several different matrices of comparable quality can be built using the same large starting matrix.

The calculation of the phases

To calculate the phases a multi-solution approach is used. De Graaff and Vermin used the phase-permutations provided by Multan as the starting point. The idea of these authors was to use KH-matrices to provide Fastan with a better start as could be obtained using the usual, fairly small, starting sets. The results presented by them showed clearly that for some Multan-failures enlarging the starting sets in this way is not good enough. The research in Leyden is now concentrated on the development of a phase-determination procedure which is based on KH-matrices only. For this reason the permutations generated by Multan are no longer used. Instead a number, usually a hundred, of trials are generated, using random phases for ing the procedure suggested by all reflections as a starting point, and for each trial the phases are refined following the procedure suggested by de Graaff & Vermin (1982) (see below).

Phase refinement

Numerical libraries, which are now widely available, generally contain routines written to maximize a certain function with respect to a number of unknown parameters. Maximizing detA as a function of the phases, using a conventional method such as a 'quasi Newton' algorithm is an m^5 order process. The method discussed below is of the order three to four in m, which is considerably more efficient.

The derivative of detA with respect to a given phase α_{ij} is given by :

$$\partial \, \mathrm{det}\mathbf{A}/\partial \, \alpha_{ij} = 2|a_{ij}||b_{ij}| \sin(\beta_{ij} - \alpha_{ij}) \, \mathrm{det}\mathbf{A} \qquad (i,j=1,2,\ldots\ldots,m) \qquad (1)$$

In the maximum $\partial \, \mathrm{det}\mathbf{A}/\partial \alpha_{ij} = 0$ or

$$|a_{ij}||b_{ij}| \sin(\alpha_{ij} - \beta_{ij}) \, \mathrm{det}\mathbf{A} \approx 0 \qquad (i,j=1,2,\ldots\ldots,m) \qquad (2)$$

These partial derivatives cannot be zero exactly because of symmetry relations existing between various α_{ij}'s. In 1977 Knossow published an interesting relation between the actual structure and the elements of the inverse of $\mathbf{A}$: if the matrix $\mathbf{A}$ contains the correct phases then

$$\tau(r) = -\sum_{i,j} b_{ij} \exp\left[-2\pi \mathrm{i}(\mathbf{H}_i - \mathbf{H}_j)r\right] \qquad (i,j=1,2,\ldots\ldots,m) \qquad (3)$$

$\tau(r)$, the negative Fourier transform of the elements of the inverse of $\mathbf{A}$, contains maxima on the atomic positions (Knossow, de Rango, Mauguen, Sarrazin & Tsoucaris, 1977). The Fourier transform of the matrix elements a_{ij} has maxima on the same positions, implying for the correct phases the relations

$$\beta_{ij} \approx \alpha_{ij} + \pi. \qquad\qquad (i,j=1,2,\ldots\ldots\ldots,m) \qquad (4)$$

These relations between the phases of the KH-matrix and its inverse are consistent with (1) under the condition that **A** contains a sufficient number of terms and, of course, that the Generalized Maximum Determinant Rule holds.

The foregoing considerations led to the formulation of an iterative scheme for the maximization of det**A** with respect to the phases : $\alpha_{new} = \alpha_{old} + \Delta\alpha$ when there are two possibilities:

[1] $\Delta\alpha = \alpha - \beta - \pi$ or [2] $\Delta\alpha = \beta - \alpha + \pi$, both resulting in $\Delta\alpha = 0$ if (4) holds for all elements of A. Close to the maximum only [1] is consistent with expression (1), giving

$$\Delta\alpha \approx \sin(\Delta\alpha) = \sin(\alpha - \beta + \pi) = \sin(\beta - \alpha) \qquad (5)$$

In practice the elements of the shift vector $\Delta\alpha$ are calculated based on (5), the consistency of the contributions of symmetry related reflections is taken into account, while large shifts are toned down by multiplying the shift obtained with the value of $\sin(\beta_i - \alpha_i)/2$. For a more detailed explanation the reader is referred to the literature.

The method of maximization given above converges to essentially the same points in parameter space as a Newton method.

<u>Recognizing the best set of phases</u>

Obvious choices for a figure of merit are the final value of det**A** as well as a measure of the way in which the relation $\beta_i - \alpha_i = \pi$ holds. Another possibility is the value of the highest eigenvalue of **A**. So far no really satisfactory figure of merit has been found. The discrimination between sets of phases obtained from a typical 20x20 KH-matrix is no easy matter. Early experiments with larger sets of phases (see below), sufficiently large to enable the calculation of (partially) interpretable maps, pointed to the product of the highest eigenvalue and the value of the determinant of **A**, as a very discriminating figure of merit.

<u>The multi matrix method</u>

The value of N limits the size of the determinant, which can be used in the phase determination process, thereby limiting the number of phases which can be determined as well. The value of the determinant is a structure invariant, *i.e.* det**A** is independent of the choice of origin. The combination of phase sets obtained from different KH-matrices is therefore non-trivial. One way of circumventing this problem is to maximize several determinants concurrently. Sufficiently large sets of phases can be obtained this way, bypassing the tangent refinement.

RESULTS

Using the recently developed construction algorithm the construction of starting sets of phases of 20-40 reflections with very low average errors (typically between ten and twenty degrees) is routine. As was remarked earlier, this does not always mean that extension of these sets using tangent refinement is automatic. Of the five Multan-failures tested two were solved easily, one with some difficulty and two not at all.

One of these structures could be determined using the multi matrix method. The twelve highest peaks in the map obtained from the second best solution - according to the figure of merit discussed above - could be extended automatically by repeated Fourier cycles to the full structure.

REFERENCES

Graaff, R.A.G de & Vermin, W.J. (1982). Acta Cryst. A**38**, 464-470.

Heinerman, J.J.L., Kroon, J. & Krabbendam, H. (1979). Acta Cryst. A**35**, 101-105.

Karle, J. & Hauptman, H. (1950). Acta Cryst. **3**, 181-187.

Kitaigorodsky, A.I. (1950). X-ray Structure Analysis, Vol. III, p. 61. Moscow: Gostekhizdat.

Knossow,M., Rango, C. de, Mauguen, Y., Sarrazin, M. & Tsoucaris, G. (1977). Acta Cryst. A**33**, 119-125.

Main, P. (1975). Crystallographic Computing Techniques, edited by F.R. Ahmed, pp. 165-175. Copenhagen: Munksgaard.

Taylor, D.J., Woolfson, M.M. & Main, P. (1978). Acta Cryst. A**34**, 870-883.

Tsoucaris, G. (1970). Acta Cryst. A**26**, 492-494.

Vermin, W.J. & Graaff, R.A.G. de (1978). Acta Cryst. A**34**, 892-894.

MAXIMUM DETERMINANT REVISITED

G. TSOUCARIS

CNRS, UPR 180, Lab. de Physique, Centre Pharmaceutique

Université Paris-Sud, 92290 Châtenay-Malabry, France

1 INTRODUCTION

Since the first publications on the Maximum Determinant Method (MDM) several papers have dealt with further developments of the method or with topics found to be related to it, including the following :
- Generalized Tangent Formula
- Heuristic derivation of determinental jpd's
- Negative Quartet Invariants
- Heuristic Derivation of MDM from Inequalities
- D4 and Higher Order jpd's
- Information Theory
- Maximum Entropy Method, MEM
- Practical Applications including Ab Initio Phase Determination
- Maximization of Eigenvalues with Use of Magic Integers
- Phase Extension in Macromolecular Structures
- Extension to GOEDGOOP Determinants Involving Irreductible Representations
- etc.

We recall that there are two different notions under the MDM term :

(1) The "$\underline{\text{correlation MDM}}$" associated with a m-dimensional Gaussian law, of random variables (r.v.) $E_1 .. E_m$, conditional to the knowledge of $m(m-1)/2$ correlation coefficients U_{pq} with p, q =1, ... m.

$$(1) \qquad p(E_1 .. E_m / U_{pq}\text{'s}) = (1/p^m D_m) \, e^{-Q} \propto e^{N\Delta_{m+1}/D_m}$$

$$\text{with} \qquad Q = N \, \frac{\Delta_{m+1} - D_m}{D_m} = \sum_{pq} D_{pq} E_p E_q^*$$

$$D_m = \text{Det} (U_{pq}) \qquad U_{pq} = U_{Hp\text{-}Hq}$$

$$E_p = E_{L\text{-}Hp}$$

$$\sqrt{N}\ U_H = E_H = \sum_{j=1}^{N} q_j\, e^{2\pi i H_j r_j} \qquad \sum_{j=1}^{N} q_j^{\,2} = 1$$

$$N = \text{Number of atoms}$$

Δ_{m+1}, obtained from D_m by bording with a $(m+1)^{th}$ row and column containing the m r.v. $E_1 \ldots E_m$, is maximum with respect to these m^{th} row elements.

(2) The general MDM rule

Δ_{m+1} determinant is maximum, in the statistical sense with respect to all structure factors (s.f.) it contains, considered here as r.v.'s.

The ab initio derivation of explicit expressions for jpd's of more that 3 or 4 r.v. is a lengthy task. Obviously, in the cases where useful jpd's are already explicitely given by the correlation or the general MDM, much work is saved and jpd of several ten's or hundred's of s.f. can be used in practice.

The set of all structure factors which are elements of D_m is uniquely defined by the choice of m reciprocal lattice vectors $H_1 \ldots H_m$. Clearly, this restricts the class of r.v's whose j.p.d.'s are readily obtained. We will show below that despite this intrinsic limitation, new results can be derived by using specially designed determinants. An example is given for D_5 by eq. 4b, section 5.

2 THE DETERMINANT AS AN ALTERNATING MULTILINEAR FORM

A determinant in linear algebra is <u>defined</u> to be an <u>alternating multilinear form</u> D_m, i.e. a mapping of m vectorial variables into $\Re$, linear with respect to each variable, invariant under any even permutation of the m variables, and changing under any odd permutation. For KARLE-HAUPTMAN determinants where U_{pq}'s are unitary structure factors ($U_{pp} = 1$), the expression of D_m has the form :

$$(2) \qquad D_m = 1 - \sum_{pq} U_{pq}U_{qp} + \sum_{pqr} U_{pq}U_{qr}U_{rp} - \sum_{pqrs} U_{pq}U_{qr}U_{rs}U_{sp} + \cdots$$

all sums ranking from 1 to m, with $p \neq q \neq r \ldots$, i.e. all indices being distinct for each summation

By regrouping terms, we have

$$(2a) \quad D_m = 1 - \sum_{pq} |U_{pq}|^2 + 2\sum_{pqr} |U_{pq}U_{qr}U_{rp}|\cos\Phi_3 - 2\sum_{pqrs} |U_{pq}U_{qr}U_{rs}U_{sp}|\cos\Phi_4 + \cdots$$

where Φ_3 represents the phase of triplets., Φ_4, that of quartets, and so on, the indices being taken with restrictions not given here.

In order to introduce a more uniform notation in eq.(1) we set n o w
$m+1=n$, $\quad L=H_n$, $\quad E_p=E_{L-H_p}=E_{H_n-H_p}=E_{np}$ $\quad$ and $E_{pq}=\sqrt{N}\,(U_{H_p-H_q})$

(1a) $\qquad p(..E_{np}..\mid E_{pq}) \propto \exp\left[\dfrac{1}{D_m}\left(-\displaystyle\sum_{p=1}|E_{np}|^2 + \dfrac{1}{\sqrt{N}}\sum_{pq}E_{pq}E_{qn}E_{np}-\right.\right.$

$$\left.\left.\dfrac{1}{N}\sum_{pqr}E_{pq}E_{qr}E_{rn}E_{np} + \dfrac{1}{N^{3/2}}\sum_{pqrs}E_{pq}E_{qr}E_{rs}E_{sn}E_{np} - \dots\right)\right]$$

all summations ranking from 1 to m and $n=m+1$ being fixed

Hence the physical meaning of the triplets, quintuplets, etc, is different from that of the quartets, sextets, etc. But, in a linear combination, the simplest way of distinguishing terms, which are non zero, is the sign of their coefficient.

Summarizing, we can write schematically :

(1b) $\qquad p(..E_{pn}..\mid E_{pq}) \propto \exp\left[\displaystyle\sum - \text{intensities} + \dfrac{1}{\sqrt{N}}\sum \text{triplets}\right.$

$$\left. - \dfrac{1}{N}\sum \text{quartets} + .. \right]$$

The expression (1a) shows that triplets, quartets, etc., contribute to the jpd with alternating signs and diminushing factors $\dfrac{1}{\sqrt{N}}, (\dfrac{1}{\sqrt{N}})^2, (\dfrac{1}{\sqrt{N}})^3$ etc.

3 FROM CONDITIONAL TO GENERAL jpd

Employing (m times) the formula of compound probabilities
$\qquad P(A,B) = P(A)\,P(B/A) = P(B)\,P(A/B)$
and starting from the conditional MDM distribution (1), one infers, with plausible approximation, the jpd for the general MDM rule

(3) $\qquad\qquad p\text{ (all elements)} \propto e^{N\Delta_{m+1}}$

The approximation is based on the fact that for m we have $m \ll N$, $D_m \approx 1$.

(3.a) $\qquad p\text{ (all elements)} \qquad \propto$

$$\exp\left[\left(-\sum_{pq}|E_{pq}|^2 + \dfrac{1}{\sqrt{N}}\sum_{pqr}E_{pq}E_{qr}E_{pr}- \dfrac{1}{N}\sum_{pqrs}E_{pq}E_{qr}E_{rs}E_{ns} + \dots\right)\right]$$

We arrive thus at the important conclusion that triplets, quintets, etc., i.e. odd order invariants, contribute to the value of Δ_{m+1} and the associated probability expression (1a) and (3a) with a plus sign ; quartets, sextets, etc., i.e. even order invariants, with a minus sign. [Note that these signs are intrinsic to the determinant, independant of the moduli]. Comparison with HAUPTMAN's and GIACOVAZZO's distributions, up to the sextet Φ_6 invariants, reveals the same alternation of sign based on the parity of the order of the invariant, although the mathematical apparatus is somewhat formally different. The one employs ab initio the classical machinery of mathemacical probability for each particular set of r.v. and the other the MDM, valid for any determinant such that m << N. We recall, however that the MDM is also established by using the classical characteristic function and the GRAM-CHARLIER (or EDGEWORTH) expansion (de RANGO et al. 1973. Appendix). But the final form of the characteristic function C(u), given the covariance matrix, leads at once to the hermitian form Q, of eq(1). Independantly, on the other hand the MEM formalism also leads to the MDM.

It is natural to ask whether a more fundamental of physical point of view would shed additional light on the question of sign alternation. We note here that the definition and use in eq.(1) of the form Q implies, in general, the structure factors corresponding to the square of the electron density (TSOUCARIS, 1970, appendix). A closer examination of known probability formula shows that the sign alternation disappears in the extreme case of atoms such that $\displaystyle\sum_{j=1}^{N} Z_j^{\,3} = 0$

4 THE REGRESSION EQUATION AS THE BASIS OF A GLOBAL DETERMINATION OF THE PHASES

The regression equation, associated with the gaussian law (1) provides the expected value of one structure factor, given that of all others included in Δ_{m+1}. Therefore, it applies as well to the general MDM.

$$(4) \qquad \overline{E}_{n,q} = -\frac{1}{D_{qq}} \sum_{p \neq q} D_{pq} E_{n,p} \qquad\qquad n = m+1$$

where D_{pq} is an element of the inverse matrix U^{-1}. It is convenient to change the notation by setting

$$H_n - H_q = H \qquad H_n - H_p = H - K$$

We have then the expansion, valid for m << N

$$(4a) \qquad \overline{E}_H = \frac{1}{\sqrt{N}} \sum_K E_K E_{H-K} - \frac{1}{N} \sum_{K,L} E_{H-K} E_L E_{K-L} + \cdots$$

We emphasize the fact that the alternation is an intrinsic property of the general probability expressions, and is therefore independant of the question of positive or negative invariants. In other words the minus sign associated with quartets (second

term in the eq. (4a) expansion) is to be introduced in the calculation of $\overline{E}_H$, whether the sign of the quartet $\overline{E}_H \, \overline{E}_{H-K} \, \overline{E}_L \, \overline{E}_{K-L}$ is positive or negative (in c - s structures) or whether its phase is close to 0 or π. However, it will be shown in the next section that the coefficient of this second term is to be modified in the case where additional information is provided about the value of other structure factors.

We write now the eq. (4) or (4a) in its "phase form", for the expected value of the phase :

$$(5) \qquad \overline{\phi}_H \;=\; \text{arc tg} \frac{\overline{B}_H}{\overline{A}_H} \;=\; \text{phase of} \left(- \sum_{p \neq q} D_{pq} E_{np} \right)$$

where $\overline{A}_H$ and $\overline{B}_H$ are the real and imaginary part of the right hand member of (4) or (4a). Note that $\overline{\phi}_H$ is independant of $D_{qq} \geq 0$

Now we select, for the purpose of computation, a set of reflections to which we apply the regression equation (5). Let us call S the corresponding reciprocal lattice vectors. The phase discreapancies of these equations will be used to construct the global minimization function of the selected phases :

$$(6) \qquad R_\phi = \sum_{H \in S} w_H \left[(\phi_H - \overline{\phi}_H) \bmod 2\Pi \right]^2$$

The actual process of the minimisation $\dfrac{\partial R_\phi}{\partial \phi_K} = 0$ relies upon the following remarkable equation (given for space group P1) :

$$(7a) \qquad \frac{\partial \overline{\phi}_H}{\partial \phi_K} \;=\; \frac{|E_K \, E_{H-K}|}{|\overline{E}_H|} \; \cos (\overline{\phi}_{-H} + \phi_K + \phi_{H-K})$$

$$(7b) \qquad \frac{\partial \overline{\phi}_H}{\partial \phi_L} \;=\; \frac{|E_{H-K} \, E_K \, E_{K-L}|}{|\overline{E}_H|} \; \cos(\overline{\phi}_{-H} + \phi_{K-L} + \phi_{H-K} + \phi_L) + .. , \text{etc}$$

For a semi-empirical weighting factor $w_H = |E_H \cdot \overline{E}_H|$ the factor in (7) depending on moduli alone, is simply equal to the usual triplet or quartet moduli $|E_H \, E_K \, E_{H-K}|, \ldots$

We emphasize the fact that the derivative of $\phi_H (\phi_1 \ldots \phi_S)$ with respect to the ϕ_K's is a "structure invariant" i.e. independant of the choice of the origin in direct space. It is remarkable however that it depends precisely on the usual phase invariants $\Phi_3, \Phi_4 \ldots$

An alternative way to exploit the "phase regression formula" (5) is the maximization of the following function which is closely connected to the determinant itself

$$(8) \qquad M (\phi_1 \ldots \phi_S) = \sum_H w_H \cos(\phi_H - \overline{\phi}_H)$$

The functions (6) and (8), appropriate for a global determination of the phases (or of a certain bloc of phases) are closely interrelated.

By expending the cosinus, we note that the LS function (6) constitutes, with a minus sign the first term of an expansion of (8).

$$M \approx Cte - \frac{1}{2} R \phi^2 + ..$$

The function (8) is closely related to eq. (1)

$$(9) \qquad Q_m = N \left(1 - \frac{\Delta_{m+1}}{D_{m+1}}\right) = \sum_{q=1}^{m} D_{qq} \left[|E_{nq}|^2 - E_{nq}^* \bar{E}_{nq} | \right]$$

Indeed, if we take the real part in (9), it yields $\sum_{q} E_q^* \bar{E}_q \in \Re$, hence

$$(9a) \qquad Q_m = \sum_{q=1}^{m} D_{qq} \left[|E_{nq}|^2 - E_{nq}\bar{E}_{nq} | \cos(\varphi_{nq} - \bar{\varphi}_{nq}) \right]$$

In the particular case where $S = (H_{n1} \dots H_{nn})$ with $n = m+1$, i.e. all selected reflexions span the last row of D_{m+1}, there is a simple relation between the hermitian form Q_m and the maximization function M defined by eq.(8) :

$$(9c) \qquad Q_m = Cte - M$$

where

$$w_H = |E_H \bar{E}_H| D_{qq} \qquad (cf. (8))$$

$$Cte = \sum_{q=1}^{m} D_{qq} |E_H|^2$$

and q is given by

$$H_q = H_n - H$$

It is a remarkable fact that the maximization function M, inferred by a Least Square Method through eq.(6), is related to the MDM.

Indeed, we have :

$$(9d) \qquad N \frac{D_{m+1}}{D_m} = Cte + M$$

For $m \ll N$, $D_{qq} \approx 1$; therefore, with the simplified notation

$$Q_m = \sum_{H} |E_H|^2 - M$$

Whether two, three or more terms of eq.(4a) are to be used in order to built up the minimization / maximization functions (6) or (8), is a question of "computer generation". Correlatively, the regression equation (4) which provides an expected value $\bar{E}_H$ for E_H, becomes a strict equality for $m = N+2$ (de RANGO et al. 1973 ; and

TSOUCARIS, 1970 Appendix) where the left hand member of (4) becomes now the exact value of the structure factor. **Therefore, the regression equation, established in the statistical case for m<<N, becomes increasingly reliable as m increases and approach N+2.**

For unequal atoms, or even atoms with both positive and negative atomic scattering factors, a somewhat surprising fact is to be observed : although, the definition of the correlation coefficient

$$Gpq \;\; = \;\; < E^{*}_{L\text{-}H_p} \;\; E_{L\text{-}H_q} >$$

involves always a (positive) ρ^2 function, in the regression eq.(4) the E_{n_p}'s may correspond to a ρ function taking positive or negative value

5 THE Δ_5 REGRESSION EQUATION AND THE A PRIORI KNOWN INFORMATION

The following redundant Δ_5 determinant, written below with different notations, will be used in order to illustrate the power of the jpd expression and the associated regression equation.

$$ND_5 = \begin{vmatrix} 1 & & & & \\ U_{H-K} & 1 & & & \\ U_{H-L} & U_{K-L} & 1 & & \\ U_{H-K+L} & U_L & U_{2l-K} & 1 & \\ E_H & E_K & E_L & E_{K-L} & 1 \end{vmatrix} = \begin{vmatrix} 1 & & & & \\ U_h & 1 & & & \\ U_{h+k} & U_k & 1 & & \\ U_{h+l} & U_l & U_{l-k} & 1 & \\ E_{h+k+l} & E_{k+l} & E_l & E_k & 1 \end{vmatrix}$$

We finally obtain a new expression of the regression equation valid in the case where **two** cross terms are zero

$$(4b) \qquad < E_H \mid E_{H\text{-}L} = E_{H\text{-}L+K} = 0 > \; = \; \frac{1}{\sqrt{N}} \, E_{H\text{-}K} \, [E_K - \frac{2}{\sqrt{N}} \sum_L E_L \, E_{K\text{-}L}]$$

In conclusion, the redundant D_5 determinant clearly shows - through the regression formula - how the classical triplet formula can be refined without considering the exact assignment of the probability for a negative quartet :

  - if no assumption on the cross terms is made, the factor of the quartet contribution is -1 (eq. 4a)

  - if we assume that **two** cross terms are zero, this factor becomes -2 (eq. 4b)

References

1   Rango C. de, Tsoucaris, G. & Zelwer, C. (1974). Acta Cryst. **A30**, 342 - 353

2   Tsoucaris, G. (1970). Acta Cryst. **A26**, 492 - 499

EXACT CONDITIONAL PROBABILITY DENSITY FUNCTION OF THREE-PHASE INVARIANT IN P1: SOME THEORETICAL AND PRACTICAL CONSIDERATIONS

Uri Shmueli[a] and George H. Weiss[b]

School of Chemistry[a]
Tel Aviv University
69 978 Tel Aviv, Israel

Physical Sciences Laboratory[b]
Division of Computer Sciences and Technology
National Institutes of Health
Bethesda, Maryland 20892, U.S.A.

INTRODUCTION

The applicability of probabilistic approaches to crystallography was first realized by Wilson (1949) in his study of intensity statistics, and found its place as the fundamental mathematical avenue to direct methods of phase determination (*e.g.* Hauptman & Karle, 1953; Cochran & Woolfson, 1955; Cochran, 1955; see also the Chapters by M. M. Woolfson on "The Cochran Distribution" and by S. Fortier and I. R. Castleden on "Some Applications of Probability Theory in Direct Methods", in this book). The actual probabilistic formalisms, used until 1984, have been based either on low-order approximations afforded by the central limit theorem, as invoked, *e.g.*, by Wilson (1949) and Cochran (1955), or constitute higher approximations (*e.g.* Hauptman & Karle, 1953; Klug, 1958; Naya, Nitta & Oda, 1964,1965 - to mention a few early ones) which can be cast into the formalism of Gram-Charlier or Edgeworth expansions. These expansions are rather lengthy and have poor convergence properties. In fact, most structure-solving packages are still based on the central limit theorem approximations, which perform well for nearly equal-atom structures, with a not too small number of atoms in the asymmetric unit and with no significant population of special positions or other sites imparting rational dependence (Karle & Hauptman, 1959). There remain, of course, small- and intermediate-molecule cases in which these approximations are not adequate. This presentation includes a description of some of our work on the development of exact probabilistic approaches, which properly account for the space-group symmetry and (arbitrarily heterogeneous) atomic composition.

The principle of our method is based on the fact that the magnitude of the normalized structure factor, *E,* or of its real or imaginary part, *A* or *B,* cannot exceed the sum of the normalized scattering factors, taken over the unit cell. These quantities are therefore bounded and therefore the probability density function (pdf) of any of them can differ from zero only over a finite interval. This is true also of the joint

Direct Methods of Solving Crystal Stuctures
Edited by H. Schenk, Plenum Press, New York, 1991

probability density function (jpdf) of several structure factors or their components. Such jpdf can be expanded in a Fourier series, the coeffients of which are calculable in terms of trigonometric structure factors and which turns out to have very good convergence properties. The Fourier-series expansion of the jpdf can thus be exactly formulated as well as accurately computed. The first application of this method to crystallographic statistics was given by Shmueli, Weiss, Kiefer & Wilson (1984) who derived the exact pdfs of $|E|$ for the two triclinic space groups. Some of its other applications, of immediate relevance to direct methods, are our studies of the exact Σ_1 (Shmueli & Weiss, 1985) and Σ_2 (Shmueli & Weiss, 1986) relationships and the study of the exact conditional pdf of the three-phase invariant (Shmueli, Rabinovich & Weiss, 1989a, 1989b). The latter study will be summarized in the following section.

CONDITIONAL DENSITY OF THE THREE-PHASE INVARIANT

The three-phase invariant is the sum of the phases associated with the product of the three normalized structure factors $E(\mathbf{h})$, $E(\mathbf{k})$ and $E(-\mathbf{h}-\mathbf{k})$, *i.e.*

$$\Phi = \varphi(\mathbf{h}) + \varphi(\mathbf{k}) + \varphi(-\mathbf{h}-\mathbf{k}). \tag{1}$$

We shall first describe the exact expression for the conditional pdf of Φ, given the magnitudes $|E(\mathbf{h})|$, $|E(\mathbf{k})|$ and $|E(-\mathbf{h}-\mathbf{k})|$, the symmetry of the crystal and the composition of the asymmetric unit, as obtained by Shmueli, Rabinovich & Weiss (1989a, 1989b) for the space group $P1$. The following commonly employed abbreviations will be used:

$$|E(\mathbf{h})| \equiv E_1 \, , \quad |E(\mathbf{k})| \equiv E_2 \, , \quad |E(-\mathbf{h}-\mathbf{k})| \equiv E_3 \, ,$$

$$\varphi(\mathbf{h}) \equiv \varphi_1, \quad \varphi(\mathbf{k}) \equiv \varphi_2 \, , \quad \varphi(-\mathbf{h}-\mathbf{k}) \equiv \varphi_3 \, ,$$

$$E(\mathbf{h}) \equiv A_1 + iB_1 \equiv E_1(\cos\varphi_1 + i\sin\varphi_1), \text{ etc. for } \mathbf{k} \text{ and } -\mathbf{h}-\mathbf{k} \, .$$

The main stages of the derivation are :

1. The expansion of the jpdf of A_1, B_1, A_2, B_2, A_3 and B_3 in the sixfold Fourier series

$$p(\mathbf{E}) = \left(\tfrac{\alpha}{2}\right)^6 \sum_{\mathbf{u}} C_{\mathbf{u}} \exp\left[-\pi i \alpha \sum_{k=1}^{3} \left(u_{2k-1}A_k + u_{2k}B_k\right)\right] \tag{2}$$

$$= \left(\tfrac{\alpha}{2}\right)^6 \sum_{\mathbf{u}} C_{\mathbf{u}} \exp\left[-\pi i \alpha \sum_{k=1}^{3} |E_k|\left(u_{2k-1}^2 + u_{2k}^2\right)^{1/2} \cos(\varphi_k - \Delta_k)\right] \tag{3}$$

where

$$\mathbf{E}^T = (A_1,\, B_1,\, A_2,\, B_2,\, A_3,\, B_3) \, , \quad \mathbf{u}^T = (u_1,\, u_2,\, \dots ,\, u_6) \, , \quad \Delta_k = \tan^{-1}(u_{2k}/u_{2k-1}) \, ,$$

and α is the reciprocal of the sum of the normalized scattering factors, taken over the unit cell.

2. Replacement of the phase φ_3 in (3) by $\Phi - \varphi_1 - \varphi_2$ and integration of the jpdf (3) over the phases φ_1 and φ_2.

3. Calculation of the Fourier coefficient $C_{\mathbf{u}}$ for the crystal symmetry and composition of interest. We assume here that (i) the atomic contributions to the structure factor are independent, and (ii) the atoms are uniformly distributed throughout the unit cell.

The above steps lead to the required exact conditional pdf, which assumes the general form

$$p_{ex}(\Phi|\ E_1,\ E_2,\ E_3) = K\sum_{\mathbf{u}} C_{\mathbf{u}} Z_{\mathbf{u}}\ , \tag{4}$$

where $Z_{\mathbf{u}}$ depends on the three-phase invariant, the magnitudes of the normalized structure factors involved and *does not* depend on symmetry and composition, and K is a normalization constant. Detailed expressions for the quantities appearing in (4), with Fourier coefficients $C_{\mathbf{u}}$ corresponding to the space group $P1$, are given by Shmueli, Rabinovich and Weiss (1989a). It is also shown in the latter reference that equation (4) reduces to the Cochran (1955) conditional pdf when only lowest order terms of the characteristic function C are retained.

Since $C_{\mathbf{u}}$ and $Z_{\mathbf{u}}$ are also infinite - albeit quickly converging - series, the numerical evaluation of the conditional pdf (4) poses non-trivial computational problems. In fact, it was possible to reduce the computing time to manageable proportions only by avoiding repetitive computations and fully exploiting the symmetry of the summation indices (Shmueli, Rabinovich & Weiss, 1989a, 1989b).

RESULTS

The conditional pdf (4) has been computed for a variety of choices of the composition of the unit-cell contents of $P1$, for heterogeneous models, and for varying number of the atoms in the unit cell, for equal-atom models. For all the parameters so far chosen, the exact conditional pdf has a single peak at $\Phi = 0$ and is invariably more sharply peaked than the corresponding Cochran conditional pdf. This means that the Cochran pdf underestimates the exact one in the (important) neighbourhood of the peak. It is also seen that the discrepancy between the exact and approximate pdfs increases as the number of the atoms in an equal-atom model in the unit cell of $P1$ decreases, and the magnitude of the E values involved increases. Increasing atomic heterogeneity has the same effect as a decreasing number of atoms in an equal-atom model.

A careful inspection of the exact conditional densities, so far computed, indicates that they have a similar behaviour to that of the corresponding Cochran conditional pdfs, and may therefore have similar functional forms. We have therefore assumed that the exact densities can be approximated by a Cochran-like function of the form

$$p_{app}(\Phi|\ \kappa') = \frac{1}{2\ \pi I_0(\kappa')}\exp(\kappa'\cos\Phi) \tag{5}$$

where $I_0(x)$ is a modified Bessel function and κ' is obtained, *e.g.*, by equating the exact conditional pdf (4) at $\Phi = 0$ to the right-hand side of (5) at this point, *i.e.*

$$p_{ex}(0|\ E_1,\ E_2,\ E_3) \equiv \frac{1}{2\ \pi I_0(\kappa')}\exp(\kappa'). \tag{6}$$

The parameter κ' has been found for all the available data on exact conditional densities and equation (5) so far appears to be a very good approximation to the exact conditional pdf. It is, of course, interesting to see if additional calculations of conditional densities may lead to some useful relationship between κ' and the conventional κ, which appears in the Cochran (1955) equation or its analogs. When such a relationship is found, a convenient approximation - remarkably close to the exact conditional pdf - should be most readily accessible. This problem is now being studied.

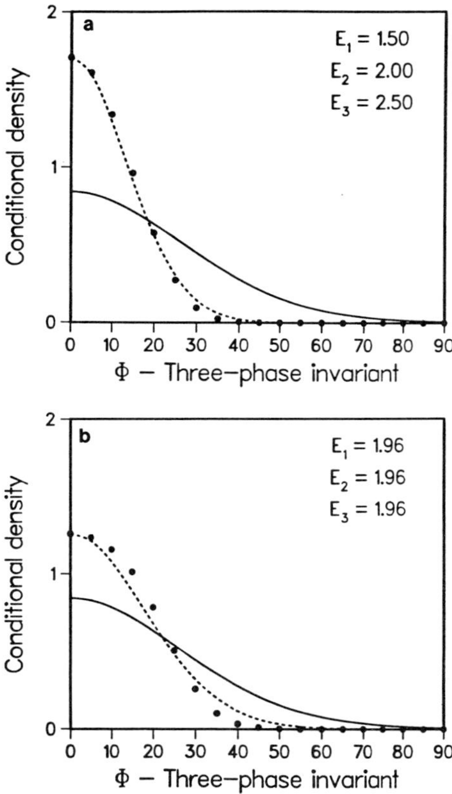

Figure 1. *Exact, modified-kappa and Cochran conditional pdfs of the three-phase invariant for the space group P1 and ten equal atoms in the unit cell.*

The centers of the filled circles correspond to values of the exact conditional density, the dashed curves are plots of the modified-kappa approximation, discussed in text, and the solid curves correspond to the central limit theorem approximation for the conditional density (Cochran, 1955). The drawings refer to two closely similar triple products, with (a) unequal E's and (b) equal E's.

TABLE 1

$E_1 = 1.50$, $E_2 = 2.00$, $E_3 = 2.50$			$E_1 = 1.96$, $E_2 = 1.96$, $E_3 = 1.96$		
N	$R(\text{ex } vs\ \kappa')$	$R(\text{ex } vs\ \text{Cochran})$	N	$R(\text{ex } vs\ \kappa')$	$R(\text{ex } vs\ \text{Cochran})$
10	0.0364	0.5042	10	0.0947	0.3591
15	0.0286	0.2963	15	0.0261	0.2389
20	0.0174	0.2191	20	0.0168	0.1826
25	0.0128	0.1760	25	0.0124	0.1492
30	0.0100	0.1479	30	0.0098	0.1268
35	0.0082	0.1280	35	0.0081	0.1104
40	0.0070	0.1129	40	0.0069	0.0979

We conclude this note with some illustrations. Figure 1 shows the exact and approximate (Cochran, 1955) conditional pdfs, along with the corresponding modified-kappa approximations.   The discrepancies between the exact and approximate conditional densities are most significant.   The agreement beween the exact and modified-kappa densities is excellent in Fig. 1(a) and somewhat worse in Fig. 1(b). Note, that the triple products in 1(a) and 1(b) are closely similar (7.50 and 7.52 respectively) but the triplets of E values involved are different and so are the peaks of the exact conditional pdfs.   This shows that the conditional density of a three-phase invariant depends on the individual magnitudes of the E values involved and not on the magnitude of their triple product alone, as implied by the Cochran pdf.

Table 1 shows a comparison of the exact, modified-kappa and Cochran (1955) densities, for the triples used in Fig. 1 and several values of N, in terms of R factors defined as

$$
R(\text{ex } vs \ A) = \left[\frac{\sum_i \left[p_{ex}(\Phi_i| \cdots) - p_A(\Phi_i| \cdots) \right]^2}{\sum_i \left[p_{ex}(\Phi_i| \cdots) \right]^2} \right]^{1/2}
\tag{7}
$$

where Φ_i ranges from 0^0 to 90^0 in steps of 5^0 and p_{ex} is the exact conditional density computed from (4). The Table supports the results of our computations insofar as the R value comparing the exact and Cochran densities decreases with increasing number of atoms in the unit cell of $P1$, and further illustrates the behaviour of the modified-kappa approximation.

ACKNOWLEDGEMENTS

We wish to thank Zafra Stein and Yael Posner for helpful computing assistance at various stages of this work.   This study was supported in part by research grant No. 88-00210 from the Binational Science Foundation United States - Israel (BSF), Jerusalem, Israel.   Some computations related to this study were carried out on a CDC Cyber 180-990 mainframe and most of them on an IRIS 4D/120GTX workstation, at the Tel Aviv University.

REFERENCES

Cochran, W., 1955, Relations Between Phases of Structure Factors,
      Acta Cryst., 8 : 473.
Cochran, W. & Woolfson, M.M., 1955, The Theory of Sign Relations Between Structure
      Factors, Acta Cryst., 8 : 1.
Hauptman, H. & Karle, J., 1953, "Solution of the Phase Problem I. The
      The Centrosymmetric Crystal". A.C.A. Monograph No. 3. Pittsburgh:
      Polycrystal Book Service.
Hauptman, H. & Karle, J., 1959, Rational Dependence and Renormalization of
      Structure Factors for Pase Determination, Acta Cryst., 12 : 846.
Klug, A., 1958, Joint Probability Distributions of Structure Factors and the Phase
      Problem, Acta Cryst., 11 : 515.
Naya, S., Nitta, I. & Oda, T. 1964, A Study on the Statistical Method for
      Determination of Signs of Structure Factors, Acta Cryst., 17 : 421.
Naya, S., Nitta, I. & Oda, T. 1965, A Theory of the Joint Probability Distribution
      of Complex-Valued Structure Factors, Acta Cryst., 19 : 734.
Shmueli, U., Weiss, G.H., Kiefer, J.E. & Wilson, A.J.C., 1984, Exact Random-Walk
      Models in Crystallographic Statistics. I. Space Groups $P\bar{1}$ and $P1$,
      Acta Cryst., 40 : 559.

Shmueli, U. & Weiss, G.H., 1985, Exact Joint Probability Distribution for Centrosymmetric Structure Factors. Derivation and Application to the Σ_1 Relationship in the Space Group $P\bar{1}$, Acta Cryst., 41 : 401.

Shmueli, U. & Weiss, G.H., 1986, Joint Distribution of $E_\mathbf{h}$, $E_\mathbf{k}$ and $E_\mathbf{h+k}$, and the Probability for the Positive Sign of the Triple Product in the Space Group $P\bar{1}$, Acta Cryst., 42 : 240.

Shmueli, U., Rabinovich, S. & Weiss, G.H., 1989a, Exact Conditional Distribution of a Three-Phase Structure Invariant in the Space Group $P1$. I. Derivation and Simplification of the Fourier Series, Acta Cryst., 45 : 361.

Shmueli, U., Rabinovich, S. & Weiss, G.H., 1989b, Exact Conditional Distribution of a Three-Phase Structure Invariant in the Space Group $P1$. II. Calculation and Comparison with the Cochran Approximation, Acta Cryst., 45 : 367.

Wilson, A.J.C., 1949, The Probability Distribution of X-ray Intensities, Acta Cryst., 2 : 318.

SOME CONSIDERATIONS CONCERNING THE PHYSICAL INTERPRETATION OF

SAYRE'S EQUATION AND PHASE TRIPLETS IN DIRECT METHODS

A. F. Mishnev

Institute of Organic Synthesis
Latvian SSR Academy of Sciences
226006 Riga, USSR

Despite of the permanent progress in the theory of direct methods a problem of physical interpretation of phase relations is still not clear enough. Schenk (1981) has given a graphic explanation of phase triplets and quartets on the basis of electron density considerations. Such a geometrical approach is very convenient and viseable, however it does not take into account a real process of X–ray scattering by crystal. In direct methods a crystal is considered as an ideal infinite periodic structure. Therefore, direct methods are also valid for a large perfect crystal and one can compare phase relations of direct methods with inferences of the dynamic theory of X–ray diffraction. For this purpose we use Ewald's dynamical theory in the case of three strong coplanar beams (Ewald & Heno, 1968; Post, 1979). The condition of compatibility of the dynamical equations has the form

$$
\begin{vmatrix}
F(0) + \dfrac{2\epsilon_0}{\Gamma} & F(-H) & F(-K) \\[2em]
F(H) & F(0) + \dfrac{2\epsilon_0}{\Gamma} & F(H-K) \\[2em]
F(K) & F(K-H) & F(0) + \dfrac{2\epsilon_0}{\Gamma}
\end{vmatrix} = 0
\tag{1}
$$

where ϵ_0 is a resonance error, $\Gamma \simeq e^2\lambda^2/mc^2\pi V$, V is the volume of the unit cell. Determinant (1) (except the diagonal terms) is identical to the Karle – Hauptman's one. If we had taken more waves into consideration we would have got a determinant of a higher order. Unfortunately, ϵ_0 cannot be measured or calculated for an unknown structure. Nevertheless, we can use the determinant (1) for some illustrations. Expansion of the latter yields the dispersion equation

$$
\tau_0{}^3 = \tau_0 \left(\sum_{i=1}^{3} |F_i|^2 \right) + 2|F(H)F(K)F(K-H)|\cos\Phi_3 = 0
\tag{2}
$$

where

$$
\tau_0 = F(0) + \frac{2\epsilon_0}{\Gamma} \quad \text{and} \quad \Phi_3 = \Phi(H)+\Phi(-K)+\Phi(K-H).
$$

If we fix τ_0 and $|F(H)|$ to some arbitrary values and calculate $\cos\Phi_3$ from Eq.(2) we shall find that for large structure factors $|F(K)|$ and $|F(K-H)|$, $\cos\Phi_3 = 1$. This result is consistent with direct methods theory.

Let us now examine some more strict approach based on the unitary equation for scattering amplitude (Gerber & Karplus, 1972)

$$A(\vec{k}',\vec{k}) \, - \, A^+(\vec{k}',\vec{k}) \; = \; \frac{ik}{2\pi} \int A^+(\vec{k}'',\vec{k}')A(\vec{k}'',\vec{k})\,d\Omega_{k''} \tag{3}$$

Eq.(3) being a consequence of energy flux conservation describes the process of multiple beam diffraction. Here $A(k',k)$ is the scattering amplitude tensor for the scattering from the direction of incident wave vector k to the direction of final vector k'. In the linear polarization representation the following approximation for the amplitude tensor is assumed (Gerber & Karplus, unpublished).

$$A(\vec{k}',\vec{k}) \; = \; \begin{bmatrix} \dfrac{e^2\cos\theta}{m(1+iq)c^2} & 0 \\[2ex] 0 & -\dfrac{e^2}{m(1+iq)c^2} \end{bmatrix} F(\vec{k}',\vec{k}) \; , \tag{4}$$

where $q = 2e^2\omega/3mc^2$, ω = the frequency of the scattered radiation. Substituting this relation into (3) we get the following equation

$$F(\vec{k}',\vec{k}) \; = \; \frac{3}{8\pi} \int F^*(\vec{k}'',\vec{k}')F(\vec{k}'',\vec{k})\,d\Omega_{k''} \tag{5}$$

In the case of X-ray diffraction by crystals the integration in Eq.(5) turns into a summation over a discrete set of reciprocal lattice vectors K and after renormalization Eq.(5) assumes the form

$$F(H) \; = \; \frac{F(0)}{\sum\limits_{K}|F(K)|^2} \; \sum\limits_{K} F(H-K)F(K) \tag{6}$$

Thus, Eq.(6) is a special form of the general unitary equation in the case of X-ray diffraction by crystal and represents a generalized form of Sayre's equation. So far as the unitary equation describes the process of multiple scattering one can ascribe to $F(H)$ the sense of primary, $F(K)$ = secondary and $F(H-K)$ = cooperative reflections in a process of three-beam diffraction. The summing up in (6) is performed over all possible simultaneous reflections for the primary reflection $F(H)$.

REFERENCES

Ewald, P.P., and Heno, Y., 1968, X-ray diffraction in the case of three strong rays. I. Crystal composed of non-absorbing point atoms, <u>Acta Cryst.</u>, A24:5.

Gerber, R.B., and Karplus, M., 1972, On the determination of the phases of electromagnetic scattering amplitudes from experimental data, <u>J. Chem. Phys.</u>, 56:1921.

Gerber, R.B., and Karplus, M., 1972, Derivation of phase-determining relations from the unitary theorem of electromagnetic scattering (unpublished).

Post, B., 1979, A solution of the X-ray 'phase problem', <u>Acta Cryst.</u>, A35:17.

Schenk, H., 1981, The negative-quartet relation from electron-density considerations, <u>Acta Cryst.</u>, A37:573.

DISTRIBUTION FITTING METHODS

Jindrich Hašek and Henk Schenk

Institute of Macromolecular Chemistry, Academy of Sciences, Prague
Tsjechoslowakia and Laboratory for Crystallography, University of
Amsterdam, The Netherlands

The measured intensities alone are not sufficient to determine the structure. The
solubility of the phase problem of a structure determination is dependent on the Amount
of a Priori Structure Information (APSI) − i.e. atomicity, the known number of atoms,
known molecular fragment, etc. Therefore the optimal procedure for the structure
determination can be formulated as an optimal transmission of APSI among all phases.
The mathematical condition for "best" utilization op APSI can be found in [1] as:

$$S = V^{-1} \int_V (\rho_{calc} - \rho_{exact})^2 \, dV = V^{-3} \sum (|E_H|^2 \, var \, \varphi_H) = minimum$$

where variances of phases can be derived from the recurent formula:

$$var \, \varphi_H = (\sum (b_j \, V_j + \sum a_{ij} \, var \, \varphi_{ij})^{-1})^{-1}$$

following from the Graph of Phase Relations (GPR). The V_i is the variance of the i−th
seminvariant used for calculation of the phase φ_H in the (n+1)−th level, $var\varphi_H$ is a
variance of the phase φ_H and $var\varphi_{ij}$ are variances of known phases in preceding levels.
The Theory of Seminvariant Graphs described in [2] gives a procedure how to find the
optimal choice of seminvariants, their optimal utilizations, and how to find the best
starting set supposing given set of seminvariants [3,4,5].

In Direct Methods the APSI is expressed as a distribution of seminvariants. Roughly
speaking, the atomicity and the partial structure knowledge determine the distribution
mean, the structure complexity the distribution width. The classical direct methods are
based on the so called phase relationships stating that every seminvariant equals to the
corresponding distribution mean value. Of course, this requirement ignore the distribution
shape (related to the complexity of the structure). When suitable constraints are not
present (contradictory phase indications, large starting set, etc), it can result in infinitely
sharp distributions of seminvariants corresponding to the simple (mostly wrong) structure
model consisting of one or of a small number of "atoms" in the unit cell. It implies that
the classical direct methods can be efficient only for small structures. When the structure
contains many atoms or the atoms are not well resolved, the distributions are wide and
therefore the approximations under which the phase relations are derived are not valid
and as a result also the classical direct methods fail.

The Distribution Fitting Methods (DFM) [1,2,6,7] are applicable to all structures not
regarding their nature. These methods directly minimize the differences between the
theoretical and empirical distributions (histograms) of seminvariants. There are more ways
to describe the distributions mathematically − probability density, cumulative distribution,

Direct Methods of Solving Crystal Stuctures
Edited by H. Schenk, Plenum Press, New York, 1991

distribution moments. This enables different formulation of the objective function according to which the methods can be devided into four groups:
1) Minimization based on a weighted sum of differences between the corresponding values of both distributions of seminvariants.
2) Minimization based on the Kolmogorov test (minimization of the maximal difference between cumulative-distributions)
3) Minimization of a weighted sum of differences between suitable characteristics of distributions (1st, 2nd moments etc.)
4) Minimization of differences between the uniform distribution and the distribution of the variable:

$$z(x) = \int_0^x P(y) \, dy$$

where $P(y)$ is the theoretical probability density of the seminvariant value and x is the actual value of the seminvariant.

There are more ways how to measure the differences between the trial and theoretical distributions. However, since no analysis of the distribution of differences between these distributions has been made, it is not evident what the best function form of the minimized objective function is. The optimal weights w_i and the exponent n function in the objective expression used for distribution fitting:

$$\sum w_i \left| P_i^{theor} - P_i^{emp} \right|^n = minimum$$

The mathematically simpliest way is to take n=2 and to derive weights from statistical considerations not regarding the specific features of probability distributions of different types of seminvariants as it has been made in [1,6,7]. Another approach is to use a more robust method with n=1 and to lower the weights for seminvariants having higher expected deviations from the theoretically derived distributions. The problem of the choice of seminvariants and their sampling can be solved using the theory of seminvariant graphs mentioned above.

All the standard direct methods can be considered as different special approximations of DFM when the distribution mean values are fitted only and all the higher moments except of the first one are neglected. This statement is evident from the fact, that the condition of equality between mean values of empirical and theoretical distributions gives the phase relationships on which the standard direct methods are build. It implicates that the DFM offer large variety of different procedures including the traditional direct methods which are preferable for their simplicity in all cases when they give an automatic solution.

The main advantage of the distribution fitting methods in comparison with standard direct methods is their universal use including the macromolecular structures with a poor atomic resolution supposing that the corresponding theoretical distributions of seminvariants are known. The formalism of seminvariant graphs makes the utilization of APSI contained in seminvariants more efficient in all methods the phase problem in reciprocal space.

References

[1] Hašek J: Acta Cryst. A40.338-352 (1984)
[2] Hašek J.: X-ray and Neutron Structure Analysis in Materials Science p.p. 293-306, Plenum Press 1989
[3] Peschar R. & Schenk H.: Acta Cryst. A.43, 751-763(1987)
[4] Schagen J.D. & Schenk H.: Zeitschrift für Kristallogr. 179, 97-111 (1987)
[5] Germain G., Main P. & Woolfson M.M.: Acta Cryst. 26, 274-285 (1970)
[6] Hašek J., Schenk H., Kiers C.Th. & Schagen J.D.: Acta Cryst. A41, 333-340(1985)
[7] Hašek J & Schenk H.: Acta Cryst.A44, 482-485(1988)

THE MINIMAL PRINCIPLE SOLVES SOME CRYSTAL STRUCTURES

Herbert Hauptman*, D. Velmurugan* and Han Fusen**

*Medical Foundation of Buffalo, Inc.
73 High Street
Buffalo, N.Y. 14203

**The Upjohn Company
Kalamazoo, MI 49001

It is assumed throughout that a crystal structure S in the space group G and consisting of N identical atoms in the asymmetric unit is fixed, but unknown, that the magnitudes $|E|$ of the normalized structure factors E are known, and that a sufficiently large base of phases, corresponding to the largest magnitudes $|E|$, is specified. The magnitudes $|E|$ determine a function R(I), called the minimal function and defined on the space of those structure invariants I which are generated by the specified base of phases:

$$R(I) = \frac{\sum_{H,K} A_{HK} \left\{ \cos \phi_{HK} - \frac{I_1(A_{HK})}{I_0(A_{HK})} \right\}^2 + \sum_{L,M,N} |B_{LMN}| \left\{ \cos \phi_{LMN} - \frac{I_1(B_{LMN})}{I_0(B_{LMN})} \right\}^2}{\sum_{H,K} A_{HK} + \sum_{L,M,N} |B_{LMN}|} , \tag{1}$$

where

$$A_{HK} = \frac{2}{(nN)^{1/2}} \left| E_H E_K E_{H+K} \right|, \tag{2}$$

$$B_{LMN} = \frac{2}{nN} \left| E_L E_M E_N E_{L+M+N} \right| \left(|E_{L+M}|^2 + |E_{M+N}|^2 + |E_{N+L}|^2 - 2 \right), \tag{3}$$

$$\phi_{HK} = \phi_H + \phi_K + \phi_{-H-K} \tag{4}$$

$$\phi_{LMN} = \phi_L + \phi_M + \phi_N + \phi_{-L-M-N} , \tag{5}$$

Direct Methods of Solving Crystal Stuctures
Edited by H. Schenk, Plenum Press, New York, 1991

n is the order of the space group and I_1 and I_0 are the Modified Bessel
Functions. In view of (4) and (5), the ϕ_{HK} are seen to be triplets
(three-phase structure invariants) and the ϕ_{LMN} are quartets (four-phase
structure invariants). From (2), $A_{HK} > 0$ but, from (3), B_{LMN} may be
greater than zero or less than zero depending on whether the three
"cross-terms", $|E_{L+M}|$, $|E_{M+N}|$, $|E_{N+L}|$ are all large or all small,
respectively. The negative quartets, those with $B_{LMN} < 0$, are
particularly important in the applications. The ratios of the Bessel
Functions $\dfrac{I_1}{I_0}$ appearing in (1) are known to be the conditional expected

values of the cosines, $\cos \phi_{HK}$ or $\cos \phi_{LMN}$, given A_{HK} or B_{LMN},
respectively.

From (4) and (5) it is clear that the phases ϕ determine the
structure invariants ϕ_{HK} and ϕ_{LMN} so that R(I) may be regarded as a
function R(ϕ) of the phases. If one denotes by ϕ_S the phases
corresponding to the structure S and by ϕ_R a set of phases chosen at
random, then

$$R(\phi_S) < R(\phi_R), \tag{6}$$

the minimal principle. (It should be noted that, since R(I) is
initially defined as a function of the cosine invariants, its value, as
a function of phases, is independent of the choice of origin, and even
of enantiomorph).

Next, if the base of phases is sufficiently large, the number of
structure invariants exceeds by far the number of phases. Hence there
must exist a large number of identities which the structure invariants
must, of necessity, satisfy. Then the minimal principle implies that
the true values of the structure invariants are those which minimize
R(I) subject to the constraint that they satisfy also the identities
among them. Alternatively, the minimal principle implies that, for
fixed origin and enantiomorph, the true values of the phases are those
which minimize R(ϕ), regarded now as a function of the phases; in this
case all identities among the structure invariants will automatically be
satisfied.

However, if the base of phases is sufficiently large, the phases
themselves may not be regarded as independent variables because their
number will exceed the number of parameters, 3N, needed to define the
structure. In such a case the phases themselves must satisfy a set of
identities. Since, for fixed origin, the crystal structure determines

the values of the phases, the minimal function may be regarded as a
function R(T) defined on the space of all crystal structures T in G
consisting of N identical atoms in the asymmetric unit. It has the
property that, if T ≠ S, then R(S) < R(T), the minimal principle again.
Thus the problem of determining the crystal structure S when magnitudes
$|E|$ are given is reduced to the problem of determining that crystal
structure which minimizes the minimal function R(T). The value of R(T)
depends on the N atomic position vectors r of T, or the 3N atomic
coordinates. Thus the problem of finding the global minimum of R, even
for simple structures, is formidable indeed. Strategies for its
solution are suggested by the following generalized minimal principle
which has statistical validity only: In the space of all crystal
structures T in G consisting of M identical atoms in the asymmetric
unit, where M ≤ N is fixed, those which minimize R are the (M-atom) substructures of S.

For most space groups the case M = 1 is of particular importance.
(For the remaining space groups, e.g. P1, when the origin may be
arbitrarily specified, it is the case M = 2 which assumes a special
importance. Because of limitations of space these cases are not further
considered here.) For such structures the minimal function is a
function of only one position vector r, or three coordinates, and the
positions of its global minima, which, in view of the generalized
minimal principle, often coincide with atomic position vectors of S for
some choice of permissible origin and enantiomorph, are easily found.
Having in this way located one atomic position vector and having fixed
the origin and enantiomorph, successive iterations may be used to
determine the others at an accelerating rate, or, better still, initial
values for the phases may be calculated and these used to find the
nearby local minimum of the minimal function, regarded now as a function
of phases. In this way improved values for the phases are determined,
the Fourier Synthesis identifies additional atomic position vectors, and
the process may be continued until the whole structure has been
revealed.

The method has been used to solve the two (previously known)
structures in the space group $P2_12_12_1:C_{18}H_{28}O_{11}$, Z=4, 2138 reflections,
300 phases in the base, number of triplets = 7110, number of negative
quartets = 139,290, number of positive quartets = 8805, error free data;
and $C_{60}N_6O_{18}H_{102}$, Z=4, 5024 reflections, 400 phases in the base, number
of triplets = 6514, number of negative quartets = 90,001, number of
positive quartets = 2880, experimental data.

In this work the only structure invariants used were the triplets and the (mostly negative) quartets the probabilistic theories of which are well known. It is not yet known whether the method described here will be useful for structures of much greater complexity because it is still not clear how rapidly the size of the phase base must increase as a function of increasing N.

ACKNOWLEDGEMENTS

This work is supported by grant number CHE8822296 from the National Science Foundation.

ESTIMATION OF STRUCTURE SEMINVARIANTS

Davide Viterbo

Dipartimento di Chimica
Universitá della Calabria
I-87030 Arcavacata di Rende (Cs), Italy

Introduction

One-phase seminvariants have been in use for many years, but it is common experience that their estimation by means of the Σ_1 formulae (Hauptman & Karle, 1953) is often unreliable.

The use of two-phase seminvariants was first proposed by Grant, Howell & Rogers (1957), who introduced the idea of *coincidence*. Later Debaerdemaeker & Woolfson (1972) extended this idea to non-centrosymmetric space groups, but these phase relationships have not been extensively used in practice, because their estimates were not sufficiently accurate.

Recently the use of the representation theory (Giacovazzo, 1977) has allowed the derivation of more powerful techniques for estimating one and two-phase seminvariants. In the following we will examine the algebraic foundations of these techniques.

One-phase seminvariants of first rank

The first representation of a one-phase s.s. of first rank, Φ_H is given by the collection of triplets

$$\{\psi_1\} = \{\Phi_H - \phi_h + \phi_{hR_n}\} \tag{1}$$

where R_n is a rotation matrix and h can be found by solving the equation

$$H = h(I - R_n) \tag{2}$$

The general algebraic procedure (Cascarano & Giacovazzo, 1983) for identifying s.s.'s of first rank and solving equation (2) is outlined in the Appendix. Here we shall just consider, as an example, the space group *P312*, the equivalent positions of which are

$$x,y,z \ ; \ -y,x-y,z \ ; \ y-x,-x,z \ ; \ -y,-x,-z \ ; \ x,x-y,-z \ ; \ y-x,y,-z$$

The corresponding rotation matrices are

$$R_1 = I = \begin{pmatrix} 1 & 0 & 0 \\ 0 & 1 & 0 \\ 0 & 0 & 1 \end{pmatrix}; \quad R_2 = \begin{pmatrix} 0 & -1 & 0 \\ 1 & -1 & 0 \\ 0 & 0 & 1 \end{pmatrix}; \quad R_3 = \begin{pmatrix} -1 & 1 & 0 \\ -1 & 0 & 0 \\ 0 & 0 & 1 \end{pmatrix}$$

$$R_4 = \begin{pmatrix} 0 & -1 & 0 \\ -1 & 0 & 0 \\ 0 & 0 & -1 \end{pmatrix}; \quad R_5 = \begin{pmatrix} 1 & 0 & 0 \\ 1 & -1 & 0 \\ 0 & 0 & -1 \end{pmatrix}; \quad R_6 = \begin{pmatrix} -1 & 1 & 0 \\ 0 & 1 & 0 \\ 0 & 0 & -1 \end{pmatrix}$$

It can be easily seen that $R_3 = R_2^{-1}$, while R_4, R_5, R_6 coincide with their inverses. Furthermore, $R_5 = R_2^{-1}R_4R_2$ and $R_6 = R_3^{-1}R_4R_3$, i.e. R_5 and R_6 are conjugate with R_4. The working table will contain two classes: one with matrix R_2 and the other with R_4, R_5, R_6. We then need to solve equation (2) only for $n = 2$ and 4, for which

$$A_2 = I - R_2 = \begin{pmatrix} 0 & -1 & 0 \\ 1 & 2 & 0 \\ 0 & 0 & 1 \end{pmatrix}; \quad A_4 = \begin{pmatrix} 1 & 1 & 0 \\ 1 & 1 & 0 \\ 0 & 0 & 2 \end{pmatrix}$$

Their generalized inverses are given by

$$A_2^* = \begin{pmatrix} 2/3 & 1/3 & 0 \\ -1/3 & 1/3 & 0 \\ 0 & 0 & 0 \end{pmatrix}; \quad A_4 = \begin{pmatrix} 1 & 0 & 0 \\ 0 & 0 & 0 \\ 0 & 0 & 1/2 \end{pmatrix}$$

Denoting by $\mathbf{H} = (H_1\ H_2\ H_3)$ and $\mathbf{h} = (h_1\ h_2\ h_3)$, the two conditions (3a) and (5a), $\mathbf{AA^*H} = \mathbf{H}$ and $\mathbf{A^*H} \equiv 0mod(1)$, will give

$$\left|\begin{aligned} A_2A_2^*\mathbf{H} &= \begin{pmatrix} 1 & 0 & 0 \\ 0 & 1 & 0 \\ 0 & 0 & 0 \end{pmatrix} \cdot \begin{pmatrix} H_1 \\ H_2 \\ H_3 \end{pmatrix} = \begin{pmatrix} H_1 \\ H_2 \\ 0 \end{pmatrix} \\[2em] A_2^*\mathbf{H} &= \left(\frac{2H_1 + H_2}{3}, \frac{-H_1 + H_2}{3}, 0 \right) \equiv 0mod(1,1,1) \\[2em] &\text{from which } (H_1 - H_2, H_3) \equiv 0mod(3,0) \ ; \end{aligned}\right.$$

and

$$\left|\begin{aligned} A_4A_4^*\mathbf{H} &= \begin{pmatrix} 1 & 0 & 0 \\ 1 & 0 & 0 \\ 0 & 0 & 1 \end{pmatrix} \cdot \begin{pmatrix} H_1 \\ H_2 \\ H_3 \end{pmatrix} = \begin{pmatrix} H_1 \\ H_1 \\ H_3 \end{pmatrix} \\[2em] A_4^*\mathbf{H} &= \left(H_1, 0, \frac{H_3}{2} \right) \equiv 0mod(1,1,1) \\[2em] &\text{from which } (H_1 - H_2, H_3) \equiv 0mod(0,2) \qquad (\text{i.e. } H_1 = H_2) \end{aligned}\right.$$

Thus $\Phi_{\mathbf{H}}$ is a first-rank s.s. in *P312* if $\mathbf{H}$ satisfies the conditions

$$(H_1 - H_2, H_3) \equiv 0mod(3,0) \text{ or } (0,2) \qquad\qquad (3)$$

In general any linear combination of phases will be a s.s. of first rank in this space group, if the corresponding combination of the indices satisfies one of the conditions (3).

Finally, the solution of (2) gives

$$\mathbf{h} = \mathbf{A}_2^*\mathbf{H} + (\mathbf{I} - \mathbf{A}_2^*\mathbf{A}_2)\mathbf{k} = \left(\frac{2H_1 + H_2}{3}, \frac{-H_1 + H_2}{3}, 0\right) + \begin{pmatrix} 0 & 0 & 0 \\ 0 & 0 & 0 \\ 0 & 0 & 1 \end{pmatrix} \cdot \begin{pmatrix} k_1 \\ k_2 \\ k_3 \end{pmatrix}$$

$$= \left(\frac{2H_1 + H_2}{3}, \frac{-H_1 + H_2}{3}, k_3\right) \qquad (k_3 \text{ is a free index}), \qquad (4)$$

and

$$\mathbf{h} = \left(H_1, 0, \frac{H_3}{2}\right) + \begin{pmatrix} 0 & -1 & 0 \\ 0 & 1 & 0 \\ 0 & 0 & 0 \end{pmatrix} \cdot \begin{pmatrix} k_1 \\ k_2 \\ k_3 \end{pmatrix}$$

$$= \left(H_1 - k_2, k_2, \frac{H_3}{2}\right) \qquad (k_2 \text{ is a free index}). \qquad (5)$$

In practice the computer program will initially derive, from the symmetry operators, the working table, the matrices **A** and their generalized inverses $\mathbf{A}^*$, which in turn define the seminvariance conditions (3). Then, for each **H** satisfying (3), (4) and (5) it will provide the solutions of (2) and define the set of vectors $\{\mathbf{h}\}$, which, together with **H**, characterise the first representation of $\Phi_{\mathbf{H}}$.

Two-phase seminvariants of first rank

Two triplet invariants

$$\phi_U = \phi_{\mathbf{h}_1 - \mathbf{h}_2} \approx \phi_{\mathbf{h}_1} - \phi_{\mathbf{h}_2}$$

$$\phi_V = \phi_{-\mathbf{h}_1\mathbf{R}_n + \mathbf{h}_2\mathbf{R}_s} \approx -\phi_{\mathbf{h}_1\mathbf{R}_n} + \phi_{\mathbf{h}_2\mathbf{R}_s} , \qquad (6)$$

Can be combined to give

$$\phi_U + \phi_V = \phi_{\mathbf{h}_1} - \phi_{\mathbf{h}_2} - \phi_{\mathbf{h}_1\mathbf{R}_n} + \phi_{\mathbf{h}_2\mathbf{R}_s} = 2\pi(\mathbf{h}_2\mathbf{T}_s - \mathbf{h}_1\mathbf{T}_n) \qquad (7)$$

The left hand side of (7) is a two-phase s.s. as the constant term on the right only depends on the fixed functional form of the structure factor and is independent of the choice of origin.
We have seen that

$$\Phi = \phi_U + \phi_V \qquad (8)$$

is a two-phase s.s. of first rank if two matrices $\mathbf{R}_n$ and $\mathbf{R}_s$ and at least a vector **h** exist, such that

$$\psi_1 = \Phi' + \phi_{\mathbf{h}\mathbf{R}_n} + \phi_{\mathbf{h}\mathbf{R}_s} \qquad (9)$$

is a universal structure invariant. In (9) Φ' is a symmetry equivalent of Φ.

The first representation of Φ is the collection of all invariants

ψ_1 obtained when $\mathbf{h}$ ranges over the reciprocal space and $\mathbf{R}_n, \mathbf{R}_s$ range over the set of rotation matrices.

The first representation of the s.s. (7) is given by the collection of pairs of quartets of type

$$\psi_1 = \phi_V + \phi_{UR_s} + \phi_{h_1R_n} - \phi_{h_1R_s}$$

$$\psi'_1 = \phi_V + \phi_{UR_n} + \phi_{h_2R_n} - \phi_{h_2R_s} \tag{10}$$

Substituting the values of $\mathbf{U}$ and $\mathbf{V}$ given in (6) we can verify that ψ_1 and ψ'_1 are quartet invariants. Their cross-vectors are respectively

$$\mathbf{h}_1(\mathbf{R}_s - \mathbf{R}_n) \; ; \quad \mathbf{h}_2\mathbf{R}_s \; ; \quad \mathbf{V} + \mathbf{h}_1\mathbf{R}_n \quad \text{and}$$

$$\mathbf{h}_2(\mathbf{R}_s - \mathbf{R}_n) \; ; \quad \mathbf{h}_1\mathbf{R}_n \; ; \quad \mathbf{V} + \mathbf{h}_2\mathbf{R}_n \tag{11}$$

The normalized amplitudes corresponding to the independent vectors contributing to ψ_1 and ψ'_1 are

$$|E_1| = |E_{h_1}| ; \quad |E_2| = |E_{h_2}| ; \quad |E_3| = |E_U| ; \quad |E_4| = |E_V| ;$$

$$|E_5| = |E_{h_1(R_s-R_n)}| ; \quad |E_6| = |E_{h_2(R_s-R_n)}| ; \tag{12}$$

$$|E_7| = |E_{V+h_1R_n}| ; \quad |E_8| = |E_{V+h_2R_n}| ;$$

and constitute a set of phasing magnitudes on which Φ depends. Because of symmetry, for a given Φ several couples of quartets may exist. In general the invariance condition is in fact satisfied, not only by the pair of vectors $\mathbf{h}_1$ and $\mathbf{h}_2$, but also by all vectors

$$\mathbf{h}'_1 = \mathbf{h}_1 + \mathbf{k} \quad \text{and} \quad \mathbf{h}'_2 = \mathbf{h}_2 + \mathbf{k} \quad \text{for which } \mathbf{k}(\mathbf{R}_s-\mathbf{R}_n) = 0.$$

The number of phasing magnitudes contributing to Φ can be rather large.

The seminvariance conditions for $\mathbf{U}+\mathbf{V}$ are derived from the symmetry operators of the space group, as we have seen in the case of one-phase s.s.'s. In order to define the first representation, we have to derive the vectors $\mathbf{h}_1$ and $\mathbf{h}_2$ from $\mathbf{U}$ and $\mathbf{V}$. From (6) we obtain

$$\mathbf{h}_1 = \mathbf{U} + \mathbf{h}_2 \tag{13}$$

and

$$\mathbf{V} = -\mathbf{h}_1\mathbf{R}_n + \mathbf{h}_2\mathbf{R}_s = -(\mathbf{U}+\mathbf{h}_2)\mathbf{R}_n + \mathbf{h}_2\mathbf{R}_s = -\mathbf{U}\mathbf{R}_n + \mathbf{h}_2(\mathbf{R}_s-\mathbf{R}_n)$$

giving

$$\mathbf{h}_2(\mathbf{R}_s-\mathbf{R}_n) = \mathbf{V} + \mathbf{U}\mathbf{R}_n \tag{14}$$

The solution of (14) to obtain $\mathbf{h}_2$ can be performed with the algebraic procedure seen for one-phase s.s.'s. We shall just consider here two examples in space groups $P\bar{1}$ and $P2_12_12_1$ [for space groups up to orthorhombic the off-diagonal elements of the $\mathbf{R}_j$ and $\mathbf{D}_{ij} = (\mathbf{R}_i - \mathbf{R}_j)$ matrices are always zero and are not given].

Space group P$\bar{1}$

$$\mathbf{R}_1 = (1,1,1) \; ; \quad \mathbf{R}_2 = (\bar{1},\bar{1},\bar{1}) \; ; \quad \mathbf{D}_{12} = (2,2,2) \; .$$

We have

$$\Phi = \phi_{\mathbf{h}_1+\mathbf{h}_2} + \phi_{\mathbf{h}_1-\mathbf{h}_2}$$

The first representation of Φ is given by

cross vectors

$$\psi_1 = \phi_{\mathbf{h}_1+\mathbf{h}_2} + \phi_{\mathbf{h}_1-\mathbf{h}_2} - \phi_{\mathbf{h}_1} - \phi_{\mathbf{h}_1} \qquad 2\mathbf{h}_1, \; \mathbf{h}_2$$

$$\psi'_1 = \phi_{\mathbf{h}_1+\mathbf{h}_2} + \phi_{\mathbf{h}_1-\mathbf{h}_2} - \phi_{\mathbf{h}_2} - \phi_{\mathbf{h}_2} \qquad 2\mathbf{h}_2, \; \mathbf{h}_1$$

Φ depends on the <u>six</u> phasing magnitudes

$$|E_{\mathbf{h}_1+\mathbf{h}_2}|, \quad |E_{\mathbf{h}_1-\mathbf{h}_2}|, \quad |E_{\mathbf{h}_1}|, \quad |E_{\mathbf{h}_2}|, \quad |E_{2\mathbf{h}_1}|, \quad |E_{2\mathbf{h}_2}|$$

Space group P$2_12_12_1$

$$\mathbf{R}_1 = (1,1,1); \quad \mathbf{R}_2 = (1,\bar{1},\bar{1}); \quad \mathbf{R}_3 = (\bar{1},1,\bar{1}); \quad \mathbf{R}_4 = (\bar{1},\bar{1},1);$$

$$\mathbf{D}_{12} = (0,2,2); \quad \mathbf{D}_{13} = (2,0,2); \quad \mathbf{D}_{14} = (2,2,0);$$

$$\mathbf{D}_{34} = (0,2,\bar{2}); \quad \mathbf{D}_{24} = (2,0,\bar{2}); \quad \mathbf{D}_{23} = (2,\bar{2},0);$$

$$\omega_{\mathbf{s}} = (2,2,2).$$

$\mathbf{D}_{12}$, $\mathbf{D}_{13}$ and $\mathbf{D}_{14}$ define the first-rank seminvariance conditions.

Let

$$\mathbf{U} = (1\ 2\ 3) \quad \text{and} \quad \mathbf{V} = (7\ 2\ 5) \; ,$$

and consider the two different seminvariants

$$\Phi = \phi_{\mathbf{U}} - \phi_{\mathbf{V}} = \phi_{\mathbf{U}} + \phi_{-\mathbf{V}} \qquad \text{and}$$

$$\Phi' = \phi_{\mathbf{U}} + \phi_{\mathbf{V}}$$

For the first, using equation (14), we have

$$\mathbf{h}_2 \cdot \mathbf{D}_{13} = \mathbf{h}_2 \cdot (2,0,2) = (\bar{7}\ \bar{2}\ \bar{5}) + (\bar{1}\ 2\ \bar{3}) = (\bar{8}\ 0\ \bar{8})$$

from which

$$\mathbf{h}_2 = (\bar{4}\ k\ \bar{4}) \quad \text{and} \quad \mathbf{h}_1 = (1\ 2\ 3) + (\bar{4}\ k\ \bar{4}) = (\bar{3}\ k{+}2\ \bar{1})$$

The first representation is given by the set of quartets

$$\psi_1 = \phi_{-7,-2,-5} + \phi_{1,2,3} + \phi_{3,k+2,1} + \phi_{3,-k-2,1}$$

$$\psi'_1 = \phi_{-7,-2,-5} + \phi_{-1,2,-3} + \phi_{4,k,4} + \phi_{4,-k,4}$$

with cross vectors

$$(\bar{6}\ 0\ \bar{2}); \quad (\bar{4}\ k\ \bar{4}); \quad (\bar{4}\ \bar{k}\text{-}4\ \bar{4}) \quad \text{for} \quad \psi_1 \quad \text{and}$$

$$(\bar{8}\ 0\ \bar{8}); \quad (\bar{3}\ k\text{-}2\ \bar{1}); \quad (\bar{3}\ \bar{k}\text{-}2\ \bar{1}) \quad \text{for} \quad \psi'_1.$$

It can be a useful exercise to verify that relation (14) is also satisfied by the vectors (1 2 3) and (7 -2 5) , but the quartets are equivalent to those already derived and no new phasing magnitude is introduced.

For Φ' we have

$$\mathbf{h}_2 \cdot \mathbf{D}_{24} = \mathbf{h}_2 \cdot (2,0,\bar{2}) = (7\ 2\ 5) + (\bar{1}\ \bar{2}\ 3) = (6\ 0\ 8)$$

from which

$$\mathbf{h}_2 = (3\ k\ \bar{4}) \quad \text{and} \quad \mathbf{h}_1 = (1\ 2\ 3) + (3\ k\ \bar{4}) = (4\ k\text{+}2\ \bar{1})$$

The first representation is defined by

$$\psi_1 = \phi_{7,2,5} + \phi_{1,-2,-3} + \phi_{-4,-k-2,-1} + \phi_{-4,k+2,-1}$$

$$\psi'_1 = \phi_{7,2,5} + \phi_{-1,-2,3} + \phi_{-3,-k,-4} + \phi_{-3,k,-4}$$

with cross vectors

$$(8\ 0\ 2); \quad (3\ \bar{k}\ 4); \quad (3\ k\text{+}4\ 4) \quad \text{for} \quad \psi_1 \quad \text{and}$$

$$(6\ 0\ 8); \quad (4\ \bar{k}\text{+}2\ 1); \quad (4\ k\text{+}2\ 1) \quad \text{for} \quad \psi'_1.$$

We can notice that as k varies, the vectors $\mathbf{h}_1$ and $\mathbf{h}_2$ lie on two reciprocal lattice rows and it can be shown that, not only in the examples considered, but in general, also the vectors $\mathbf{V} + \mathbf{h}_1\mathbf{R}_n$ and $\mathbf{V} + \mathbf{h}_2\mathbf{R}_n$ lie on the same reciprocal lattice rows. The latter two vectors will not give rise to any new phasing magnitude and will not appear in the probability formulae.

A p p e n d i x

Given $\mathbf{H}$ and $\mathbf{R}_n$, we have to solve equation (2) to obtain $\mathbf{h}$. The matrix $\mathbf{A}$ = $\mathbf{I} - \mathbf{R}_n$ may be singular and in order to solve (2) we have to use its *reflexive generalized inverse*. This is defined (see Ben-Israel & Greville, 1974) as the matrix $\mathbf{A}^*$ satisfying the relations

$$\mathbf{A}\mathbf{A}^*\mathbf{A} = \mathbf{A} \quad \text{and} \quad \mathbf{A}^*\mathbf{A}\mathbf{A}^* = \mathbf{A}^* \tag{1a}$$

A system of linear equations

$$\mathbf{A}\mathbf{x} = \mathbf{b} \tag{2a}$$

has a solution if

$$\mathbf{A}\mathbf{A}^*\mathbf{b} = \mathbf{b} \tag{3a}$$

The solution is

$$\mathbf{x} = \mathbf{A}^*\mathbf{b} + (\mathbf{I} - \mathbf{A}^*\mathbf{A})\mathbf{z} \tag{4a}$$

where z is an arbitrary vector. In our case $\mathbf{A} = (\mathbf{I}-\mathbf{R}_n)^T$ and $\mathbf{b} = \mathbf{H}$ have

412

integer elements and also the solution $x = h$ must be integer.

According to Hurd & Waid (1970) theorem, the solution has integer elements if

$$A^*b \equiv 0\,mod(1) \tag{5a}$$

In order to explain the procedure for identifying s.s.'s of first rank we have to recall some group-theory definitions:

a) Let A,B,C be three elements of a group. If $B = C\,A\,C^{-1}$, then we say that B is the *transform* of A by C or that A and B are *conjugate* to each other.
b) The complete set of elements conjugate to each other forms a *class of the group*.
c) Any element can not belong to two different classes.

We now prove that the solution h' of

$$h'(I - R_m) = HR_t \tag{6a}$$

is symmetry equivalent to h from (2), if $R_n = R_t R_m R_t^{-1}$. In fact (6a) can be rewritten as

$$H = h'(I - R_m)R_t^{-1} = h'R_t^{-1}(R_t - R_m R_t^{-1}) = h'R_t^{-1}(I - R_t R_m R_t^{-1}) \tag{7a}$$

When R_t varies over all rotation matrices, $R_t R_m R_t^{-1}$ describes the class of matrices conjugate to R_m and $h'R_t^{-1}$ spans over all symmetry equivalents to h. It is also easy to prove that if h satisfies (2) then $h' = -hR_n$ satisfies

$$h'(I - R_n^{-1}) = H \tag{8a}$$

Therefore, once the solution of (2) has been found, there is no need for solving also (6a) and (8a). The following procedure will allow us to obtain only the unique solutions:

1) Exclude I from the list of the rotation matrices;
2) Exclude all R^{-1} matrices;
3) Divide the remaining matrices into classes to obtain the *working table*;
4) Solve (2) for one (first) matrix in each class of the working table.

R e f e r e n c e s

- Ben-Israel A. & Greville T.N.E. (1974). `Generalized Inverses: Theory and Application '. J. Wiley, New York.

- Cascarano G. & Giacovazzo C. (1983). Z. Kristallogr., **165**, 169.

- Debaerdemaeker T. & Woolfson M.M. (1972). Acta Cryst., **A28**, 477.

- Giacovazzo C. (1977). Acta Cryst., **A33**, 933.

- Grant D., Howells R. & Rogers D. (1957). Acta Cryst., **10**, 489.

- Hauptman H. & Karle J. (1953). "Solution of the Phase Problem. I. The Centrosymmetric Crystal". ACA Monograph N.3, Pittsburgh: Polycrystal Book Service.

- Hurt M.F. & Waid C. (1970). SIAM (Soc. Ind. Appl. Math.), J. Appl. Math., **19**, 547.

Exercises on the representation method

1 - Derive the first representation and the corresponding first phasing shell of the quartet in the space group $P2_1/c$

$$\phi_{123} + \phi_{5\bar{1}4} + \phi_{\bar{1}\bar{4}3} + \phi_{\bar{5}\bar{5}\bar{4}}$$

Which normalized amplitudes have to be large and which small for the quartet to be positive? Answer the same question for the quartet

$$\phi_{123} + \phi_{5\bar{1}4} + \phi_{\bar{1}\bar{3}3} + \phi_{\bar{5}\bar{4}4}$$

2 - Derive the first representation and the corresponding first phasing shell of the quintet in the space group $P2_1$

$$\phi_{123} + \phi_{\bar{1}\bar{4}3} + \phi_{3\bar{7}2} + \phi_{59\bar{6}} + \phi_{8\bar{8}4}$$

The equivalent positions are: $x,y,z;$ $\bar{x},y+\frac{1}{2},\bar{z}.$

3 - In the space group $P2_1$ derive the first and second representations and the corresponding phasing shells for the one phase s.s. $\phi_{6010}.$

4 - In the same space group derive the first representation and the corresponding phasing shell of the two-phase s.s.

$$\phi_{235} + \phi_{\bar{6}\bar{3}\bar{9}}$$

FIGURES OF MERIT IN THE SIR PROGRAM

Davide Viterbo

Dipartimento di Chimica
Universitá della Calabria
I-87030 Arcavacata di Rende (Cs), Italy

Introduction

Given several sets of phases it would be rather time consuming to compute and interpret all the corresponding electron density maps to see which yield the correct structure. It is instead easier to compute some appropriate functions, called _figures of merit_ (FOM); their values are expected to be extreme for the correct solution and thus allow an _a-priori_ estimate of the goodness of each phase set. The most commonly used FOM's are:

1) _ABS_ (absolute FOM) represents a measure of the internal consistency of the employed triplet relationships in estimating the phases. It is defined as

$$ ABS = \frac{\Sigma_h \alpha_h}{\Sigma_h <\alpha_h>} = \frac{A}{A_e} \tag{1} $$

for a correct structure A should be close to the theoretically estimated A_e and $ABS \approx 1.0$. In practice it has been found that often for the correct set of phases $A > A_e$ and ABS values between 0.9 and 1.3 indicate a promising phase set. Higher values indicate an overconsistency and are typical of some troublesome structures.

2) R_α FOM: it is a measure of how much triplets deviate from their expected statistical behavior and is defined as

$$ R_\alpha = \frac{\Sigma_h |\alpha_h - <\alpha_h>|}{\Sigma_h <\alpha_h>} \tag{2} $$

it should be minimum for the correct set of phases. In (1) and (2)

$$ \alpha_h = [(\sum_{j=1}^{r} G_j \cos\Theta_j)^2 + [(\sum_{j=1}^{r} G_j \sin\Theta_j)^2]^{\frac{1}{2}} \tag{3} $$

$$ \alpha_h = \sum_{j=1}^{r} G_j D_1(G_j) \tag{4} $$

where $D_i(x) = I_i(x)/I_o(x)$ is the ratio of the modified Bessel functions of i-th and zero order,

$$G_j = 2\sigma_3\sigma_2^{-3/2}|E_h E_{k_j} E_{h-k_j}| \quad \text{and} \quad \Theta_j = \phi_{k_j} + \phi_{h-k_j} \tag{5}$$

3) ψ_o FOM: it is defined as

$$\psi_o = \frac{\Sigma_1[|\Sigma_k E_k E_{1-k}|]}{\Sigma_1[\Sigma_k|E_k E_{1-k}|^2]^{\frac{1}{2}}} = \frac{\Sigma_1\alpha'_1}{\Sigma_1 v_1^{\frac{1}{2}}} \tag{6}$$

where the outer summations are over a certain number (50÷150) of reflections with lowest $|E_1|$ value. The triplet generating routine will have to set up also the relationships (ψ_o triplets) linking each of these 1 reflections with pairs of reflections k and $1-k$ with large $|E|$, the phases of which have been determined. The inner summation at the numerator corresponds to Sayre's equation, written in terms of normalized structure factors, relative to each reflection 1. Since $|E_1| \approx 0$, the large terms in the summation must tend to cancel each other out and the correct set of phases will correspond to a minimum ψ_o value. This function differs from the previous FOM's because it is not a self-consistency figure among the determined reflections only, but it establishes the coherence of the obtained phases with some weak reflections not used for their determination. In (6)

$$\alpha'_h = [(\sum_{j=1}^{r'} A_j \cos\Theta_j)^2 + [(\sum_{j=1}^{r'} A_j \sin\Theta_j)^2]^{\frac{1}{2}} \tag{7}$$

where

$$A_j = 2\sigma_3\sigma_2^{-3/2}|E_{k_j} E_{1-k_j}| \quad \text{and} \quad \Theta_j = \phi_{k_j} + \phi_{1-k_j} \tag{8}$$

Then

$$v_1 = \sum_{j=1}^{r'} A_j^2 \tag{9}$$

In the SIR program (Burla et Al., 1989) several types of structure invariants (i.s.) and seminvariants (s.s.) are estimated by the representation method and all these phase relationships may be used to compute a variety of FOM's. Besides also the above mentioned traditional FOMS have beeen reconsidered (Cascarano, Gicovazzo & Viterbo, 1987) under a more rigorous probabilistic point of view and new functions have been derived.

FOM's directly based on s.i. and s.s. estimates

The two functions SS1FOM, SS2FOM, based on one- and two-phase s.s.'s are defined as:

$$\text{SS1FOM} = \frac{\Sigma\ w_1 G_1 \cos(\Phi_1 - \Theta_1)}{\Sigma\ w_1 G_1 D_1(G_1)} = \frac{T_1}{B_1} \tag{10}$$

$$\text{SS2FOM} = \frac{\Sigma\ w_2 G_2 \cos(\Phi_2 - \Theta_2)}{\Sigma\ w_2 G_2 D_1(G_2)} = \frac{T_2}{B_2} \tag{11}$$

where Θ_1 and Θ_2 are the estimated values of the s.s.'s, with distribution parameter G_1 and G_2 respectively, Φ_1 and Φ_2 are the computed values and $w_1=w_2=0.5$ for the actively used s.s.'s and $=1.0$ for the others.

Negative triplets (estimated by the P10 formula using the second representation) and quartets (estimated by their first representation) give the FOM's:

$$\text{NTRFOM} = \frac{\Sigma \ G_3 \cos\Phi_3}{\Sigma \ |G_3| D_1(|G_3|)} = \frac{T_3}{B_3} \tag{12}$$

$$\text{NQUEST} = \frac{\Sigma \ G_4 \cos\Phi_4}{\Sigma \ |G_4| D_1(|G_4|)} = \frac{T_4}{B_4} \tag{13}$$

where the summations are extended to all negatively estimated ($G_j<0.0$) triplets and quartets respectively. In (10)-(13) $D_1(G_j)$ is the expected value of $\cos\Phi_j$; all these functions are in the range $-1\div1$ and are expected to be extreme and positive for the correct solution.

Since the various phase relationships defining a given FOM are singly evaluated, the effectiveness of each FOM depends on the number and on the reliability of the relations concurring to its value. This suggests to combine the four FOM's in the following way

$$\text{CPHAS} = \frac{T_1 + T_2 + T_3 + T_4}{B_1 + B_2 + B_3 + B_4} \tag{14}$$

Enantiomorph sensitive functions - Triplets estimated by the P10 formula with $G_3\approx0.0$ can not be reliably fixed, but on an exclusion basis, we may assume that they will preferably lie around $\pm\pi/2$. The following enantiomorph sensitive function is therefore justified

$$\text{ENTR} = \frac{1}{t} \sum_{j=1}^{t} |\sin\Phi_{3j}| = \frac{T_3^e}{t} \tag{15}$$

The same idea applied to quartets gives

$$\text{ENQU} = \frac{1}{q} \sum_{j=1}^{q} |\sin\Phi_{4j}| = \frac{T_4^e}{q} \tag{16}$$

Values of the combined function

$$\text{ENANT} = (T_3^e + T_4^e)/(t + q) \leq \sin 35°$$

indicate that the enantiomorph has probably been lost and a warning is issued by the program.

Revision of the traditional FOM's

A better understanding of the statistical meaning of the ABS, R_α and ψ_0 FOM's was recently achieved by a more rigorous probabilistic approach, which also allowed to devise some new effective functions. At the basis of this approach is the use of the probability distributions $P(\alpha_h)$ of α_h defined in (3) and $P(\alpha'_h)$ of α'_h defined in (7), derived by

Burla et Al. (1987). From $P(\alpha_h)$ (having the form of a centric distribution) and $P(\alpha'_h)$ (having the form of an acentric distribution) it is possible to derive various theoretical moments, which may then be compared with the corresponding "experimental" values computed using the different sets of phases. ABS and R_α are related to moments of $P(\alpha_h)$, but also the expected values of other appropriate functions of α_h may be derived and used as FOM's. In SIR four such functions are combined to give

$$\text{ALCOM} = \tfrac{1}{4}\Big\{\Big[\frac{1}{q_1}\frac{1}{n}\sum_h\Big(\frac{|K\alpha_h-<\alpha_h>|}{\sigma_h}\Big)\Big] + \Big[\frac{1}{q_2}\frac{\sum_h(K\alpha_h-<\alpha_h>)^2}{\sum_h\sigma_h^2}\Big]^{\tfrac{1}{2}} +$$

$$\Big[\frac{1}{q_3}\frac{1}{n}\sum_h\Big(\frac{K\alpha_h-<\alpha_h>}{\sigma_h}\Big)^2\Big]^{\tfrac{1}{2}} + \Big[\frac{1}{q_4}\frac{1}{n}\sum_h\Big|\Big(\frac{K\alpha_h-<\alpha_h>}{\sigma_h}\Big)^2-1\Big|\Big]^{\tfrac{1}{2}}\Big\} \qquad (17)$$

where the q_i's are the theoretical expected values of the four functions, the summations are over the <u>n</u> strong phased reflexions, K=1/ABS is used to rescale the computed α's on the expected values and

$$\sigma_h^2 = \tfrac{1}{2}\sum_{j=1}^{r}G_j^2[1+D_2(G_j)-2D_1^2(G_j)] \qquad (18)$$

All terms in (17) are expected to be equal to 1 for the correct solution and in this case also ALCOM$\approx$1; higher values indicate incorrect sets of phases.

The $_o$ FOM is related to a moment of $P(\alpha'_h)$, but also the expected values of other appropriate functions of α'_h may be derived and used as FOM's. Also in this case four such moments are used in SIR and combined to give

$$\text{PSCOM} = \frac{1}{4\cdot\text{ABS}}\Big\{\Big[\frac{1}{q_o}\frac{\sum_h\alpha'_h}{\sum_h v_h^{\tfrac{1}{2}}}\Big] + \Big[\frac{1}{q_1}\frac{1}{n}\sum_h\Big(\frac{\alpha'^2_h}{v_h}\Big)\Big]^{\tfrac{1}{2}} +$$

$$\Big[\frac{1}{q_2}\frac{1}{n}\sum_h\Big|\frac{\alpha'^2_h}{v_h}-1\Big|\Big]^{\tfrac{1}{2}} + \Big[\frac{1}{q_3}\frac{1}{n}\sum_h\Big|\Big(\frac{\alpha'_h}{v_h^{\tfrac{1}{2}}}-q_o\Big)^2\Big|\Big]^{\tfrac{1}{2}}\Big\} \qquad (19)$$

where again q_o, q_1, q_2 and q_3 are the theoretical expected values, <u>v</u> is defined in (9) and the division by ABS allows to recognize those wrong solutions for which small values of ABS are associated to small values of the different terms in (19) (low consistency both for Σ_2 and ψ_o triplets). Also PSCOM should be close to 1 and minimum for the correct set of phases.

Once the different FOM's have been calculated for each phase set it is useful to derive an overall <u>combined figure of merit</u>, CFOM; its capability of discriminating the correct set of phases will in general be higher than that of the individual functions. In SIR CFOM is obtained as

$$\text{CFOM} = \Big(1/\sum_{i=1}^{4}w_i\Big)\Big\{w_1[1-|\text{ABS}-1|] + w_2\exp[-(1-\text{CPHAS})^{1.5}] +$$

$$w_3\exp[-(\text{PSCOM}-1)^{1.5}] + w_4\exp[-(\text{ALCOM}-1)^{1.5}]\Big\} = \qquad (20)$$

$$\Big(1/\sum_{i=1}^{4}w_i\Big)\Big\{w_1\text{MABS} + w_2\text{CPHASE} + w_3\text{PSCOMB} + w_4\text{ALCOMB}\Big\}$$

418

It was found empirically that the most effective values for the weights are w_1=0.2, W_2=1.0, w_3=1.4, w_4=1.0. A correct set of phases should lead to CFOM$\approx$1.0, while CFOM$\ll$1.0 should indicate a wrong solution.

The electron density map(s) corresponding to the most promising phase set(s) (with highest CFOM) will then be calculated first.

References

- Burla M.C., Camalli M., Cascarano G., Giacovazzo C., Polidori G., Spagna R. & Viterbo D. (1989), J. Appl. Crystallogr., **22**, 389.

- Burla M.C., Cascarano G., Giacovazzo C., Nunzi A. & Polidori G. (1987), Acta Crystallogr., **A43**, 370.

- Cascarano G., Giacovazzo C. & Viterbo D. (1987), Acta Crystallogr., **A43**, 22.

SQUASH: A PROCEDURE FOR PHASE REFINEMENT

AND EXTENSION OF PROTEIN STRUCTURES

Kevin Cowtan

Department of Physics
University of York
Heslington, York, YO1 5DD. ENGLAND

INTRODUCTION

Direct methods are now well established as an automatic method for the
solution of small molecule structures. Features expected of the electron
density are used to constrain the values of the unknown phases. The result
is a map at atomic resolution from which atomic co-ordinates can be
extracted by peak search routines.

These methods are generally dependent on the availability of data at
atomic resolution. This presents difficulties in the solution of protein
structures, where data resolution is severely limited by disorder and high
thermal motion. Such structures are usually solved by the Multiple
Isomorphous Replacement (MIR) method, which typically gives phased data to
a resolution of 3.0 Å. The resultant map will not give atomic co-ordinates
directly, but a model can usually be fitted into the density using the
chemical knowledge of the user.

There are however many proteins for which MIR phases do not lead to an
interpretable map. This may result from correlation between heavy atom
positions between derivatives, lack of isomorphism, or simply a lack of
isomorphous derivatives. In these circumstances it would be useful to apply
procedures from direct methods to refine MIR phases and generate phases for
unphased data at higher resolutions.

To achieve this aim we must look for constraints on the electron
density that we can use to improve our electron density map. Sayre's
equation is widely used in existing direct methods procedures, and we apply
this to the problem of phase extension. Density modification techniques
have already been used by protein crystallographers for phase extension,
and these techniques are expanded and applied. In addition Sayre's
equation, which has been widely used in direct methods, is applied and
produces useful results at below atomic resolution.

Three constraints on the electron density are therefore applied to
obtain extended phases: Sayre's eQUAtion, Solvent flattening, and Histogram
matching, giving rise to the acronym SQUASH. The constraints are applied as
follows:

<u>Sayre's Equation</u>

Sayre (1952) derived an equation which would be obeyed by any electron density map made solely of equal, resolved atoms. This equation can therefore be used to include a great deal of useful information into the problem, and through various simplifications forms the basis of many direct methods packages in use today.

Sayre's equation can be written in real space as:

$$\rho(x) = \Sigma_y \; \rho(y) \; \psi(x-y)$$

where ψ is a function dependent on atomic shape.

By taking the Fourier transform the reciprocal space form of Sayre's equation is obtained:

$$F(h) = \frac{\Theta(|h|)}{V} \; \Sigma_k \; F(k) \; F(h-k)$$

This equation relates the magnitudes <u>and</u> phases of structure factors. This equation can then be used to derive phases for observed structure factors given an initial set of phases, say for the low resolution data.

In practice for protein structures we are working at well below atomic resolution, and so Sayre's equation is not strictly valid. However we find that it can still be applied in a modified form. Instead of using the shape function ψ for a single atom some sort of shape function for a polypeptide group must be used. This function is determined empirically from the current map, the only constraint being that it must be spherically symmetric.

<u>Solvent Flattening</u>

Protein molecules are often globular or irregular in shape, and thus when they are packed together to form a crystal there are gaps left between them. These spaces are usually filled by a region of disordered solvent. As the x-ray diffraction pattern comes from an average over the whole crystal of the electron density in the unit cell these solvent regions are approximately flat in the final electron density map. If we know the position of the molecule in the unit cell we can therefore smooth out the density in the rest of the cell to improve the electron density map. This leads to an improvement in the calculated phases (Wang, 1985).

How can the molecule be located? If the solvent is water, it will have a mean electron density of around 0.32 e/$\mathring{A}^3$. Typically the mean density of a protein is around 0.43 e/$\mathring{A}^3$. Thus an approximate molecular envelope can be obtained by convoluting an MIR map with some averaging function, typically a sphere of radius 8.0$\mathring{A}$.

This method has been shown to give good results when a good estimate of the molecular envelope can be obtained. Its effectiveness increases as the solvent content of the unit cell increases.

<u>Histogram Matching</u>

A density modification technique which can be applied to the protein region in a similar way that solvent flattening is applied to solvent would be useful. In fact such a technique already exists in the field of image processing.

For any electron density map we can calculate a histogram of density
values. This will probably show a large fraction of the map with electron
density around zero, with a smaller fraction of high density representing
the atomic peaks. In fact we can predict what the density histogram should
look like for a perfect electron density map at any particular resolution,
and it turns out that the histograms are similar for a wide range of
structures.

How can this information be applied? We can alter the density
histogram of a map simply by rescaling the electron density. Thus the map
may be altered in a systematic way to obtain the correct electron density
histogram. Again this gives rise to improved phases and to increased
resolution (Zhang, Main, 1990).

METHOD

The above methods give rise to a system of equations written
in terms of the electron density as follows:

$$\rho(x) = \Sigma_y \; \rho(y) \; \psi(x-y) \quad \text{(Sayre's equation)}$$

$$\rho(x) = H(x) \quad \text{(Solv. flattening/Hist. Matching)}$$

Note that there are two equations and one unknown for each pixel in
the electron density map, and that the equations are non-linear. Thus the
solution represents a considerable computational problem. Using the
conventional least-squares Newton Raphson method a matrix of approximately
10^{10} elements would have to be stored, and the calculation would require
about 50 years on a typical mainframe.

To turn this into a feasible computational problem two steps have to
be taken. Firstly the conjugate gradient algorithm is used to solve
approximately the system linear equations obtained by Newton's method.
Secondly it is noted that the product of the Jacobian of the system with an
arbitrary vector can be calculated by fourier transforms without explicitly
storing the Jacobian. Thus the whole Newton Raphson calculation can be
performed by around 30 FFTs (Main, 1990).

RESULTS

The SQUASH procedure has been successfully applied to 2 Zn Insulin
data using MIR phases to 3.0 Angstroms and structure factor magnitudes to
2.15 Angstroms. From this data phases were derived for the all the
structure factors to 2.15 Å. The resulting phases were compared with
calculated phases from the known structure. A correlation coefficient was
also calculated between both the MIR and SQUASH maps and the calculated
electron density. The results of these tests are shown in Table 1.

Table 1. Application of SQUASH to 2 Zn Insulin, 3.0Å MIR phases.

	Mean Phase Error (°)		
	Initial Phases	Extended Phases	Map Correlation
3.0 Å MIR data	46.2	-	0.55
2.15 Å SQUASH data	39.3	61.4	0.70

The SQUASH map clearly shows the solvent and protein regions of the
crystal. The resolution of the map has been considerably improved. Detailed
examination of three dimensional plots has shown that sections of the
backbone which are broken in the MIR map have been restored in the SQUASH
map.

As a result the SQUASH map is more easily interpretable than the MIR
map. It incorporates unphased high-resolution data which could not by
conventional techniques have been included in the calculation of the
electron density.

Tests are currently under way on other structures for which the input
phases are far worse. Initial results are encouraging, although there are
problems in detecting when there has been an improvement in the map.

CONCLUSIONS

A procedure has been developed which has the potential to allow the
solution of protein structures from less or poorer quality data.
Consequently less time need be spent in the preparation of isomorphous
derivatives, and less diffractometer time is required collecting data. More
importantly, it should become possible to solve larger structures than have
been practical up to now.

ACKNOWLEDGEMENTS

I would like to thank Glaxo for use of the Alliant FX40
mini-supercomputer at York, without which the development of the SQUASH
procedure to its current state would have been extremely difficult. I would
also like to thank the Rigaku Corporation, of Japan, for providing my
research studentship.

REFERENCES

Main, P., 1990, The Use of Sayre's Equation with Constraints for Direct
    Determination of Phases, Acta Cryst. (to be published)
Sayre, D., 1952, Acta Cryst. 5:60
Wang, B.C., 1985, Methods in Enzymology 115:90
Zhang, K.Y.J., 1989, "On the Phase Refinement and Extension of
    Macromolecular Structures using both Real and Reciprocal Space
    Approaches." (Thesis), University of York
Zhang, K.Y.J., Main, P., 1990, Histogram Matching as a New Density
    Modification Technique for Phase Refinement and Extension of Protein
    Molecules, Acta Cryst. A46:41

SAYTAN: THE SAYRE-EQUATION TANGENT FORMULA

Cecil Tate

Department of Physics
University of York
York YO1 5DD UK

SAYTAN is a computer-programmed tangent formula for phase determination, employed in a multi-trial approach in the MULTAN tradition. It has been described in two publications by Debaerdemaeker, Tate & Woolfson (1985, 1988). In the first of these papers it was called the "Sayre Tangent Formula", but a referee for the second paper pointed out that Sayre did not have a tangent formula and so, since then, we use the more appropriate description "Sayre-equation Tangent Formula".

The basic idea of SAYTAN is that a good set of phases ϕ_h belonging to structure factors $|E_h|$ should satisfy a system of Sayre equations

$$E_h = \frac{K}{g_h} \sum_h E_k E_{h-k} \quad ; \quad E_h = |E_h| e^{i\phi_h} \qquad (1)$$

where K is an overall scaling factor which is meant to allow for the fact that the set $\langle|E_h|\rangle$ is not a complete set of data, and g_h is the scattering factor for the "squared" structure.

The algebraic derivation of SAYTAN starts from the following residual for a system of Sayre equations

$$R = \sum_h |g_h E_h - KG_h|^2 \quad ; \quad G_h = \sum_k E_k E_{h-k} \qquad (2)$$

In the two publications mentioned above, the scattering factors g_h appear on the right-hand side, as in equation (1), but in later work we have multiplied each Sayre equation by g_h, as in equation (2), because this simplifies the program slightly and seems to work equally well.

The residual (2) can be written in the form

$$R = U - 2KT + K^2 Q$$

$$U = \sum_h g_h^2 |E_h|^2$$

$$T = R_e \sum_h g_h E_h{}^* G_h$$

$$= \sum_{\underset{\sim}{h}} \sum_{\underset{\sim}{k}} g_h |E_k E_{h-k} E_h| \cos(\Phi_k + \Phi_{h-k} - \Phi_h)$$

$$Q = \sum_{\underset{\sim}{h}} |G_h|^2 \qquad\qquad (3)$$

$$= \sum_{\underset{\sim}{h}} \sum_{\underset{\sim}{h'}} \sum_{\underset{\sim}{h''}} |E_{h'} E_{k-h'} E_{h-h''} E_{h''}| \cos(\Phi_{h'} + \Phi_{h-h'} - \Phi_{h-h''} - \Phi_{h''})$$

The phases which minimize R should make

$$\frac{\partial R}{\partial \Phi_\ell} = 0$$

for all $\underset{\sim}{\ell}$. Differentiating (3),

$$- 2K \frac{\partial T}{\partial \Phi_\ell} + K^2 \frac{\partial Q}{\partial \Phi_\ell} = 0 \qquad\qquad (4)$$

since U is independent of phases. It can easily be shown that

$$\frac{\partial T}{\partial \Phi_\ell} = \sum (g_\ell + g_h + g_{\ell-h}) |E_\ell E_h E_{\ell-h}| \sin(\Phi_h + \Phi_{\ell-h} - \Phi_\ell)$$

$$\frac{\partial Q}{\partial \Phi_\ell} = 4 \sum_{\underset{\sim}{h}} \sum_{\underset{\sim}{k}} |E_\ell E_{h-\ell} E_{h-k} E_k| \sin(\Phi_k + \Phi_{h-k} - \Phi_{h-\ell} - \Phi_\ell) \qquad (5)$$

Substituting (5) in (4) and rearranging we have

$$\Phi_\ell = \text{phase of } \left\{ \sum_{\underset{\sim}{h}} (g_\ell + g_h + g_{\ell-h}) E_h E_{\ell-h} - 2K \sum_{\underset{\sim}{h}} E_{\ell-h} \sum_{\underset{\sim}{k}} E_k E_{h-k} \right\} \quad (6)$$

This may be a suitable point at which to note that in R we can include
Sayre equations for structure factors on the left-hand side which are very
small, ideally zero, for which we do not calculate a phase. These weak
reflections are labelled E_{hz}: they are the "psi-zero" reflections which
have been used in older versions of MULTAN only passively for the "psi-
zero" figure of merit; but in SAYTAN they play an active role in phase
determination although they are not phased themselves.

The part of R arising from these "zero" terms simply requires that

$$R_z = K^2 \sum_{\underset{\sim}{h_z}} \left| \sum_{\underset{\sim}{k}} E_k E_{h_z-k} \right|^2 \qquad\qquad (7)$$

should be small, which is like saying that the psi-zero figure of merit
should be small.

It should also be noticed that R_z can be put into the total residual
with its own weight which we'll call w_z. This is made explicit by
re-writing equation (6) as follows

$$\Phi_\ell = \text{phase of } \left\{ \sum_{\underset{\sim}{h}} (g_\ell + g_h + g_{\ell-h}) E_h E_{\ell-h} - 2K(q_{B,\ell} + w_z q_{z,\ell}) \right\} \quad (8)$$

where

$$q_{B,\underset{\sim}{\ell}} = \sum_{\underset{\sim}{h}} E_{\underset{\sim}{\ell}-\underset{\sim}{h}} \sum_{\underset{\sim}{k}} E_{\underset{\sim}{k}} E_{\underset{\sim}{h}-\underset{\sim}{k}} \tag{9}$$

the indices $\underset{\sim}{h}$, $\underset{\sim}{k}$ and $\underset{\sim}{\ell}$ all referring to strong reflections which are to be phased, and

$$q_{z,\underset{\sim}{\ell}} = \sum_{\underset{\sim}{h}_z} E_{\underset{\sim}{\ell}-\underset{\sim}{h}_z} \sum_{\underset{\sim}{k}} E_{\underset{\sim}{k}} E_{\underset{\sim}{h}_z-\underset{\sim}{k}} \tag{10}$$

where the indices $\underset{\sim}{h}_z$ are for weak reflections which are not to be phased. We usually call $q_{B,\underset{\sim}{\ell}}$ the "big" quartets and $q_{z,\underset{\sim}{\ell}}$ the "zero" quartets.

We also separate the triplet terms in (8) into two parts as follows

$$t_{1,\underset{\sim}{\ell}} = g_{\underset{\sim}{\ell}} \sum_{\underset{\sim}{h}} E_{\underset{\sim}{h}} E_{\underset{\sim}{\ell}-\underset{\sim}{h}}$$

$$\tag{11}$$

$$t_{2,\underset{\sim}{\ell}} = \sum_{\underset{\sim}{h}} (g_{\underset{\sim}{h}} + g_{\underset{\sim}{\ell}-\underset{\sim}{h}}) E_{\underset{\sim}{h}} E_{\underset{\sim}{\ell}-\underset{\sim}{h}}$$

It is then interesting to note a relationship between $t_{2,\underset{\sim}{\ell}}$ and $q_{B,\underset{\sim}{\ell}}$. Consider the expression

$$t_{2,\underset{\sim}{\ell}} - 2Kq_{B,\underset{\sim}{\ell}} = \sum_{\underset{\sim}{h}} \langle E_{\underset{\sim}{\ell}-\underset{\sim}{h}} (g_{\underset{\sim}{h}} E_{\underset{\sim}{h}} - KG_{\underset{\sim}{h}})$$

$$+ E_{\underset{\sim}{h}} (g_{\underset{\sim}{\ell}-\underset{\sim}{h}} E_{\underset{\sim}{\ell}-\underset{\sim}{h}} - KG_{\underset{\sim}{\ell}-\underset{\sim}{h}}) \rangle \tag{12}$$

We see that this expression has Sayre equations as factors throughout and, therefore, for a set of phases which approximately satisfy the Sayre equations, we have

$$t_{2,\underset{\sim}{\ell}} - 2Kq_{B,\underset{\sim}{\ell}} \simeq 0 \tag{13}$$

Furthermore, in the initial stages of phase development, when the phases are still fairly random, it is best to rely mainly on the classical triplet term $t_{1,\underset{\sim}{\ell}}$ for phase refinement and so we can assume equation (13) to be satisfied in that case also. This suggests the following form for SAYTAN

$$q_{\underset{\sim}{\ell}} = \text{phase of } \langle t_{1,\underset{\sim}{\ell}} + w_B (t_{2,\underset{\sim}{\ell}} - 2Kq_{B,\underset{\sim}{\ell}}) - 2Kw_z q_{z,\underset{\sim}{\ell}} \rangle \tag{14}$$

where w_B and w_z are parameters which can be input to the computer program by means of keywords. If these keywords are not used, the program provides default values for w_B and w_z. By default, the program sets $w_B = 0$ and this is found to give good results in practice. It also has the advantage that computation of the terms $q_{B,\underset{\sim}{\ell}}$, which are relatively costly to compute, can be skipped.

The question now arises as to what to use for the scaling factor K. One possibility for K is to use the value give by

$$\frac{\partial R}{\partial K} = -2T + 2KQ = 0 \tag{15}$$

which minimizes R. This gives

$$K = T/Q \tag{16}$$

However, at any given stage of the refinement process, T and Q are functions of the phases and, in the beginning, when the phases are random, T and Q will both be small and random, and T/Q will then fluctuate wildly. We have found in experiments with SAYTAN on various structures that satisfactory results are obtained with

$$K = T/\langle Q \rangle \tag{17}$$

where $\langle Q \rangle$ is an estimate of the value for Q that would be expected for good phases.

$\langle Q \rangle$ can be separated into two parts

$$\langle Q \rangle = \langle Q_B \rangle + \langle Q_Z \rangle \tag{18}$$

where $\langle Q_B \rangle$ is estimated from the probability distribution for Q_B and $\langle Q_Z \rangle$ is given by a "random-walk" estimate of Q_Z. The value for K given by K = T/$\langle Q \rangle$ seems to be particularly suitable because, at the start of refinement, when the phases are random, K is quite small and the phase Φ_Q in (14) is then mainly determined by the triplets. After a few cycles of phase development, however, K becomes appreciable and thus brings in the quartets. If it happens that a particular trial approaches a good solution, K will then have a value nearly the same as the true T/Q.

So far, no indication has been given of how to evaluate the parameter w_Z. It seems reasonable to suppose that the contribution to SAYTAN (equation (14)) from the "zero" quartets $q_{Z,\varrho}$, should be comparable to the triplet term $t_{1,\varrho}$. For any given w_Z the ratio of these contributions depends upon the ratio.

$$r = \text{NSRPSI/NRST} \tag{19}$$

where NSRPSI and NSRT are respectively the number of "weak", or psi-zero relationships and the number of strong relationships in the Sigma-2 list of MULTAN. If r is decreased, w_Z should be increased, and vice-versa. SAYTAN seems to work best when the ratio r is about 0.5 but this condition is not critical. After experimenting on a number of known, small structures we settled on the following formula

$$w_Z = 5/(1 + 8r) \tag{20}$$

The formula (20) has been incorporated into the program as the default option, but this can be overwritten by using the appropriate keyword. It has been found that the effectiveness of SAYTAN in finding a good solution is fairly sensitive to w_Z, particularly for larger structures, and therefore, when difficulty is encountered in solving a structure, it is often advantageous to try several values for w_Z.

If w_B and w_Z are both zero, of course, SAYTAN reduces to the classical tangent formula of Karle & Hauptman. It is interesting to observe how the quality of the results from SAYTAN changes as w_Z is increased from zero. In general, as w_Z is increased, at first the quality improves: solutions are found more frequently and each of the good solutions tends to show more atoms in the structure. But when some optimum value has been passed, the quality deteriorates.

TRITAN - OR, RECYCLED FAILURE

Laila Refaat

Department of Physics
University of York
York YO1 5DD UK

Even when direct methods fail to give a clear solution, E maps from some of the phase sets obtained contain correctly oriented fragments. Actually a number of workers have made use of structural information in the application of direct methods - for example, Hauptman (1964), Oda, Naya and Nitta (1967), Kroon and Krabbendam (1970) and Theissen and Busing (1974). The most systematic approach has been that of Main (1976) who has shown that knowledge of a correctly oriented structural fragment modifies the Cochran distribution for a three-phase invariant.

We wish to report a simple and automatic procedure for using this concealed partial structure information without the need specifically to look at E maps and give estimates of the values of three-phase invariants. These estimates are incorporated into a modified tangent formula which is used in a new run of a multi solution direct-methods procedure, the total process called TRITAN.

THE TRITAN PROCEDURE

The scenario that we are presenting is that there have been runs of, say, MULTAN87 (Debaerdemaeker, Tate & Woolfson (1988)) in all its possible modes but that no correct solution has been found. Entering the TRITAN procedure initiates the following sequence of calculations.

1. For each of n (usually six to ten) 'best' phase sets, as judged by the combined figure of merif CFOM, the E map is calculated and the largest chemically sensible fragment is automatically found with the SEARCH routine of MULTAN.

2. For each of the fragments partial structure factors are calculated

$$S_j(h) = |S_j(h)|\exp[i\psi_j(h)] \quad (j = 1 \text{ to } n), \qquad (1)$$

for h corresponding to the set of large E(h) for which phases are required.

3. It is now required to find the fragment that bears probably the greatest resemblance to the complete structure. This is done by finding for each fragment a correlation coefficient.

$$r_j = \frac{\langle |S_j(h)E(h)| \rangle_h - \langle |S_j(h)| \rangle_h \, \langle |E(h)| \rangle_h}{\sigma_j{}^S \, \sigma_E} \qquad (2)$$

where

$$\sigma_j{}^S = \left[\langle |S_j(h)|^2 \rangle_h - \langle |S_j(h)| \rangle_h^2 \right]^{\frac{1}{2}} \qquad (3a)$$

and

$$\sigma_E = \left[\langle |E(h)^2| \rangle - \langle |E(h)| \rangle_h^2 \right]^{\frac{1}{2}} \qquad (3b)$$

If the largest value of r is for $j = i$ then the first estimates of the three phase invariants are obtained from

$$\Phi^1_{3,e}(h,k) = \Psi_i(h) - \Psi_i(k) - \Psi_i(h-k). \qquad (4)$$

4. For each other fragment $(j \neq i)$ there are calculated

$$R_j = \left[\langle \sin \Phi^1_{3,e}(h,k) \sin \Psi_{3,j}(h,k) \rangle_{h,k} \right.$$

$$\left. - \langle \sin \Phi^1_{3,e}(h,k) \rangle_{h,k} \, \langle \sin \Psi_{3,j}(h,k) \rangle_{h,k} \right] / \mu_e \mu_j$$

$$(5)$$

where

$$\Psi_{3,j}(h,k) = \Psi_j(h) - \Psi_j(k) - \Psi_j(h-k) \qquad (6)$$

and μ_e and μ_j are the standard deviations of the quantities $\sin\Phi^1_{3,e}(h,k)$ and $\sin\Psi_{3j}(h,k)$ respectively. The value of R_j, which is a linear correlation coefficient, is theoretically constrained by the form of (5) to be in the range

$$-1 \leqslant R_j \leqslant 1. \qquad (7)$$

If $R_j = 1$ then it indicates perfect positive correlation of the quantities $\sin\Phi^1_{3,e}(h,k)$ and $\sin\Psi_{3j}(h,k)$. On the other hand $R_j = -1$ indicates a perfect negative correlation which can be transformed into a perfect positive correlation by taking the enantiomorph of the jth fragment which reverses all the values of $\Psi_j(h)$ and hence of $\Psi_{3,j}(h,k)$.

5. For the largest value of $|R_j|$ calculate a second estimate of each invariant from

$$\tan\Phi^2_{3,e}(h,k) =$$

$$\frac{K^1(h,k) \sin\Phi^1_{3,e}(h,k) + s_j |R_j|^{\frac{1}{2}} K_j(h,k) \sin\Psi_{3,j}(h,k)}{K^1(h,k) \cos\Phi^1_{3,e}(h,k) + |R_j|^{\frac{1}{2}} K_j(h,k) \cos\Psi_{3,j}(h,k)}$$

$$= T^1(h,k)/B^1(h,k), \qquad (8)$$

where

$$K^1(h,k) \;=\; 2\sigma_3\sigma_2^{-3/2}\;|\,S_i(h)S_i(k)S_i(h-k)\,| \qquad\qquad (9a)$$

$$K_j(h,k) \;=\; 2\sigma_3\sigma_2^{-3/2}\;|\,S_j(h)S_j(k)S_j(h-k)\,| \qquad\qquad (9b)$$

$\sigma_n \;=\; \Sigma_{j=1}^{N}\; N_j^n$ (z_j is the atomic number of the jth atom)

and s_j is the sign of R_j.

Equation (8) is a tangent-formula combination of two estimates of the three-phase invariants. The presence of S_j ensures that the estimates are from the same enantiomorph and $|R_j|^{1/2}$ is an arbitrary, but empirically effective way of influencing the degree to which the new estimate should modify the previous one. The K values for these second estimates of the three-phase invariants are derived from

$$K^2(h,k) \;=\; [T^1(h,k)^2 \;+\; B^1(h,k)^2]^{1/2} \qquad\qquad (10)$$

6. A chain process is now entered in which at each stage triple-phase estimates from each residual fragment are compared with the current combined estimates to yield values of R_j, as indicated in (5) followed by the estimates corresponding to the largest $|R_j|$ being combined with the current estimate as shown by (8).

7. A modified tangent formula, similar to that proposed by Main (1976) and also used by Olthof, Sint & Schenk (1979) and in the program MITHRIL by Gilmore (1984) is used in place of the normal one in a rerun of MULTAN. This tangent formula has the form

$$\Phi(h) \;=\; \text{phase of } \sum_k K^n(h,k)$$

$$\times\; \exp\{i[\Phi(k) \;+\; \Phi(h-k) \;-\; \Phi_{3,e}^n(h,k)]\}. \qquad\qquad (11)$$

SOME PRACTICAL TRIALS

The TRITAN procedure has been tried on several known structures which can be solved by direct methods but which present difficulties for some direct-methods procedures. In these trials sets of phases were selected which did not individually reveal the structures in order to see whether an amalgamation of information from them would give a solution. The structures are referred to by code names for brevity.

CORTISONE [Declercq, Germain & Van Meerssche (1972). $C_{21}H_{28}O_5$, $P2_12_12_1$, $Z = 4$]

Although this structure can be solved by any of the five procedures available in MULTAN87 there is no perfect solution in 100 trials with MULTAN80. None of these sets of phases gave an E map showing a five-membered ring although several produced linked six-membered rings.

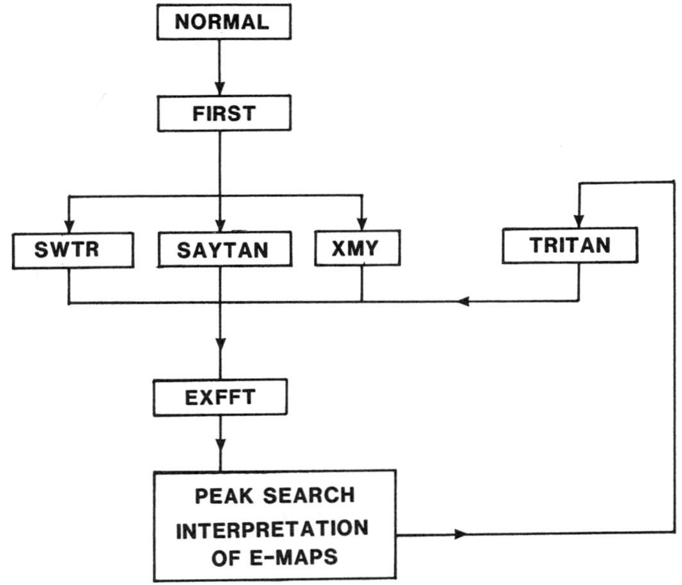

Figure 1. Tritan recycle.

The ten sets of phases with the highest values of CFOM were subjected to the TRITAN procedure. The amalgamated three-phase invariant estimates, used with the modified tangent formula (15), then gave six perfect solutions in 30 trials.

MUCCAR

For this P1 structure both MULTAN80 and MULTAN87 using the statistically weighted tangent formula (SWTF) gave poor results although the SAYTAN mode in MULTAN87 solves it quite easily. MULTAN80, with the SWTF, was run for 100 trials. The maps corresponding to the two best CFOMs revealed less than one-half of the structure. Nine phase sets, corresponding to the third to eleventh best CFOMs were input into TRITAN. In 30 trials there were found four perfect solutions and three others with only one atom missing.

MUNICH4 [Szeimes-Seebach, Harnisch, Szeimes, Van Meerssche, Germain & Declercq (1978). $C_{27}H_{22}O$, Cc, Z = 4]

With the SWTF no solution was found in 200 trials of MULTAN 80 and 40 trials of MULTAN87. The six phase sets with the highest values of CFOM from MULTAN80 were input into TRITAN. In 30 trials there were two perfect maps and four with only one atom missing.

CINOBUFAGIN [Declercq, Germain & King (1977). $C_{26}H_{27}O_6$, $P2_12_12_1$, Z = 4]

This was originally solved with MULTAN, using the SWTF, with great difficulty after generating 5000 phase sets. We carried out 200 trials with the SWTF mode of MULTAN87 but the figures of merit were so poor that we did not event try to use them. Instead we did 50 trials with the SAYTAN mode of MULTAN87, which readily solves the structure, and input the worst six solutions in terms of CFOM into TRITAN. In 30 trials there were six maps produced lacking only one atom.

AZET [Colens, Declercq, Germain, Putzeys & Van Meerssche (1974) $C_{21}H_{16}ClNO$, $Pca2_1$, Z = 4]

Use of MULTAN87 in the SWTF mode gave nothing promising but for 50 trials in the SATTAN mode the best map showed one-half of the structure. After amalgamating the information in the five phase sets with the highest CFOMs, in 30 trials TRITAN gave one map with four atoms missing and seven maps with eight or nine atoms missing.

REFERENCES

Colens, A., Declercq, J.P., Germain, G., Putzeys, J.P. & Van Meerssche, M. 1974, Cryst Struct Commun <u>3</u>, 119-122.

Debaerdemaeker, T., Tate, C. & Woolfson, M.M., 1988, Acta Cryst A<u>44</u>, 353-357.

Declercq, J.P., Germain, G. & King, G.S.D., 1977, Fourtth Eur Crystallogr Meet., Oxford. Abstracts A, pp 279-280.

Declercq, J.P., Germain, G. & Van Meerssche, M., 1972, Cry Struct Commun <u>1</u>, 13-15.

Gilmore, C.J., 1984, J Appl Cryst <u>17</u>, 42-46.

Hauptman, H., 1964, Acta Cryst <u>17</u>, 1421-1433.

Kroon, J. & Krabbendam, H., 1970, Acta Cryst B$\underline{26}$, 312-314.

Main, P., 1976, Crystallographic Computing Techniques, edited by
    F.R. Ahmed, pp 97-105. Copenhagen: Munksgaard.

Oda, T., Naya, S. & Nitta, I., 1967) Me Osaka Univ $\underline{16}$, 19-36.

Olthof, G.J., Sint, L. & Schenk, H., 1979, Acta Cryst A$\underline{35}$, 941-946.

Szeimes-Seebach, U., Harnisch, J., Szeimes, G., Van Meerssche, M.,
    Germain, G. & Declercq, J.P., 1978, Angew Chem Int Ed Engl $\underline{17}$,
    848-850.

Thiessen, W.E. & Busing, W., 1974, Acta Cryst A$\underline{30}$, 914.

THE DESIGN OF A KNOWLEDGE-BASED SYSTEM FOR CRYSTAL STRUCTURE DETERMINATION

Suzanne Fortier[*], Janice Glasgow[#] and Frank H. Allen[##]

Dept. of Chemistry[*] and Dept. of Computing and
Information Science[#]
Queen's University, Kingston
Canada K7L 3N6

Cambridge Crystallographic Data Centre[##]
University Chemical Laboratory
Lensfield Road, Cambridge, CB2 1EW
England

1 INTRODUCTION

The determination of crystal structures from their diffraction data
belongs to the general class of image reconstruction exercises from
incomplete and/or noisy data. In the crystallographic example, recovery
of a three-dimensional image requires knowledge of both the amplitude
and the phase of the diffracted rays. Since it is not usually possible
to measure the phases experimentally, this exercise is not
straightforward. Rather it involves the retrieval of some phase
information from either additional isomorphous data sets, partial
structure information or, in a direct method approach, from the
amplitude data themselves through the application of probability theory.
In the simplest cases, the process of image recovery is one of image
processing. In more complex cases and, in particular in the case of
macromolecular crystal structures, pattern recognition is an important
component of the image reconstruction exercise. Chemical and structural
rules are used to guide the iterative structure determination process.

Pattern recognition, in most crystal structure determination
projects, comes from the user, through his/her individual, mental,
recollection of structural patterns. Clearly the individual memory of
possible structural templates can be greatly expanded by using the
information available in the crystallographic databases. The proposed
knowledge-based system for crystal structure determination uses an
artificial intelligence infrastructure to embed current phasing
strategies - primarily direct methods, but also Patterson based methods
and densitiy modification techniques - with crystallographic database
information. The process of crystal structure determination is modelled
as the iterative resolution of a three-dimensional image. Thus, solving
a crystal structure becomes a more fluid process in which the search
space is scanned by mathematical tools afforded by direct methods, while
it is guided by pattern recognition techniques derived from chemical and
crystallographic reasoning.

Direct Methods of Solving Crystal Stuctures
Edited by H. Schenk, Plenum Press, New York, 1991

2 THE ARTIFICIAL INTELLIGENCE INFRASTRUCTURE

At the core of the system is the concept of *imagery*, that is the ability to construct and reason about images in creative and unique ways. Thus both the knowledge representation model and the search techniques are based on current theories of cognitive psychology on the mental processes involved in imagery.

2.1 The phase problem as a search problem

The crystallographic phase problem can be thought of as a general search problem. In the initial state, only a handful of phases are known. In spanning the search space, it is hoped that a final state, or goal state, will be reached in which enough phases will have been determined with sufficient accuracy to compute an interpretable electron density map. In this context, the multisolution approach, commonly used in direct methods, can be described as a simple generate-and-test search procedure. The morphology of the search tree is unusual, though. The tree has a single depth level with a large branching factor, the number of nodes in such a tree is usually between 32 and 128. What characterizes and limits the search is the fact that the heuristic evaluation functions used (the figures of merit) are not effective in pruning partially developed solutions. This is due, in part, to the fact that these heuristic functions do not test the actual goal of the search, the interpretability of the reconstructed image. Rather, they provide ranking numbers that are derived from the underlying probabilistic model. The *a priori* model, assumed by direct methods, is that the repeating atomic motif in the crystal can be represented by a collection of atoms randomly distributed through space. The only chemical information that is used is the chemical formula. A great deal of additional information is actually available, if not currently used, at the outset of a crystal structure determination project. The crystallographic databases now contain information on over 100,000 fully determined crystal structures. Any new structure/molecule will almost certainly contain chemical fragments for which reasonable guesses can be made concerning their likely three-dimensional shapes. Indeed, such guesses are made quite automatically by individual crystallographers in their attempts to interpret features of the image reconstructed from the diffraction data.

In the knowledge-based approach to crystal structure determination, active use is made of the available structural information. In particular, initial chemical information such as two-dimensional chemical diagrams, sequences, etc., together with expert knowledge of related chemistry is used to extract possible three-dimensional structural templates from the crystallographic databases. These are then used to guide the search. The phasing search space is expanded by using a hierarchical approach whereby the process of structure determination flows through resolution levels and is modelled as *resolving* the image of the crystal structure. Phases are thus expanded in shells of increasing resolution using a direct methods approach. At each stage, an electron density image is computed. As part of the reconstruction, standard image preprocessing techniques such as noise removal, local averaging, ensemble averaging, etc., under which fall the density modification procedures, are applied. The image is then segmented and analyzed for feature identification and classification. A general-sphere representation model, following the approach of Marr and Nishihara (1978), is used for shape description. Concurrently, possible structural patterns are identified in the databases. These are then reconstructed at a resolution matching that of the current electron density image, through the selection of appropriate atomic displacement factors. Symbolic array representations for the fragments retrieved from the databases are then compared with the depictions of the features identified in the electron density image. In the first stage of pattern matching, Patterson-based techniques are used to focus attention in the most promising regions of the electron density map. A template matching approach is then applied and the degree of fit assessed. In addition, the matches are checked for consistency with world knowledge which, in the crystallographic context, derives mainly from the constraints imposed by symmetry and packing considerations.

At each phasing stage, any partial structure information identified through pattern matching is used to update the probability distribution phasing tools. Recent direct methods investigations (Bricogne, 1988; Fortier and Nigam, 1989) have shown that it is possible to integrate in a general joint probability distribution framework various kinds of information, e.g. diffraction data from isomorphous derivatives, Friedel pair data and partial structure identification. Furthermore, this work has provided the theoretical basis needed for the computer generation of such distributions. It thus becomes possible to think in terms of a dynamic system in which distributions tailored to the knowledge base are generated as needed. Such a framework provides the tools required for incorporating and updating all available sources of phase information and allows the structure determination process to be modelled as an iterative recognition process, as depicted in Figure 1 which shows the main modules of the knowledge-based structure determination algorithm.

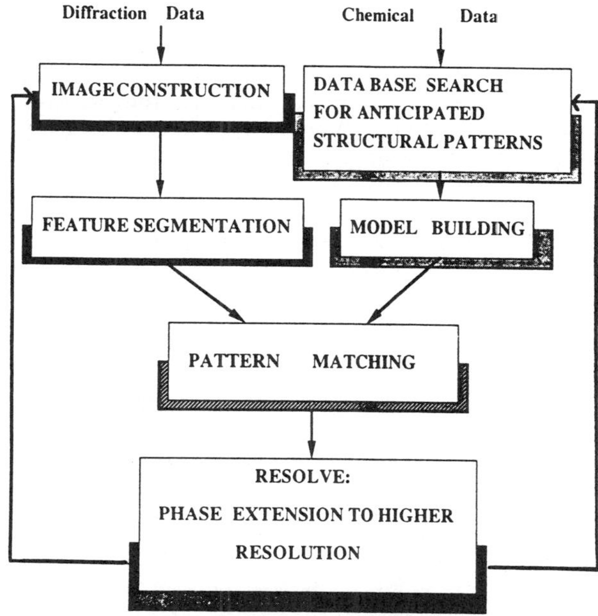

Figure 1 The knowledge-based crystal structure determination
algorithm

2.2 Knowledge acquisition from database analysis

The generation of suitable three-dimensional templates for a given two-dimensional chemical fragment is itself a pattern recognition (classification) problem applied to the accumulated crystal structure information. Within the Cambridge Structural Database system, an initial two-dimensional search locates all examples of the fragment, with restrictions placed on its chemical environment if appropriate. Suitable three-dimensional shape descriptors, e.g. torsion angles, are then generated for each fragment and form the basis for auto-classification into conformational subgroups via specially modified cluster analysis algorithms (Allen, Doyle and Taylor, 1990). The most representative fragment for each subgroup is selected as a possible template for pattern matching with the electron density features. Templates may be used automatically, using the subgroup population as a ranking measure, or may be selected by the user on the basis of expert knowledge of the relevant three-dimensional structural chemistry. The system will build a three-dimensional template library as part of its knowledge base, a library which has obvious uses in other modelling applications, particularly in rational drug design. In all applications, including ours, it is important to extend the template definitions to a prediction of the full three-dimensional molecular structure wherever possible. We intend to combine molecular similarity analysis and molecular segmentation algorithms to elucidate rules which govern fragment combination. These rules represent a context-sensitive grammar which links the semantic units (templates) within a formal linguistic framework which describes three-dimensional molecular structures. The problem of full structure prediction from a fragmentary basis has already received some attention using artificial intelligence (Dolata, Leach and Prout, 1987) and analogy techniques (Wippke and Hahn, 1988). Similar rule-based approaches to protein structure prediction are also being developed (Blundell, Sibanda, Sternberg and Thornton, 1987).

Database analyses also contribute to world knowledge in the crystallographic context, in providing systematic information about the intermolecular interactions which define packing constraints. This information covers preferred H-bond geometries (Taylor and Kennard, 1984) and the topology of extended H-bonded motifs (Etter, MacDonald and Bernstein, 1990). It also covers the geometry of non-bonded interactions between atoms (Nyburg and Faerman, 1986) and between chemical functional groups (Rowland, Allen, Carson and Bugg, 1990). This type of knowledge is applicable to analyses of both small molecules and proteins.

2.3 Knowledge representation

A formal scheme for representing imagery has been developed (Glasgow, 1990). This scheme is based on array theory, the mathematics of embedded, rectangularly arranged collections of data (More, 1981). Similar to set theory, array theory is concerned with the concepts of aggregation, membership and nesting. Array theory is also concerned with the concept of data objects having a spatial position relative to other objects in a collection. As such it provides a metalanguage that can be used for describing and implementing the representations and processes that correspond to mental imagery. This metalanguage has three components: a data structure for depicting images symbolically, a set of array functions that operate on such representations, and a domain of interpretation. The embedded array structure is used to represent both the spatial and hierarchical structure of an·image. The functions defined for representation correspond to the cognitive processes that are involved in constructing, transforming and accessing images. These functions are applied in the context of world knowledge that includes a database representation of long-term memory.

Theories of cognition suggest that imagery involves both descriptive and depictive information. They also suggest that, in long-term memory, depictive knowledge may be stored as a literal description of the locations of the parts within an image. Thus, the data structure used is one which represents an image descriptively, yet allows a symbolic array depiction to be generated when needed.

The embedded array data structure provides a multi-dimensional realizaton of an image. Although the representation is viewer independent, functions such as rotate, translate and project allow an image to be viewed from a variety of perspectives. The embedded nature of the array also allows for a hierarchical depiction of an image. Detailed information (lower levels of the hierarchy) can either be hidden or made explicit in such a representation. Hierarchical organization is a fundamental aspect of imagery. Theories of selected attention suggest the need for an integrated spatial/hierarchical representation: when attention is focussed on a particular feature, the brain is still partially aware of other features and their spatial relation to the considered feature. Our symbolic array representation supports such theories by considering an image as a *recursive* data structure.

2.4 Implementation

We are presently implementing a prototype system using the programming language Nial (Jenkins, Glasgow and McCrosky, 1986). This implementation makes use of specialized software developed in Nial for applications in artificial intelligence and takes advantage of the underlying embedded array data structure on which Nial is based. Provisions are made for reimplementation, at a later stage, of parts of the code in a compiler based or parallel language. The prototype system is initially being developed for and tested on small organic structures. The final goal, however, is to implement a similar system for elucidating protein crystal structures. Because of their hierarchical structural organization, proteins are ideally suited to the phasing strategies used in the knowledge-based approach to crystal structure determination.

3 CONCLUDING REMARKS

The proposed knowledge-based system offers a comprehensive approach to crystal structure determination which accomodates a variety of phasing tools and takes advantage of the structural information already accumulated in the databases. The phasing search is conducted in a flexible way with a search tree that is dynamically generated and explored using guiding heuristics derived from the continual updating of structural information. The simplicity of a general algorithmic solution is thus replaced by the flexibility of a context driven solution process which attempts to make use of all available information. By rephrasing the crystal structure determination problem in the general context of image reconstruction, a dynamic approach results. In this approach, *resolving* the crystal structure image is analogous to a scene analysis in which structural information of increasing resolution takes shape in the frame of the unit cell.

ACKNOWLEDGEMENTS

Finanical assistance from the Natural Sciences and Engineering Research Council of Canada and from Queen's University is gratefully acknowledged.

REFERENCES

Allen, F.H., Doyle, M.J. and Taylor, R. (1990) Acta Cryst., B46, in press.

Blundell, T.L., Sibanda, B.L., Sternberg, M.J.E. and Thornton, J.M. (1987) Nature, 326, 347.

Bricogne, G. (1988) Acta Cryst., A44, 517.

Dolata, D.P., Leach, A.R. and Prout, K. (1987) J. Computer-Aided Mol. Design, 1, 73.

Etter, M.C., MacDonald, J.C. and Bernstein, J. (1990) Acta Cryst., B46, 256.

Fortier, S. and Nigam, G.D. (1989) Acta Cryst., A45, 247.

Glasgow, J. (1990) IEEE Trans. Patt. Anal. Mach. Intell., submitted.

Jenkins, M.A., Glasgow, J.I. and McCrosky, C. (1986) IEEE Software, Vol. 3, No. 1, 46.

Marr, D. and Nishihara, H.K. (1977) Proc. Roy. Soc. Lond., B.200, 269.

Moore, T. (1981) In *Structures and Operations in Engineering & Management Systems*, O. Bjorke and O. Fanksen, eds., Raper Publishers, Trondheim, Norway.

Nyburg, S.C. and Faerman, C.H. (1985) Acta Cryst., B41, 274.

Rowland, R.S., Allen F.H., Carson, W.M. and Bugg, C.E. (1990) In *Crystallographic Applications in Molecular Modelling*, S.E. Ealick and C.E. Bugg, eds., Springer Verlag, New York, in press.

Taylor, R. and Kennard, O. (1984) Acc. Chem. Res., 17, 320.

Wippke, W.T. and Hahn, M.A. (1988) Tetrahedron Computer Methodology, 1, 141.